AF861953

Homogenisation: Averaging Processes in Periodic Media

Mathematics and Its Applications (*Soviet Series*)

Managing Editor:

M. HAZEWINKEL
Centre for Mathematics and Computer Science, Amsterdam, The Netherlands

Editorial Board:

A. A. KIRILLOV, *MGU, Moscow, U.S.S.R.*
Yu. I. MANIN, *Steklov Institute of Mathematics, Moscow, U.S.S.R.*
N. N. MOISEEV, *Computing Centre, Academy of Sciences, Moscow, U.S.S.R.*
S. P. NOVIKOV, *Landau Institute of Theoretical Physics, Moscow, U.S.S.R.*
M. C. POLYVANOV, *Steklov Institute of Mathematics, Moscow, U.S.S.R.*
Yu. A. ROZANOV, *Steklov Institute of Mathematics, Moscow, U.S.S.R.*

Volume 36

Homogenisation: Averaging Processes in Periodic Media

Mathematical Problems in the Mechanics of Composite Materials

by

N. BAKHVALOV

and

G. PANASENKO

Mechanics and Mathematics Faculty,
Moscow University, U.S.S.R.

KLUWER ACADEMIC PUBLISHERS

DORDRECHT / BOSTON / LONDON

Library of Congress Cataloging in Publication Data

Bakhvalov, N. S.
[Osrednenie protsessov v periodicheskikh sredakh. English]
Homogenisation : averaging processes in periodic media : mathematical problems in the mechanics of composite materials / by N. Bakhvalov and G. Panasenko.
p. cm. -- (Mathematics and its applications (Soviet series))
Translation of: Osrednenie protsessov v periodichskikh sredakh.
Includes bibliographical references.
ISBN 0-7923-0049-1
1. Composite materials--Mathematical models. I. Panasenko, G. P. (Grigoriĭ Petrovich) II. Title. III. Series: Mathematics and its applications (Kluwer Academic Publishers). Soviet series.
TA418.9.C6B3513 1989
620.1'183--dc20 89-19747

ISBN 0-7923-0049-1

Published by Kluwer Academic Publishers,
P.O. Box 17, 3300 AA Dordrecht, The Netherlands.

Kluwer Academic Publishers incorporates
the publishing programmes of
D. Reidel, Martinus Nijhoff, Dr W. Junk and MTP Press.

Sold and distributed in the U.S.A. and Canada
by Kluwer Academic Publishers,
101 Philip Drive, Norwell, MA 02061, U.S.A.

In all other countries, sold and distributed
by Kluwer Academic Publishers Group,
P.O. Box 322, 3300 AH Dordrecht, The Netherlands.

This is the expanded and revised edition of the original work
ОСРЕДНЕНИЕ ПРОЦЕССОВ В ПЕРИОДИЧЕСКИХ СРЕДАХ
Published by Nauka Publishers, Moscow, © 1984.

Translated from the Russian by D. Leites.
Typeset by Rosenlaui Publishing Services, Inc.

All Rights Reserved
This English edition © 1989 by Kluwer Academic Publishers
No part of the material protected by this copyright notice may be reproduced or utilized in any form or by any means, electronic or mechanical including photocopying, recording, or by any information storage and retrieval system, without written permission from the copyright owner.

Printed in The Netherlands

SERIES EDITOR'S PREFACE

'Et moi, ..., si j'avait su comment en revenir, je n'y serais point allé.'

Jules Verne

The series is divergent; therefore we may be able to do something with it.

O. Heaviside

One service mathematics has rendered the human race. It has put common sense back where it belongs, on the topmost shelf next to the dusty canister labelled 'discarded non-sense'.

Eric T. Bell

Mathematics is a tool for thought. A highly necessary tool in a world where both feedback and non-linearities abound. Similarly, all kinds of parts of mathematics serve as tools for other parts and for other sciences.

Applying a simple rewriting rule to the quote on the right above one finds such statements as: 'One service topology has rendered mathematical physics ...'; 'One service logic has rendered computer science ...'; 'One service category theory has rendered mathematics ...'. All arguably true. And all statements obtainable this way form part of the raison d'être of this series.

This series, *Mathematics and Its Applications*, started in 1977. Now that over one hundred volumes have appeared it seems opportune to reexamine its scope. At the time I wrote

> "Growing specialization and diversification have brought a host of monographs and textbooks on increasingly specialized topics. However, the 'tree' of knowledge of mathematics and related fields does not grow only by putting forth new branches. It also happens, quite often in fact, that branches which were thought to be completely disparate are suddenly seen to be related. Further, the kind and level of sophistication of mathematics applied in various sciences has changed drastically in recent years: measure theory is used (non-trivially) in regional and theoretical economics; algebraic geometry interacts with physics; the Minkowsky lemma, coding theory and the structure of water meet one another in packing and covering theory; quantum fields, crystal defects and mathematical programming profit from homotopy theory; Lie algebras are relevant to filtering; and prediction and electrical engineering can use Stein spaces. And in addition to this there are such new emerging subdisciplines as 'experimental mathematics', 'CFD', 'completely integrable systems', 'chaos, synergetics and large-scale order', which are almost impossible to fit into the existing classification schemes. They draw upon widely different sections of mathematics."

By and large, all this still applies today. It is still true that at first sight mathematics seems rather fragmented and that to find, see, and exploit the deeper underlying interrelations more effort is needed and so are books that can help mathematicians and scientists do so. Accordingly MIA will continue to try to make such books available.

If anything, the description I gave in 1977 is now an understatement. To the examples of interaction areas one should add string theory where Riemann surfaces, algebraic geometry, modular functions, knots, quantum field theory, Kac-Moody algebras, monstrous moonshine (and more) all come together. And to the examples of things which can be usefully applied let me add the topic 'finite geometry'; a combination of words which sounds like it might not even exist, let alone be applicable. And yet it is being applied: to statistics via designs, to radar/sonar detection arrays (via finite projective planes), and to bus connections of VLSI chips (via difference sets). There seems to be no part of (so-called pure) mathematics that is not in immediate danger of being applied. And, accordingly, the applied mathematician needs to be aware of much more. Besides analysis and numerics, the traditional workhorses, he may need all kinds of combinatorics, algebra, probability, and so on.

In addition, the applied scientist needs to cope increasingly with the nonlinear world and the extra mathematical sophistication that this requires. For that is where the rewards are. Linear models are honest and a bit sad and depressing: proportional efforts and results. It is in the nonlinear world that infinitesimal inputs may result in macroscopic outputs (or vice versa). To appreciate what I am hinting at: if electronics were linear we would have no fun with transistors and computers; we would have no TV; in fact you would not be reading these lines.

There is also no safety in ignoring such outlandish things as nonstandard analysis, superspace and anticommuting integration, p-adic and ultrametric space. All three have applications in both electrical engineering and physics. Once, complex numbers were equally outlandish, but they frequently proved the shortest path between 'real' results. Similarly, the first two topics named have already provided a number of 'wormhole' paths. There is no telling where all this is leading fortunately.

Thus the original scope of the series, which for various (sound) reasons now comprises five subseries: white (Japan), yellow (China), red (USSR), blue (Eastern Europe), and green (everything else), still applies. It has been enlarged a bit to include books treating of the tools from one subdiscipline which are used in others. Thus the series still aims at books dealing with:

- a central concept which plays an important role in several different mathematical and/or scientific specialization areas;
- new applications of the results and ideas from one area of scientific endeavour into another;
- influences which the results, problems and concepts of one field of enquiry have, and have had on the development of another.

The idea of a natural scale for all kinds of phenomena is an old and valuable one in physics, and if only one natural scale is involved, that certainly tends to simplify things; when all scales are involved and interact, things are apt to get complicated and messy; there it is hoped that the systematic study of fractals and their kin will provide a good solid foundational framework. There is an intermediate case, naturally occurring, and important from the viewpoint of applications in engineering and elsewhere. This concerns the case where there are just a few clearly distinguishable natural scales present. One case is that of composite materials consisting of alternate layers of various materials. Here the fact that the periodic layer structure is at a completely different scale than the molecular scales of the constituent materials is at the root of the very different physical and engineering properties of these materials.

Mathematically, we are concerned here, by and large, with differential equations with rapidly oscillating coefficients and the ideas of averaging and corresponding asymptotics. The book is concerned with the applications of this branch of mathematics to the case of composite materials and the development and description of special mathematical tools to be applied to the design of such materials (with pregiven desired properties).

At least one of the authors, Bakhvalov, appears to have a natural bend in the direction of thinking in terms of multiple scales. Indeed, he is the original inventor (around 1965), mostly at the theoretical level, of the so-called multigrid methods in numerical analysis, later independently rediscovered by Hackbousch (and implemented by him (1976)), a set of methods which is rapidly becoming the most popular one for numerically solving linear and nonlinear boundary-value problems.

There is much I like about the present volume, but perhaps the main point is that it goes a long way in the direction of illustrating the power of mathematical analysis in improving our understanding of what goes on in composites, and in predicting the properties of such materials.

The shortest path between two truths in the real domain passes through the complex domain.

J. Hadamard

La physique ne nous donne pas seulement l'occasion de résoudre des problèmes ... elle nous fait pressentir la solution.

H. Poincaré

Never lend books, for no one ever returns them; the only books I have in my library are books that other folk have lent me.

Anatole France

The function of an expert is not to be more right than other people, but to be wrong for more sophisticated reasons.

David Butler

Bussum, September 1989

Michiel Hazewinkel

TABLE OF CONTENTS

Notations xi

Preface xiii

Preface to English Edition xvii

Introduction xix

§1. Medium With Periodic Structure xix

§2. What Are Effective Coefficients? xxii

§3. A Scheme for Calculating Effective Coefficients xxiii

§4. Investigating the Microscopic Structure of a Field xxv

§5. What is the Asymptotic Averaging Method? xxvi

§6. Where can these Methods Help? xxxi

§7. Advantages and Disadvantages of the Averaging Method xxxii

§8. A Brief Review of the Literature xxxiii

Chapter 1. Formulation of Elementary Boundary Value Problems 1

§1. The Concept of the Classical Formulation of a Boundary Value Problem for Equations with Discontinuous Coefficients 1

§2. The Concept of Generalized Solution 2

§3. Generalized Formulations of Problems for the Basic Equations of Mathematical Physics 4

Chapter 2. The Concept of Asymptotic Expansion. A Model Example to Illustrate the Averaging Method 12

§1. Asymptotic expansion. A Formal Asymptotic Solution 12

§2. Asymptotic Expansion of a Solution of the Equation $u = 1 + \varepsilon u^3$ 14

§3. Asymptotic Expansion of a Solution of the Equation $(K(x/\varepsilon)u')' = f(x)$ by the Averaging Method 16

§4. Generalization of the Averaging Method in the Case of a Piecewise Smooth Coefficient 26

§5. Averaging the System of Differential Equations 29

Chapter 3. Averaging Processes in Layered Media 32

§1. Problem of Small Longitudinal Vibrations of a Rod 34

§2. Nonstationary Problem of Heat Conduction 42
§3. Averaging Maxwell Equations 43
§4. Averaging Equations of a Viscoelastic Medium 45
§5. Media with Slowly Changing Geometric Characteristics 54
§6. Heat Transfer Through a System of Screens 57
§7. Averaging a Nonlinear Problem of the Elasticity Theory in an Inhomogeneous Rod 61
§8. The System of Equations of Elasticity Theory in a Layered Medium 68
§9. Considerations Permitting Reduction of Calculations in Constructing Averaged Equations 77
§10. Nonstationary Nonlinear Problems 80
§11. Averaging Equations with Rapidly Oscillating Nonperiodic Coefficients 81
§12. Problems of Plasticity and Dynamics of Viscous Fluid as Described by Functions Depending on Fast Variables 89

Chapter 4. Averaging Basic Equations of Mathematical Physics 102

§1. Averaging Stationary Thermal Fields in a Composite 104
§2. Asymptotic Expansion of Solution of the Stationary Heat Conduction Problem 125
§3. Stationary Thermal Field in a Porous Medium 132
§4. Averaging a Stationary System of Equations of Elasticity Theory in Composite and Porous Materials 137
§5. Nonstationary Systems of Equations of Elasticity and Diffusion Theory 142
§6. Averaging Nonstationary Nonlinear System of Equations of Elasticity Theory 147
§7. Averaging Stokes and Navier–Stokes Equations. The Derivation of the Percolation Law for a Porous Medium (Darcy's Law) 151
§8. Averaging in case of Short-Wave Propagation 157
§9. Averaging the Transition Equation for a Periodic Medium 165
§10. Eigenvalue Problems 168

Chapter 5. General Formal Averaging Procedure 171

§1. Averaging Nonlinear Equations 171
§2. Averaged Equations of Infinite Order for a Linear Periodic Medium and for the Equation of Moment Theory 175
§3. A Method of Describing Multi-Dimensional Periodic Media that does not Involve Separating Fast and Slow Variables 191

Chapter 6. Properties of Effective Coefficients. Relationship Among Local and Averaged Characteristics of a Solution 196

§1. Maintaining the Properties of Convexity and Symmetry of the Minimized Functional in Averaging 196

§2. On the Principle of Equivalent Homogeneity 205

§3. The Symmetry Properties of Effective Coefficients and Reduction of Periodic Problems to Boundary Value Problems 212

§4. Agreement Between Theoretically Predicted Values of Effective Coefficients and Those Determined by an Ideal Experiment 222

Chapter 7. Composite Materials Containing High-Modulus Reinforcement 224

§1. The Stationary Field in a Layered Material 225

§2. Composite Materials with Grains for Reinforcement 229

§3. Dissipation of Waves in Layered Media 236

§4. High-Modulus 3D Composite Materials 239

§5. The Splitting Principle for the Averaged Operator for 3D High-Modulus Composites 245

Chapter 8. Averaging of Processes in Skeletal Structures 257

§1. An Example of Averaging a Problem on the Simplest Framework 262

§2. A Geometric Model of a Framework 272

§3. The Splitting Principle for the Averaged Operator for a Periodic Framework 273

§4. The Splitting Principle for the Averaged Operator for Trusses and Thin-walled Structures 295

§5. On Refining the Splitting Principle for the Averaged Operator 298

§6 Asymptotic Expansion of a Solution of a Linear Equation in Partial Derivatives for a Rectangular Framework 299

§7 Skeletal Structures with Random Properties 310

Chapter 9. Mathematics of Boundary-Layer Theory in Composite Materials 312

§1. Problem on the Contact of Two Layered Media 314

§2. The Boundary Layer for an Elliptic Equation Defined on a Half-Plane 317

§3. The Boundary Layer Near the Interface of Two Periodic Structures 324

§4. Problem on the Contact of Two Media Divided by a Thin Interlayer 331

§5. The Boundary Layer for the Nonstationary System of Equations of Elasticity Theory 334

§6. On the Ultimate Strength of a Composite 339

§7. Boundary Conditions of Other Types 340
§8. On the Averaging of Fields in Layer Media with Layers of Composite Materials 343
§9. The Time Boundary Layer for the Cauchy Parabolic Problem 344

Supplement: Existence and Uniqueness Theorems for the Problem on a Cell 346

Bibliography 351

Subject Index 365

NOTATIONS

ε, μ, ω^{-1}	–small parameters
R^s	–s-dimensional vector space over the field of real vectors
G	–a domain in R^s, i.e., an open connected set (usually bounded)
$\partial \cdot G$	–boundary of the domain G
$\overline{G}$	–closure of the domain G, i.e., $\overline{G} = G \cup \partial G$
$\forall$	–for all
$\exists$	–there exist
$x = (x_1, \dots, x_s)$	–point in R^s, slow (macroscopic) variable
$\xi = (\xi_1, \dots, \xi_s)$	–fast (microscopic) variable, $\xi = x/\varepsilon$
$L_2(G)$	–space of functions with bounded norm $\| u \|_{L_2(G)} = \sqrt{\int_G u^2(x)dx}$
$C(\overline{G})$	–space of functions continuous in $\overline{G}$, $\| u \|_e = \sup_{x \in \overline{G}} \|u(x)\|$
$C^\infty(\overline{G})$	–space of functions infinitely differentiable in $\bar{G}$
$W_P^1, \overset{\circ}{W}{}_P^1$	–are the classical Sobolev spaces
$W_2^{1\ \mathrm{loc}}$	–space of functions $\{f \mid \forall \text{ compact } K \quad f \in W_2^1(K)\}$
$u\vert_{\partial G}$	–trace on ∂G
$[u]_\Sigma$	–means the difference of values of the function u on the two sides of the surface Σ

$\langle f(x_1, \dots, x_s, \xi_1, \dots, \xi_s) \rangle$ is a mean over the period, i.e.,

$$\int_\varnothing^1 \cdots \int_\varnothing^1 f(x_1, \dots, x_s, t, \xi_1, \dots, \xi_s) d\xi_1 \dots d\xi_s$$

$\{f\}\|_{x,t}$	–is a local averaging of function f, i.e., $\varepsilon^{-s} \int_{\sqcap_x^\varepsilon} f(y,t)dy$, where $\sqcap_x^\varepsilon = \{y \in R^s \mid y_q - x_q\| \le \varepsilon/2 \ \forall\, q = 1,\dots,s\}$
$\langle f(x)\rangle_T$	$= \int_\varnothing^T \cdots \int_\varnothing^T f(x)dx$
(a,b)	–scalar product of the vectors a and b
$A \times B$	–Cartesian product of the sets A and B, i.e., the set of pairs $\{a,b\}$, where $a \in A$ and $b \in B$
$\mathrm{div}(\mathrm{div}_x), \mathrm{div}_\xi$	–divergence with respect to the variables x or ξ
$\mathrm{grad}(\mathrm{grad}_x), \mathrm{grad}_\xi$	–gradient with respect to the variables x or ξ
$\nabla(\nabla_x), \nabla_\xi$	–the same
$\Delta(\Delta_x), \Delta_\xi$	–Laplacian with respect to the variables x or ξ

$$L_{\alpha\beta}U(x,\xi) = \sum_{k,j=1}^{S} \frac{\partial}{\partial\alpha_k}\left(A_{k_j}(\xi)\frac{\partial u(x,\xi)}{\partial\beta_j}\right)$$

$i = (i_1,\dots,i_s)$	is a multiindex, $\|i\| = l$ is its length, $i_j \in \{1,\dots,s\}$
$D^i = \frac{\partial^l}{\partial x_{i_1}\dots\partial x_{i_l}}$	(partial derivative)
$D^{q,i} = \frac{\partial^{q+l}}{\partial t^q \partial x_{i_1}\dots\partial x_{i_l}}$	
a. e.	–asymptotic expansion
f. a. s.	–formal asymptotic solution
$f(x,\varepsilon) \sim \sum_{i=l}^{\infty} \varepsilon^i g_i(x,\varepsilon)$	is a. e. of $f(x,\varepsilon)$, i.e.,

$$\forall N \exists M_0 : \forall m \ge M_0 \quad f(x,\varepsilon) - \sum_{i=l}^{S_m} \varepsilon^i g_i(x,\varepsilon) = O(\varepsilon^n)$$

u	–exact solution of the problem
u^∞	–f. a. s. of the problem
$\vee$	–f. a. s. of the averaged problem
$\wedge$	–effective coefficient symbol
δ_{i_j}	–Kronecker symbol

We use the convention of repeated indexes, i.e.,

$$a_i b_i = \sum_{i=1}^{s} a_i b_i$$

averaging = homogenisation.

PREFACE

Many branches of science and engineering today have adopted materials which consist of alternating volumes of compounds with different properties. An elementary example is the simple felt-like material that has long been used for heat insulation and soundproofing. To a great extent, these are composite materials or composites having periodic structures. The largely new properties of the composites are due mostly to the fact that the volumes of uniform compounds alternate on a level other than the molecular one. The linear dimensions of the heterogeneities are much smaller than the linear dimensions of the body as a whole but are considerably larger than those of molecules—so much so, in general, that the state of a compound in each of these volumes can be described by equations of continuum mechanics. We use the term "microlevel" below, meaning the size level of these volumes.

The progress made in design and application of such materials is mainly due to improvements in their production procedures. However, after the first fast successes in engineering a great number of new composites, it became necessary to get a better idea of what had to go into them. It turned out that the behaviour of a composite material on its microlevel under external loads is often quite different from that of a uniform material with the same effective characteristics. Cracking, for instance, starts and develops in this material in a different way.

Many semiempiric theories have been proposed to describe processes in composites, and a little later, efforts got under way to study these processes under the rigorous standards of mathematics. A strict mathematical treatment often proved to be instrumental in better understanding the processes that take place in composites under loading. The mathematical constructions made it possible to predict properties of a composite material even before it is engineered and, consequently, to consider the problem of developing a composite with predetermined or optimum (in terms of some particular characteristics) properties.

Averaged (effective) characteristics of composite materials can also be determined by testing prototypes rather than by calculations, but mathematical treatment, in addition to the averaged characteristics, can also help to reveal the microstructure of stress fields or thermal fields in a material.

All of the above, of course, is not to disparage the role of a nonrigorous treatment in the theory of composite materials, since the problem is quite complex and many details cannot be investigated by mathematical methods. For example, the theory of averaging the processes in random media is yet to be developed mathematically. Adhesion phenomena and other circumstances, when disregarded, may reduce to zero the results of strictly mathematical as well as other theoretical investigations. However, we hope that the mathematical constructions made in the present book can be useful to engineers engaged in research and development of composite materials.

The processes in composites are generally described by differential equations with rapidly oscillating coefficients. We use standard asymptotic methods to investigate solutions of the equations. The book deals mainly with applications of these methods to investigations of various processes in media with periodic structures.

We selected the material by taking into consideration only the simplest problems and methods and sometimes restricted ourselves to a formal asymptotic expansion of a solution. Rigorous mathematical proofs of complex problems are given mostly so that the reader may have no doubts about the correctness of the results produced by formal asymptotic constructions.

Knowledge acquired after the first three years of advanced mathematical training at a college is sufficient to read most of the book. Presentation of the material is based on the principle "from simple to complicated," every chapter beginning with an elementary example to illustrate the main idea of the method to be described. We have often rejected a shorter mathematical justification of a result in favour of one which may be longer but more graphic.

The first part of the book investigates fields and processes in layered media. Chapter 1 defines a generalized solution and formulates theorems which prove the existence and uniqueness of solutions of basic equations in mathematical physics. Chapter 2 introduced 0-symbols and the concepts of asymptotic expansion and of formal asymptotic solution; a solution of a simple model problem is expanded asymptotically, as is that of the simplest equation with rapidly oscillating coefficients, to demonstrate the averaging method. Chapter 3 deals with the investigation of stationary and nonstationary problems of mathematical physics in layered media.

The second part of the book contains largely spatial many-dimensional problems. Chapter 4 is devoted to the investigation of the equations of mathematical physics, which describe principal processes in periodic media. Chapter 5 describes general schemes of the averaging method. Chapter 7 discusses composite materials whose components have widely different properties. Chapter 8 sets forth asymptotic methods for calculating skeletal or latticed structures and gives explicit formulas to effectively calculate characteristics of this type of structure. Chapter 9 discusses the theory of boundary layer in composite materials, which takes into consideration the effects that occur in the vicinity of a specimen's boundary.

The book is based on a course of lectures delivered by the authors at the Faculty of Mathematics and Mechanics and at the Faculty of Computing Mathematics and Cybernetics at Moscow University.

Separate numbering is adopted for sections, with separate numbering for formulas within each section. The number of a section is added to references within each chapter. The number of a chapter is added to references concerning formulas from another chapter. The same system is used for theorems.

The authors wish to express gratitude to M. E. Eglit and S. M. Tsidilin who read the manuscript and greatly contributed to its improvement.

The authors repeatedly discussed some topics included in the book with M. R. Korotkina, A. G. Kulikov, B. E. Pobedrya, and V. L. Berdichevskii. Our sincere thanks are due to all of them.

The authors want to thank N. V. Novikova, E. A. Yakovleva, D. B. Volkova, A. A. Zlotnik, A. Yu. Kosarev, S. V. Mozolin, and M. V. Reztsov for their great help in preparing the manuscript.

N. S. Bakhvalov and G. P. Panasenko

PREFACE TO THE ENGLISH EDITION

This book was first published in the U. S. S. R. in 1984. Since then, however, some new results have emerged, and these are included in this English edition.

The most complicated step in averaging (homogenisation) method algorithm is the so-called "cell-problem" (considered in the third section of the Introduction). A program package EFMODUL to solve this problem and to compute the effective coefficients of composites has been developed by the authors. This package contains both fast programs based on an asymptotic technique to compute the effective coefficients of some special types of composites and, in general, more complicated ones based on numerical methods.* Some numerical results of EFMODUL computations have already been published in a paper by Panasenko [1].

Chapter 7 deals with the model of 3-D reinforced composites having two parameters; the first is a "geometrical" parameter ε which is equal to the ratio of the period of a composite to the peculiar size of the specimen ($\varepsilon \ll 1$). The second is a "physical" parameter ω which is equal to the ratio of the physical property of the reinforcement characteristic to that of the matrix ($\omega \gg 1$). In Chapter 7, the case $\omega\varepsilon^2 \ll 1$ is investigated; other cases have been considered. In the cases $\omega\varepsilon^2 \gg 1$ and $\omega\varepsilon^2 = \text{const.}$, a multicontinuum averaged model was obtained. This result and the method involved appear in another paper by Panasenko [2].

[1] Panasenko G. P., "Numerical solution of cell problems in homogenisation," Zh. Byč. Mat. i Mat. Fis. (ZHVM and MF) **28** (1988), pp. 281–286.

[2] Panasenko G. P., "Homogenisation processes in highly non-homogeneous structures," Doklady Akademia Nauk SSSR (DAN SSSR) **298** (1988), pp. 76–79.

In addition, the asymptotic theory of processes in a thin non-homogeneous plate has been developed in the following papers:

Panasenko, G. P., "Boundary layer in homogenisation problems for non-homogeneous media," Proc. of the 4th Int. Conf. on Boundary and Interior Layers, Boole-Press, Dublin (1986), pp. 398–402.

Panasenko G. P. and Reztsov, M. V., "Averaging the 3-D elasticity problem in non-homogeneous plates," DAN SSSR **294** (1987), pp. 1061–1065.

---

* *Publisher's Note*: For further information concerning this software please contact the publisher. (Kluwer Academic Publishers, SciTech Division, P.O. Box 17, 3300 AA Dordrecht, The Netherlands).

Some new results are also presented in the reviews:

Bakhvalov, N. S., Panasenko, G. P., Shtaras, A. L., Eglit, M. E., "Numerical-asymptotic methods," Asymptotic Methods in Mathematical Physics, Naukova Dumka, Kiev, (1987).

Bakhvalov, N. S., Panasenko, G. P., Shtaras, A. L., "Homogenisation partial differential equations," Modern Mathematical Problems. Fundamental Directions. 34, Nauka, Moscow (1988) (English translation: Encyclopedia Mathematica, 34, Springer).

N. Bakhvalov and G. Panasenko

INTRODUCTION

Composite materials have mainly periodic or nearly periodic structures, and therefore we shall primarily investigate processes in periodic media.

§1 Medium with Periodic Structure

We define here a medium with a periodic structure as one which is composed of recurrent elements (cells). Many composite materials have this structure. Thus, a unidirectionally reinforced fibrous composite comprises a periodic system of unidirectional fibers of one compound and the matrix of another compound which fills the space between the fibers. The thickness of a fiber is of the same order as the fiber spacing. Figure 1 shows a structure of a composite cut across the fibers (the fiber sections are shaded).

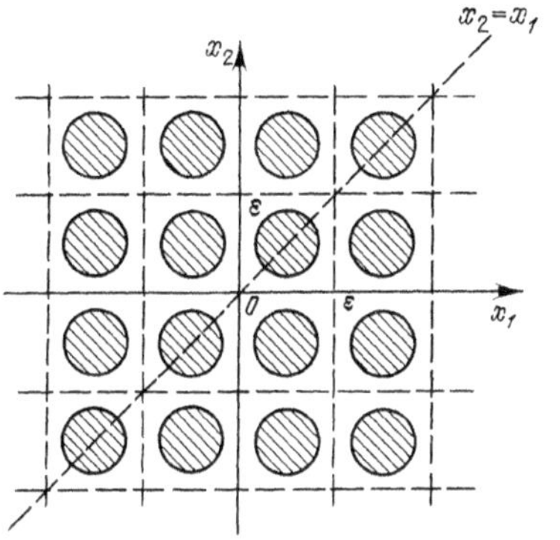

Figure 1

Another example of the periodic structure is a composite made up of a three-dimensional periodic system of grains that may be irregular in shape, and a material which fills the space between the grains (see Fig. 2).

Let L be the characteristic size of a composite specimen and ε the side of the recurrent cell. It is assumed throughout that the dimensions of the periodic cell are much smaller than the characteristic size of the specimen: $\varepsilon \ll L$. Let λ be the characteristic spatial size of a problem (e.g., the wavelength when wave propagation is investigated), and $l = \min(L\lambda)$. We also assume that $\varepsilon \ll l$ except in §4.10. If we take $l = 1$, then $\varepsilon \ll 1$, and ε can be considered a small dimensionless parameter.

Various skeletal structures, such as the trusses of bridges, cranes, and industrial installations, the frameworks of buildings, power transmission line supports, etc., are periodic. A model of a continuous medium with periodic arrangement of voids can describe such porous media as powders, foam plastics, wood, soil, materials with systems of cracks, perforated plates, etc.

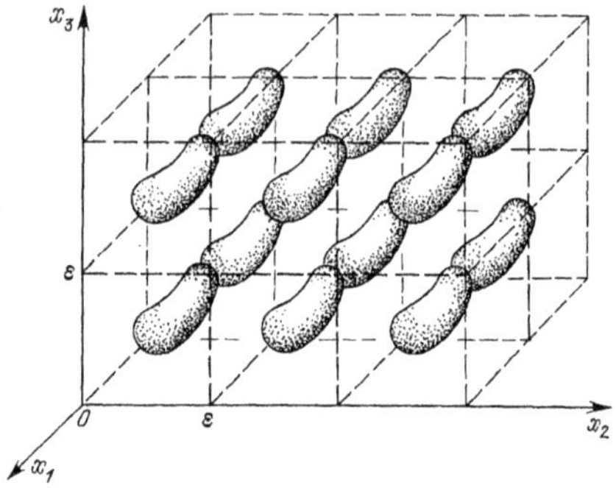

Figure 2

Fields and processes in periodic media are described by partial differential equations with periodic coefficients or by systems thereof. Thus, if the temperature of the fibrous composite shown in Fig. 1 is constant along the orientation of the fibers, the two-dimensional thermal field in the composite is described by the Poisson equation [151]

$$\frac{\partial}{\partial x_1}\left(K_\varepsilon(x_1, x_2)\frac{\partial u}{\partial x_1}\right) + \frac{\partial}{\partial x_2}\left(K_\varepsilon(x_1, x_2)\frac{\partial u}{\partial x_2}\right) = f(x_1, x_2) \tag{1}$$

everywhere outside the matrix–fiber contact lines. Here $K_\varepsilon(x_1, x_2)$ is the conductivity coefficient at a point (x_1, x_2): if this point falls on a fiber, K_ε takes the value K_B, the fiber conductivity coefficient; otherwise K_ε is the conductivity coefficient of the matrix K_M; the function $f(x_1, x_2)$ is the density of heat sources in the composite (it is supposed to be smooth); and $u(x_1, x_2)$ is the temperature at the point (x_1, x_2).

Continuity conditions are satisfied on the matrix–fiber contact lines for temperature:

$$[u] = 0 \tag{2}$$

and for the heat-flux density:

$$\left[K_\varepsilon \frac{\partial u}{\partial n}\right] = 0; \tag{3}$$

the square brackets denote the jump of the function in transition through the interface; and $\frac{\partial u}{\partial n}$ is the normal derivative to the interface.

The coefficient of conductivity $K_\varepsilon(x_1, x_2)$ changes by $|K_B - K_M|$ when (x_1, x_2) changes by a value of the order of $\varepsilon \ll 1$. Thus the function $K_\varepsilon(x_1, x_2)$ oscillates rapidly (see Fig. 3) This makes a numerical solution of such an equation practically impossible, because too small a mesh has to be taken for at least several nodes of the difference scheme to fall on each fiber. However, since $\varepsilon \ll 1$, problems (1)–(3) can be investigated asymptotically as $\varepsilon \to 0$.

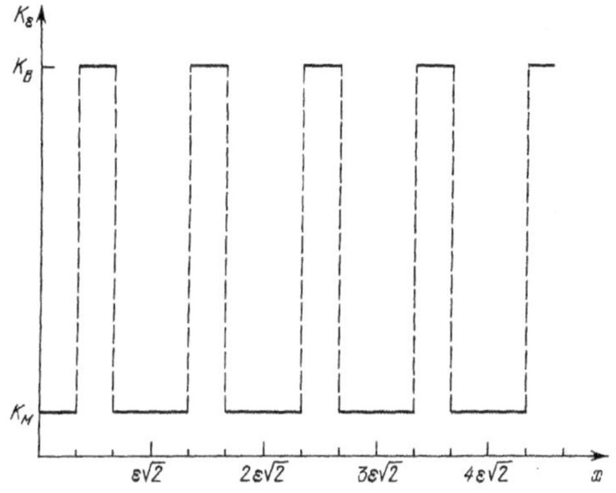

Figure 3

In the same way, a problem of static elasticity theory can be formulated for the composite material shown in Fig. 2 (see [1, 143, 155]). The equilibrium state of a linear elastic composite is described by the system of equations

$$\sum_{i,j=1}^{3} \frac{\partial}{\partial x_i}\left(A_{ij}^{\varepsilon}(x_1, x_2, x_3)\frac{\partial u}{\partial x_j}\right) = f(x_1, x_2, x_3) \qquad (4)$$

everywhere outside the grain surfaces. Here the 3×3 matrix-valued functions

$$A_{ij}^{\varepsilon}(x_1, x_2, x_3) = \left\| a_{ij}^{\varepsilon kl}(x_1, x_2, x_3) \right\|$$

determine the rigidity tensor at (x_1, x_3, x_3): if the point belongs to the grain, then the tensor $a_{ij}^{\varepsilon kl}(x_1, x_2, x_3)$ coincides with the rigidity tensors a_{Bij}^{kl} of the grain material; otherwise $a_{ij}^{\varepsilon kl}(x_1, x_2, x_3)$ equals the matrix rigidity tensor a_{Mij}^{kl}; the three-dimensional vector-valued function $u(x_1, x_2, x_3)$ defines the translation of the point which had the coordinates (x_1, x_2, x_3) in equilibrium; and $f(x_1, x_2, x_3)$ is a mass-force vector.

The continuity conditions are satisfied on the grain surfaces for the translation vector and normal component of the stress tensor;

$$[u] = 0, \tag{5}$$

$$\left[\sum_{i,j=1}^{3} A_{ij}^{\varepsilon} \frac{\partial u}{\partial x_j} n_i \right] = 0, \tag{6}$$

where n_i are the direction cosines of the normal to the grain surface.

We assume for simplicity that the recurrent element of the structure was transformed into a square (cube) by linear substitution of the independent variables. For example, the case of a cellular structure (see Fig. 4) is reduced to the considered case if the parallelogram $ABCD$ is taken as the periodicity domain and then is transformed into a square by a change of space variables.

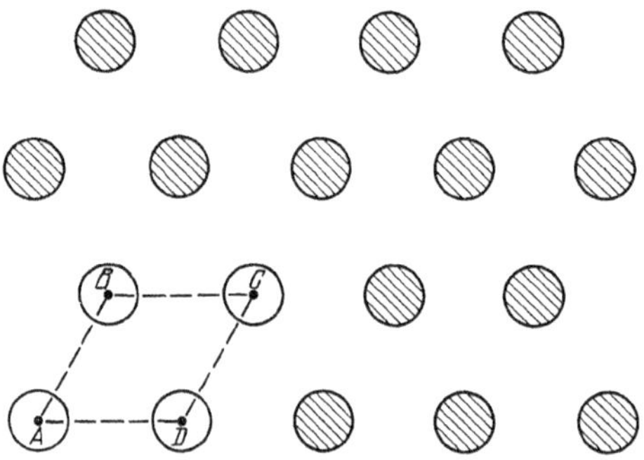

Figure 4

§2 What are Effective Coefficients?

The purpose of our reasoning is to obtain equations whose coefficients are not rapidly oscillating while their solutions are close to those of the original equations (for appropriate boundary conditions). These new equations are called *averaged equations*, and their coefficients are the *effective coefficients* of a composite material. Sometimes averaging yields equations of a type quite different from the original ones; for example, in §3.4 the averaging of a system of differential equations results in a system in integral-differential equations.

In problems (1.1)–(1.3), let zero temperature be maintained on the boundary Γ of a specimen occupying a domain G; then a solution of the problem $u(x_1, x_2)$ is close to the solution $v_0(x_1, x_2)$ of the averaged equation

$$\frac{\partial}{\partial x_1}\left(\widehat{K} \frac{\partial v_0}{\partial x_1}\right) + \frac{\partial}{\partial x_2}\left(\widehat{K} \frac{\partial v_0}{\partial x_2}\right) = f(x_1, x_2) \tag{1}$$

provided

$$v_0 = 0 \tag{2}$$

on the boundary of the whole specimen. Here $\widehat{K}$ is the composite's effective coefficient of conductivity, with an algorithm for the calculation of $\widehat{K}$ being given in §3. It will be shown in Chapter 4 that the error $\max_{x\to\bar{G}}|u - v_0|$ is of order ε. Thus, the smaller the side of a periodic cell as compared to the characteristic size of the specimen, the more accurate is the approximation of $v_0(x_1, x_2)$. Unlike the original problem, (1)–(2) can be solved numerically on a large mesh.

The system of equations of elasticity theory (1.4)–(1.6) has a similar property, i.e., if the boundary of a specimen is rigidly fixed, the solution of (1.4)–(1.6) is close to that of the averaged system of equations

$$\sum_{i,j=1}^{3} \frac{\partial}{\partial x_i}\left(\widehat{A}_{ij}\frac{\partial v_0}{\partial x_j}\right) = f(x_1, x_2, x_3)$$

provided $v_o = 0$ on the boundary of the specimen. Here $\hat{A}_{ij}$ is the effective elasticity tensor ($\hat{A}_{ij}$ are constant matrices). So it is much simpler to investigate the original problem when it is replaced by an averaged one.

Note that some works hypothesize the possibility of replacement of a non-homogeneous medium by an equivalent homogeneous one (see, for example, the equivalent homogeneity hypothesis [71]). The propositions set forth below are free from such assumptions. Moreover, the results given in §6.2 can be regarded as justification for the equivalent homogeneity hypothesis.

§3 A Scheme for Calculating Effective Coefficients

The theoretical constructions given below yield the following algorithm for calculating the coefficients of an averaged equation:

1. In the case of (2.1)–(2.2), we use the "fast" variables $\xi_1 = x_1/\varepsilon$, $\xi_2 = x_2/\varepsilon$ to find the effective coefficient $\widehat{K}$. Then the elementary cell—the square $(0,\varepsilon)\times(0,\varepsilon)$—turns into the square $(0,1)\times(0,1)$ in the variables ξ_1, ξ_2 (see Fig. 5).

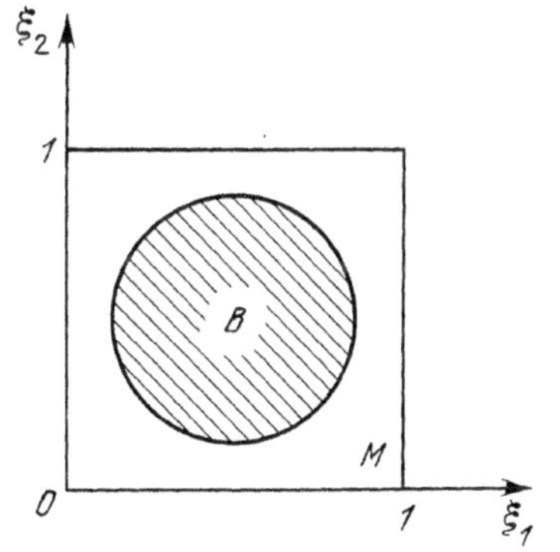

Figure 5

Let $\mathcal{B}$ be the domain in the unit square (a circle with point $(1/2, 1/2)$ as the centre) which corresponds to the fiber cross-section, and $\mathcal{M}$ the domain occupied by the matrix. We set

$$K(\xi_1, \xi_2) = \begin{cases} K_B & \text{for} \quad (\xi_1, \xi_2) \in \mathscr{B}, \\ K_M & \text{for} \quad (\xi_1, \xi_2) \in \mathscr{M}. \end{cases}$$

In this particular case, when the function $K(\xi_1, \xi_2)$ is even, relative to the straight lines $\xi_1 = 1/2$, $\xi_2 = 1/2$, and $\xi_1 = \xi_2$, namely $K(\xi_1, \xi_2) \equiv K(1 - \xi_1, \xi_2) \equiv K(\xi_1, 1 - \xi_2)$, $K(\xi_1, \xi_2) \equiv K(\xi_2, \xi_1)$, the averaged equation has the form

$$\frac{\partial}{\partial x_1}\left(\widehat{K}\frac{\partial v}{\partial x_1}\right) + \frac{\partial}{\partial x_2}\left(\widehat{K}\frac{\partial v}{\partial x_2}\right) = f$$

and the algorithm for calculating $\widehat{K}$ is as follows.

Let Π be an open square: $\Pi = \{\xi_1, \xi_2 | 0 < \xi_1, \xi_2 < 1/2\}$. We find the function M that satisfies Laplace's equation in the domain $\Pi \setminus \partial\mathcal{B}$,

$$\frac{\partial}{\partial \xi_1}\left(K(\xi_1, \xi_2)\frac{\partial M}{\partial \xi_1}\right) + \frac{\partial}{\partial \xi_2}\left(K(\xi_1, \xi_2)\frac{\partial M}{\partial \xi_2}\right) = 0 \tag{1}$$

the relations on the coefficient discontinuity line

$$[M] = 0, \quad (\xi_1, \xi_2) \in \partial\mathscr{B}, \tag{2}$$

$$\left[K(\xi_1, \xi_2)\frac{\partial M}{\partial n}\right] = 0, \quad (\xi_1, \xi_2) \in \partial\mathscr{B}, \tag{3}$$

and the boundary conditions

$$M(0, \xi_2) = 0, \quad M(1/2, \xi_2) = 1/2, \tag{4}$$

$$\frac{\partial M}{\partial \xi_2}(\xi_1, 0) = \frac{\partial M}{\partial \xi_2}\left(\xi_1, \frac{1}{2}\right) = 0. \tag{5}$$

The problem (1)–(5) can be solved numerically, by the finite element method, [55, 96, 148], or by finite difference methods [142].

In Chapter 4 we prove that the effective coefficient of conductivity is

$$\widehat{K} = 4\int_0^{1/2}\int_0^{1/2} K(\xi_1, \xi_2)\frac{\partial M}{\partial \xi_1}(\xi_1, \xi_2)\, d\xi_1 d\xi_2. \tag{6}$$

Note that with the volume fractions of the fiber and matrix being equal, (6) produces a result which differs from the arithmetic mean coefficient $(K_B + K_M)/2$, and the harmonic mean, $\left((K_B^{-1} + K_M^{-1})/2\right)^{-1}$, of K_B and K_M. The greater the difference between K_B and K_M, the more K and these means differ.

2. In the case of the three-dimensional structure shown in Fig. 2, the cube of periodicity $(0, \varepsilon) \times (0, \varepsilon) \times (0, \varepsilon)$ turns into the unit cube $Q = (0, 1) \times (0, 1) \times (0, 1)$ under the change of variables $\xi_1 = x_1/\varepsilon$, $\xi_2 = x_2/\varepsilon$, $\xi_3 = x_3/\varepsilon$. Let $\mathcal{B}$ be a subregion of the interior of the unit cube occupied by the grain and $\mathcal{M}$, a region

which corresponds to the matrix. We set

$$A_{ij}(\xi_1, \xi_2, \xi_3) = \begin{cases} A_{Bij} & \text{for} \quad (\xi_1, \xi_2, \xi_3) \in \mathcal{B}, \\ A_{Mij} & \text{for} \quad (\xi_1, \xi_2, \xi_3) \in \mathcal{M}. \end{cases}$$

Consider separately the systems of equations with the unknown 3×3 matrix-valued functions $N_1(\xi_1, \xi_2, \xi_3)$, $N_2(\xi_1, \xi_2, \xi_3)$, and $N_3(\xi_1, \xi_2, \xi_3)$:

$$\sum_{i,j=1}^{3} \frac{\partial}{\partial \xi_i} \left(A_{ij}(\xi_1, \xi_2, \xi_3) \frac{\partial N_l}{\partial \xi_j} \right) = 0, \quad (\xi_1, \xi_2, \xi_3) \in Q \setminus \partial \mathcal{B}, \tag{7}$$

for $(l = 1, 2, 3)$ with the relation on the grain-matrix interface $\partial \mathcal{B}$

$$[N_l] = 0, \quad (\xi_1, \xi_2, \xi_3) \in \partial \mathcal{B}, \tag{8}$$

$$\left[\sum_{i,j=1}^{3} A_{ij}(\xi_1, \xi_2, \xi_3) \frac{\partial (N_l + \xi_l E)}{\partial \xi_j} n_i \right] = 0, \quad (\xi_1, \xi_2, \xi_3) \in \partial \mathcal{B}, \tag{9}$$

and with the periodic boundary conditions

$$N_l|_{\xi_r = 0} = N_l|_{\xi_r = 1}, \quad \left. \frac{\partial N_l}{\partial \xi_r} \right|_{\xi_r = 0} = \left. \frac{\partial N_l}{\partial \xi_r} \right|_{\xi_r = 1}, \quad r = 1, 2, 3; \tag{10}$$

Here E is the unit 3×3 matrix, n_i are the direction cosines of the normal to $\partial \mathcal{B}$. These matrix problems can be split into independent systems of equations for the columns of the matrices N_l.

Assuming that all the problems (7)–(10) for $(l = 1, 2, 3)$ have been solved, we calculate the averaged elasticity tensor by the formula (see Chapter 4):

$$\widehat{A}_{ij} = \int_0^1 \int_0^1 \int_0^1 \sum_{q=1}^{3} A_{iq}(\xi_1, \xi_2, \xi_3) \frac{\partial}{\partial \xi_q} (N_j + \xi_j E) \, d\xi_1 d\xi_2 d\xi_3. \tag{11}$$

Problems (7)–(10) can be solved by using the finite difference methods or the method of finite elements [55, 96, 125, 142, 148].[1] When the region of periodicity is symmetric in a certain way, problems (7)–(10) are reduced to standard boundary value problems (§6.3).

§4 Investigating the Microscopic Structure of a Field

In §2 we noted that an exact solution of the original problem is close to that of the averaged one. Nevertheless, we shall see that the difference between derivatives of the exact solution and those of the averaged problem are great, while the

[1]Methods of the complex analysis are quite effective in the two-dimensional case [37, 45]. For further development see [225, 226].

derivatives of the exact solution are parts of expressions for heat fluxes or the stress tensor. Therefore, it does not suffice just to know effective coefficients to be able to determine correctly fluxes and stresses in a composite. However, if problems (7)–(10) are solved, the formula [8]

$$\sigma_i = \sum_{j=1}^{3} A_{ij}^{\varepsilon}(x_1, x_2, x_3)\frac{\partial u}{\partial x_j} \approx$$
$$\approx \sum_{q,j=1}^{3} A_{iq}^{\varepsilon}(x_1, x_2, x_3)\left(\frac{\partial \widetilde{N}_j}{\partial \xi_q}\left(\frac{x_1}{\varepsilon}, \frac{x_2}{\varepsilon}, \frac{x_3}{\varepsilon}\right) + \delta_{qj}E\right)\frac{\partial v_0}{\partial x_j}(x_1, x_2, x_3),$$
$$i = 1, 2, 3, \quad (1)$$

enables stresses to be determined with accuracy of order ε in the interior points of a composite material. Here $\widetilde{N}_j(\xi_1, \xi_2, \xi_3)$ are periodic extensions of the matrix-valued functions $N_j(\xi_1, \xi_2, \xi_3)$ over the whole space R^3 (with unit periodicity cube),δ_{ij} the Kronecker symbol (i.e., $\delta_{ij} = 1$ for $i = j$, and $\delta_{ij} = 0$ for $i \neq j$), and $v_0(x_1, x_2, x_3)$ a solution of the averaged problem.

Formula (1) takes into account the macroscopic as well as microscopic structure of the stress field and provides an approach to solving problems of strength calculations for composite materials. Similar formulas can be obtained for heat flows problems [(1.1)–(1.3)].

All the above can be explained, by using Fig. 14–16 of §4.1. The difference between the solutions u and v_0 turns out to be a rapidly oscillating function of order ε. The derivative of the difference is a value of the same order as that of the solution.

§5 What is the Asymptotic Averaging Method?

1. We said in §1 that fields and processes in periodic media are described by partial differential equations with rapidly oscillating coefficients.

To investigate solutions of the equations, we use the method of many scales [36, 73, 88, 101, 103] combined with techniques applied by Krylov, Bogoliubov, and Mitropolskii [27, 72, 100]. Solution is sought in the form of series in powers of a small parameter ε with coefficients depending both on the variables x_i (usually referred to as *slow* or *macroscopic variables*) and $\xi_i = x_i/\varepsilon$ (*fast* or *microscopic variables*):

$$u \sim \sum_{i=0}^{\infty} \varepsilon^i u_i(x_1, x_2, x_3, \xi_1, \xi_2, \xi_3).$$

The slow variables correspond to the global structure of the fields and the fast variables to their local structure. Such a series is substituted into the original equations. By equating the coefficients of the powers of ε to zero, we get equations in the functions u_i. The function u_0 and the coefficients of an equation in u_0 often turn out to be independent of the fast variables. We call this equation the *averaged equation*. The numerical solving of the averaged problem is considerably less difficult than that of the original one. It is assumed throughout, except in Chapter 9 dealing with boundary effects, that the u_i are 1-periodic functions of ξ.

A *T-periodicity* henceforth means period T in the variables indicated; but if T or the variables are not specified, period 1 is assumed for periodicity of all the ξ_j.

2. To illustrate the averaging method by a simple example, consider the equation

$$\frac{d}{dx}\left(K\left(\frac{x}{\varepsilon}\right)\frac{du}{dx}\right)-f(x)=0, \tag{1}$$

where $K(\xi)$ is a periodic function with period 1 and $\varepsilon \ll 1$. Equation (1) describes, for example, a stationary thermal field (with K being the conductivity coefficient and u the temperature) in a rod with rapidly oscillating conductivity coefficient. For simplicity, first consider the case where f and K are infinitely differentiable.

We seek an asymptotic solution of the problem in the form

$$u^{(\infty)}=u_0(x,\ \xi)+\varepsilon u_1(x,\ \xi)+\varepsilon^2 u_2(x,\ \xi)+\dots,\quad \xi=x/\varepsilon, \tag{2}$$

where $u_i(x,\xi)$ are periodic functions in ξ with period 1. Substitute expansion (2) into equation (1) following the chain rule

$$\frac{d}{dx}F\left(x,\ \frac{x}{\varepsilon}\right)=\left(\frac{\partial F(x,\ \xi)}{\partial x}+\varepsilon^{-1}\frac{\partial F(x,\ \xi)}{\partial \xi}\right)\bigg|_{\xi=x/\varepsilon}.$$

By simplifying, we get

$$\begin{aligned}\Bigg(\varepsilon^{-2}\frac{\partial}{\partial\xi}\left(K(\xi)\frac{\partial u_0(x,\ \xi)}{\partial\xi}\right)&+\varepsilon^{-1}\Bigg(\frac{\partial}{\partial\xi}\left(K(\xi)\frac{\partial u_1(x,\ \xi)}{\partial\xi}\right)+\\&+\frac{\partial}{\partial\xi}\left(K(\xi)\frac{\partial u_0(x,\ \xi)}{\partial x}\right)+\frac{\partial}{\partial x}\left(K(\xi)\frac{\partial u_0(x,\ \xi)}{\partial\xi}\right)\Bigg)+\\+\varepsilon^0\Bigg(\frac{\partial}{\partial\xi}\left(K(\xi)\frac{\partial u_2(x,\ \xi)}{\partial\xi}\right)&+\frac{\partial}{\partial\xi}\left(K(\xi)\frac{\partial u_1(x,\ \xi)}{\partial x}\right)+\frac{\partial}{\partial x}\left(K(\xi)\frac{\partial u_1(x,\ \xi)}{\partial\xi}\right)+\\&+\frac{\partial}{\partial x}\left(K(\xi)\frac{\partial u_0(x,\ \xi)}{\partial x}\right)-f(x)\Bigg)+\dots\Bigg)\bigg|_{\xi=x/\varepsilon}\sim 0.\end{aligned} \tag{3}$$

Here the symbol $\sim$ means the asymptotic equality (see §2.1). By equating the terms of order $\varepsilon^{-2}, \varepsilon^{-1}, \varepsilon^0$ to zero in (3), we obtain

$$\frac{\partial}{\partial \xi}\left(K(\xi)\frac{\partial u_0(x,\xi)}{\partial \xi}\right)=0, \tag{4_0}$$

$$\frac{\partial}{\partial \xi}\left(K(\xi)\frac{\partial u_1(x,\xi)}{\partial \xi}\right)+\frac{\partial}{\partial \xi}\left(K(\xi)\frac{\partial u_0(x,\xi)}{\partial x}\right)+\frac{\partial}{\partial x}\left(K(\xi)\frac{\partial u_0(x,\xi)}{\partial \xi}\right)=0, \tag{4_1}$$

$$\frac{\partial}{\partial \xi}\left(K(\xi)\frac{\partial u_2(x,\xi)}{\partial \xi}\right)+\frac{\partial}{\partial \xi}\left(K(\xi)\frac{\partial u_1(x,\xi)}{\partial x}\right)+$$
$$+\frac{\partial}{\partial x}\left(K(\xi)\frac{\partial u_1(x,\xi)}{\partial \xi}\right)+\frac{\partial}{\partial x}\left(K(\xi)\frac{\partial u_0(x,\xi)}{\partial x}\right)-f(x)=0. \tag{4_2}$$

By formally assuming x and ξ to be independent variables, we consider the relations (4_i) for $i = 0.1.2$ as a recurrent chain of equations differential in ξ with an unknown function $u_i(x,\xi)$ and a parameter x.

3. It follows from (4_0) that $K(\xi)\partial u_0(x,\xi)/\partial\xi$ is independent of ξ, i.e., $K(\xi)\partial u_0(x,\xi)/\partial\xi = C(x)$. The assumption that $K(\xi) > 0$, which is natural for the applications, implies

$$\frac{\partial u_0(x,\xi)}{\partial \xi}=\frac{C(x)}{K(\xi)}.$$

We adopt the following notation for the *mean over period* in the one-dimensional as well as many-dimensional case,

$$\langle f(x_1,\ldots,x_s,t,\xi_1,\ldots,\xi_s)\rangle =$$
$$=\int_0^1\ldots\int_0^1 f(x_1,\ldots,x_s,t,\xi_1,\ldots,\xi_s)\,d\xi_1\ldots d\xi_s;$$

and x and ξ are considered independed variables in the last integral.

By applying the operator $\langle\,\cdot\,\rangle$ to the equality, we see from the periodicity of $u_0(x,\xi)$ in ξ that

$$\left\langle\frac{\partial u_0(x,\xi)}{\partial \xi}\right\rangle=\int_0^1\frac{\partial u_0(x,\xi)}{\partial \xi}\,d\xi=u_0(x,\xi)\Big|_{\xi=0}^{\xi=1}=0.$$

Thus $0 = \langle K^{-1}(\xi)\rangle C(x)$, and consequently, $C(x) = 0$, $\partial u_0/\partial\xi = 0$, and $u_0(x,\xi)$ is independent of ξ, i.e.,

$$u_0(x,\ \xi)=v_0(x). \tag{5}$$

4. Considering (5), we rewrite (4_1) as

$$\frac{\partial}{\partial \xi}\left(K(\xi)\left(\frac{\partial u_1(x,\xi)}{\partial \xi}+\frac{dv_0}{dx}\right)\right)=0.$$

This implies that

$$
\begin{aligned}
&K(\xi)\left(\frac{\partial u_1}{\partial \xi}+\frac{dv_0}{dx}\right)=C_1(x),\\
&\frac{\partial u_1}{\partial \xi}+\frac{dv_0}{dx}=C_1(x)\,K^{-1}(\xi).
\end{aligned}
\tag{6}
$$

By applying the operator $\langle\,.\,\rangle$, we get $d\nu_0/dx = C_1(x)\langle 1/K(\xi)\rangle$, i.e., $c_1(x) = \widehat{K}d\nu_0/dx$, where $\widehat{K}$ is the harmonic mean of $K(\xi)$:

$$
\widehat{K} = \langle 1/K(\xi)\rangle^{-1}. \tag{7}
$$

Therefore, $\partial u_1/\partial\xi = (\widehat{K}/K(\xi) - 1)d\nu_0/dx$, hence

$$
u_1(x,\xi) = N_1(\xi)\frac{dv_0}{dx} + D_1(x),
$$

where

$$
N_1(\xi) = \int_0^{\xi}\left(\frac{\widehat{K}}{K(\xi)} - 1\right)d\xi,
$$

and $D_1(x)$ is an arbitrary function, for example $D_1(x) \equiv 0$.

The same expression for $u_1(x,\xi)$ can be obtained from other considerations. The equation in $u_1(x,\xi)$ has the form

$$
\frac{\partial}{\partial\xi}\left(K(\xi)\frac{\partial u_1(x,\xi)}{\partial\xi}\right) = -\left(\frac{d}{d\xi}K(\xi)\right)\frac{dv_0(x)}{dx}; \tag{8}
$$

and so we will seek a solution by the Fourier method of separating variables in the form $u_1(x,\xi) = N(\xi)d\nu_0(x)/dx$. By substituting u_1 into (8), we get an equation in N:

$$
\frac{d}{d\xi}\left(K(\xi)\frac{d}{d\xi}(N(\xi)+\xi)\right) = 0. \tag{9}
$$

$N = n_1(\xi)$ is a particular solution of equation (9) which is a one-dimensional analogue of the problems on a cell (3.1)–(3.5) and (3.7)–(3.10).

5. Integrating (4_2) with respect to ξ over $[0,1]$ and considering the periodicity of $u_1(\xi)$ and $K(\xi)$ yields

$$
\int_0^1\left(\frac{\partial}{\partial x}\left(K(\xi)\frac{\partial u_1(x,\xi)}{\partial\xi}\right) + \frac{\partial}{\partial x}\left(K(\xi)\frac{\partial u_0(x,\xi)}{\partial x}\right)\right)d\xi = f(x). \tag{10}
$$

Since this equation follows from (4_2), it is a necessary condition for (4_2) to be solvable in the class of 1-periodic functions in ξ. It will be proved in Chapter 2 that (10) is also a sufficient condition.

In view of the relations (5)–(7), we rewrite (10) as

$$\widehat{K}\frac{d^2 v_0}{dx^2} = f(x). \tag{11}$$

It is natural to call (11) an *averaged equation* and $\widehat{K}$ an *effective coefficient*. If $v_0(x)$ is a solution of equation (11), then problem (4_2) is solvable with respect to u_2 in the class of periodic functions in ξ. Consequently, due to (3) and (4), expansion (2) satisfies (1) at least up to terms of order ε.

Thus, if an original inhomogeneous rod is replaced with a homogeneous one having a constant effective characteristic $\widehat{K}$, the properties of the latter will be close, in a sense, to those of the former. Formula (7), which defines $\widehat{K}$, is a one-dimensional analogue of (3.6) and (3.11). The closeness of the original model to the averaged one is investigated more rigorously below.

Note that (2) and (5)–(9) imply an approximate formula for the flow, which is analogous to (4.1):

$$K\left(\frac{x}{\varepsilon}\right)\frac{du}{dx} \approx K(\xi)\frac{d(N_1+\xi)}{d\xi}\frac{dv_0}{dx}.$$

Further expansion (2) in powers of ε can yield a function satisfying (1) up to any predetermined order ε^j. The appropriate construction is given in §2.3. It will be shown in Chapter 2 that the above approximation to the solution also holds in the case of the discontinuity coefficient $K(\xi)$.

Consider a boundary value problem for (5.1) on the segment $[0,1]$ for $f \equiv 0$ and the boundary values $u(0) = 0$ and $u(1) = 1$. We set

$$K(\xi) = \begin{cases} K_1 & \text{for } \xi \in (0, 1/2), \\ K_2 & \text{for } \xi \in (1/2, 1). \end{cases}$$

The problem has an analytical solution

$$\begin{aligned} u(x) &= \int_0^x \frac{1}{K(x/\varepsilon)}\,dx \left(\int_0^1 \frac{1}{K(x/\varepsilon)}\,dx\right)^{-1} = x + \varepsilon\psi\left(\frac{x}{\varepsilon}\right), \\ \psi(\xi) &= \int_0^\xi \left(\frac{\widehat{K}}{K(\xi)} - 1\right) d\xi. \end{aligned} \tag{1}$$

It turns out that for $f \equiv 0$, two terms of the asymptotic expansion $n_0(x) + \varepsilon u_1(x,\xi), \xi = x/\varepsilon$, give an approximation which coincides with an exact solution of (1) (provided the boundary values were chosen appropriately for v_0 and u_1, see §2.4).

The following estimate of accuracy for the solution will be produced in §2.4: $\|u - v_0(x)\|_{c([0,1])} = 0(\varepsilon)$, i.e., $\max|u(x) - v_0(x)| \leqslant c_0\varepsilon$. It can be verified in the case under consideration that

$$c_0 = \max_{\xi \in [0,1]} |\psi(\xi)| = \frac{1}{2} \frac{|K_1 - K_2|}{K_1 + K_2} \leqslant \frac{1}{2}.$$

Since the total of layers $\mathcal{N}$ is $2/\varepsilon$, we have

$$\max_{[0,1]} |u(x) - v_0(x)| \leqslant 1/\mathcal{N}. \tag{2}$$

Estimate (2) can give an idea of the magnitude of error of the averaging method, and it shows that there is an inverse relationship between a maximum value of the absolute difference between solutions of the original and averaged problems and the number of layers in a composite. Therefore, the accuracy 10^{-2} is attained for a composite consisting of 100 layers. In more complex multidimensional problems, for example, the factors of proportionality between error and ε depend of many parameters and often cannot be estimated as explicitly as in (2). In such a situation, one has to be content with the order of smallness of error as $\varepsilon \to 0$, and the estimates such as (2) can be used as models.

REMARK 1: It is generally assumed in the book that the initial data of a problem including boundary (initial) conditions, right-hand sides, and coefficients are infinitely differentiable with respect to the variables x_q and t. Actually it suffices to assume that there exist variables of some finite order of differentiability to obtain asymptotics in ε of a given order of accuracy.

REMARK 2: We also assume that the coefficients of the equation are piecewise smooth and infinitely differentiable with respect to the ζ_q outside the discontinuity surfaces. Actually, it suffices in most cases [8, 53, 54, 188] for the coefficients to be measurable with respect to ζ_q. However, the presentation under so weak an assumption is less clear.

REMARK 3: In real problems, as a rule, the right-hand side f in equations either identically equals zero or oscillates rapidly, i.e.,

$$f = f(x_i, \ldots, x_s, \xi_i, \ldots, \xi_s)|_{\xi = x/\varepsilon},$$

with function f being 1-periodic in the ξ_q. For simplicity, we often consider the case of $f = f(x)$ when the right-hand side is independent of the fast variables ξ_q. The case of $f = f(x,\xi)|_{\xi=x/\varepsilon}$ yields the same averaged zero-order equations, but they contain $\langle f(x), \xi \rangle$ in the right-hand side instead of f. This fact can be seen for T-periodic problems from the general averaging scheme described in §5.1. The case of boundary value problems in the simplest domains is discussed in [112].

§6 Where Can These Methods Help?

When a specimen made of a composite material is under some loads, it is of interest to determine its stresses. It would appear initially that the effective characteristics of the composite can be first determined by bed-testing, and then the stresses in the material with such characteristics can be computed using numerical integration.

In this way, we get a true picture of translations. The true translations u are close to the average translations v_0 obtained by the above-described method: $u \approx v_0$. But, as has been mentioned above, there is closeness between actual stresses and those produced by calculations for a specimen with average (effective) characteristics. The trouble is that the stresses are calculated in terms of derivatives a solution, and in general it does not follow from $u \approx v_0$ that $\partial u/\partial x_i \approx \partial v_0/\partial x_i$.

Additional investigations are needed to get the correct picture of stresses. The complexity of the constructions made in §3 and 4, as compared to those hitherto adopted in mechanics, is the "price of accuracy," in the words of Ilyushin. We obtain an asymptotically correct representation of a solution, and the complexity of the representation and the method by which it is constructed is due to the fact that it is a correct approximation to the solution; there is no other accurate approximation within the selected model.

Mathematical investigations of processes in periodic media are also of interest because mathematical theory can often predict properties of a composite, whereas the latter is being engineered and thus the problem of producing a material with predetermined properties can be considered, and optimization of the composite's performance in some particular aspects can be realized.

It should be noted that mathematical modelling of processes in composites is often difficult due to the lack of a satisfactory physical model. For example, there is a transition zone at the contact point of two materials where they are glued or welded together or where diffusion occurs. This zone can be allowed for in our investigations, but more often than not, the material properties in this zone are not sufficiently known.

The procedure described in this book is applicable to composites with periodic structures where the components are prestressed. But in certain cases, its real value may also be limited due to lack of information about the state of matter in such a composite.

§7 Advantages and Disadvantages of the Averaging Method

1. The asymptotic averaging method is an effective technique for investigating both macroscopic and microscopic properties of periodic structures. Even the first approximation yields an error of order ε/L, where ε is the side of a periodic cell and L the characteristic macroscopic size of the problem. This value is within thousandths of one percent for many composites. However, this is not the limit. An asymptotic expansion of the solution enables accuracy of order $(\varepsilon/L)^n$ to be reached for any integer $n > 0$.

2. The averaging method is universal. It is applicable to all kinds of processes that might occur in periodic media, such as elastic vibrations, heat propagation, diffusion, fluid percolation, electromagnetic oscillations, and radiation, among others. It is possible to consider both linear and nonlinear models, differential as well as operator equations. The procedure described in Chapter 9 can also take into account boundary effects, in some cases to any degree of accuracy.

3. The averaging method is rigorously justified for many models. Theorems have been proved about the solvability of the problems which are elementary steps of an algorithm for constructing an asymptotic solution. Estimates of the method's accuracy have been obtained to determine the order of error in a particular approximation and, consequently, to draw distinct limits for application of this approximation.

4. The main disadvantage of the method is that the numerical solution of cell problems of types (3.1)–(3.5) and (3.7)–(3.10) is required. Three-dimensional problems on a cell take a lot of computer time and memory capacity.

It should be borne in mind, however, that the many existing "forks," mixing rules, and engineering design formulas frequently produce very rough approximations to effective characteristics. Thus, for composites with widely different characteristics of their components, the error of the Voight–Reuss–Hill approximations is comparable to the effective characteristics themselves and may exceed 100 percent (see Figures in §4.1).

The above-mentioned design formulas have rigid limits of applicability, and moreover, the question of the error value and the applicability limits often has not been studied sufficiently well. At the same time, the methods described below yield simple explicit formulas for the leading term of the effective coefficients in the case, being quite important for the applications of composite materials reinforced with high-modulus fibers (Chapter 7 and 8).

5. The traditional methods do not describe the macroscopic characteristics of processes, such as local strains and stresses, and do not take into account effects in the boundary layer. The averaging method permits the limits of applicability to be established for the approximate engineering methods used now for design and to considerably improve their accuracy when necessary. Thus, a unified approach becomes possible to a number of problems that seem to be dissimilar at first sight.

§8 A Brief Review of the Literature

The question of calculating effective characteristics of inhomogeneous media with periodic structures was already raised in the classical works by Poisson, Maxwell, Rayleigh, Voight, and Reuss. Thus, Voight [220] suggested that parameters of polycrystals, such as the stiffness tensor, be calculated by averaging the appropriate values over volume and orientation. For the same case, Reuss [212] used averaging of the components of the reverse tensor (compliance). It was shown later [59, 159, 194, 217] that Voight's method gives the upper bound of effective parameters, whereas the method proposed by Reuss gives their lower bound.

As has been mentioned above (see also [182, 190]), the "fork" between these approximations (Hill's fork) can be quite wide. Therefore, devices are often used in practice to establish averaged relations whose applicability range is very difficult to estimate and where there is no way to improve the accuracy.

An asymptotically exact scheme for calculating the effective characteristics of laminated media was first proposed in [85, 86].

Methods of the complex analysis have been used in [37, 45] for exact solutions of two-dimensional problems of elasticity theory for composite materials and perforated plates and shells. Further development of the mathematical modelling of processes in homogeneous media with periodic structures is based on the theory of averaging differential equations with rapidly oscillating coefficients (see §1).

Works on averaging partial differential equations in domains with fine-grained boundaries first appeared in the 1960s [93, 94].

First results from averaging partial differential equations of divergent type were produced in [8–10, 22, 106, 180, 186, 198, 213]. In particular, averaging of parabolic and hyperbolic equations as well as those of stationary and nonstationary simultaneous equations of the elasticity theory were obtained in [8]. An asymptotic expansion of solutions of nonstationary second-order simultaneous equations and averaged systems were constructed in [9]. The results of [8, 9] were generalized into nonlinear second-order [10, 17] and higher-order [10] equations, as well as into operator equations [11]. The author of [22] proposed a method of averaging variational problems at the physical level of rigour that was later developed in [21].

Further progress of the theory of averaging processes in periodic structures is associated with [187, 188, 206, 207] (variational inequalities, equations with coefficients oscillating faster in the time variable than in the space variable, etc.).

A number of authors studied convergence of solutions of equations with rapidly oscillating coefficients to a solution of an averaged equation that does not require the coefficients to be periodic, including a nearly periodic case [62], sufficient [144,156] and necessary [106] conditions for the convergence of solutions with a weak convergence of the equation coefficients, an averaging of equations with random coefficients [17, 63, 64, 95, 157, 161, 175, 185, 211], questions concerning G-convergence of elliptic [53] and parabolic [54] differential operators, and of nonlinear operators [135]. Note, however, that the mathematical questions of averaging the media with random properties have not yet been looked into well enough. However, there are some works in mechanics devoted to this problem [30, 87, 160, 169].

Problems relating to the theory of the boundary layer in inhomogeneous media were first treated by [110–113]. Boundary effects of contact of two periodic structures are investigated in [111–113]. A boundary layer is built in time for a parabolic equation in [132]. The authors of [39, 40, 60, 134, 176, 205] also consider some mathematical questions of the boundary-layer theory for inhomogeneous media.

Averaging of composite media with widely different properties of the components is studied in [114, 115, 119, 120, 154]. Methods of averaging the processes in skeletal structures (see Chapter 8) are treated by [116–118, 121, 137]. The principle of splitting an averaged operator formulated in [118] makes it possible to obtain explicit formulas for averaging the characteristics of skeletal structures. Other approaches to averaging latticed structures are set forth in [2, 131].

Methods for the calculation of inhomogeneous shells (ribbed, folded, corrugated, perforated, and other types) are developing rapidly on the basis of the averaging theory; see [3–7, 50, 65, 83, 92, 124, 195].

Problems in eigenvalues for equations with rapidly oscillating coefficients are investigated in [115, 168, 200, 205, 218, 219]. High-frequency asymptotics of a wave equation in a periodic structure is presented in [12].

Some works are devoted to averaging the Stokes and Navier–Stokes equation [20, 23, 24, 39, 40, 214], the Maxwell equation [214], the systems of equations of elasticity theory [8, 23, 26, 50, 59, 60, 68, 109, 125–127, 177–179, 191–193, 210, 214] and viscoelasticity theory [11, 16, 201, 202, 214], the transfer equation [15, 102, 187], the systems of equations of plasticity [18] and of radiation heat transfer [13], and Stefan's problem [197]. Processes in porous media are averaged in [13, 22, 26, 39, 40, 60, 64, 93–95, 109, 118, 121, 161–165, 191, 192, 201, 203–205, 208, 210, 214, 218, 219].

Regretfully, we had to leave out a great number of interesting works which studied microinhomogeneous media at the physical level of rigour; we just note works [32, 34, 35, 48, 49, 52, 71, 91, 138, 140, 158].

In particular, formulas such as (3.1)–(3.11) were obtained in [34, 48, 49, 138] for averaged coefficients (in the case of a symmetric cell), which coincide with those derived by the averaging methods.

CHAPTER 1

FORMULATION OF ELEMENTARY BOUNDARY VALUE PROBLEMS

Sections 1 and 2 deal with the classical and generalized formulations of boundary value problems for the heat equation; §3 presents the generalized formulations of problems for the basic equations of mathematical physics and states the existence and uniqueness theorems for generalized solutions. These questions are dealt with in more detail in the monographs [74–76, 78–80, 147, 151, 155] and articles [43, 51, 77, 104, 105, 171].

§1 The Concept of the Classical Formulation of a Boundary Value Problem for Equations with Discontinuous Coefficients

The process of heat transfer in a one-dimensional medium is described by the heat equation

$$C(x)\frac{\partial u}{\partial t} = \frac{\partial}{\partial x}\left(K(x)\frac{\partial u}{\partial x}\right) + f(x, t), \tag{1}$$

where $C(x) \geqslant 0$ is the heat capacity per unit volume, $K(x) > 0$ is a conductivity coefficient, and f is the density of the heat sources. For definiteness, consider a boundary value problem in the domain $a < x < b$, $0 < t$ for the boundary and initial conditions

$$u(x, 0) = u_0(x), \quad u(a, t) = u_1(t), \quad u(b, t) = u_2(t). \tag{2}$$

In particular, the coefficients C and K may be discontinuous. If C and K are continuous on $[a, b]$, with a possible exception of the points $x^1, \dots, x^{q-1}$ where they may have first-order discontinuities, and if they are continuously differentiable at all other points, then (1) holds only in the smoothness domains of the coefficients, $x^i < x < x^{i+1} (i = 0, \dots, q-1)$; here $x^0 = a, x^q = b$. In the absence of concentrated sources, with $x = x^i (i = 1, \dots, q-1)$, the continuity conditions for the temperature $u(x, t)$ and the heat flow $K(x)\partial u/\partial x$ should be satisfied:

$$[u]|_{x=x^i} = 0, \quad \left[K(x)\frac{\partial u}{\partial x}\right]\Big|_{x=x^i} = 0;$$

here the symbol $[\bullet]$ means the difference between the right-hand and left-hand limit values of a function at a given point:

$$[g]|_{x=y} = g(y+0, t) - g(y-0, t), \quad g(y \pm 0, t) = \lim_{\eta \to 0, \eta > 0} g(y \pm \eta, t).$$

The continuity condition of the heat flow follows from the law of heat conservation written in the integral form [142].

Consider a problem on heat conduction in a multi-dimensional anisotropic inhomogeneous medium. Let the heat capacity per unit volume be $C(x)$ and the elements of the conductivity tensor $A_{ij}(x)(x \in R^s)$ be smooth functions everywhere except on sufficiently smooth $(s-1)$-dimensional surfaces where they suffer a first-order discontinuity. Then, assuming that $u(x,t)$ and the initial values are piecewise smooth, we can derive the classical formulation of the problem from the law of heat conservation, namely, that due to the discontinuity surfaces of the coefficients, $u(x,t)$ satisfies everywhere the heat equation

$$C(x)\frac{\partial u}{\partial t} - \sum_{i,j=1}^{s} \frac{\partial}{\partial x_i}\left(A_{ij}(x)\frac{\partial u}{\partial x_j}\right) = f(x, t), \tag{3}$$

and the conjugation conditions

$$[u]|_{\Gamma_k} = \left[\frac{\partial u}{\partial \nu}\right]\Big|_{\Gamma_k} = 0 \tag{4}$$

are satisfied on the discontinuity surfaces Γ_k. Here $[\varphi]|_\Gamma$ means the difference between the limit values of $\varphi(x,t)$ on Γ in the approach from different sides of Γ at x and $\frac{\partial u}{\partial v} = \sum_{i,j=1}^{s} A_{ij}(x)\frac{\partial u}{\partial x_j} n_i$, n_i are the direction cosines of the normal to Γ.

By analogy with problems (3) and (4), the classical formulation can be stated for elliptic or hyperbolic equations with discontinuous coefficients, namely, an equation (system of equations) is satisfied because of the discontinuity surfaces, and two conjugation conditions of type (4) are satisfied on the discontinuity surfaces (see Introduction).

§2 The Concept of Generalized Solution

The concept of a generalized solution often becomes useful in considering equations with discontinuous coefficients. The function $u(x,t)$ will be called a *generalized solution* of problems (1.1) and (1.2) if it is continuous in $[a,b] \times [0,\infty)$, satisfies conditions (1.2), has the derivative $\partial u/\partial x$ which is continuous in each of the sets $[x^{i-1}, x^i] \times [0,\infty)$, and if for any infinitely differentiable function $\varphi(x,t)$ vanishing when $x = a$ and $x = b$, it satisfies the *integral* identity

$$\int_a^b C(x)\,u(x,\tau)\,\varphi(x,\tau)\,dx - \int_a^b C(x)\,u_0(x)\,\varphi(x,0)\,dx + \\ + \int_0^\tau\int_a^b \left(-C(x)\,u\frac{\partial \varphi}{\partial t} + K(x)\frac{\partial u}{\partial x}\frac{\partial \varphi}{\partial x}\right)dx\,dt = \int_0^\tau\int_a^b f(x,t)\varphi\,dx\,dt, \tag{1}$$

where $\int_a^b K(x)\frac{\partial u}{\partial x}\frac{\partial \varphi}{\partial x}dx$ is understood in the sense $\sum_{i=1}^{N}\int_{x^{i-1}}^{x^i} K(x)\frac{\partial u}{\partial x}\frac{\partial \varphi}{\partial x}dx$, and τ is an arbitrary positive number. Integral identity (1) is obtained from equation

(1.1) by multiplying by $\varphi(x,t)$, integrating over $[a,b] \times [0,\tau]$, and by subsequent integration by parts with respect to x and t.

It has been proved in [78] that the classical formulation of the problem is equivalent to the definition of the solution by means of the integral identity, under the assumption that the coefficients are piecewise smooth and the integral and boundary conditions agree in smoothness. The formulation of a problem in the form of an integral identity will henceforth be called the *generalized formulation.* When the initial data are not piecewise smooth functions (for example, $K(x)$ is a bounded, measurable, positive function), the classical formulation of the problem may not make sense, whereas the generalized formulation remains meaningful for broad classes of nonsmooth, discontinuous initial data.

In the same way as identity (1), integral identities are derived from classical formulations for multi-dimensional elliptic, parabolic, and hyperbolic equations. Where the initial and boundary conditions are smooth and in agreement and the coefficients of the equations are smooth functions everywhere except on some (s–1)–dimensional smooth surfaces on which the coefficients have a first-order discontinuity, the classical formulation is equivalent to the generalized one. Rigorous statements of theorems concerning the relationship between the classical and generalized formulation can be found in [51, 74, 76, 78–80, 104, 105].

It is possible to give another definition for generalized solution of a problem with discontinuous data (boundary conditions and coefficients of equations), namely, by considering the sequence of boundary value problems with smooth data converging (in the required norms) to the initial data. If there is a limit of the sequence of solutions of such problems in the required norm, it will be called a *generalized solution* of the original problem. It has been proved for broad classes of linear boundary value problems that these definitions of generalized solution are equivalent.

Starting with different definitions of generalized solution, different approaches for the construction and justification of the average procedure can be proposed in the case of discontinuous coefficients.

One of the possible ways of justifying asymptotics, which we sometimes use, is to substitute an asymptotic expansion into the integral identity and to estimate the accuracy through its discrepancy.

This can also be done on the basis of the second definition of the generalized solution. Having obtained asymptotics in the case of smooth coefficients, we pass to the limit of discontinuous coefficients both in the solution itself and in the asymptotic expansion. This method is more complicated technically because it requires control over the dimensionality of the estimate of accuracy in the limiting process, but the idea is much simpler.

We prefer the following scheme for the presentation of the material. We average the problem assuming the coefficients to be smooth, extend the resulting asymptotics to the case of discontinuous coefficients, taking into account the conjugation conditions on the discontinuity surfaces, and justify the asymptotics by means of a priori estimates.

§3 Generalized Formulations of Problems for the Basic Equations of Mathematical Physics

As indicated above, the generalized solution of a problem may have different definitions. This section presents a universally accepted definition of the concept of a generalized solution, which we shall use in the following.

Let G be a bounded domain in R^s with a piecewise smooth boundary ∂G, and $u(x)$ a function defined on G. Denote by $L_2(G)$ the space of functions with a bounded norm

$$\| u \|_{L_2(G)} = \sqrt{\int_G u^2(x)\, dx} .$$

(Strictly speaking, the integral is understood throughout the book as the Lebesque integral. However, the generalized solution coincides with the classical one in an overwhelming majority of the problems with piecewise smooth coefficients and smooth data and the integrals in the norms of the solutions, the right-hand sides of the equations and of the initial and boundary conditions are the Riemann integral. Therefore, the reader does not need to know the Lebesque integral theory. By the same token, the reader is not required to know the theories of distributions and generalized solutions.)

We say that $u(x)$ belongs to $W_2^1(G)$ if there exist functions $u^{(1)}(x), \ldots, u^{(s)}$ and a sequence of continuously differentiable functions $u_k(x)$ in $\overline{G}(k = 1, 2, \ldots)$ such that

$$\| u_k(x) - u(x) \|_{L_2(G)} \to 0, \quad \left\| \frac{\partial u_k(x)}{\partial x_i} - u^{(i)}(x) \right\|_{L_2(G)} \to 0 \ \text{ for } \ k \to \infty.$$

The functions $u^{(i)}(x) (i = 1, \ldots, s)$ will be called the *generalized derivatives* of u and denoted by $\partial u / \partial x_i$.

On $W_2^1(G)$ we define the norm

$$\| u \|_{W_2^1(G)} = \sqrt{\int_G \left(u^2 + \sum_{i=1}^{s} \left(\frac{\partial u}{\partial x_i} \right)^2 \right) dx}.$$

If each of the functions $u_k(x)$ vanishes in a neighbourhood of the boundary of G, we say that $u(x)$ belongs to $\overset{\circ}{W}{}_2^1(G)$.

Consider the cylinder $G \times (0,T)$. The definition of a function of W_2^1 also encompasses obviously the case of an $(s+1)$-dimensional domain $G \times (0,T)$. Let us denote by $W_{2,0}^1(G \times (0,T))$ the subspace $W_2^1(G \times (0,T))$ which consists of the functions $u(x,t)$ such that there exists a sequence of functions $u_k(x,t)$ continuously differentiable in $\overline{G} \times [0,T]$ that vanish in a neighbourhood of $\partial G \times [0,T]$ and satisfy the relation $\lim_{k \to \infty} \| u_k(x,t) - u(x,t) \|_{W_2^1(G \times (0,T))} = 0$, where

$$\| u \|_{W_2^1(G \times (0,T))} = \sqrt{\int_{G \times (0,T)} \left(u^2 + \sum_{i=1}^{s} \left(\frac{\partial u}{\partial x_i} \right)^2 + \left(\frac{\partial u}{\partial t} \right)^2 \right) dx\, dt}.$$

We say that $u(x,t)$ belongs to $W_2^{1,0}(G \times (0,T))$ if there exist functions $u^{(1)}(x,t), \ldots, u^{(s)}(x,t)$ and a sequence of functions $v_k(x,t)$ continuously differentiable in $\overline{G} \times [0,T]$ such that

$$\| v_k(x, t) - u(x, t) \|_{L_2(G\times(0,T))} \to 0$$
$$\left\| \frac{\partial v_k}{\partial x_i}(x, t) - u^{(i)}(x, t) \right\|_{L_2(G\times(0,T))} \to 0 \quad \text{for } k \to \infty,$$

where $\|u\|_{L_2(G\times(0,T))} = \sqrt{\int_{G\times(0,T)} u^2 dx dt}$. The functions $u^{(i)}(x,t)(i = 1, \ldots, s)$ will be denoted by $\partial u(x,t)/\partial x_i$ as above.

On $W_2^{1,0}(G \times (0,T))$, we introduce the norm

$$\| u \|_{W_2^{1,0}(G\times(0,T))} = \sqrt{\int\limits_{G\times(0,T)} \left(u^2 + \sum_{i=1}^{s} \left(\frac{\partial u}{\partial x_i} \right)^2 \right) dx\, dt}.$$

If the sequence $v_k(x,t)$ vanishes in a neighbourhood of $\partial G \times [0,T]$, we say that $u(x,t)$ belongs to $\overset{\circ}{W}{}_2^{1,0}(G \times (0,T))$.

Let us denote by $C(\overline{G})$ a set of functions continuous in $\overline{G}$ with the norm $\|u\|_{C(\overline{G})} = \max_{x\in\overline{G}} |u(x)|$, and by $C^\infty(\overline{G})$ we denote the set of functions infinitely differentiable in $\overline{G}$.

If all the components Ψ_i of a vector Ψ belong to a space of functions V, we say that Ψ also belongs to V, assuming that $\|\Psi\|_V = \sqrt{\sum_i \|\Psi\|_V^2}$.

1. *Elliptic equations*

Let $G, G_i (i = 1, 2, \ldots, N)$ be unbounded domains in R^s; $\partial G, \partial G_i$ their piecewise smooth boundaries satisfying the cone condition; $A_{ij}(x)$, $A_i(x)$, $B_i(x)$, $A(x)$, $f(x)$, and $f_i(x)$ functions infinitely differentiable everywhere except on $\bigcup_{i=1}^N \partial G_i$ up to $\bigcup_{i=1}^N \partial G_i \bigcup \partial G$, whereas on $\bigcup_{i=1}^N \partial G_i$ these functions and all their derivatives may have a first-order discontinuity. (It is implied here that the surfaces ∂G_i split G into a finite number of disjoint domains $\widetilde{G}_q$, and the coefficients as well as f and f_i are infinitely differentiable in each of the $\overline{\widetilde{G}}_q$.) The equation

$$\sum_{i,j=1}^{s} \frac{\partial}{\partial x_i} \left(A_{ij}(x) \frac{\partial u}{\partial x_j} \right) + \sum_{i=1}^{s} \frac{\partial}{\partial x_i} (A_i(x) u) + \\ + \sum_{i=1}^{s} B_i(x) \frac{\partial u}{\partial x_i} + A(x) u = f(x) + \sum_{i=1}^{s} \frac{\partial f_i(x)}{\partial x_i} \tag{1}$$

is called an *elliptic equation* if for any fixed $x \in G \backslash \bigcup_{i=1}^N \partial G_i$ and for any

$(\xi_1, \dots, \xi_s)$, the relations

$$\sum_{i,j=1}^{s} A_{ij}(x)\,\xi_i\xi_j \geqslant \varkappa \sum_{i=1}^{s} \xi_i^2, \quad A_{ij}(x) = A_{ji}(x), \tag{2}$$

are true, with the constant $\kappa > 0$ being independent of x.

If the summation over one or several repeated indices is from 1 to s, the summation sign will be generally omitted. Thus we write $A_{ij}(x)\xi_i\xi_j \geqslant \kappa\xi_i\xi_i$ instead of (2).

Elliptic equations often describe stationary fields. Thus, the equation

$$\frac{\partial}{\partial x_i}\left(A_{ij}(x)\frac{\partial u}{\partial x_j}\right) = f(x) + \frac{\partial f_i(x)}{\partial x_i} \tag{3}$$

may describe a temperature field in a medium with the conductivity tensor $A_{ij}(x)$, a concentration field in a medium with the diffusion tensor $A_{ij}(x)$, and a function of groundwater lowering in a medium with the percolation tensor $A_{ij}(x)$. The Dirichlet condition

$$u|_{\partial G} = 0. \tag{4}$$

is supposed to be satisfied on the boundary ∂G.

The function $u(x) \in \overset{\circ}{W}{}_2^1(G)$ which satisfies the integral identity

$$\int_G \left(A_{ij}(x)\frac{\partial u}{\partial x_j}\frac{\partial \varphi}{\partial x_i} + A_i(x)\,u\frac{\partial \varphi}{\partial x_i} - B_i(x)\frac{\partial u}{\partial x_i}\varphi - A(x)\,u\varphi\right)dx =$$

$$= \int_G \left(-f(x)\,\varphi + f_i(x)\frac{\partial \varphi}{\partial x_i}\right)dx. \tag{5}$$

for any function $\varphi(x)(\varphi(x)|_{\partial G} = 0)$ infinitely differentiable in $\overline{G}$ is called a generalized Sobolev solution of problems (1) and (4) in $W_2^1(G)$.

Remark 1: For the piecewise smooth coefficients and the right-hand sides (f, f_i), it can be shown that the generalized solution satisfies equations (1) at all points of smoothness, and we have $[u] = 0$, $(A_{ij}\partial u/\partial x_j + A_i u - f_i)n_i = 0$ on the discontinuity surfaces (lines), where n_i are direction cosines of the normal. The same conditions occur on the discontinuity surfaces also in the case of nonstationary problems (8) and (9).

Three Fredholm theorems [76] hold for problems (1) and (4). Let us formulate one of them.

Theorem 1. *Problems (1) and (4) cannot have more than one generalized solution in $W_2^1(G)$. Then this solution exists and the following estimate is valid*

$$\|u\|_{W_2^1(G)} \leqslant c\left(\|f\|_{L_2(G)} + \sum_{i=1}^{s}\|f_i\|_{L_2(G)}\right), \tag{6}$$

where c is a positive constant independent of the right-hand side of (1). *In particular, problem* (3), (4) *has a unique solution in* $W_2^1(G)$ *and estimate* (6) *holds.*

2. *Parabolic and hyperbolic equations*

Further we set $\sum = \bigcup_{i=1}^N \partial G_i \times [0,\infty)$. Let $R(x,t)$, $S(x,t)$, $A_{ij}(x,t)$, $A_i(x,t)$, $B_i(x,t)$, $A(x,t)$, $f(x,t)$, $f_i(x,t)$ be functions infinitely differentiable in $(G \times [0,\infty))\setminus\sum$ up to $\sum$, and on $\sum$ these functions and their derivatives have a first-order discontinuity. Let then the following conditions be valid for any $x \in G\setminus\bigcup_{i=1}^N \partial G_i$ and $t > 0$:

$$\begin{gathered} R(x,t) > R_0, \quad S(x,t) > S_0, \\ A_{ij}(x,t)\xi_i\xi_j \geqslant \varkappa\xi_i\xi_i \quad \forall(\xi_1,\ldots,\xi_s) \in \mathbf{R}^s, \\ A_{ij}(x,t) = A_{ji}(x,t), \end{gathered} \tag{7}$$

where R_0, S_0, and κ are positive constants independent of x, t, ξ. The equation

$$S(x,t)\frac{\partial u}{\partial t} - \frac{\partial}{\partial x_i}\left(A_{ij}(x,t)\frac{\partial u}{\partial x_j} + A_i(x,t)u\right) + \\ + B_i(x,t)\frac{\partial u}{\partial x_i} + A(x,t)u = f(x,t) + \frac{\partial f_i(x,t)}{\partial x_i} \tag{8}$$

is called a *parabolic equation.*

Parabolic equations can describe nonstationary thermal and diffusion processes, ground-water percolation, etc. The physical meaning of A_{ij} is the same as in the elliptic case, and $S(x,t)$ is the heat capacity per unit volume equal to specific heat times density (the thermal conductivity problem). It is natural for the applications, as is usually done, to write the term containing differentiation with respect to t in the form of $\frac{\partial}{\partial t}(Su)$, but in the case when S is smoothly dependent on t, this equation reduces to (8). Henceforth, as a rule, the case is considered where S does not depend on t at all.

Let the following initial conditions be given for $t = 0$:

$$u(x,0) = \psi(x), \quad \psi \in \overset{\circ}{W}{}_2^1(G). \tag{9}$$

By the generalized solution of (8), (9), and (4) from $W_{22}^{1,0}(G \times (0,T)$, we mean the function $u(x,t) \in \overset{\circ}{W}{}_2^{1,0}(G \times (0,T))$, satisfying the integral identity

$$\int_0^T\int_G\left(-Su\frac{\partial\varphi}{\partial t} + A_{ij}\frac{\partial u}{\partial x_j}\frac{\partial\varphi}{\partial x_i} + A_iu\frac{\partial\varphi}{\partial x_i} + B_i\frac{\partial u}{\partial x_i}\varphi + \left(A - \frac{\partial S}{\partial t}\right)u\varphi\right)dx\,dt = \\ = \int_G S(x,0)\,\psi(x)\,\varphi(x,0)dx + \int_0^T\int_G\left(f\varphi + f_i\frac{\partial\varphi}{\partial x_i}\right)dx\,dt$$

for any function $\varphi(x,t)(\varphi(x,t)_{\partial G} = 0,\ \varphi(x,T) = 0)$ infinitely differentiable in

$\overline{G} \times [0,T]$.

The following theorem is valid for problems (8), (9), and (4).

THEOREM 2. *A generalized solution of problems (8) and (9) exists and is unique; and for this solution the following estimate holds:*

$$\|u\|_{W_2^{1,0}(G\times(0,T))} \leqslant$$
$$\leqslant c(T)\left[\|\psi(x)\|_{L_2(G)} + \|f\|_{L_2(G\times(0,T))} + \sum_{i=1}^{s}\|f_i\|_{L_2(G\times(0,T))}\right].$$

The proof of the theorem repeats almost literally those of Theorems 3.1 and 4.1 in Chapter 3 of [76].

Consider, further, the equation

$$R(x,t)\frac{\partial^2 u}{\partial t^2} + S(x,t)\frac{\partial u}{\partial t} - \frac{\partial}{\partial x_i}\left(A_{ij}(x,t)\frac{\partial u}{\partial x_j}\right) +$$
$$+ B_i(x,t)\frac{\partial u}{\partial x_i} + A(x,t)u = f(x,t) + \frac{\partial f_i(x,t)}{\partial x_i} \tag{10}$$

for the same conditions (7) on R and A_{ij}. It is called a *hyperbolic equation*; it can model wave processes in media [151].

Typically in applications, the term containing two differentiations with respect to t is of the form $\frac{\partial}{\partial t}(R\frac{\partial u}{\partial t})$, but by analogy with the parabolic case for smooth dependence of R on t, the equation reduces to (10). As a rule, R no longer depends on t in what follows.

Two initial conditions are set at $t = 0$:

$$\begin{aligned} u(x,0) &= \psi_1(x), \quad \psi_1 \in \overset{\circ}{W}{}_2^1(G), \\ \frac{\partial u}{\partial t}(x,0) &= \psi_2(x), \quad \psi_2 \in L_2(G). \end{aligned} \tag{11}$$

By a *generalized solution* of problems (10), (10), and (4) of $W_2^1(G\times(0,T))$, we mean the function $u(x,t) \in W_{2,0}^1(G\times(0,T))$ that satisfies the integral identity

$$\int_0^T\int_G\left(-R\frac{\partial u}{\partial t}\frac{\partial \varphi}{\partial t} + A_{ij}\frac{\partial u}{\partial x_j}\frac{\partial \varphi}{\partial x_i} + B_i\frac{\partial u}{\partial x_i}\varphi + Au\varphi +\right.$$
$$\left. + \left(S - \frac{\partial R}{\partial t}\right)\frac{\partial u}{\partial t}\varphi\right)dx\,dt - \int_G R(x,0)\psi_2(x)\varphi(x,0)\,dx =$$
$$= \int_0^T\int_G\left(f\varphi - f_i\frac{\partial \varphi}{\partial x_i}\right)dx\,dt$$

and the condition $u(x,0) = \psi_1(x)$ for any function $\varphi(x,t)(\varphi(x,t)|_{\partial G} = 0, \varphi(x,T) = 0)$ infinitely differentiable in $\overline{G} \times [0,T]$.

THEOREM 3. *A generalized solution of problems (10), (11), and (4) of $W_2^1(G\times(0,T))$ exists and is unique, and for this solution the following estimate holds:*

$$\|u\|_{W_2^1(G\times(0,T))} \leqslant c(T)\left(\|\psi_1\|_{W_2^1(G)} + \|\psi_2\|_{L_2(G)} + \|f\|_{L_2(G\times(0,T))}\right) + \\ + \sum_{i=1}^{s}\left(\left\|\frac{\partial f_i}{\partial t}\right\|_{L_2(G\times(0,T))} + \max_{t\in[0,T]}\|f_i\|_{L_2(G)}\right).$$

The theorem is proved in the same way as Theorems 3.1 and 3.2 in Chapter 4 of [76].

3. *The system of equations of elasticity theory*

Let $R(x), A_{ij}(x)$ be $s\times s$ matrix-valued functions with elements infinitely differentiable everywhere in G except on $\bigcup_{i=1}^N \partial G_i$ up to $\bigcup_{i=1}^N \partial G_i \bigcup \partial G$, whereas on $\bigcup_{i=1}^N \partial G_i$ these functions and all their derivatives may have a first-order discontinuity. Let $f(x,t), f_i(x,t)$ be s-dimensional vectors with similar smoothness everywhere from $\bigcup_{i=1}^N \partial G_i \times [0,T] (T>0)$ and let $R(x), A_{ij}(x) = (a_{ij}^{kl}(x))$ satisfy the inequalities

$$R(x)\geqslant \rho_0 E, \qquad \sum_{i,j,k,l=1}^{s} a_{ij}^{kl}(x)\,\xi_i^k\xi_j^l \geqslant \varkappa \sum_{i,k=1}^{s}\xi_i^k\xi_i^k$$

$$\forall \xi_i^k \in \mathbf{R}, \quad \xi_i^k = \xi_k^i, \quad x\in G\setminus\bigcup_{i=1}^N \partial G_i, \quad a_{ij}^{kl} = a_{kj}^{il} = a_{il}^{kj} = a_{ji}^{lk},$$

where ρ_0, κ are positive constants and E is the unit matrix.

The system of equations

$$R(x)\frac{\partial^2 u}{\partial t^2} - \frac{\partial}{\partial x_i}\left(A_{ij}(x)\frac{\partial u}{\partial x_j}\right) = f(x,t) + \frac{\partial f_i}{\partial x_i}(x,t) \tag{12}$$

is called a *nonstationary system* of equations of the linear elasticity theory. Here $a_{ij}^{kl}(x)$ is the stiffness tensor, $R(x) = \rho(x)E, \rho(x)$ is the density, and u is the displacement vector.

Let f, f_i be independent of t. The system of equations

$$\frac{\partial}{\partial x_i}\left(A_{ij}(x)\frac{\partial u}{\partial x_j}\right) = f(x) + \frac{\partial f_i}{\partial x_i} \tag{13}$$

is called a *stationary system of equations* of the linear elasticity theory. We assume the boundary condition to be homogeneous:

$$u|_{\partial G} = 0. \tag{14}$$

We set initial values (11) for system (12), where $\psi_1(x), \psi_2(x)$ are vector-functions from $\overset{\circ}{W}{}_2^1(G)$ and $L_2(G)$, respectively.

By a *generalized solution* of $W_2^1(G)$ of problems (13) and (14), we mean a vector-function $u(x) \in \overset{\circ}{W}{}_2^1(G)$ that satisfies the integral identity

$$\int_G \left(A_{ij}(x)\frac{\partial u}{\partial x_j}, \frac{\partial \varphi}{\partial x_i}\right)dx = \int_G \left(-(f(x), \varphi) + \left(f_i(x), \frac{\partial \varphi}{\partial x_i}\right)\right)dx,$$

for any vector-function $\varphi(x), (\varphi(x)|_{\partial G} = 0)$ indefinitely differentiable in $\overline{G}$. Here (a, b) is the scalar product of vectors $a, b \in R^s$, $|a| = \sqrt{(a,a)}$.

THEOREM 4. *There exists a unique generalized solution of $W_2^1(G)$ of problem (13), (14). For this solution, the following estimate holds:*

$$\|u\|_{W_2^1(G)} \leqslant c\left(\|f\|_{L_2(G)} + \sum_{i=1}^{s}\|f_i\|_{L_2(G)}\right).$$

By a generalized solution of problems (12), (14), and (11) of $W_2^1(G\times(0,T))$, we mean a vector-function $u(x,t) \in W_{2,0}^1(G\times(0,T))$ that satisfies the integral identity

$$\int_0^T\int_G \left(\left(-R\frac{\partial u}{\partial t}, \frac{\partial \varphi}{\partial t}\right) + \left(A_{ij}\frac{\partial u}{\partial x_j}, \frac{\partial \varphi}{\partial x_i}\right)\right)dx\,dt -$$

$$-\int_G R(x)\psi_2(x)\varphi(x,0)\,dx = \int_0^T\int_G \left((f, \varphi) - \left(f_i, \frac{\partial \varphi}{\partial x_i}\right)\right)dx\,dt$$

and the condition $u(x,0) = \psi_1(x)$ for any vector function

$$\varphi(x,t)(\varphi(x,t)|_{\partial G\times[0,T]}-, \varphi(x,T) = 0)$$

infinitely differentiable in $\overline{G}\times[0,T]$.

By analogy with Theorems 3.1 and 3.2 in Chapter 4 of [76] and by using Corn's inequality [155], we can prove the following theorem.

THEOREM 5. *There exists a unique generalized solution of $W_2^1(G\times(0,T))$ of problems (12), (14), and (11), and for this solution the following estimate holds:*

$$\|u\|_{W_2^1(G\times(0,T))} \leqslant c(T)\left(\|\psi_1\|_{W_2^1(G)} + \|\psi_2\|_{L_2(G)} + \|f\|_{L_2(G\times(0,T))} + \right.$$

$$\left. + \sum_{i=1}^{s}\left(\left\|\frac{\partial f_i}{\partial t}\right\|_{L_2(G\times(0,T))} + \max_{t\in[0,T]}\|f_i\|_{L_2(G)}\right)\right).$$

The methods of [74–80, 104, 105] can be used to state generalized formulations also for problems with the second-order and third-order boundary conditions as well as with other types of conditions. Instead of condition (4), one may require that the function u be T-periodic in each of the variables x_j or in some of them (see Supplement). We shall later consider individual and simultaneous equations with first-order boundary conditions. However, most of these methods and results can be employed to solve boundary value problems of other types.

Some difficulties arise in constructing asymptotic expansions of the solution of boundary value problems because of the need to take into consideration the effects of the boundary layer (see Chapter 9). Therefore, we often consider equations in the entire space $\mathbf{R}^s$ so as not to complicate the presentation by discussing these subtle effects; and, instead of boundary conditions, the requirement was made that the solution be T-periodic in the x_i; here T is a number independent of ε. If the effective and local characteristics of a medium have been defined correctly, they must coincide with the same characteristics for the periodic problem.

CHAPTER 2

The Concept of Asymptotic Expansion. A Model Example to Illustrate the Averaging Method

This Chapter introduces definitions of an asymptotic expansion of a function, operations with asymptotic series, and discusses elementary examples of the methods of asymptotic expansions applied to the problem of averaging the processes in periodic media. Many model examples of asymptotic methods used for investigating problems of mechanics are discussed in [101, 103].

§1 Asymptotic Expansion. A Formal Asymptotic Solution

1. The notation $\varphi(\varepsilon) = 0(\psi(\varepsilon))$ as $\varepsilon \downarrow 0$ (i.e., as $\varepsilon \rightarrow 0$ over a subset E of the values of $\varepsilon > 0$) means that there exists a constant $A > 0$ such that $|\varphi(\varepsilon)| \leqslant A|\psi(\varepsilon)|$ for any sufficiently small $\varepsilon > 0$. To be more exact, this notation means that there exist values A, $\varepsilon_0 > 0$ such that $|\varphi(\varepsilon)| \leqslant A|\psi(\varepsilon)|$ for $0 < \varepsilon < \varepsilon_0$. In particular, $\varphi(\varepsilon) = 0(\varepsilon^N)$ is equivalent to $|\varphi(\varepsilon) \leqslant A\varepsilon^N$ for a sufficiently small $\varepsilon > 0$. Hereafter, ε belongs to E.

2. Let a normed space B_ε of functions $f(x,\varepsilon)(x \in M \subset \mathbf{R}^s)$ with the norm $\|f(x,\varepsilon)\|_{B_\varepsilon(M)}$ be specified for each sufficiently small $\varepsilon > 0$. The notation $f(x,\varepsilon) = 0(\psi(x,\varepsilon)$ as $\varepsilon \downarrow 0$ in the norm of B_ε means that there exist A, $\varepsilon_0 > 0$ such that $\|f(x,\varepsilon)\|_{B_\varepsilon(M)} \leqslant A\|\psi(x,\varepsilon)\|_{B_\varepsilon(M)}$ for $0 < \varepsilon < \varepsilon_0$. In particular, $f(x,\varepsilon) = 0(\varepsilon^N)$.

Similar symbols may be introduced as $\varepsilon \rightarrow \pm\infty$. Thus, $f(x,\varepsilon) = 0(\psi(x,\varepsilon)$ as $\varepsilon \rightarrow +\infty(\varepsilon \rightarrow -\infty)$ in the norm of B_ε if there exist constants A, $\varepsilon_0 > 0$ such that $\|f(x,\varepsilon)\|_{B_\varepsilon(M)} \leqslant A\|\psi(x,\varepsilon)\|_{B_\varepsilon(M)}$ for all $\varepsilon > \varepsilon_0$, $(\varepsilon < -\varepsilon_0)$.

3. Let the normed spaces $B_{\varepsilon,\mu}$ of the functions $f(x,\varepsilon,\mu)(x \in M \subset R^s)$ be specified for any sufficiently small $M \subset \mathbf{R}^3)$ be specified for any sufficiently small $e > 0$ and $\mu > 0$. The notation $f(x,\varepsilon,\mu) = 0(\psi(x,\varepsilon,\mu)$ as $\varepsilon \downarrow 0$, $\mu \downarrow 0$ in the norm of $B_{\varepsilon,\mu}$ means that there exist constants A, ε_0, $\mu_0 > 0$ such that $\|f(x,\varepsilon,\mu)\|_{B_{\varepsilon,\mu}(M)} \leqslant A\|\psi(x,\varepsilon,\mu)\|_{B_{\varepsilon,\mu}(M)}$ for $0 < \varepsilon < \varepsilon_0$, $0 < \mu < \mu_0$. In particular, $f(x,\varepsilon,\mu) = O(\varepsilon^{N_1} + \mu^{N_2})$ in the norm of B_ε if $\|f(x,\varepsilon,\mu\|_{B_{\varepsilon,\mu}(M)} \leqslant A(\varepsilon^{N_1} + \mu^{N_2})$ for sufficiently small ε,μ.

4. By an *asymptotic expansion* (a.e.) of a function $f(x,\varepsilon)$, we mean the asymptotic series $\sum_{i=l}^{\infty} \varepsilon^i g_i(x,\varepsilon)$, probably divergent, that satisfies the following conditions: for any N there exists a number M_0, such that with each $m \geqslant M_0$

$$f(x, \varepsilon) - \sum_{i=l}^{m} \varepsilon^i g_i(x, \varepsilon) = O(\varepsilon^N)$$

in the norm of $B_\varepsilon(\varepsilon \downarrow 0)$. For a.e., we set

$$f(x, \varepsilon) \sim \sum_{i=l}^{\infty} \varepsilon^i g_i(x, \varepsilon). \tag{1}$$

We now introduce operations with the asymptotic series. The notation

$$\sum_{i=l}^{\infty} \varepsilon^i g_i(x, \varepsilon) \sim \sum_{i=l}^{\infty} \varepsilon^i h_i(x, \varepsilon)$$

means that for any $N > 0$ there exists a number M such that with each $m > M$, the relation

$$\sum_{i=l}^{m} \varepsilon^i (g_i(x, \varepsilon) - h_i(x, \varepsilon)) = O(\varepsilon^N)$$

in the norm of $B_\varepsilon(\varepsilon \downarrow 0)$ is true. The sum, difference, and product of the asymptotic series and other operations are defined in the same way as for ordinary series as follows:

$$\sum_{i=l}^{\infty} \varepsilon^i g_i(x, \varepsilon) \pm \sum_{i=l}^{\infty} \varepsilon^i h_i(x, \varepsilon) = \sum_{i=l}^{\infty} \varepsilon^i (g_i(x, \varepsilon) \pm h_i(x, \varepsilon)),$$

$$\left(\sum_{i=l}^{\infty} \varepsilon^i g_i(x, \varepsilon)\right)\left(\sum_{i=l}^{\infty} \varepsilon^i h_i(x, \varepsilon)\right) = \sum_{r=2l}^{\infty} \varepsilon^r \sum_{i=l}^{r-l} g_i(x, \varepsilon) h_{r-i}(x, \varepsilon),$$

$$K(x, \varepsilon) \sum_{i=l}^{\infty} \varepsilon^i g_i(x, \varepsilon) = \sum_{i=l}^{\infty} \varepsilon^i K(x, \varepsilon) g_i(x, \varepsilon).$$

Notice that the series do not necessarily converge.

5. The functions of g_i in (1) may be asymptotically expanded, in general, in powers of another small parameter μ : $f(x, \varepsilon, \mu) \sim \sum_{i=l}^{\infty} \varepsilon^i g_i(x, \varepsilon, \mu)$, where $g_i(x, \varepsilon, \mu) \sim \sum_{j=q_i}^{\infty} \mu^j h_{ij}(x, \varepsilon, \mu)$, or

$$f(x, \varepsilon, \mu) \sim \sum_{i=l}^{\infty} \varepsilon^i \sum_{j=q_i}^{\infty} \mu^j h_{ij}(x, \varepsilon, \mu).$$

The latter relation means that for any $N_1, N_2 > 0$ there exist numbers M_1 and M_2 such that for any pair $m_1 \geqslant M_1$ and $m_2 \geqslant M_2$, the conditions

$$f(x, \varepsilon, \mu) - \sum_{i=l}^{m_1} \varepsilon^i \sum_{j=q_i}^{m_2} \mu^j h_{ij}(x, \varepsilon, \mu) = O\left(\varepsilon^{N_1} + \mu^{N_2}\right)$$

in the norm of $B_{\varepsilon,\mu}(\varepsilon, \mu \downarrow 0)$ hold.

Operations with double asymptotic series can be introduced by analogy with Subsection 4 above.

6. Given the equation

$$L_\varepsilon u = f, \tag{2}$$

where L_ε is an operator from $B_{1\varepsilon}$ into $B_{2\varepsilon}(B_{1\varepsilon}$ and $B_{2\varepsilon}$ are normed spaces), by the *formal asymptotic solution* (f.a.s.) of equation (2), we shall call the asymptotic series $u^{(\infty)} = \sum_{i=l}^{\infty} \varepsilon^i g_i(x, \varepsilon)$, $g_i(x, \varepsilon) \in B_{1\varepsilon}$ such that for any natural N, there exists a number M for which the relation

$$L_\varepsilon u^{(m)} - f = O(\varepsilon^N) \tag{3}$$

holds for all $m \geqslant M$ as $\varepsilon \downarrow 0$ in the norm of $B_{2\varepsilon}$, where $u^{(m)} = \sum_{i=l}^{m} \varepsilon^i g_i(x, \varepsilon)$.

If L_ε is a linear operator and there is an a priori estimate $\|u\|_{B_{1\varepsilon}} \leqslant c_1 \varepsilon^{c_2} \|f\|_{B_{2\varepsilon}}$, where $c_1 > 0$, and c_2 are constants independent of ε, then it follows from (3) that for any natural N there exists a number M such that $\|u^{(m)} - u\|_{B_{1\varepsilon}} = O(\varepsilon^N), \varepsilon \downarrow 0$, for all $m \geqslant M$, and consequently, the f.a.s. of $u^{(\infty)}$ is an a.e. of the exact solution of the problem: $u \sim u^{(\infty)}$.

The f.a.s. of the problem $L_{\varepsilon,\mu} u = f, \varepsilon \downarrow 0, \mu \downarrow 0$ is found in much the same way.

§2 Asymptotic Expansion of a Solution of the Equation $u = 1 + \varepsilon u^3$

In this section, we shall demonstrate principal steps in constructing and justifying an a.e. solution by using a cubic equation as an example.

1. *Formulation of the problem*

Construct a f.a.s. as $\varepsilon \downarrow 0$ of a solution of the equation

$$u - 1 - \varepsilon u^3 = 0, \tag{1}$$

satisfying the condition

$$0 \leqslant u \leqslant 2. \tag{2}$$

Condition (2) acts as a "class" where the solution is sought because there are also other roots (1) that do not satisfy this condition. (After substituting $v = u\sqrt{\varepsilon}$, we obtain the equation $v - v^3 - \sqrt{\varepsilon} = 0$. All three of its f.a.s. are sought in the form of series in powers of $\sqrt{\varepsilon}$, as in the above-mentioned examples.)

2. *Construction of f.a.s.*

We will seek f.a.s. of the form

$$u^{(\infty)} = \sum_{i=0}^{\infty} \varepsilon^i u_i. \tag{3}$$

By substituting series (3) into the left-hand side of equation (1) and simplifying, we get

$$\sum_{i=0}^{\infty} \varepsilon^i u_i - 1 - \varepsilon\left(\sum_{i=0}^{\infty} \varepsilon^i u_i\right)^3 \sim$$
$$\sim (u_0 - 1) + \sum_{i=1}^{\infty} \varepsilon^i \left(u_i - \sum_{j=0}^{i-1}\sum_{k=0}^{i-j-1} u_j u_k u_{i-j-k-1}\right). \tag{4}$$

We require that all the terms of the formal series (4) be equal to zero in order to have the recurrent chain of equalities

$$u_0 - 1 = 0, \quad u_i - \sum_{j=0}^{i-1}\sum_{k=0}^{i-j-1} u_j u_k u_{i-j-k-1} = 0, \qquad i \geqslant 1. \tag{5}$$

We determine $u_0, u_1, \ldots$ consecutively from (5). Then (3) satisfies the relation $u^{(\infty)} - 1 - \varepsilon(u^{(\infty)})^3 \sim 0$. It means that if the series (3) is truncated at the Nth term, and the partial sum $u^{(N)} = \sum_{i=0}^{N} \varepsilon^i u_i$ is substituted in the left-hand side of (1), the discrepancy is of order $O(\varepsilon^{N+1})$. Indeed, we have

$$r\left(u^{(N)}\right) \overset{\text{def}}{=} u^{(N)} - 1 - \varepsilon\left(u^{(N)}\right)^3 =$$
$$= (u_0 - 1) + \sum_{i=1}^{N} \varepsilon^i \left(u_i - \sum_{j=0}^{i-1}\sum_{k=0}^{i-j-1} u_j u_k u_{i-j-k-1}\right) -$$
$$- \varepsilon^{N+1}\left(\sum_{j=0}^{N}\sum_{k=0}^{N-j} u_j u_k u_{N-j-k} + \sum_{i=N+1}^{3N} \varepsilon^{i-N} \sum_{j=0}^{N}\sum_{k=0}^{i-j} u_j u_k u_{i-j-k}\right);$$

here the u_l, with $l > N$, are assumed to be zero. By taking into account the equalities (5), we get $r(u^{(N)}) = O(\varepsilon^{N+1})$. Thus we have constructed f.a.s. $u^{(\infty)}$ of (1) whose partial sums $u^{(N)}$ reduce the left-hand side of (1) to $O(\varepsilon^{N+1})$.

It is easy to see that for any N there exists a number $\varepsilon_N > 0$ such that $u^{(N)} \in [0, 2]$ for all ε $(0 < \varepsilon < \varepsilon_N)$. Therefore ε_0 is any number, $\varepsilon_1 = 1$, and $\varepsilon_2 = 1/3$.

3. *Existence and uniqueness theorem for a solution of problem* (1), (2). *An a priori estimate*

THEOREM 1. *Let $\varepsilon < 1/12$. Then there exists a unique solution of problem (1), (2).*

PROOF: Assume $f(u) = u - 1 - \varepsilon u^3$, then $f'(u) = 1 - 3\varepsilon u^2$. If $\varepsilon < 1/12$, then $3\varepsilon u^2 < 1$ for $u \in [0, 2]$. Hence we have $f'(u) > 0$. Therefore $f(u)$ is monotonously increasing on $[0, 2]$, and since $f(0) = -1 < 0$, $f(2) = 1 - 8\varepsilon > 0$, it follows from the continuity of $f(u)$ that a solution of equation (1) on $[0, 2]$ exists and is unique.

THEOREM 2. *Let v satisfy the equation*

$$v - 1 - \varepsilon v^3 = g, \quad v \in [0,\ 2], \tag{6}$$

and u be an exact solution of problem (1), (2). *Then $|u - v| \leqslant c|g|$, where* $c = 1/(1 - 12\varepsilon)$, $\varepsilon < 1/12$.

PROOF: By subtracting (6) from (1), we get $u - v = \varepsilon(u - v)\ (u^2 + uv + v^2) - g$ or $(u - v)(1 - \varepsilon(u^2 + uv + v^2)) = -g$, whence

$$|u - v| = \frac{|g|}{|1 - \varepsilon(u^2 + uv + v^2)|} \leqslant \frac{|g|}{1 - 12\varepsilon}.$$

We have given these two elementary theorems, because they are the simplest analogues of more complex existence and uniqueness theorems for solutions of the differential equations described in Chapter 1.

4. *Justification of f.a.s. An estimate of accuracy*

Theorems 1 and 2 make it possible to derive the closeness of the exact solution u and the asymptotic solution $u^{(N)}$ of problems (1) and (2) from the smallness of the discrepancy $r(u^{(N)})$ for a sufficiently small ε. Namely, $|u - u^{(N)}| \leqslant r(u^{(N)})/(1 - 12\varepsilon)$, whence $|u - u^{(N)}| = O(\varepsilon^{N+1})$.

To illustrate the accuracy of the asymptotic method, we introduced a table of the discrepancies and errors for $N = 0,\ 1,\ 2$ and $\varepsilon = 0.01,\ 0.001$.

N	ε	$u^{(N)}$	$r(u^{(N)})$	q: $\lvert u^{(N)} - u\rvert < q$
0	ε	1	$-\varepsilon$	$\varepsilon/(1-12\varepsilon)$
	0,01	1	—0,01	0,01136
	0,001	1	—0.001	0.001012
1	ε	$1+\varepsilon$	$-\varepsilon^2(3+3\varepsilon+\varepsilon^2)$	$\varepsilon^2(3+3\varepsilon+\varepsilon^2)/(1-12\varepsilon)$
	0,01	1.01	—0,00030301	0,0003443
	0.001	1.001	—0.000003003001	0,000003039
2	ε	$1+\varepsilon+3\varepsilon^2$	$-\varepsilon^3(12+19\varepsilon+36\varepsilon^2+27\varepsilon^3+27\varepsilon^4)$	$\varepsilon^3(12+19\varepsilon+36\varepsilon^2+27\varepsilon^3+27\varepsilon^4)/(1-12\varepsilon)$
	0,01	1,0103	—0,00001219362727	0,00001386
	0,001	1,001003	—0,000000012019036027027	0,00000001217

Table 1

§3 Asymptotic Expansion of a Solution of the Equation $(K(x/\varepsilon)u')' = f(x)$ by the Averaging Method

1. Let $K(\xi)$ be an infinitely differentiable 1-periodic function of $\xi \in R$ satisfying the inequality $0 < \kappa_1 \leqslant K(\xi) \leqslant \kappa_2$ and $f(x)$ be an infinitely differentiable function determined on a segment $[0, 1]$. We construct the f.a.s. of a solution of the problem

$$Pu \equiv (K(x/\varepsilon)u')' = f(x), \quad x \in (0, l), \tag{1}$$

$$u(0) = g_1, \quad u(l) = g_2, \tag{2}$$

where ε is a small parameter ($\varepsilon = l/n$, n an integer).

Problem (1), (2) describes a stationary temperature field in an inhomogeneous rod (layered medium) with a rapidly changing conductivity coefficient $K(x/\varepsilon)$ in the presence of sources spread according to the law $f(x)$. Figure 6 shows the curve of the conductivity coefficient.

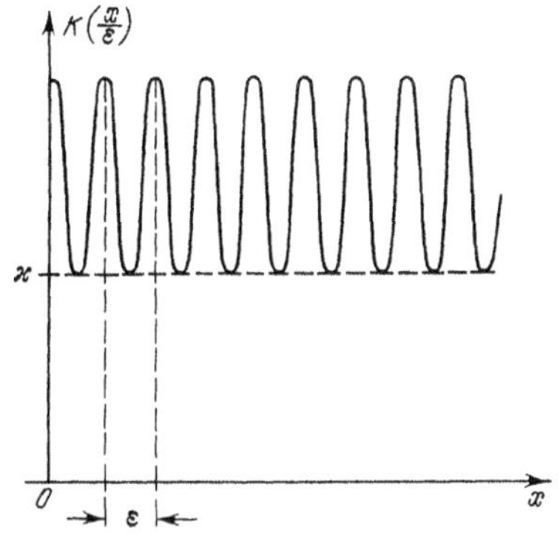

Figure 6

2. In §5 of the Introduction, we have sought the a.e. of a solution of this equation in the form

$$u^{(\infty)} = u_0(x, \xi) + \varepsilon u_1(x, \xi) + \ldots, \quad \xi = x/\varepsilon,$$

and saw that it is worth taking $u_0(x,\xi) = v(x)$ and $u_1 = N_1(\xi)dv/dx$. This causes us to suppose that in this case one can try to seek an a.e. of the solution in the form

$$u^{(\infty)} = \sum_{i=0}^{\infty} \varepsilon^i N_i\left(\frac{x}{\varepsilon}\right) \frac{d^i v\,(x)}{dx^i}, \tag{3}$$

where the $N_i(\xi)$ are 1-periodic functions and $v(x)$ is a solution of some averaged problem with constant coefficients, which will be constructed below; with v being independent of ξ, and $v(x)$ having an a.e.

$$v\,(x) = \sum_{j=0}^{\infty} \varepsilon^j v_j\,(x), \tag{4}$$

where $v_j(x)$ does not depend on ε.

Let us substitute series (3) into the left-hand side of (1) and use the chain rule

$$Pu^{(\infty)} \sim \left(K\left(\frac{x}{\varepsilon}\right) \sum_{i=0}^{\infty} \varepsilon^{i-1} \frac{dN_i(\xi)}{d\xi}\bigg|_{\xi=\frac{x}{\varepsilon}} \frac{d^i v}{dx^i} + \right.$$

$$\left. + K\left(\frac{x}{\varepsilon}\right) \sum_{i=0}^{\infty} \varepsilon^i N_i\left(\frac{x}{\varepsilon}\right) \frac{d^{i+1} v}{dx^{i+1}} \right)' \sim \sum_{i=0}^{\infty} \varepsilon^{i-2} \frac{d}{d\xi}\left(K(\xi) \frac{dN_i(\xi)}{d\xi}\right) \frac{d^i v}{dx^i} +$$

$$+ \sum_{i=0}^{\infty} \varepsilon^{i-1} K(\xi) \frac{dN_i(\xi)}{d\xi} \frac{d^{i+1} v}{dx^{i+1}} + \sum_{i=0}^{\infty} \varepsilon^{i-1} \frac{d}{d\xi}(K(\xi) N_i(\xi)) \frac{d^{i+1} v}{dx^{i+1}} +$$

$$+ \sum_{i=0}^{\infty} \varepsilon^i K(\xi) N_i(\xi) \frac{d^{i+2} v}{dx^{i+2}}, \quad \xi = \frac{x}{\varepsilon}.$$

After simplification, we get

$$Pu^{(\infty)} \sim \varepsilon^{-2} \frac{d}{d\xi}\left(K(\xi) \frac{dN_0}{d\xi}\right) v + \varepsilon^{-1}\left(\frac{d}{d\xi}\left(K(\xi) \frac{dN_1}{d\xi}\right) + K(\xi) \frac{dN_0}{d\xi} + \right.$$

$$\left. + \frac{d}{d\xi}(K(\xi) N_0)\right) \frac{dv}{dx} + \sum_{i=2}^{\infty} \varepsilon^{i-2}\left(\frac{d}{d\xi}\left(K(\xi) \frac{dN_i}{d\xi}\right) + \right.$$

$$\left. + K(\xi) \frac{dN_{i-1}}{d\xi} + \frac{d}{d\xi}(K(\xi) N_{i-1}) + K(\xi) N_{i-2}\right) \frac{d^i v}{dx^i}. \quad (5)$$

Let us try to select the functions v_0 and N_i so that terms of order ε^{-2} and ε^{-1} vanish and all the terms of higher order are independent of ξ. We obtain a chain of relations

$$LN_0 = 0, \quad (6)$$

$$LN_1 + K(\xi) \frac{dN_0}{d\xi} + \frac{d}{d\xi}(K(\xi) N_0) = 0, \quad (7)$$

$$LN_i + K(\xi) \frac{dN_{i-1}}{d\xi} + \frac{d}{d\xi}(K(\xi) N_{i-1}) + K(\xi) N_{i-2} = h_i, \quad i > 1. \quad (8)$$

Here $L = \frac{d}{d\xi}(K(\xi)\frac{d}{d\xi})$ and h_i are constants.

Assume $N_0 = 1$. Then (6) is satisfied automatically, and (7) takes the form

$$LN_1 + \frac{d}{d\xi} K(\xi) = 0, \quad (9)$$

or

$$\frac{d}{d\xi}\left(K(\xi)\left(\frac{dN_1}{d\xi} + 1\right)\right) = 0. \quad (10)$$

Since $N_0 \equiv 1$, then (3) starts with $v(x)$:

$$u^{(\infty)}(x) = v(x) + \sum_{i=1}^{\infty} \varepsilon^i N_i\left(\frac{x}{\varepsilon}\right) \frac{d^i v(x)}{dx^i}. \tag{11}$$

By substituting (11) into condition (2), we get

$$\begin{aligned} &v(0) + \sum_{i=1}^{\infty} \varepsilon^i N_i(0) \frac{d^i v(0)}{dx^i} \sim g_1, \\ &v(l) + \sum_{i=1}^{\infty} \varepsilon^i N_i\left(\frac{l}{\varepsilon}\right) \frac{d^i v(l)}{dx^i} \sim v(l) + \sum_{i=1}^{\infty} \varepsilon^i N_i(0) \frac{d^i v(l)}{dx^i} \sim g_2, \end{aligned} \tag{12}$$

because $N_i(l/\varepsilon) = N_i(n) = N_i(0)$ due to the periodicity of $N_i(\xi)$. Then, by selecting N_i, we impose the condition

$$N_i(0) = 0 \quad \text{for} \quad i \geqslant 1. \tag{13}$$

In this case, relations (192) take the form

$$v(0) \sim g_1, \quad v(l) \sim g_2. \tag{14}$$

Note that (8) and (9) can be rewritten as

$$LN_i - T_i(\xi) = h_i, \quad i \geqslant 1, \text{ where } \ h_1 = 0, \tag{15}$$

$$\begin{aligned} &T_1(\xi) = -\frac{d}{d\xi} K(\xi), \\ &T_i(\xi) = -K(\xi)\left(\frac{dN_{i-1}}{d\xi} + N_{i-2}\right) - \frac{d}{d\xi}(K(\xi) N_{i-1}), \quad i > 1, \end{aligned} \tag{16}$$

with $T_i(\xi)$ being expressed through N_j with $j < i$. This circumstance can be used for the recurrent definition of N_i from (15). The constants h_i are chosen under the assumption that there exist 1-periodic solutions of equation (15).

Lemma 1. *Let $F(\xi)$, $K(\xi)$ be 1-periodic differentiable functions, with $K(\xi)$ being greater than zero. A necessary and sufficient condition for a 1-periodic solution of the equation*

$$LN = F(\xi) \tag{17}$$

to exist is that

$$\langle F(\xi) \rangle \equiv \int_0^1 F(\xi)\, d\xi = 0. \tag{18}$$

Proof: Necessity. Let $N(\xi)$ be a 1-periodic solution of (17). After integrating identity (17) from 0 to 1, we get $\langle LN\rangle = \langle F\rangle$, that is,

$$K(1)\frac{dN}{d\xi}(1) - K(0)\frac{dN}{d\xi}(0) = \langle F\rangle.$$

The periodicity of $K(\xi)$ and $N(\xi)$ implies (18).

Sufficiency. Find a general solution of the ordinary differential equation (17). The latter is equivalent to the relation

$$K(\xi)\frac{dN}{d\xi} = \int_0^\xi F(t)\,dt + C_1,$$

i.e., $\frac{dN}{d\xi} = K^{-1}(\xi)\int_0^\xi F(t)\,dt + K^{-1}(\xi)C_1$, or

$$N(\xi) = \int_0^\xi K^{-1}(y)\left(\int_0^y F(t)\,dt + C_1\right)dy + C_2. \qquad (19)$$

It follows from the condition $\langle F\rangle = 0$ and periodicity of $F(t)$ that

$$\int_0^{y+1} F(t)\,dt - \int_0^y F(t)\,dt = \int_y^{y+1} F(t)\,dt = 0,$$

that is, $\int_0^y F(t)dt$ is 1-periodic in y, and so the whole expression $K^{-1}(y)\left[\int_0^y F(t)dt + C_1\right]$ is a function 1-periodic in y. Then

$$N(\xi+1) - N(\xi) = \int_\xi^{\xi+1} K^{-1}(y)\left[\int_0^y F(t)\,dt + C_1\right]dy =$$

$$= \left\langle K^{-1}(y)\left[\int_0^y F(t)\,dt + C_1\right]\right\rangle. \qquad (20)$$

Choose the constant C_1 in the general solution of (19) so that the integral in (20) vanishes, that is,

$$C_1 = -\langle K^{-1}\rangle^{-1}\left\langle K^{-1}(y)\int_0^y F(t)\,dt\right\rangle. \qquad (21)$$

Then $N(\xi+1) = N(\xi)$, i.e., the function $N(\xi)$ is 1-periodic. Thus, we have found the desired periodic solution of equation (17). The lemma is proved.

Note that the solution of $N(\xi)$ is determined by (19) and (21) up to an arbitrary constant C_2. If we require also that a solution of (17) satisfy the condition

$$N(0) = 0, \tag{22}$$

then the constant C_2 must be equal to zero, and the solution of problems (19) and (22) with the condition (18) exists and is unique.

According to the above-proved lemma, the constants h_i in equations (15) should be chosen from the condition (see (18)) $\langle T_i(\xi) + h_i \rangle = 0$, that is,

$$h_i = -\langle T_i \rangle. \tag{23}$$

In particular, for $i = 1$, we obtain $h_1 = -\langle T_1 \rangle = \left\langle \frac{d}{d\xi} K \right\rangle = K(1) - K(0) = 0$, which implies that for $h_i = 0$ there exists a periodic solution of (9).

Thus, the functions N_i are constructed by using the following recurrent procedure: Let all the functions N_j be found for $j \leqslant i-1$. Determine $T_i(\xi)$ from the formulas (16), calculate h_i from (23), and then find $N_i(\xi)$ as a 1-periodic solution of (15) that satisfies condition (13). The basis of the recursion is $N_0 = 1$. Problems (15) and (13) are called the *periodic cell problem.*

Since $N_i(\xi)$ satisfies (6)–(8), series (5) takes the form

$$\left(K\left(\frac{x}{\varepsilon}\right) u^{(\infty)\prime}\right)' \sim \sum_{i=2}^{\infty} \varepsilon^{i-2} h_i \frac{d^i v}{dx^i}.$$

If v is a f.a.s. satisfying the relations

$$\sum_{i=2}^{\infty} \varepsilon^{i-2} h_i \frac{d^i v}{dx^i} \sim f(x), \tag{24}$$

$$v(0) \sim g_1, \quad v(l) \sim g_2, \tag{25}$$

then $u^{(\infty)}$ from (3) is the f.a.s. of problem (1), (2). We call (24) the *formal averaged equation* of infinite order of accuracy. Its coefficients h_i are constant. Let us calculate h_2, the coefficient of the principal term $h_2 d^2v/dx^2$ of (24) (h_2 will be called the *averaged conductivity coefficient of the rod* and it will be denoted by $\widehat{K}$):

$$h_2 = -\langle T_2 \rangle = \left\langle K(\xi)\left(\frac{dN_1}{d\xi} + N_0\right) + \frac{d}{d\xi}\left(K(\xi) N_0\right) \right\rangle = \\ = \left\langle K(\xi)\left(\frac{dN_1}{d\xi} + 1\right)\right\rangle.$$

From (10) we get

$$K(\xi)(1 + dN_1/d\xi) = C = \text{const},$$
$$dN_1/d\xi = K^{-1}(\xi)C - 1,$$
$$\langle dN_1/d\xi \rangle = \langle K^{-1}(\xi)C - 1 \rangle,$$
$$N_1(1) - N_1(0) = \langle K^{-1} \rangle C - 1.$$

Since $N_1(1) = N_1(0)$, then $C = \langle K^{-1}\rangle^{-1}$. Consequently, $h_2 = \widehat{K} = \langle C\rangle = C = \langle K^{-1}\rangle^{-1} > 0$.

By substituting the f.a.s. (4) into equation (24) and conditions (25), we get

$$\sum_{i=2}^{\infty} \varepsilon^{i-2} h_i \frac{d^i v}{dx^i} \sim \sum_{i=2}^{\infty} \varepsilon^{i-2} h_i \sum_{j=0}^{\infty} \varepsilon^j \frac{d^i v_j}{dx^i} \sim$$

$$\sim \text{[change of the indices: } q = i + j - 2,\ j = j] \sim$$

$$\sim \sum_{q=0}^{\infty} \varepsilon^q \left(h_2 \frac{d^2 v_q}{dx^2} + \sum_{j=0}^{q-1} h_{q-j+2} \frac{d^{q-j+2} v_j}{dx^{q-j+2}} \right) \sim f(x),$$

$$v(0) \sim \sum_{q=0}^{\infty} \varepsilon^q v_q(0) \sim g_1, \quad v(l) \sim \sum_{q=0}^{\infty} \varepsilon^q v_q(l) \sim g_2.$$

It follows that the $v_q(x)$ must satisfy the equations

$$h_2 \frac{d^2 v_q}{dx^2} = f_q(x), \quad q \geqslant 0, \tag{26}$$

and boundary conditions

$$v_q(0) = g_1^q, \quad v_q(l) = g_2^q, \quad q \geqslant 0, \tag{27}$$

where

$$f_q(x) = \begin{cases} f(x) & \text{for} \quad q = 0, \\ -\displaystyle\sum_{j=0}^{q-1} h_{q-j+2} \frac{d^{q-j+2} v_j}{dx^{q-j+2}} & \text{for} \quad q > 0, \end{cases}$$

$$g_i^q = g_i \delta_q^0 = \begin{cases} g_i & \text{for} \quad q = 0, \\ 0 & \text{for} \quad q > 0. \end{cases}$$

Thus all the v_q can consecutively be found following the recurrent chain of boundary value problems (26) and (27). Moreover, it can be proved by induction that the $f_q(x)$ are infinitely differentiable and so are the $v_q(x)$.

Problems (26) and (27) for $q = 0$ have the form

$$\widehat{K} d^2 v_0 / dx^2 = f(x), \quad v_0(0) = g_1, \quad v_0(l) = g_2$$

and is called the *averaged zeroth-order problem*. It describes a thermal field in a homogeneous medium whose properties are close in a sense to the effective properties of the original inhomogeneous medium. The solution of the chain of problems (26), (27) for all $q \geqslant 0$ completes the procedure of constructing the f.a.s. of problem (1), (2).

Remark 1: Let $u^{(\infty)}$ be sought now in the form [188],

$$u^{(\infty)} \sim \sum_{i=0}^{\infty} \varepsilon^i u_i\left(x, \frac{x}{\varepsilon}\right), \tag{28}$$

where $u_i(x,\xi)$ is a function 1-periodic in ξ; it belongs to $C^\infty([0,1] \times R)$ and is regularly dependent on ε, i.e., it has an asymptotic expansion of the form

$$u_i(x, \xi) \sim \sum_{j=0}^{\infty} \varepsilon^j u_i^j(x, \xi)$$

(u_i^j are independent of ε).

By substituting (28) in the left side of (1), by applying the chain rule, and by simplifying, we get

$$Pu^{(\infty)} \sim \sum_{i=0}^{\infty} \varepsilon^{i-2}\left(\frac{\partial}{\partial\xi}\left(K(\xi)\frac{\partial u_i}{\partial\xi}\right) + \frac{\partial}{\partial\xi}\left(K(\xi)\frac{\partial u_{i-1}}{\partial x}\right) + \right.$$
$$\left. + \frac{\partial}{\partial x}\left(K(\xi)\frac{\partial u_{i-1}}{\partial\xi}\right) + \frac{\partial}{\partial x}\left(K(\xi)\frac{\partial u_{i-2}}{\partial x}\right)\right)\Big|_{\xi=x/\varepsilon},$$

where $u_i = 0$ for $i < 0$. The fact that u_i depends on ε is ignored. We set $L_{\alpha\beta}\varphi(x,\xi) = \frac{\partial}{\partial\alpha}\left(K(\xi)\frac{\partial\varphi(x,\xi)}{\partial\beta}\right)$. We require that the terms of order ε^{-2} and ε^{-1} be zero and that all the terms of higher order be independent of ξ. Hence,

$$L_{\xi\xi}u_0(x, \xi) = 0, \tag{29}$$

$$L_{\xi\xi}u_1 + L_{\xi x}u_0 + L_{x\xi}u_0 = 0, \tag{30}$$

$$L_{\xi\xi}u_i + L_{\xi x}u_{i-1} + L_{x\xi}u_{i-1} + L_{xx}u_{i-2} = \hbar_i(x), \quad i > 1. \tag{31}$$

It follows from (29) that $u_0(x,\xi)$ is independent of ξ, i.e., $u_0(x,\xi) = v(x)$. Then (30) is transformed into

$$L_{\xi\xi}u_1 + \frac{dK(\xi)}{d\xi}\frac{dv}{dx} = 0. \tag{32}$$

By using the method of separation of variables, we conclude that a solution of (32) can be represented as a product

$$u_1(x, \xi) = N_1(\xi)dv/dx,$$

where $N_i(\xi)$ is a 1-periodic solution of (20). Then we shall reason by induction. Let all the $u_k(x,\xi)$ with $k = 0, 1, \ldots, i-1$ be chosen in the form

$$u_k(x,\ \xi) = N_k(\xi) d^k v/dx^k,$$

where the $N_k(\xi)$ are 1-periodic functions. By integrating (31) with respect to ξ over $[0, 1]$ and by considering the periodicity of u_k in ξ, we get

$$\widetilde{h}_i(x) = \langle L_{x\xi} u_{i-1} + L_{xx} u_{i-2} \rangle =$$

$$= \left\langle K(\xi) \frac{dN_{i-1}}{d\xi} \frac{d^i v}{dx^i} + K(\xi) N_{i-2} \frac{d^i v}{dx^i} \right\rangle = h_i \frac{d^i v}{dx^i},$$

where

$$h_i = \left\langle K(\xi) \frac{dN_{i-1}}{d\xi} + K(\xi) N_{i-2} \right\rangle = \text{const.}$$

Then (31) can be rearranged as follows:

$$L_{\xi\xi} u_i = \frac{d^i v}{dx^i} \left(h_i - \left(\frac{d}{d\xi} (K(\xi) N_{i-1}) + K(\xi) \frac{dN_{i-1}}{d\xi} + K(\xi) N_{i-2} \right) \right).$$

By using the method of separation of variables again, we seek $u_i(x, \xi)$ in the form

$$u_i(x,\ \xi) = N_i(\xi) d^i v/dx^i,$$

where $N_i(\xi)$ is a 1-periodic solution of equation (8) with condition (13).

Thus, by reasoning from the fact that the f.a.s. of problem (1), (2) can be found in the form of series (28), we have obtained expansion (3), whereas for the rapidly oscillating coefficients of the expansion, we have obtained equations (6)–(8).

3. Justification of this asymptotic expansion is based on the standard technique. Denote by $u^{(J)}$ the partial sum of series (3):

$$u^{(J)} = \sum_{i=0}^{J+1} \varepsilon^i N_i \left(\frac{x}{\varepsilon} \right) \frac{d^i v^{(J)}}{dx^i}, \tag{33}$$

where

$$v^{(J)} = \sum_{j=0}^{J} \varepsilon^j v_j(x). \tag{34}$$

By substituting $u^{(J)}$ into the equation and the boundary conditions as in Subsection 2, we see that

$$Pu^{(J)} = f(x) + r(u^{(J)}), \tag{35}$$

$$u^{(J)}(0) = g_1, \quad u^{(J)}(l) = g_2, \quad |r(u^{(J)})| \leqslant c_J \varepsilon^J, \tag{36}$$

where c_J is a positive constant independent of ε. Thus, the error of the partial sum of the asymptotic series $R^{(J)} = u^{(J)} - u$ satisfies the relations

$$PR^{(J)} = r(u^{(J)}), \quad |r(u^{(J)})| \leqslant c_J \varepsilon^J, \quad R^{(J)}(0) = R^{(J)}(l) = 0. \quad (37)$$

The generalized maximum principle for solutions of differential equations implies directly the estimate $\|u^{(J)} - u\|_{C([0,1])} = O(\varepsilon^J)$. The definition of $u^{(J)}$ (see (33), (34)) implies

$$\|u^{(J)} - \widetilde{u}^{(J)}\|_{C([0,l])} = O(\varepsilon^J),$$

where

$$\widetilde{u}^{(J)} = \sum_{s=0}^{J-1} \varepsilon^s \sum_{j=0}^{s} N_{s-j}(\xi) \frac{d^{s-j} v_j}{dx^{s-j}}.$$

In particular, for $J = 1$, we get

$$\|v_0 - u\|_{C([0,l])} = O(\varepsilon).$$

The estimates of the closeness of the flows $K(x/\varepsilon)u^{(J)\prime}$ and $K(x/\varepsilon)u'$ may also be obtained. Indeed, the solution $R^{(J)}$ of (37) satisfies the inequality

$$\left\| K\left(\frac{x}{\varepsilon}\right) R^{(J)\prime} \right\|_{C([0,l])} \leqslant \text{const} \max_{x \in [0,l]} |r(u^{(J)})| = O(\varepsilon^J),$$

whence

$$\left\| K\left(\frac{x}{\varepsilon}\right) \widetilde{u}^{(J)\prime} - K\left(\frac{x}{\varepsilon}\right) u' \right\|_{C([0,l])} = O(\varepsilon^{J-1}).$$

In particular,

$$\left\| K\left(\frac{x}{\varepsilon}\right)\left(1 + \frac{dN_1}{d\xi}\right)\frac{dv_0}{dx} - K\left(\frac{x}{\varepsilon}\right) u' \right\|_{C([0,l])} = O(\varepsilon),$$

or (cf. (12), (13))

$$\left\| \widehat{K} \frac{dv_0}{dx} - K\left(\frac{x}{\varepsilon}\right) u' \right\|_{C([0,l])} = O(\varepsilon).$$

We have described in detail the procedure of constructing the averaged equation of infinite order of accuracy in ε the way it would look in the general case. It can be simplified considerably for our case, due to the following. The function N_1 is found from (10):

$$K(\xi)\left(\frac{dN_1}{d\xi} + 1\right) = h_1 = 0.$$

The equation in N_2 has the form

$$\frac{d}{d\xi}\left(K(\xi)\left(\frac{dN_2}{d\xi}+N_1\right)\right)+K(\xi)\left(\frac{dN_1}{d\xi}+1\right)=h_2,$$

or

$$\frac{d}{d\xi}\left(K(\xi)\left(\frac{dN_2}{d\xi}+N_1\right)\right)=0.$$

The function N_1 is determined up to a summand. Select this summand so as to have the equality $\langle N_1\rangle = 0$, with each N_i being consecutively determined from the conditions $dN_i/d\xi + N_{i-1} = 0$, $\langle N_i\rangle = 0$. Then for any $i > 2$, we get

$$h_i = \frac{d}{d\xi}\left(K(\xi)\left(\frac{dN_i}{d\xi}+N_{i-1}\right)\right)+K\left(\frac{dN_{i-1}}{d\xi}+N_{i-2}\right)=0$$

and the averaged equation of infinite order of accuracy in ε has the simplest form

$$h_2 d^2v/dx^2 = f(x). \qquad (38)$$

Notice, however, that with the above-indicated choice of N_i, the boundary conditions are satisfied up to terms of order ε. A more complex procedure needed to construct an asymptotic solution satisfying the boundary conditions up to a greater order of accuracy is described in Chapter 9.

§4 Generalization of the Averaging Method in the Case of a Piecewise Smooth Coefficient

Consider problem (4.1), (4.2), assuming that $K(\xi)$ is an infinitely differentiable function on the segments $[0,\xi_1]$, $[\xi_1,\xi_2], \dots, [\xi_p,1]$, where $0<\xi_1<\xi_2<\dots<\xi_p<1$, and u is a generalized solution. We also suppose that $K(\xi)$ is a 1-periodic function satisfying the inequality $0<\kappa_1 \leqslant K(\xi) \leqslant \kappa_2$, and $f(x)$ is an infinitely differentiable function defined on $[0,1]$. For simplicity, we consider $K(\xi)$ to be infinitely differentiable at the point $\xi = 0$ and consequently, on the segment $[\xi_p, 1+\xi_1]$.

As was shown in Chapter 1, the generalized formulation of problem (4.1), (4.2) is equivalent to the following classical formulation:

$$Pu \equiv (K(x/\varepsilon)u')' = f(x), \quad x \in (0,l), \quad x \neq x_{mj}, \qquad (1)$$

$$u(0) = g_1, \quad u(l) = g_2, \qquad (2)$$

$$[u]\,|_{x=x_{mj}} = 0, \qquad (3)$$

$$[K(x/\varepsilon)u']\,|_{x=x_{mj}} = 0, \qquad (4)$$

where the x_{mj} are the discontinuity points of the coefficient $K(x/\varepsilon)$,

$$x_{mj} = (\xi_m + j)\varepsilon, \quad m = 1,\dots,p, \quad j = 0,\dots,n-1, \quad n = l/\varepsilon.$$

The procedure of constructing $u^{(\infty)}$ is fully applicable to problems (1)–(4), with the difference being that the functions $N_1(\xi)$ are generalized solutions of equations (3.15). In the same way as Lemma 4.1, the following lemma is proved.

Lemma 1. *Let $F_0(\xi)$, $F_1(\xi)$, $K(\xi)$ be 1-periodic functions infinitely differentiable on each of the segments $[0,\xi_1]$, $[\xi_1,\xi_2],\dots,[\xi_p,1]$ and at the point $\xi=0$, and $K(\xi)>0$. Then a necessary and sufficient condition for a 1-periodic solution of the problem*

$$LN = F_0(\xi) + \frac{d}{d\xi}F_1(\xi), \quad \xi \neq \xi_m, \tag{5}$$

$$[N]|_{\xi_m} = 0, \tag{6}$$

$$\left[K(\xi)\frac{dN}{d\xi} - F_1(\xi)\right]\bigg|_{\xi_m} = 0, \tag{7}$$

$$N(0) = 0 \tag{8}$$

to exist and be unique is $\langle F_0\rangle = 0$.

The solution of problems (5)–(8) is a generalized solution of the equation $LN = F_0(\xi) + \frac{d}{d\xi}F_1(\xi)$, provided $N(0)=0$.

The functions N_i are constructed consecutively by the following algorithm. Let all the functions N_j be found for $j \leqslant i-1$.

1) Define $T_i(\xi)$ from (3.16), the terms $-K(\xi)(dN_{i-1}/d\xi + N_{i-2})$ operating as $F_0(\xi)$ and the terms $\frac{d}{d\xi}(K(\xi)N_{i-1})$ as $\frac{d}{d\xi}F-1(\xi)$.

2) Calculate h_i by the formula

$$h_i = \left\langle K(\xi)\left(\frac{dN_{i-1}}{d\xi} + N_{i-2}\right)\right\rangle. \tag{9}$$

3) Find a 1-periodic generalized solution of the equation $LN_i = T_i(\xi) + h_i$, $N_i(0) = 0$, i.e., the classical solution of the problem

$$LN_i = T_i(\xi) + h_i, \quad \xi \neq \xi_m, \quad m = 1,\dots,p, \tag{10}$$

$$[N_i]|_{\xi_m} = 0, \tag{11}$$

$$\left[K(\xi)\left(\frac{dN_i}{d\xi} + N_{i-1}\right)\right]\bigg|_{\xi_m} = 0, \tag{12}$$

$$N_i(0) = 0. \tag{13}$$

The existence and uniqueness of the solution of problems (10)–(13) follow from Lemma 1 and (9). The basis of induction is $N_0 \equiv 1$.

The procedure of constructing f.a.s. and the problem for the functions $v_j(x)$ repeats literally the corresponding procedure in §3. Let $u^{(J)}$ and $v^{(J)}$ be partial sums of the form (3.33), (3.34), where N_i is the solution of problems (10)–(13). Then, repeating word for word the reasoning of §3 (for $x = x_{mi}$), we see that $u^{(J)}$ satisfies (3.35) and (3.36) with $|r(u^{(J)})| \leqslant 0\varepsilon^J$ as above.

Next, with the $N_i(\xi)$ being continuous and the $v_j(x)$ being infinitely differentiable due to (11), the function of $u^{(J)}$ in (3.33) and (3.34) is continuous, and so $[u^{(J)}]|_{x_{mj}} = 0$. Let us prove that

$$\left[K\left(\frac{x}{\varepsilon}\right)u^{(J)\prime}\right]\Big|_{x_{mj}} = O\left(\varepsilon^{J+1}\right).$$

Set

$$\left[K\left(\frac{x}{\varepsilon}\right)u^{(J)\prime}\right]\Big|_{x_{mj}} =$$
$$= \left[K\left(\frac{x}{\varepsilon}\right)\left(\sum_{i=0}^{J+1}\varepsilon^{i-1}\frac{dN_i}{d\xi}\frac{d^i v^{(J)}}{dx^i} + \sum_{i=0}^{J+1}\varepsilon^i N_i \frac{d^{i+1}v^{(J)}}{dx^{i+1}}\right)\right]\Big|_{x_{mj}}.$$

Since $v^{(J)}(x)$ is infinitely differentiable and the periodic functions $N_i(\xi)$ satisfy (11) and (12) on the discontinuities,

$$\left[K\left(\frac{x}{\varepsilon}\right)u^{(J)\prime}\right]\Big|_{x_{mj}} =$$
$$= \sum_{i=1}^{J+1}\varepsilon^{i-1}\left[K(\xi)\left(\frac{d}{d\xi}N_i + N_{i-1}\right)\right]\Big|_{\xi_m}\frac{d^i v^{(J)}}{dx^i}(x_{mj}) + \varepsilon^{J+1}\left[K(\xi)\right]|_{\xi_m}\times$$
$$\times N_{J+1}(\xi_m)\frac{d^{J+2}v^{(J)}}{dx^{J+2}}(x_{mj}) = \varepsilon^{J+1}\left[K(\xi)\right]|_{\xi_m}N_{J+1}(\xi_m)\times$$
$$\times \frac{d^{J+2}v^{(J)}}{dx^{J+2}}(x_{mj}) = O\left(\varepsilon^{J+1}\right). \quad (14)$$

If the flow continuity condition $K(x/\varepsilon)u^{(J)}$ had held exactly, it would have been possible to use the generalized maximum principle and to obtain the estimate of the closeness of u and $u^{(J)}$ as in §3. Since, according to (14), the flow has a discontinuity of order ε^{J+1}, we add a piecewise smooth function $\varepsilon^{J+1}\Phi(x)$ to $u^{(J)}$, which satisfies the following conditions at the points x_{mj}:

$$[\Phi]|_{x_{mj}} = 0, \quad [\varepsilon^{J+1}K(x/\varepsilon)\Phi']|_{x_{mj}} = -[K(x/\varepsilon)u^{(J)\prime}]|_{x_{mj}}.$$

Then the conjugation conditions (3) and (4) are exactly true for the sum $\overline{u}^{(J)} = u^{(J)} + \varepsilon^{J+1}\Phi(x)$; however, when $\overline{u}^{(J)}$ is substituted into (1), an additional discrepancy $p\overline{u}^{(J)} = f(x) + r(u^{(J)}) + \varepsilon^{J+1}P\Phi$ occurs. If $\Phi(x)$ is such that

$$\varepsilon^{J+1}\|P\Phi\|_{C([0,l])} = O(\varepsilon^{J-1}), \quad \|\Phi\|_{C([0,l])} = O(1),$$

then we use the generalized maximum principle to get

$$\|u - u^{(J)} - \varepsilon^{J+1}\Phi(x)\|_{C([0,l])} = O(\varepsilon^{J-1}),$$

and consequently

$$\|u - u^{(J)}\|_{C([0,\,l])} = O(\varepsilon^{J-1}).$$

Since $\|u^{(J)} - \tilde{u}^{(J)}\|_{c([0,l])} = O(\varepsilon^J)$, we get

$$\|u - \tilde{u}^{(J-1)}\|_{C([0,\,l])} = O(\varepsilon^{J-1}),$$
$$\|u - \tilde{u}^{(J)}\|_{C([0,\,l])} = O(\varepsilon^{J}). \tag{15}$$

In particular, the following can be taken as $\Phi(x)$:

$$\chi\left(\frac{x-l}{\varepsilon}\right)\int_0^x \Psi(x)\,dx,$$

where $\chi(\xi)$ is an infinitely differentiable function which is equal to 1 for $|\xi| \geqslant \bar{\xi}$ and to zero for $|\xi| \leqslant \bar{\xi}/2$.

$$\Psi(x) = -\sum_{m=1}^{p}\sum_{j=0}^{n-1}\left(1 - \chi\left(\frac{x - x_{mj}}{\varepsilon}\right)\right) N_{J+1}(\xi_m)\,\frac{d^{J+2}v^{(J)}(x_{mj})}{dx^{J+2}},$$
$$\bar{\xi} = \frac{1}{2}\min\left\{(1-\xi_p),\, \xi_1,\, \min_{1\leqslant m\leqslant p-1}|\xi_m - \xi_{m+1}|\right\}.$$

Estimate (15) for $J = 0$ implies that $\|v_0 - u\|_{C[(0,l)]} = O(\varepsilon)$.

Thus, the difference between the procedures of constructing f.a.s. of solutions for smooth and piecewise smooth coefficients consists of replacing the classical formulation of the "cell problems" such as (3.15) by a generalized formulation and of easy verification that the asymptotic solution of the original problem $u^{(J)}$ satisfies the conjugation conditions at the discontinuity points of the coefficients to a sufficiently high order of accuracy in ε.

§5 Averaging the System of Differential Equations

Let $K(\xi)$ be an $n \times n$ matrix whose elements are 1-periodic functions and $K(\xi) > 0$, i.e.,

$$\varkappa_1 \eta_i \eta_i \leqslant k_{ij}(\xi)\,\eta_i\eta_j, \quad \varkappa_1 > 0 \qquad \forall\,(\eta_1,\ldots,\eta_n) \in \mathbf{R}^n;$$

here summation over repeated indices i, j is between 1 and n.

Consider a boundary value problem

$$\frac{d}{dx}\left(K\left(\frac{x}{\varepsilon}\right)\frac{du}{dx}\right) = f(x), \quad x \in (0,\, l),$$

for the boundary conditions $u(0) = g_1$, $u(l) = g_2$; here $u(x)$ and $f(x)$ are vector-valued functions, g_1 and g_2 are vectors, and $f(x)$ is infinitely differentiable. Asymptotics of the solution are sought in the form

$$u^{(\infty)} = u_0(x, \xi) + \varepsilon u_1(x, \xi) + \ldots, \quad \xi = x/\varepsilon,$$

where u_1 is a vector-valued function (columns) of dimension n.

All the constructions and relations in §§3 and 4 are applicable to this case with the following changes. The scalars K, N_i, h_i are replaced by matrices and the scalars v and v_i by vectors, $N_0 = E$ is the unit matrix, and so (3.10) has an analogue in the form of the matrix equation

$$\frac{d}{d\xi}\left(K(\xi)\left(\frac{dN_1}{d\xi} + E\right)\right) = 0;$$

and when h_2 is calculated, 1 should also be replaced by the unit matrix.

Consider the system of equations of elasticity theory for a layered composite with isotropic layers:

$$\begin{gathered} \left(\left(2\mu\left(\frac{x}{\varepsilon}\right) + \lambda\left(\frac{x}{\varepsilon}\right)\right) u_1'\right)' = 0, \\ \left(\mu\left(\frac{x}{\varepsilon}\right) u_2'\right)' = 0, \quad x \in (0, 1), \\ \left(\mu\left(\frac{x}{\varepsilon}\right) u_3'\right)' = 0, \\ u_i(0) = 0, \quad u_i(1) = 1, \qquad i = 1, 2, 3, \end{gathered} \tag{1}$$

where the vector $u = (u_1, u_2, u_3)^T$ is a generalized solution of the systems of equations (1) and $\lambda(\xi), \mu(\xi)$ are 1-periodic functions

$$\lambda(\xi) = \begin{cases} \lambda_1 & \text{for } \xi \in (0, 1/2), \\ \lambda_2 & \text{for } \xi \in (1/2, 1), \end{cases}$$

$$\mu(\xi) = \begin{cases} \mu_1 & \text{for } \xi \in (0, 1/2), \\ \mu_2 & \text{for } \xi \in (1/2, 1); \end{cases}$$

here

$$\lambda = \frac{\nu E}{(1+\nu)(1-2\nu)}, \quad \mu = \frac{E}{2(1+\nu)}. \tag{2}$$

The pairs of numbers (λ_1, μ_1) and $\lambda_2, \mu_2)$ are Lamé constants of the layers. They are related to Young's modulus E and the Poisson ratio ν by formulas (2). We assume that $\lambda_i > 0$, $\mu_i > 0$, and u is the displacement vector.

In the classical formulation, equations (1) are given only off the discontinuity points, while the conjugation conditions hold at these points (see Chapter 1)

$$[u_1] = [u_2] = [u_3] = 0,$$

$$\left[\left(2\mu\left(\frac{x}{\varepsilon}\right) + \lambda\left(\frac{x}{\varepsilon}\right)\right) u_1'\right] = \left[\mu\left(\frac{x}{\varepsilon}\right) u_2'\right] = \left[\mu\left(\frac{x}{\varepsilon}\right) u_3'\right] = 0.$$

Problem (1) splits into three independent problems of type (3.1)–(3.4), and thus the averaged problem has the form

$$\left((2\widehat{\mu}+\widehat{\lambda})\, v_1'\right)'=0, \quad \left(\widehat{\mu} v_2'\right)'=0, \quad \left(\widehat{\mu} v_3'\right)'=0, \quad x\in(0,1),$$
$$v_i(0)=0, \quad v_i(1)=1, \qquad i=1,2,3,$$

where
$$\widehat{\mu}=2\left(\frac{1}{\mu_1}+\frac{1}{\mu_2}\right)^{-1}, \quad \widehat{\lambda}=2\left(\frac{1}{2\mu_1+\lambda_1}+\frac{1}{2\mu_2+\lambda_2}\right)^{-1}-2\widehat{\mu}.$$

The truth of the estimate $\max\limits_{i=1,2,3}\max\limits_{x\in[0,1]}|u_i-v_i|<\varepsilon/2$ is demonstrated in the same way as in §5.

The problems discussed in §§3–5 are solved by quadratures. We have investigated them by the asymptotic methods to make the presentation of the structure of these methods as lucid as possible.

CHAPTER 3

Averaging Processes in Layered Media

The averaging method is applied in this chapter to investigations of some spatial one-dimensional problems. The chapter considers the elastic rod, the heat equation, nonlinear equations, the process of heat transfer in a system of screens separated by interlayers, and equations with random coefficients.

The technique of averaging for linear nonstationary problems including justification of asymptotics is set forth in detail in §1. Section 9 gives justification of asymptotics for a nonlinear stationary problem, using other methods based on definitions of a solution by means of an integral identity. We describe two different ways of proving that the asymptotics are correct, because the second scheme is applicable to a greater number of problems, achieves the purpose faster, but is less transparent.

Section 1, in particular, contains the averaging of an equation that describes wave propagation in a layered periodic media; we shall note some peculiarities of this process.

Consider a hyperbolic equation,

$$R(\xi)\frac{\partial^2 u}{\partial t^2} - \frac{\partial}{\partial x}\left(K(\xi)\frac{\partial u}{\partial x}\right) = 0, \quad \xi = x/\varepsilon, \tag{1}$$

with 1-periodic coefficients $R(\xi)$, $K(\xi)$. The local perturbation speed is found from the formula $c(\xi) = \sqrt{K(\xi)/R(\xi)}$; therefore, in the case of this periodic medium, the front boundary of the perturbation's front covers the distance ε equal to the period during the time

$$\tau_\varepsilon = \int_0^\varepsilon \frac{dx}{\sqrt{K(x/\varepsilon)/R(x/\varepsilon)}}.$$

Thus, the average speed of the wave-front propagation is $\overline{c} = \varepsilon/\tau_\varepsilon = \langle\sqrt{R/K}\rangle^{-1}$. The averaged equation has the form (cf §1):

$$\widehat{R}\frac{\partial^2 v}{\partial t^2} - \frac{\partial}{\partial x}\left(\widehat{K}\frac{\partial v}{\partial x}\right) = 0, \quad \widehat{R} = \langle R\rangle, \quad \widehat{K}^{-1} = \langle K^{-1}\rangle.$$

Accordingly, the perturbation speed for this equation is $\hat{c} = \sqrt{\widehat{K}/\widehat{R}} = \sqrt{\langle 1/K\rangle^{-1}\langle R\rangle^{-1}}$. From the Cauchy–Buniakowski (Cauchy–Schwarz) inequality we see that $\langle\sqrt{R/K}\rangle^2 \leqslant \langle R\rangle\langle 1/K\rangle$, and therefore, $\hat{c} \leqslant \overline{c}$; the equality only occurs when $1/K = \text{const}\cdot R$, that is, $KR = \text{const}$.

It looks as if the inequality $\hat{c} < \bar{c}$, meaning that the speed of the leading edge in the periodic medium is greater than that in the averaged equation, contradicts common sense.

We shall explain why $\hat{c} < \bar{c}$. Consider the problem of a wave travelling across the interface of two media. We have the equation

$$R\frac{\partial^2 u}{\partial t^2} = \frac{\partial}{\partial x}\left(K\frac{\partial u}{\partial x}\right)$$

with the coefficients

$$R,\, K = \begin{cases} R_1,\, K_1 & \text{for } x < 0, \\ R_2,\, K_2 & \text{for } x > 0, \end{cases}$$

with R_i, K_i being constant. The continuity condition of u and of the flow $K(\partial u/\partial x)$ hold at the point $x = 0 : [u] = 0$, $[K\partial u/\partial x] = 0$.

Suppose the wave falls on the interface of the two media ($x = 0$) from the side of the negative x; in other words, $u = \varphi(x - c_i t)$ with $t \leqslant 0$, where $c_i\sqrt{k_i R_i} > 0$, $\varphi(y) = 0$ with $y \leqslant 0$. For $t \leqslant 0$ the solution is of the form

$$u(x,\, t) = \begin{cases} \varphi_2(x - c_2 t), & x \geqslant 0, \quad \varphi_2 - \text{ transient wave} \\ \varphi_1(x + c_1 t) + \varphi(x - c_1 t), & x \leqslant 0, \quad \varphi_1 - \text{ reflected wave} \end{cases}$$

The conditions at the point $x = 0$ lead to the equations

$$\varphi(-c_1 t) + \varphi_1(c_1 t) = \varphi_2(-c_2 t)$$
$$K_1\left(\varphi'(-c_1 t) + \varphi_1'(c_1 t)\right) = K_2 \varphi_2'(-c_2 t).$$

It implies

$$\varphi_1(y) = \frac{\sigma_1 - \sigma_2}{\sigma_1 + \sigma_2}\varphi(-y),$$
$$\varphi_2(y) = \frac{2\sigma_1}{\sigma_1 + \sigma_2}\varphi\left(\frac{c_1}{c_2}y\right), \quad \sigma_i = \sqrt{K_i R_i}. \tag{2}$$

So the reflected wave always appears except for the case when $\sigma_i = \text{const}$.

Consider the case of a medium ε-periodic in x with $x > 0$, with the medium consisting of layers, $\varepsilon/2$ thick, of compounds with the values of the coefficients of equation (1) (R_2, K_2) and (R_1, K_1), respectively. Let the half-space $x < 0$ be filled, with the medium having the values R_1, K_1 and the wave fall on the plane $x = 0$ from the side of the negative $x : u = \varphi(x - c_i t)$ with $t < 0$ and $\varphi(y) = 0$ with $y \geqslant 0$. When it crosses the first interface $x = 0$, the wave amplitude is multiplied by $2\sigma_1/(\sigma_1 + \sigma_2)$, and when it crosses the second one, $x = \varepsilon/2$, by $2\sigma_2/(\sigma_1 + \sigma_2)$. Thus, after the wave has travelled over the period of the structure, its amplitude is multiplied by

$$q = \frac{4\sigma_1\sigma_2}{(\sigma_1 + \sigma_2)^2} = 1 - \left(\frac{\sigma_1 - \sigma_2}{\sigma_1 + \sigma_2}\right)^2 \leqslant 1.$$

Consequently, for $\sigma_1 + \sigma_2$, after crossing $n = x/\varepsilon$ layers, the leading wavefront amplitude is multiplied by $q^{x/\varepsilon} \ll 1$ and becomes infinitesimal.

The reflected waves moving in the negative direction reflect from the next interface, producing waves moving again in the positive direction. It is these effects of the wave interaction that produce the wave of the final amplitude, which corresponds to the averaged equation and travels in the positive direction of the axis x with the velocity $\hat{c}$.

The situation is similar with the reflected wave. The wave reflected from the first layer is determined by formula (2). The constructions in Chapter 9 imply that the reflected wave equals approximately

$$\frac{\sigma_1 - \widehat{\sigma}}{\sigma_1 + \widehat{\sigma}} \varphi\,(x + c_1 t),$$

being generated by the waves reflected many times from the inner layers of the medium.

The two phenomena—the travel and reflection of the wave—could be studied also by tracing the waves as they consequently cross the interfaces of the media [u]. However, this treatment is more cumbersome.

Remember that we deal with the processes whose characteristic size is much greater than the period of the structure. Otherwise, as can be seen from §4.8, our approach is applicable only to periodic media whose properties change more or less smoothly.

The above reasoning suggests that a wave of small amplitude propagates in front of the principal wave moving at speed $\hat{c}$. It appears to be this wave that is recorded as a "foreshock" in earthquakes (see also [133]).

§1 Problem of Small Longitudinal Vibrations of a Rod

It is well known [151] that the process of small longitudinal linear elastic vibrations in a thin homogeneous rod can be described by the hyperbolic equation

$$R\,\frac{\partial^2 u}{\partial t^2} + S\,\frac{\partial u}{\partial t} - \frac{\partial}{\partial x}\left(K\,\frac{\partial u}{\partial x}\right) = f\,(x,\,t),$$

where R is the linear density of the rod, S the dissipation coefficient, K Young's modulus, $f(x,t)$ the distribution of the density of an external force directed along the rod, and $u(x,t)$ the displacement of the point which had the abscissa x in equilibrium.

Consider an inhomogeneous rod C composed of $2N$ homogeneous rods C_i:

$$C_i = \begin{cases} ((j-1)l/N, \ (j-1+\theta_1)\,l/N), & i = 2j-1, \\ ((j-1+\theta_1)\,l/N, \ jl/N), & i = 2j, \end{cases}$$
$$j = 1, 2, \ldots, N, \quad \theta_1 \in (0, 1).$$

The rods $C_{2j-1}(j = 1, \ldots, N)$ are characterized by the coefficients $R = R_1 > 0$, $S = S_1 \geqslant 0$, $K = K_1 > 0$, and the rods $C_{2j}(j = 1, \ldots, N)$ are characterized by the coefficients $R = R_2 > 0$, $S = S_2 \geqslant 0$, $K = K_2 > 0$. The process of longitudinal vibrations in C is described by the equations

$$\begin{aligned} R_1 \frac{\partial^2 u}{\partial t^2} + S_1 \frac{\partial u}{\partial t} - \frac{\partial}{\partial x}\left(K_1 \frac{\partial u}{\partial x}\right) &= f(x, t), \quad x \in C_{2j-1}, \\ R_2 \frac{\partial^2 u}{\partial t^2} + S_2 \frac{\partial u}{\partial t} - \frac{\partial}{\partial x}\left(K_2 \frac{\partial u}{\partial x}\right) &= f(x, t), \quad x \in C_{2j}, \\ j &= 1, \ldots, N, \end{aligned} \tag{1}$$

with the following relations of discontinuities:

$$\begin{aligned} [u]\,|_{(j+\theta_1)\varepsilon} &= 0, \quad j = 0, \ldots, N-1, \\ [u]\,|_{j\varepsilon} &= 0, \quad j = 1, \ldots, N-1, \end{aligned} \tag{2}$$

$$\begin{aligned} K_2 \frac{\partial u}{\partial x}\Big|_{(j+\theta_1)\varepsilon+0} - K_1 \frac{\partial u}{\partial x}\Big|_{(j+\theta_1)\varepsilon-0} &= 0, \\ K_1 \frac{\partial u}{\partial x}\Big|_{j\varepsilon+0} - K_2 \frac{\partial u}{\partial x}\Big|_{j\varepsilon-0} &= 0, \\ j = 1, \ldots, N-1, \quad \varepsilon &= l/N. \end{aligned} \tag{3}$$

Note: The term $S\partial u/\partial t$ corresponds to the frictional force when the rod vibrates in a viscous medium. We have introduced it to simplify the transition to the heat equation when $R \equiv 0$, $S > 0$.

Let the ends of the rod C move along given paths:

$$u(0, t) = g_1(t), \quad u(l, t) = g_2(t), \tag{4}$$

and at the initial moment the rod be at rest:

$$u(x, 0) = \frac{\partial u}{\partial t}(x, 0) = 0. \tag{5}$$

We introduce the 1-periodic functions $R(\xi)$, $S(\xi)$, $K(\xi)$ (see Figs. 7, 8):

$$(R(\xi), S(\xi), K(\xi)) = \begin{cases} (R_1, S_1, K_1), & \xi \in (j-1, \ j-1+\theta_1), \\ (R_2, S_2, K_2), & \xi \in (j-1+\theta_1, \ j), \end{cases}$$
$$j = 1, \ldots, N.$$

Then problems (1)–(5) can be rewritten as

$$R\left(\frac{x}{\varepsilon}\right)\frac{\partial^2 u}{\partial t^2}+S\left(\frac{x}{\varepsilon}\right)\frac{\partial u}{\partial t}-\frac{\partial}{\partial x}\left(K\left(\frac{x}{\varepsilon}\right)\frac{\partial u}{\partial x}\right)=f(x,\,t),\quad x\neq x^i, \tag{6}$$

$$[u]\big|_{x^i}=0, \tag{7}$$

$$\left[K\left(\frac{x}{\varepsilon}\right)\frac{\partial u}{\partial x}\right]\Bigg|_{x^i}=0,\quad i=1,\,\ldots,\,2N-1. \tag{8}$$

Here

$$x^i=\xi^i\varepsilon,\quad \xi^i=\begin{cases}(j-1+\theta_1)\,\varepsilon & \text{for } i=2j-1,\\ j\varepsilon & \text{for } i=2j,\end{cases}$$

with the boundary and initial values being (4) and (5).

Relations (6)–(8) are equivalent to the equation

$$R\left(\frac{x}{\varepsilon}\right)\frac{\partial^2 u}{\partial t^2}+S\left(\frac{x}{\varepsilon}\right)\frac{\partial u}{\partial t}-\frac{\partial}{\partial x}\left(K\left(\frac{x}{\varepsilon}\right)\frac{\partial u}{\partial x}\right)-f(x,\,t)=0,\, x\in(0,\,l), \tag{9}$$

where $u(x,t)$ is a generalized solution (see Chapter 1).

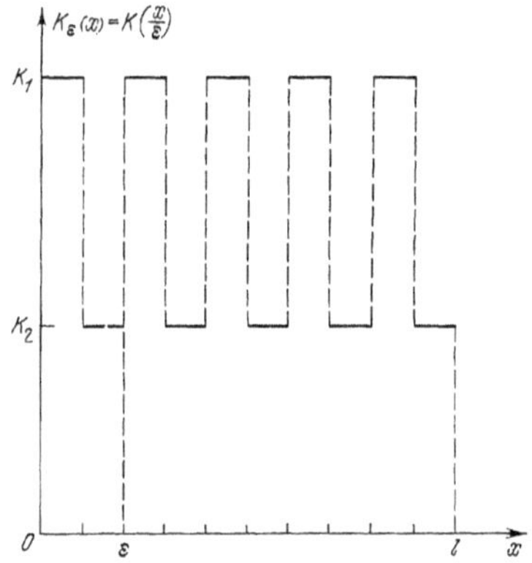

Figure 7

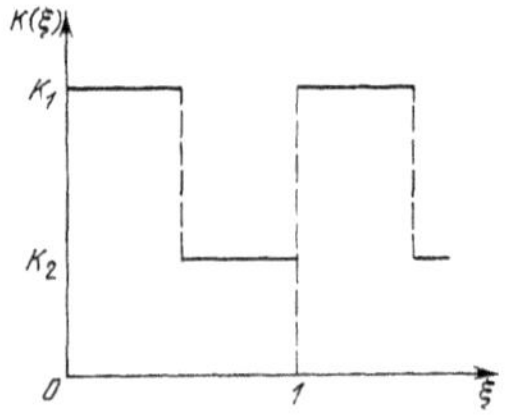

Figure 8

We assume that the initial data $f(x,t), g_i(t)$ are infinitely differentiable and agree at the points $(0,0)$, $(1,0)$, for instance:

$$\frac{\partial^{k+j}f}{\partial x^k\partial t^j}(0,\,0)=\frac{\partial^{k+j}f}{\partial x^k\partial t^j}(l,\,0)=\frac{d^j g_i}{dt^j}(0)=0$$

for all $k,j\geqslant 0$.

It is necessary for these conditions to be satisfied for all $k,j\geqslant 0$ to obtain an asymptotic expansion of the solution; a finite number of the relations is sufficient to get an approximation of a finite order of accuracy in ε.

The asymptotics of a solution of (4)–(8) is sought for in the form

$$u^{(2)}(x,t) = u_0\left(x, t, \frac{x}{\varepsilon}\right) + \varepsilon u_1\left(x, t, \frac{x}{\varepsilon}\right) + \varepsilon^2 u_2\left(x, t, \frac{x}{\varepsilon}\right), \tag{10}$$

where $u_i(x,t,\xi)$ are 1-periodic functions in ξ. By substituting (10) into the left side of (9), with $x \neq x^i$, we obtain

$$\begin{aligned} &-\varepsilon^{-2} L_{\xi\xi} u_0 - \varepsilon^{-1}(L_{\xi\xi}u_1 + L_{\xi x}u_0 + L_{x\xi}u_0) - \\ &\quad -\varepsilon^0 \Bigg(L_{\xi\xi}u_2 + L_{\xi x}u_1 + L_{x\xi}u_1 + L_{xx}u_0 - R(\xi)\frac{\partial^2 u_0}{\partial t^2} - \\ &\quad\quad - S(\xi)\frac{\partial u_0}{\partial t} + f(x,t)\Bigg) + \varepsilon r^{(2)}(x,t,\varepsilon), \quad \xi = \frac{x}{\varepsilon}. \end{aligned} \tag{11}$$

Here

$$L_{\alpha\beta} u(x,t,\xi) = \frac{\partial}{\partial \alpha}\left(K(\xi)\frac{\partial}{\partial \beta}u(x,t,\xi)\right);$$

in particular, $L_{\xi\xi} = L$,

$$r^{(2)}(x,t,\varepsilon) = r_0^{(2)}(x,t,\varepsilon) + \varepsilon \frac{\partial}{\partial x} r_1^{(2)}(x,t,\varepsilon),$$

$$\begin{aligned} r_0^{(2)}(x,t,\varepsilon) = \Bigg(&-L_{x\xi}u_2 - L_{xx}u_1 + R(\xi)\frac{\partial^2 u_1}{\partial t^2} + S(\xi)\frac{\partial u_1}{\partial t} + \\ &+ \varepsilon R(\xi)\frac{\partial^2 u_2}{\partial t^2} + \varepsilon S(\xi)\frac{\partial u_2}{\partial t}\Bigg)\Bigg|_{\xi = x/\varepsilon}, \end{aligned} \tag{12}$$

$$r_1^{(2)}(x,t,\varepsilon) = -K(\xi)\frac{\partial u_2}{\partial x}\Bigg|_{\xi = x/\varepsilon}.$$

Hereafter, $\frac{\partial}{\partial x} r_1^{(2)}$ is a full derivative with respect to x.

Let us require that the terms of orders $\varepsilon^{-2}, \varepsilon^{-1}, \varepsilon^0$ vanish. Then

$$L_{\xi\xi}u_0 = 0, \tag{13}$$

$$L_{\xi\xi}u_1 + L_{\xi x}u_0 + L_{x\xi}u_0 = 0, \tag{14}$$

$$\begin{aligned} &L_{\xi\xi}u_2 + L_{\xi x}u_1 + L_{x\xi}u_1 + L_{xx}u_0 - \\ &\quad - R(\xi)\frac{\partial^2 u_0}{\partial t^2} - S(\xi)\frac{\partial u_0}{\partial t} + f(x,t) = 0. \end{aligned} \tag{15}$$

Substituting $u(2)$ into the conjugation conditions (7) and (8) yields

$$\begin{aligned} [u^{(2)}]\big|_{x=x^i} = \big([u_0(x,t,\xi)]\big|_{\xi=\xi^i} &+ \varepsilon[u_1(x,t,\xi)]\big|_{\xi=\xi^i} + \\ &+ \varepsilon^2[u_2(x,t,\xi)]\big|_{\xi=\xi^i}\big)\big|_{\xi=x/\varepsilon}, \end{aligned} \tag{16}$$

$$\begin{aligned} \left[K\left(\frac{x}{\varepsilon}\right)\frac{\partial u^{(2)}}{\partial x}\right]\Bigg|_{x=x^i} &= \left(\varepsilon^{-1}\left[K(\xi)\frac{\partial u_0}{\partial \xi}\right]\Bigg|_{\xi=\xi^i} + \right. \\ &+ \left[K(\xi)\left(\frac{\partial u_1}{\partial \xi} + \frac{\partial u_0}{\partial x}\right)\right]\Bigg|_{\xi=\xi^i} + \varepsilon\left[K(\xi)\left(\frac{\partial u_2}{\partial \xi} + \frac{\partial u_1}{\partial x}\right)\right]\Bigg|_{\xi=\xi^i} + \\ &\left. + \varepsilon^2\left[K(\xi)\frac{\partial u_2}{\partial x}\right]\Bigg|_{\xi=\xi^i}\right)\Bigg|_{\xi=x/\varepsilon}. \end{aligned}$$

By equating the three higher terms in each of the relations (15) to zero, we see that

$$[u_0(x,\,t,\,\xi)]\,|_{\xi=\xi^i}=0, \tag{17}$$

$$\left[K(\xi)\frac{\partial u_0}{\partial\xi}\right]\Big|_{\xi=\xi^i}=0, \tag{18}$$

$$[u_1(x,\,t,\,\xi)]\,|_{\xi=\xi^i}=0, \tag{19}$$

$$\left[K(\xi)\left(\frac{\partial u_1}{\partial\xi}+\frac{\partial u_0}{\partial x}\right)\right]\Big|_{\xi=\xi^i}=0, \tag{20}$$

$$[u_2(x,\,t,\,\xi)]\,|_{\xi=\xi^i}=0, \tag{21}$$

$$\left[K(\xi)\left(\frac{\partial u_2}{\partial\xi}+\frac{\partial u_1}{\partial x}\right)\right]\Big|_{\xi=\xi^i}=0. \tag{22}$$

Let u_0 be a solution of (13), (17), and (18); u_1 of (14), (19), and (20); u_2 of (15), (21), and (22); and x,t be parameters of these problems. Then $u^{(2)}$ satisfies (7) exactly, (6) with $x\neq x^{i-c}$ to terms of order ε, and to conditions (8), with accuracy to $O(\varepsilon^2)$. Thus, the original problem was reduced to the search for generalized solutions of equations (12)–(14). As indicated in §5 of the Introduction, u_0 is independent of ξ, that is,

$$u_0(x,\,t,\,\xi)=v_0(x,\,t), \tag{23}$$

$$u_1(x,\,t,\,\xi)=N_1(\xi)\frac{\partial v_0}{\partial x}. \tag{24}$$

Here $N_1(\xi)$ is a generalized solution of the equation

$$\frac{d}{d\xi}\left(K(\xi)\left(\frac{dN_1}{d\xi}+1\right)\right)=0, \tag{25}$$

1-periodic in ξ and equal to

$$N_1(\xi)=\int_0^{\xi}(K^{-1}(\tau)\widehat{K}-1)\,d\tau+\text{const}, \tag{26}$$

where

$$\widehat{K}=\langle K^{-1}(\xi)\rangle^{-1}>0. \tag{27}$$

By applying the averaging operation $\langle\cdot\rangle=\int_0^1\cdot d\xi$ to (15), and by Lemma 2.4.1, we obtain the necessary and sufficient solvability condition for equation (15) in the class of 1-periodic functions:

$$-\left\langle\frac{\partial}{\partial x}\left(K(\xi)\left(\frac{\partial u_1}{\partial\xi}+\frac{\partial v_0}{\partial x}\right)\right)-R(\xi)\frac{\partial^2 v_0}{\partial t^2}-S(\xi)\frac{\partial v_0}{\partial t}\right\rangle=f(x,\,t).$$

By substituting u_1 from (24), we obtain

$$\widehat{R}\frac{\partial^2 v_0}{\partial t^2}+\widehat{S}\frac{\partial v_0}{\partial t}-\widehat{K}\frac{\partial^2 v_0}{\partial x^2}=f(x,t), \tag{28}$$

$$\widehat{R}=\langle R(\xi)\rangle>0,\quad \widehat{S}=\langle S(\xi)\rangle\geqslant 0. \tag{29}$$

Condition (28) is an averaged solution used below to define $v_0(x,t)$. Note that the averaged equation is of the same type as the original one, i.e., it is hyperbolic (see (27), (29)).

By substituting $u^{(2)}$ into the boundary and initial conditions (4) and (5), we get

$$v_0(0,t)+\varepsilon u_1(0,t,0)+\varepsilon^2 u_2(0,t,0)=g_1(t), \tag{30}$$

$$v_0(l,t)+\varepsilon u_1\left(l,t,\frac{l}{\varepsilon}\right)+\varepsilon^2 u_2\left(l,t,\frac{l}{\varepsilon}\right)=$$

$$=v_0(l,t)+\varepsilon u_1(l,t,0)+\varepsilon^2 u_2(l,t,0)=g_2(t), \tag{31}$$

$$u^{(2)}\big|_{t=0}=v_0(x,0)+\varepsilon u_1(x,0,\xi)+\varepsilon^2 u_2(x,0,\xi), \tag{32}$$

$$\left.\frac{\partial u^{(2)}}{\partial t}\right|_{t=0}=\frac{\partial v_0}{\partial t}(x,0)+\varepsilon\frac{\partial u_1}{\partial t}(x,0,\xi)+\varepsilon^2\frac{\partial u_2}{\partial t}(x,0,\xi), \tag{33}$$

whence we derive the boundary and initial values for v_0:

$$v_0(0,t)=g_1(t),\quad v_0(l,t)=g_2(t), \tag{34}$$

$$v_0(x,0)=\frac{\partial v_0}{\partial t}(x,0)=0, \tag{35}$$

and the conditions

$$u_1(x,t,0)=0, \tag{36}$$

$$u_2(x,t,0)=0, \tag{37}$$

which determine the unambiguous periodic solutions of (14) and (15). In particular, the constant in (26) vanishes. Formulas (30)–(37) imply that $u^{(2)}$ satisfies (4) exactly, and (5) up to the terms of order ε.

Thus, (11) and (30)–(37) imply that $u^{(2)}$ is a solution of the problem

$$R\left(\frac{x}{\varepsilon}\right)\frac{\partial^2 u^{(2)}}{\partial t^2}+S\left(\frac{x}{\varepsilon}\right)\frac{\partial u^{(2)}}{\partial t}-\frac{\partial}{\partial x}\left(K\left(\frac{x}{\varepsilon}\right)\frac{\partial u^{(2)}}{\partial x}\right)=$$

$$=f(x,t)+\varepsilon r_0^{(2)}(x,t,\varepsilon)+\varepsilon^2\frac{\partial}{\partial x}r_1^{(2)}(x,t,\varepsilon)\quad\text{for}\quad x\neq x^i,$$

$$[u^{(2)}]\big|_{x=x^i}=0,$$

$$\left[K\left(\frac{x}{\varepsilon}\right)\frac{\partial u^{(2)}}{\partial x}+\varepsilon^2 r_1^{(2)}(x,t,\varepsilon)\right]\Big|_{x=x^i}=0,$$

$$u^{(2)}(0,t)=g_1(t),\quad u^{(2)}(l,t)=g_2(t),$$

$$u^{(2)}(x,\,0)=\varepsilon u_1|_{t=0}+\varepsilon^2u_2|_{t=0},$$
$$\frac{\partial u^{(2)}}{\partial t}(x,\,0)=\varepsilon\,\frac{\partial u_1}{\partial t}\bigg|_{t=0}+\varepsilon^2\,\frac{\partial u_2}{\partial t}\bigg|_{t=0};$$

by the definition of $r_0^{(2)}$ and $r_1^{(2)}$, cf. (12). Representation (24) and the relation (35) imply that

$$u_1|_{t=0}=\partial u_1/\partial t|_{t=0}=0.$$

Two methods can be used to obtain the estimate of $w=u^{(2))}-u$. With the first method, just as in §2.4, we add to $u^{(2)}$ a term $\varepsilon^2\psi$ such that for $u^{(2)}+\varepsilon^2\psi$ the conditions on discontinuities (7) and (8) are satisfied exactly. Then the function $w-(u^{(2)}+\varepsilon^2\psi)$ can be estimated in the norm W_2^1 by using the standard technique (Chapter 1). With the second method (to be discussed later in §6.1 and Chapter 8), one should proceed to the generalized solution (§1.3).

The function $w(x,t)$ is a generalized solution of the problem

$$R\left(\frac{x}{\varepsilon}\right)\frac{\partial^2 w}{\partial t^2}+S\left(\frac{x}{\varepsilon}\right)\frac{\partial w}{\partial t}-\frac{\partial}{\partial x}\left(K\left(\frac{x}{\varepsilon}\right)\frac{\partial w}{\partial x}\right)=$$
$$=\varepsilon r_0^{(2)}+\varepsilon^2\frac{\partial}{\partial x}r_1^{(2)},\quad x\in(0,\,l),\tag{38}$$
$$w\,(0,\,t)=w\,(l,\,t)=0,\tag{39}$$
$$\left.\begin{aligned}w\,(x,\,0)&=\varepsilon^2u_2\,|_{t=0},\\ \frac{\partial w}{\partial t}(x,\,0)&=\varepsilon^2\,\frac{\partial u_2}{\partial t}\bigg|_{t=0}.\end{aligned}\right.\tag{40}$$

Equation (28) with the boundary and initial conditions (24) and (25) will be called the *averaged problem*. With the initial values being infinitely differentiable and consistent, the solution of the averaged problem is infinitely differentiable on $[0,l]\times[0,+\infty)$. The relations (25)–(27) imply that the function $u(x,t,\xi)$ is bounded for $t\leqslant T$ as are its first and second derivatives for $\xi\neq\xi^i$ whence it is easy to obtain the boundedness of $u_2(x,t,\xi)$ and of its first and second derivatives for $\xi\neq\xi^i$ (up to the discontinuity surfaces).

Thus, the $|r_0^{(2)}|,|r_1^{(2)}|,|u_2|,|\partial u_2/\partial t|$ are bounded by a constant, and consequently, the following estimate from Chapter 1 is valid for the solution w of (38)–(40):

$$\|w\|_{W_2^1((0,l)\times(0,T))}\leqslant c\,(T)\,\varepsilon,$$

that is,

$$\left\|u^{(2)}-u\right\|_{W_2^1((0,l)\times(0,T))}\leqslant c\,(T)\,\varepsilon,\tag{41}$$

where $c(T)$ is the constant independent of ε. Estimate (41) implies

$$\left\| v_0 + \varepsilon u_1\left(x, t, \frac{x}{\varepsilon}\right) - u \right\|_{W_2^1((0,l)\times(0,T))} = O(\varepsilon), \tag{42}$$

$$\| v_0 - u \|_{L_2((0,l)\times(0,T))} = O(\varepsilon). \tag{43}$$

Thus, we have proved the following theorem.

THEOREM 1. *Let u be an exact solution of problems (4)–(8), v_0 a solution of the averaged problem (28), (34), (35), $u_1(x,t,\xi)$ a 1-periodic solution in ξ of problem (14), (36), $u_2(x,t,\xi)$ a solution of problem (15), (37) of the same periodicity, and $u^{(2)}$ be determined by relation (10). Then the estimates (41)–(43) hold.*

REMARK 1: By analogy with §5 of the Introduction, we demonstrate that

$$u_2(x, t, \xi) = N_{02}^0(\xi)\frac{\partial^2 v_0}{\partial x^2} + N_{10}^1(\xi)\frac{\partial v_0}{\partial t} + N_{20}^0(\xi)\frac{\partial^2 v_0}{\partial t^2},$$

where N_{02}^0, N_{10}^1, N_{20}^0 are 1-periodic generalized solutions of the problems

$$L_{\xi\xi}N_{02}^0 = -\frac{d}{d\xi}(K(\xi)N_1(\xi)) - K(\xi)\frac{dN_1}{d\xi} - K(\xi) + \widehat{K}, \quad N_{02}^0(0) = 0;$$

$$L_{\xi\xi}N_{10}^1 = S(\xi) - \widehat{S}, \quad N_{10}^1(0) = 0;$$

$$L_{\xi\xi}N_{20}^0 = R(\xi) - \widehat{R}, \quad N_{20}^0(0) = 0.$$

Thus,

$$u^{(2)} = v_0 + \varepsilon N_1\left(\frac{x}{\varepsilon}\right)\frac{\partial v_0}{\partial x} + \varepsilon^2 N_{02}^0\left(\frac{x}{\varepsilon}\right)\frac{\partial^2 v_0}{\partial x^2} + {}$$
$$+ \varepsilon^2 N_{10}^1\left(\frac{x}{\varepsilon}\right)\frac{\partial v_0}{\partial t} + \varepsilon^2 N_{20}^0\left(\frac{x}{\varepsilon}\right)\frac{\partial^2 v_0}{\partial t^2}.$$

REMARK 2: Let R_i, S_i, K_i be $n \times n$ matrices, u, f, g_i n-dimensional vector-valued functions, $R_i > 0$, $S_i > 0$, $K_i > 0$, and (6) a system of equations. Then the asymptotics of $u(x)$ are sought for in the form (10), where the $u_i(x,\xi)$ are n-dimensional vector-valued functions 1-periodic in ξ. The vector-valued function $u(2)$ is constructed by exactly repeating the procedure described above. Consider the case

$$R(\xi) = \rho_0(\xi)E, \quad S(\xi) \equiv 0,$$

$$K(\xi) = \begin{bmatrix} 2\mu(\xi) + \lambda(\xi) & 0 & 0 \\ 0 & \mu(\xi) & 0 \\ 0 & 0 & \mu(\xi) \end{bmatrix},$$

where the $\rho_0(\xi)$, $\mu(\xi)$, $\lambda(\xi)$ are piecewise constant 1-periodic functions, and E is the unit matrix. The system of equations (9) with conditions (4) and (5) describes a wave process in a layered composite with the layers of an isotropic, linearly elastic material.

In the same way as in §§2.3 and 2.4, an asymptotic expansion of the solution of (4)–(8) can be structured as

$$u^{(\infty)}(x,\,t) = \sum_{p,q,i=0}^{\infty} \varepsilon^{q+p+i} N_{qi}^{p}(\xi)\, \frac{\partial^{q+i} v(x,t)}{\partial t^q \partial x^i}\bigg|_{\xi=x/\varepsilon},$$

where the $N_{qpi}(\xi)$ are 1-periodic functions of the fast variable $\xi, v(x,t) = \sum_{j=0}^{\infty} \varepsilon^j v_j(x,t)$ (see §4.5).

§2 Nonstationary Problem of Heat Conduction

Consider the heat equation

$$S\left(\frac{x}{\varepsilon}\right)\frac{\partial u}{\partial t} - \frac{\partial}{\partial x}\left(K\left(\frac{x}{\varepsilon}\right)\frac{\partial u}{\partial x}\right) = f(x,\,t), \qquad x \in (0,\,l), \tag{1}$$

where S is heat capacity per unit volume, and K is the thermal conductivity coefficient. Let $S(\xi)$, $K(\xi)$ be 1-periodic discontinuous functions and $u(x)$ a generalized solution of (1). Suppose that $S(\xi)$ and $K(\xi)$ are infinitely differentiable on the segments $[0,\xi_1]$, $[\xi_1,\xi_2], \ldots, [\xi_r, 1]$; $K(\xi) > 0$, $S(\xi) > 0$.

Let the boundary values be

$$u(0,\,t) = g_1(t), \quad u(l,\,t) = g_2(t) \tag{2}$$

and the initial values

$$u(x,\,0) = \varphi(x), \tag{3}$$

f, g_1, g_2, and φ being infinitely differentiable, and the conjugation conditions

$$\frac{\partial^{k+j} f}{\partial x^k \partial t^j}(0,\,0) = \frac{\partial^{k+j} f}{\partial x^k \partial t^j}(l,\,0) = \frac{d^j g_i(0)}{dt^j} = \frac{d^k \varphi}{dx^k}(0) = \frac{d^k \varphi}{dx^k}(l) = 0$$

being satisfied for all $k \geqslant 0$, $j \geqslant 0$, $i = 1, 2$.

The asymptotic solution of (1)–(3) is found in the form (1.10) The procedure of constructing $u(2)$ is derived from that in §1 if we set $R \equiv 0$. The averaged problem of heat conductivity has the form

$$\widehat{S}\frac{\partial v_0}{\partial t} - \widehat{K}\frac{\partial^2 v_0}{\partial x^2} = f(x,\,t), \quad x \in (0,\,l), \tag{4}$$

$$v_0(0,\,t) = g_1(t), \quad v_0(l,\,t) = g_2(t), \tag{5}$$

$$v_0(x,\,0) = \varphi(x). \tag{6}$$

Using the a priori estimate from Theorem 1.4.2, the following theorem can be proved by analogy with §1.

THEOREM 1. *Let u be an exact solution of problems (1)–(3), v_0 a solution of (1.13) and (1.26) periodic in ξ, $u_2(x,t,\xi)$ a solution of (1.14), the averaged problem (4)–(6), $u(x,t,\xi)$ a solution of (1.37) for $R \equiv 0$ of the same periodicity, and $u^{(2)}$ the sum of (1.10). Then the estimates*

$$\| u^{(2)} - u \|_{W_2^{1,0}((0,l)\times(0,T))} \leqslant c(T)\,\varepsilon,$$

$$\left\| v_0 + \varepsilon u_1\left(x,\, t,\, \frac{x}{\varepsilon}\right) - u \right\|_{W_2^{1,0}((0,l)\times(0,T))} \leqslant c(T)\,\varepsilon,$$

$$\| v_0 - u \|_{L_2((0,l)\times(0,T))} \leqslant c(T)\,\varepsilon,$$

hold, with $c(T)$ being independent of ε.

§3 Averaging Maxwell Equations

The problem of electromagnetic plane-wave propagation in layered periodic media can be reduced to the problem solved in §1. Consider the system of electrodynamic equations

$$e\left(\frac{x}{\varepsilon}\right)\frac{\partial E}{\partial t} = \operatorname{rot} H - \sigma\left(\frac{x}{\varepsilon}\right)E, \tag{1}$$

$$\mu\left(\frac{x}{\varepsilon}\right)\frac{\partial H}{\partial t} = -\operatorname{rot} E, \quad x \in (0,\, l), \quad t > 0, \tag{2}$$

where $E(x,t)$, $H(x,t)$ are three-dimensional vector-valued functions of the strengths of electric and magnetic fields, $e(\xi)$, $\mu(\xi)$, and $\sigma(\xi)$ (dielectric constant, magnetic permeability and electric conductivity, respectively) are 1-periodic smooth functions, and $e, \mu > 0$, $\sigma \geqslant 0$.

Let the following initial values be given at $t = 0$:

$$E(x,\, 0) = H(x,\, 0) = 0, \tag{3}$$

and the boundary values at $x = 0$, $x = 1$,

$$E^j(0,\, t) = E_1^j(t),\ E^j(l,\, t) = E_2^j(t), \tag{4_E}$$

$$H^1(0,\, t) = 0, \qquad H^1(l,\, t) = 0, \tag{4_H}$$

where the $E_i(t)$ are infinitely differentiable vector-valued functions of t,

$$\frac{d^q E_i^j}{dt^q} = 0, \quad t = 0, \quad i = 1,\, 2, \quad q = 1,\, 2,\, \ldots,\ j = 2,\ 3,$$

where E_i^j, H_i^j are the jth components of the vectors E_i and H_i. Since the components E^1 and H^1 satisfy the equations

$$e\left(\frac{x}{\varepsilon}\right)\frac{\partial E^1}{\partial t} = -\sigma\left(\frac{x}{\varepsilon}\right)E^1, \quad \mu\left(\frac{x}{\varepsilon}\right)\frac{\partial H^1}{\partial t} = 0$$

with homogeneous initial values, then $E^1 = H^1 \equiv 0$.

Let us rearrange equation (2) to the form

$$\frac{\partial H}{\partial t} = -\mu^{-1}\left(\frac{x}{\varepsilon}\right)\operatorname{rot} E \tag{5}$$

and take *rot* of (5). We get

$$\frac{\partial}{\partial t}\operatorname{rot} H = -\operatorname{rot}\left(\mu^{-1}\left(\frac{x}{\varepsilon}\right)\operatorname{rot} E\right). \tag{6}$$

Differentiating (1) with respect to t, we get

$$e\left(\frac{x}{\varepsilon}\right)\frac{\partial^2 E}{\partial t^2} = \frac{\partial}{\partial t}\operatorname{rot} H - \sigma\left(\frac{x}{\varepsilon}\right)\frac{\partial E}{\partial t}. \tag{7}$$

It follows from (6) and (7) that $E(x,t)$ is a solution of the equation

$$e\left(\frac{x}{\varepsilon}\right)\frac{\partial^2 E}{\partial t^2} = -\operatorname{rot}\left(\mu^{-1}\left(\frac{x}{\varepsilon}\right)\operatorname{rot} E\right) - \sigma\left(\frac{x}{\varepsilon}\right)\frac{\partial E}{\partial t},$$

that is,

$$e\left(\frac{x}{\varepsilon}\right)\frac{\partial^2 E^j}{\partial t^2} + \sigma\left(\frac{x}{\varepsilon}\right)\frac{\partial E^j}{\partial t} = \frac{\partial}{\partial x}\left(\mu^{-1}\left(\frac{x}{\varepsilon}\right)\frac{\partial E^j}{\partial x}\right), \tag{8}$$

$$j = 2,\ 3, \qquad E^1 \equiv 0.$$

Thus, the E^j satisfy the equation of type (1.6). Equation (1) and condition (3) yield the initial conditions for E:

$$E^j(x,\ 0) = \frac{\partial E^j}{\partial t}(x,\ 0) = 0, \qquad j = 2,\ 3. \tag{9}$$

Thus, (8), (9), and (4_E) is a corollary of problems (1)–(4). A solution of problems (8), (9) and (4_E) exists and is unique [76]. After E^j $(j = 2,3)$ has been defined from (8), (9), and (4), H can be found from the relations

$$H(x,\ t) = -\mu^{-1}\left(\frac{x}{\varepsilon}\right)\int_0^t \operatorname{rot} E(x,\ \tau)\,d\tau, \tag{10}$$

that is,

$$H^2(x,t) = \mu^{-1}\left(\frac{x}{\varepsilon}\right)\int_0^t \frac{\partial E^3}{\partial x}\,d\tau, \quad H^3(x,t) = -\mu^{-1}\left(\frac{x}{\varepsilon}\right)\int_0^t \frac{\partial E^2}{\partial x}\,d\tau.$$

The averaging of (8), (9), and (4_E) was carried out in §1.

REMARK 1: The above-described procedure of reducing problems (1)–(4) to (8), (9), (4_E), and (10) is also valid for discontinuous σ, e, and μ, which corresponds to layered media.

REMARK 2: When Maxwell equations are averaged in a multi-dimensional periodic medium, a retarding system of integral-differential equations with a delay emerges [214]. Some media are known to retain a memory of electromagnetic waves. It can be assumed from the above that the reason lies in their inhomogeneous microstructure.

§4 Averaging Equations of a Viscoelastic Medium

Differential equations are derived in developing models of mechanics and physics for homogeneous media on the basis of the laws of conservation of mass, momentum, energy, etc. However, many processes cannot be described accurately enough by the differential equations alone and so additional terms similar to those occurring in averaging Maxwell equations for a multi-dimensional case have to be added or the models become more complicated in some other way. The example given below shows that such complicated models may result from averaging one-dimensional problems.

Consider a model system of equations of motion of a periodic linear viscoelastic medium

$$l_1 \equiv \frac{\partial(a(x/\varepsilon)\,p)}{\partial t} - \frac{\partial u}{\partial x} = 0,$$
$$l_2 \equiv \frac{\partial u}{\partial t} + \frac{\partial(p - w(x/\varepsilon)\,\partial u/\partial x)}{\partial x} = 0, \tag{1}$$

where $u(x,t)$ is velocity, $p(x,t)$ pressure, x a Lagrange variable, and $a(\xi)$, $w(\xi)$ are 1-periodic functions. To make the presentation shorter, we arrange the equations as if $a(\xi)$ and $w(\xi)$ were differentiable with respect to ξ. We seek an expansion of the solution in an asymptotic series:

$$u^{(\infty)} = u_0(x,t,\xi) + \varepsilon u_1(x,t,\xi) + \ldots,$$
$$p^{(\infty)} = p_0(x,t,\xi) + \varepsilon p_1(x,t,\xi) + \ldots, \quad \xi = x/\varepsilon.$$

We have

$$l_1 \sim -\frac{1}{\varepsilon}\frac{\partial u_0}{\partial \xi} + \frac{\partial(a(\xi)\,p_0(x,t,\xi))}{\partial t} - \frac{\partial u_0}{\partial x} - \frac{\partial u_1}{\partial \xi} + O(\varepsilon) \sim 0,$$

$$l_2 \sim -\frac{1}{\varepsilon^2}\frac{\partial}{\partial \xi}\left(w\frac{\partial u_0}{\partial \xi}\right) + \frac{1}{\varepsilon}\left(\frac{\partial}{\partial \xi}\left(p_0 - w\left(\frac{\partial u_0}{\partial x} + \frac{\partial u_1}{\partial \xi}\right)\right) - \frac{\partial}{\partial x}\left(w\frac{\partial u_0}{\partial x}\right)\right) +$$
$$+\frac{\partial u_0}{\partial t} - \frac{\partial}{\partial x}\left(p_0 - w\left(\frac{\partial u_0}{\partial x} + \frac{\partial u_1}{\partial \xi}\right)\right) - \frac{\partial}{\partial \xi}\left(p_1 - w\left(\frac{\partial u_1}{\partial x} + \frac{\partial u_2}{\partial \xi}\right)\right) + O(\varepsilon) \sim 0.$$

The coefficient of ε^{-1} in the first equation and that of ε^{-2} in the second equation are zero if u_0 does not depend on ξ: $u_0 = v_0(x,t)$. We require that the coefficient of ε^{-1} vanish in the second equation:

$$\frac{\partial}{\partial \xi}\left(p_0 - w\left(\frac{\partial v_0}{\partial x} + \frac{\partial u_1}{\partial \xi}\right)\right) = 0;$$

then

$$p_0 - w\left(\frac{\partial v_0}{\partial x} + \frac{\partial u_1}{\partial \xi}\right) = h_1(x, t). \tag{2}$$

Let us divide the equation by w and apply $\langle\cdot\rangle$. Since $\langle \partial u_1/\partial \xi\rangle = 0$ due to the periodicity of u_1 in ξ, we get

$$\langle p_0/w\rangle - \partial v_0/\partial x = h_1\langle 1/w\rangle.$$

We set $\langle 1/w\rangle = 1/\hat{w}$; thus,

$$h_1 = \hat{w}\left\langle \frac{p_0}{w}\right\rangle - \hat{w}\frac{\partial v_0}{\partial x}. \tag{3}$$

By substituting h_1 into (2), we obtain a representation of $\partial v_0/\partial x + u_1/\partial \xi$ which enters the zeroth term of the first equation

$$\frac{\partial v_0}{\partial x} + \frac{\partial u_1}{\partial \xi} = \frac{p_0}{w} - \frac{\hat{w}}{w}\left\langle \frac{p_0}{w}\right\rangle + \frac{\hat{w}}{w}\frac{\partial v_0}{\partial x}. \tag{4}$$

Now we have

$$l_2 = \frac{\partial v_0}{\partial t} - \frac{\partial h_1(x, t)}{\partial x} - \frac{\partial}{\partial \xi}\left(p_1 - w\left(\frac{\partial u_1}{\partial x} + \frac{\partial u_2}{\partial \xi}\right)\right) + O(\varepsilon).$$

We define the functions $p_1(x,t,\xi)$ and $u_2(x,t,\xi)$ so that $p_1 - w(\frac{\partial u}{\partial x} + \frac{\partial u_2}{\partial \xi})$ were independent of ξ, for example, $p_1(x,t,\xi) \equiv 0$. We have

$$w(\partial u_1/\partial x + \partial u_2/\partial \xi) = h_2(x, t),$$

where $h_2(x,t)$ is a function yet unknown, that is

$$\partial u_1/\partial x + \partial u_2/\partial \xi = h_2(x, t)/w.$$

After applying the operation $\langle\cdot\rangle$, we see that

$$\partial u_1/\partial x = h_2(x,\ t)\langle 1/w\rangle;$$

$h_2 = \hat{w}\partial u_1/\partial x$, and consequently,

$$\partial u_2/\partial \xi = \partial u_1/\partial x(\widehat{w}/w - 1);$$

the mean over the period $\langle\cdot\rangle$ of the right side is zero, and so the function $u_2(x,t,\xi)$, periodic in ξ, can be found from the last equation. With this choice of p_1 and u_2, we get

$$l_2 = \frac{\partial v_0}{\partial t} - \frac{\partial h_1(x,\ t)}{\partial x} + O(\varepsilon).$$

Substitute h_1 from (3) into the expression for l_2 and $\partial v_0/\partial x + \partial u_1/\partial \xi$ from (4) into the expression for l_1. Requiring that the zeroth-order term be zero, we get a system of equations for the zeroth approximation:

$$\frac{\partial(a(\xi)p_0(x,t,\xi))}{\partial t} - \frac{\widehat{w}}{w(\xi)}\frac{\partial v_0(x,t)}{\partial x} + \frac{\widehat{w}}{w(\xi)}\left\langle\frac{p_0(x,t,\xi)}{w(\xi)}\right\rangle - \frac{p_0(x,t,\xi)}{w(\xi)} = 0, \quad (5)$$

$$\frac{\partial v_0(x,t)}{\partial t} + \frac{\partial}{\partial x}\left(\widehat{w}\left\langle\frac{p_0(x,t,\xi)}{w(\xi)}\right\rangle - \widehat{w}\frac{\partial v_0(x,t)}{\partial x}\right) = 0.$$

It can demonstrated that this system of equations is a correct averaging of the original problem (1) provided the appropriate initial conditions for the problems (1) and (5) are close. And it suffices to require that the coefficients of $a(\xi)$ and $w(\xi)$ were measurable and the following conditions were satisfied: $0 < \alpha_1 \leqslant -a(\xi) \leqslant \alpha_2 < \infty$, $0 < \beta_1 \leqslant w(\xi) \leqslant \beta_2 < \infty$.

In the original system of equations (1), the coefficients a and w had a characteristic size of variation of ε, and so their numerical solution needed an integration step with respect to the variable $x : h \ll \varepsilon$. The small parameter ε is absent in the resulting system (5), and the latter can be solved numerically with sufficiently large integration steps with respect to all the variables x, t, ξ. At the same time, unlike the above-described problems, the leading term of one of the functions $p_0(x,t,\xi)$ depends on the fast variable. Therefore (5) contains already two space variables, and its numerical solution is more laborious than that of the original system (1) with constant a and w. In actual problems, the coefficients a and w are piecewise constant, and this makes the numerical solution of (5) much simpler.

Assuming that

$$a = \begin{cases} a_1 & \text{for} \quad 0 < \xi \leqslant \theta_1, \\ a_2 & \text{for} \quad \theta_1 < \xi \leqslant 1, \end{cases} \qquad w = \begin{cases} w_1 & \text{for} \quad 0 < \xi \leqslant \theta_1, \\ w_2 & \text{for} \quad \theta_1 < \xi \leqslant 1, \end{cases}$$

$$p_0(x,\ 0,\ \xi) = \begin{cases} p_0^1(x) & \text{for} \quad 0 < \xi \leqslant \theta_1, \\ p_0^2(x) & \text{for} \quad \theta_1 < \xi \leqslant 1. \end{cases}$$

we have $1/\hat{w} = \theta_1/w_1 + \theta_2/w_2$. Then it can be shown that for all t a solution of the Cauchy problem for (5) satisfies the condition

$$p_0(x, t, \xi) = \begin{cases} p_0^1(x, t) & \text{for } 0 < \xi \leqslant \theta_1, \\ p_0^2(x, t) & \text{for } \theta_1 < \xi \leqslant 1. \end{cases}$$

By setting $\theta_2 = 1 - \theta_1$, we can rewrite (5) as

$$\begin{aligned} & a_i \frac{\partial p_0^i(x, t)}{\partial t} - \frac{\hat{w}}{w_i} \frac{\partial v_0(x, t)}{\partial x} + \\ & + \frac{\hat{w}}{w_i}\left(\frac{\theta_1}{w_1} p_0^1(x, t) + \frac{\theta_2}{w_2} p_0^2(x, t)\right) - \frac{p_0^i(x, t)}{w_i} = 0, \quad i = 1, 2, \\ & \frac{\partial v_0(x, t)}{\partial t} + \frac{\partial}{\partial x}\left(\hat{w}\left(\frac{\theta_1}{w_1} p_0^1(x, t) + \frac{\theta_2}{w_2} p_0^2(x, t)\right) - \hat{w}\frac{\partial v_0(x, t)}{\partial x}\right) = 0. \end{aligned} \tag{6}$$

The numerical solution of (6) is a little more difficult than that of (1) for constant a and w.

The problem of averaging (1) may be approached in a somewhat different way. By dividing the first equation by $a(x/\varepsilon)$ and by introducing $A(x/\varepsilon) = -1/a(x/\varepsilon)$, we obtain the system of equations

$$\frac{\partial p}{\partial t} + A\left(\frac{x}{\varepsilon}\right)\frac{\partial u}{\partial x} = 0, \quad \frac{\partial u}{\partial t} + \frac{\partial p}{\partial x} - \frac{\partial}{\partial x}\left(w\left(\frac{x}{\varepsilon}\right)\frac{\partial u}{\partial x}\right) = 0.$$

By applying the operator $\partial/\partial x$ to the first equation and the operator $-\partial/\partial t$ to the second equation and by adding the resulting expressions, we come to an equation in one function $u(x,t)$:

$$Pu \equiv -\frac{\partial^2 u}{\partial t^2} + \frac{\partial}{\partial x}\left(A\left(\frac{x}{\varepsilon}\right)\frac{\partial u}{\partial x}\right) + \frac{\partial}{\partial x}\left(w\left(\frac{x}{\varepsilon}\right)\frac{\partial}{\partial x}\frac{\partial u}{\partial t}\right) = 0. \tag{7}$$

By substituting the asymptotic series $u_{(\infty)} = u_0(x,t,\xi) + \varepsilon u_1(x,t,\xi) + \ldots$ for u, we see that

$$Pu^{(\infty)} \sim \sum_{i=-2}^{\infty} \varepsilon^i H_i.$$

For $u_0 = u_0(x,t)$ independent of ξ, the expression

$$H_{-2} = L^0 u_0 + L^1 \partial u_0/\partial t$$

vanishes; here

$$L^0 = \frac{\partial}{\partial \xi}\left(A(\xi)\frac{\partial}{\partial \xi}\right), \quad L^1 = \frac{\partial}{\partial \xi}\left(w(\xi)\frac{\partial}{\partial \xi}\right).$$

We have

$$H_{-1} = \frac{\partial A}{\partial \xi}\frac{\partial u_0}{\partial x} + \frac{\partial w}{\partial \xi}\frac{\partial^2 u_0}{\partial x\,\partial t} + L^0 u_1 + L^1 \frac{\partial u_1}{\partial t}.$$

We have considered the case of equation (7) above for $w = 0$, with the function u_1 taking the form

$$u_1(x, t, \xi) = N_1(\xi)\partial u_0/\partial x,$$

where $N_1(\xi)$ is a 1-periodic solution of the equation

$$L^0 N_1 + \partial A/\partial \xi = 0.$$

For case (7) with $w \neq 0$, we seek u_1 in the form

$$u_1 = N_1(\xi)\frac{\partial u_0}{\partial x} + z(x, t, \xi).$$

Then

$$H_{-1} = L^0 z + L^1 \frac{\partial z}{\partial t} + \frac{\partial}{\partial \xi}\left(w(\xi)\left(\frac{\partial N_1}{\partial \xi} + 1\right)\right)\frac{\partial^2 u_0}{\partial x\,\partial t}.$$

By requiring that H_{-1} be zero, we obtain an equation in $z(x, t, \xi)$:

$$L^1 \frac{\partial z}{\partial t} + L^0 z = -M(\xi)\frac{\partial^2 u_0}{\partial x\,\partial t}; \tag{8}$$

here $M(\xi) = \frac{\partial}{\partial \xi}(w(\xi)(\frac{\partial N_1}{\partial \xi} + 1))$. The variable x appears in this equation as a parameter, and therefore, in order to find a solution, it is necessary to investigate an equation of the form:

$$L^1 \frac{\partial W}{\partial t} + L^0 W = M(\xi)\, g(t); \tag{9}$$

Note that $\langle M(\xi)\rangle = 0$. We denote by $\Omega_0(\xi)$ a solution of the equation $L^1\Omega_0 = M(\xi)$ which is periodic in ξ and satisfies the condition $\langle \Omega_0\rangle = 0$. This equation is written as

$$\frac{\partial}{\partial \xi}\left(w(\xi)\left(\frac{\partial \Omega_0}{\partial \xi} + \frac{\partial N_1}{\partial \xi} + 1\right)\right) = 0.$$

It implies that $w(\xi)(\partial\Omega_0/\partial\xi + \partial N_1/\partial\xi + 1) = c$, or

$$\partial\Omega_0/\partial\xi + \partial N_1/\partial\xi + 1 = c/w(\xi).$$

After applying the operation $\langle\cdot\rangle$ to this equality, we get

$$1 = c\langle 1/w(\xi)\rangle = c/\widehat{w}, \quad c = \widehat{w}.$$

From the equation defining $N_1(\xi)$, we obtain $\partial N_1/\partial\xi = \hat{A}/A - 1$, where $1/\hat{A} = \langle 1/A\rangle$. Thus, $\partial\Omega_0/\partial\xi = \hat{w}/w - \hat{A}/A$, hence

$$\Omega_0 = \int_0^{\xi}\left(\frac{\widehat{w}}{w(\alpha)} - \frac{\widehat{A}}{A(\alpha)}\right)d\alpha + c_1,$$

where the constant c_1 is determined by the condition $\langle\Omega_0\rangle = 0$.

We have

$$\left|\frac{\partial N_1}{\partial\xi}\right|, \left|\frac{\partial\Omega_0}{\partial\xi}\right| \leqslant c_2$$

and moreover,

$$\int_0^1\left(\frac{\partial\Omega_0}{\partial\xi}\right)^2 d\xi \leqslant c_2.$$

We denote by $\Omega(t,\xi)$ a solution of the equation

$$L^2\Omega \equiv L^1\frac{\partial\Omega}{\partial t} + L^0\Omega = 0 \tag{10}$$

for the initial value

$$\Omega(0, \xi) = \Omega_0(\xi). \tag{11}$$

A solution of problems (10) and (11) is ambiguous. Let $\overline{\Omega}(t,\xi)$ be a solution of (10) and (11). Add an arbitrary differentiable function $c(t)$ to it. We get

$$L^2(\overline{\Omega} + c) = L^2\overline{\Omega} + L^2 c,$$
$$L^1\frac{\partial c}{\partial t} = \frac{\partial}{\partial\xi}\left(w(\xi)\frac{\partial}{\partial\xi}\frac{\partial c}{\partial t}\right) = 0,$$
$$L^0 c = \frac{\partial}{\partial\xi}\left(A(\xi)\frac{\partial c}{\partial\xi}\right) = 0.$$

Therefore $L^2 c = 0$. Thus, in order to get an unambiguous solution of (10) and (11), we impose the additional condition

$$\langle\Omega(t, \xi)\rangle = 0 \qquad \text{for all} \quad t. \tag{12}$$

Suppose Ω is some solution of (10); multiply (10) by Ω and integrate it over the period. Integration by parts with respect to ξ yields

$$\frac{\partial}{\partial t}\int_0^1 w(\xi)\left(\frac{\partial\Omega}{\partial\xi}\right)^2 d\xi+\int_0^1 A(\xi)\left(\frac{\partial\Omega}{\partial\xi}\right)^2 d\xi=0. \tag{13}$$

We get

$$A(\xi)\geqslant\frac{1}{\alpha_2}=\frac{1}{\alpha_2\beta_2}\beta_2\geqslant\gamma w(\xi),$$

where $\varphi = 1/\alpha_2\beta_2)$; therefore,

$$\int_0^1 A(\xi)\left(\frac{\partial\Omega}{\partial\xi}\right)^2 d\xi\geqslant\gamma\int_0^1 w(\xi)\left(\frac{\partial\Omega}{\partial\xi}\right)^2 d\xi.$$

Assume

$$I(t)=\int_0^1 w(\xi)\left(\frac{\partial\Omega(t,\xi)}{\partial\xi}\right)^2 d\xi.$$

It follows from (13) and the latter inequality that $\partial I/\partial t \leqslant \gamma I$. This implies

$$I(t)\leqslant e^{-\gamma t}I(0). \tag{14}$$

Let us prove that the solution of (10) and (11) is unique for condition (12). Let Ω be the difference between two solutions of (10) for the same initial condition. Then $I(0) = 0$, and from (14) $I(t) \equiv 0$; besides, $\langle\Omega\rangle \equiv 0$ according to (12). For $\langle\Omega\rangle = 0$ the following embedding relation holds:

$$\int_0^1 \Omega^2 d\xi\leqslant\frac{1}{\pi^2}\int_0^1\left(\frac{\partial\Omega}{\partial\xi}\right)^2 d\xi;$$

Therefore the condition $I(t) \equiv 0$ for $\langle\Omega\rangle \equiv 0$ implies that $\Omega(t,\xi) \equiv 0$. Since

$$\int_0^1 w\left(\frac{\partial\Omega_0}{\partial\xi}\right)^2 d\xi\leqslant\beta_2 c_2,$$

then $I(t) \leqslant \beta_2 c_2 e^{-\gamma t}$.

In the presence of the energy estimate (14), the existence of the solution of (10) with condition (12) can be proved using the standard scheme of Galerkin's method.

It can be verified directly for a continuously differentiable and uniformly bounded function $g(\tau)$ that

$$W(t,\xi)=\int_0^\infty \Omega(t-\tau,\xi)\,g(\tau)\,d\tau$$

is a solution of (9). Accordingly,

$$z(x, t, \xi) = \int_0^\infty \Omega(t-\tau, \xi) \frac{\partial^2 u_0(\tau, x)}{\partial \tau\, \partial x} d\tau \tag{15}$$

is a solution of (8), and with this choice of $z(x,t,\xi)$, we see that $H_{-1} = 0$. Further

$$H_0 = G_0 + L^0 u_2 + L^1 \frac{\partial u_2}{\partial t},$$

$$G_0 = -\frac{\partial^2 u_0}{\partial t^2} + A(\xi)\frac{\partial^2 u_0}{\partial x^2} + w(\xi)\frac{\partial^3 u_0}{\partial t\, \partial x^2} + \frac{\partial}{\partial \xi}\left(A(\xi)\frac{\partial u_1}{\partial x}\right) + A(\xi)\frac{\partial^2 u_1}{\partial x\, \partial \xi} + \\ + \frac{\partial}{\partial \xi}\left(w(\xi)\frac{\partial^2 u_1}{\partial t\, \partial x}\right) + w(\xi)\frac{\partial^3 u_1}{\partial t\, \partial x\, \partial \xi}.$$

We require that the equality $H_0 = 0$ hold. If it is so, then also $\langle H_0 \rangle = \langle G_0 \rangle = 0$. By substituting the expression for u_1 into H_0, we get an equation in $u_0(x,t)$:

$$\langle H_0 \rangle = -\frac{\partial^2 u_0}{\partial t^2} + \widehat{A}\frac{\partial^2 u_0}{\partial x^2} + \widehat{w}\frac{\partial^3 u_0}{\partial t\, \partial x^2} + \\ + \int_{-\infty}^{t} \left\langle A(\xi)\frac{\partial \Omega(\tau, \xi)}{\partial \xi} \right\rangle \frac{\partial^3 u_0(x, t-\tau)}{\partial t\, \partial x^2} d\tau + \\ + \int_{-\infty}^{t} \left\langle w(\xi)\frac{\partial \Omega(\tau, \xi)}{\partial \xi} \right\rangle \frac{\partial^4 u_0(x, t-\tau)}{\partial t^2 \partial x^2} d\tau = 0. \tag{16}$$

Let u_0 satisfy (16), and u_1 be of the form

$$u_1(x, t, \xi) = N_1(\xi)\partial u_0/\partial x + z(x, t, \xi),$$

where z is determined by (15); then $\langle H_0 \rangle = \langle G_0 \rangle = 0$. In this case, $H_0 = 0$ can be written as

$$G_0 - \langle G_0 \rangle + L^0 u_2 + L^1 \frac{\partial u_2}{\partial t} = 0. \tag{17}$$

The expression $G_0 - \langle G_0 \rangle$ can be presented as $\sum_i g_i(\xi) \times W_i(x,t), \langle g_i \rangle = 0$, where the W_i are defined by u_0 and u_1. The equation

$$g_i(\xi) W_i(x, t) + L^0 u_2^i + L^1 \frac{\partial u_2^i}{\partial t} = 0$$

is solved in the same way as (8), the function $u_2 = \sum_i u_2^i$ satisfying equation (17). With u_2 constructed in this way, we get

$$P(u_0 + \varepsilon u_1 + \varepsilon^2 u_2) = O(\varepsilon).$$

Thus, equation (16) can be regarded as an averaging of equation (17) or of the equivalent system (1).

In the case of the piecewise constant coefficients $a(\xi)$ and $w(\xi)$, solving (6) numerically is less laborious than that of the retarding differential equation (16). Assuming the boundedness

$$\frac{\partial^{k+2} u_0(x,t)}{\partial t^k \partial x^2}, \qquad k = 0, 1, 2,$$

and after single integration by parts with respect to the variable τ in the first integral and double integration by parts in the second integral, we obtain a simpler equation in $u_0(x,t)$:

$$\langle H_0 \rangle = -\frac{\partial^2 u_0}{\partial t^2} + A_0 \frac{\partial^2 u_0}{\partial x^2} + w_0 \frac{\partial^3 u_0}{\partial t\, \partial x^2} - \int_0^\infty g(\tau) \frac{\partial^2 u_0(x, t-\tau)}{\partial x^2}\, d\tau. \quad (18)$$

Let us demonstrate that the kernel $g(\tau)$ diminishes exponentially. We have the chain of inequalities

$$\left| \left\langle A(\xi) \frac{\partial \Omega(\tau,\xi)}{\partial \xi} \right\rangle \right| \leqslant \sup_{[0,1]} |A(\xi)| \int_0^1 \left| \frac{\partial \Omega(\tau,\xi)}{\partial \xi} \right| d\xi \leqslant$$

$$\leqslant \frac{1}{\alpha_1} \sqrt{\int_0^1 \left(\frac{\partial \Omega}{\partial \xi} \right)^2 d\xi} \leqslant \frac{1}{\alpha_1} \sqrt{\frac{I(\tau)}{\beta_1}} \leqslant c_3 e^{-\gamma\tau/2}.$$

In the same way, we get

$$\left| \left\langle w(\xi) \frac{\partial \Omega(\tau,\xi)}{\partial \xi} \right\rangle \right| \leqslant c_4 e^{-\gamma\tau/2}.$$

These relations yield $|g(\tau)| \leqslant c_5 e^{-\gamma\tau/2}$.

The averaging of a multi-dimensional analogue of the equation $Pu = 0$ was performed in [14, 16, 188]. A nonlinear analogue of (1) is dealt with in §12.

§5 Media with Slowly Changing Geometric Characteristics

Actual composite materials do not constitute a strictly periodic structure. The periodicity is disturbed due to some features of the production process (e.g., winding of fibre on a cylinder) or due to manufacturing errors. In other cases, the nonrigorous periodicity of a problem is planned in advance.

In the simplest cases, processes in such materials are described by equations whose coefficients depend smoothly on the slow variable x and periodically on the fast variables ξ, e.g., via the equation

$$\frac{d}{dx}\left(K(x,\xi)\frac{du}{dx}\right)-q(x,\xi)\,u=f(x,\xi),\quad \xi=\frac{x}{\varepsilon},$$

and the functions of K, q, and f, 1-periodic in ξ, may have first-order discontinuities in ξ.

The equation is not often written in this way in original variables, but can be reduced to this form by some change of independent variables. Consider an elementary example of a one-dimensional medium consisting of layers that vary in thickness. Let the coefficients $K(x)$, $q(x)$ and function $f(x)$ in the equation

$$\frac{d}{dx}\left(K_\varepsilon(x)\frac{du}{dx}\right)-q_\varepsilon(x)\,u=f_\varepsilon(x) \tag{1}$$

be

$$(K_\varepsilon(x),q_\varepsilon(x),f_\varepsilon(x))=\begin{cases}(K_1,q_1,f_1) & \text{for } x\in(j\varepsilon,(j+1/2+\varphi_j)\varepsilon),\\ (K_2,q_2,f_2) & \text{for } x\in((j+1/2+\varphi_j)\varepsilon,\ (j+1)\varepsilon),\end{cases}$$
$$j=0,1,\ldots,N-1,\ l=N\varepsilon,$$

where $\varphi_j=\varphi(j+1/2))$, $\varphi(y)$ is an infinitely differentiable function of y, independent of ε, and $-1/2<\varphi(y)<1/2$, $K_i>0$, $q_i\geqslant 0$, f_i are some numbers. We make the change of variables

$$x=y+\psi\left(\frac{y}{\varepsilon}\right)\varphi(y),$$

where $\psi(\eta)$ is a 1-periodic function that satisfies the conditions

$$\psi(\eta)=0 \quad \text{for} \quad |\eta|\leqslant\gamma,\quad 0<\gamma<1/2,$$
$$\psi(1/2)=1,\ \psi(\eta)\in C^\infty(R).$$

Let $\max_{[0,1]}|\psi'(\eta)|\cdot\max_{[0,l]}|\varphi(y)|<1$. Equation (1) assumes the form

$$\frac{d}{dy}\left(K\left(y,\frac{y}{\varepsilon},\varepsilon\right)\frac{du}{dy}\right)-q\left(\frac{y}{\varepsilon}\right)u=f\left(\frac{y}{\varepsilon}\right). \tag{2}$$

Here $K(y,\eta,\varepsilon)$, $q(\eta)$, $f(\eta)$ are 1-periodic functions in $\eta, K(y,\eta,\varepsilon)$ depending smoothly on y and ε:

$$K(y,\eta,\varepsilon)=\begin{cases} K_1\left(1+\dfrac{d\psi(\eta)}{d\eta}\varphi(y)+\varepsilon\psi(\eta)\dfrac{d\varphi}{dy}\right)^{-1} & \text{при} \quad \eta\in(0,1/2),\\ K_2\left(1+\dfrac{d\psi(\eta)}{d\eta}\varphi(y)+\varepsilon\psi(\eta)\dfrac{d\varphi}{dy}\right)^{-1} & \text{при} \quad \eta\in(1/2,1),\end{cases}$$

$$q(\eta),\,f(\eta)=\begin{cases} q_1\left(1+\dfrac{d\psi(\eta)}{d\eta}\varphi(y)+\varepsilon\psi(\eta)\dfrac{d\varphi}{dy}\right), & \\ f_1\left(1+\dfrac{d\psi(\eta)}{d\eta}\varphi(y)+\varepsilon\psi(\eta)\dfrac{d\varphi}{dy}\right) & \text{при} \quad \eta\in(0,1/2),\\ q_2\left(1+\dfrac{d\psi(\eta)}{d\eta}\varphi(y)+\varepsilon\psi(\eta)\dfrac{d\varphi}{dy}\right), & \\ f_2\left(1+\dfrac{d\psi(\eta)}{d\eta}\varphi(y)+\varepsilon\psi(\eta)\dfrac{d\varphi}{dy}\right) & \text{при} \quad \eta\in(1/2,1).\end{cases}$$

Equation (2) is a particular case of the equation

$$Pu\equiv\frac{d}{dx}\left(\left(\sum_{j=0}^{\infty}\varepsilon^jK^j\left(x,\frac{x}{\varepsilon}\right)\right)\frac{du}{dx}\right)-$$
$$-\sum_{j=0}^{\infty}\varepsilon^jq^j\left(x,\frac{x}{\varepsilon}\right)u-\sum_{j=0}^{\infty}\varepsilon^jf^j\left(x,\frac{x}{\varepsilon}\right)=0, \quad (3)$$

where $K^j(x,\xi)$, $q^j(x,\xi)$, $f^j(x,\xi)$ are 1-periodic functions in ε, infinitely differentiable with respect to x, piecewise smooth in ξ, $K^0(x,\xi)>\kappa_0>0$, $q_0(x,\xi)\geqslant 0$, with the functions of K^j, q^j, and f^j and their derivatives of the first and second order being uniformly bounded.

Equation (3) is averaged in complete analogy with the constructions in §5 of the Introduction and §1 of the present chapter. By substituting the function

$$u^{(2)}=u_0(x,\xi)+\varepsilon u_1(x,\xi)+\varepsilon^2u_2(x,\xi)$$

($\xi=x/\varepsilon$, $u_i(x,\xi)$ are 1-periodic in ξ) into (3) and by simplifying, we get

$$Pu^{(2)}=\varepsilon^{-2}L^0_{\xi\xi}u_0+\varepsilon^{-1}\left(L^0_{\xi\xi}u_1+L^0_{\xi x}u_0+L^0_{x\xi}u_0+L^1_{\xi\xi}u_0\right)+$$
$$+\varepsilon^0\left(L^0_{\xi\xi}u_2+L^0_{\xi x}u_1+L^0_{x\xi}u_1+L^0_{xx}u_0+L^1_{\xi\xi}u_1+\right.$$
$$\left.+L^1_{\xi x}u_0+L^1_{x\xi}u_0+L^2_{\xi\xi}u_0-qu_0-f^0\right)+O(\varepsilon),$$

$$L^j_{\alpha\beta}=\frac{\partial}{\partial\alpha}\left(K^j(x,\xi)\frac{\partial}{\partial\beta}\right).$$

Then we obtain equations to define the functions $u_i(x,\xi)$:

$$L^0_{\xi\xi}u_0=0, \quad (4)$$

$$L^0_{\xi\xi}u_1=-\left(L^0_{\xi x}u_0+L^0_{x\xi}u_0+L^1_{\xi\xi}u_0\right), \quad (5)$$

$$L^0_{\xi\xi}u_2=-\left(L^0_{\xi x}u_1+L^0_{x\xi}u_1+L^0_{xx}u_0+L^1_{xx}u_1+\right.$$
$$\left.+L^1_{\xi x}u_0+L^1_{x\xi}u_0+L^2_{\xi\xi}u_0-qu_0-f^0\right). \quad (6)$$

Here the variable x is considered as a parameter of equations in ξ.

By substituting $u^{(2)}$ into relations on discontinuities and by analogy with §1, we derive the conjugation conditions for the $u_i(x,\xi)$, the classical formulation of problems (4)–(6), with the above conjugation conditions being equivalent to the generalized formulation of equations (4)–(6).

In much the same way as in §5 of the Introduction and in §1 in this chapter, we derive $u_0(x,\xi) = v_0(x)$ from (4), which implies that $L^1_{\xi\xi}u_0 \equiv 0$ in (5). Then, by applying the method of separation of variables, we get a particular solution of (5): $u_1(x,\xi) = N_1(x,\xi)dv_0/dx$, with $N_1(x,\xi)$ being a generalized solution of the equation

$$L^0_{\xi\xi}(N_1(x,\xi)+\xi) = 0.$$

This solution is equal to

$$\int_0^{\xi}\left(\frac{\widehat{K}(x)}{K^0(x,\xi)} - 1\right)d\xi,\quad \widehat{K}(x) = \left\langle\frac{1}{K^0}\right\rangle^{-1}.$$

By applying the operator $\langle\cdot\rangle$ to (6), we get the averaged equation

$$\frac{d}{dx}\left(\widehat{K}(x)\frac{dv_0}{dx}\right) - \widehat{q}(x)v_0 - \widehat{f}(x) = 0, \tag{7}$$

where $\hat{q}(x) = \langle q^0(x,\xi)\rangle$, $\hat{f}(x) = \langle f^0(x,\xi)\rangle$. The solvability of this equation is necessary and sufficient for the solvability of equation (6) in the class of functions 1-periodic in ξ.

In complete analogy with §1, it can be demonstrated that the error $w = u - u^{(2)}$ is a generalized solution of the equation

$$Pw = \varepsilon\tilde{r}_0^{(2)}(x,\varepsilon) + \varepsilon^2\frac{d}{dx}\tilde{r}_1^{(2)}(x,\varepsilon),$$

where $|\tilde{r}_i^{(2)}| \leqslant c_i, c_i$ do not depend on ε. Therefore, if (3) is considered with the boundary values $u(0) = g_j$, $u(l) = g_2$, and v_0 is a solution of (7) with the boundary conditions $v(0) = g_1$, $v(l) = g_2$, the estimate $\|w\|_{W_2^1((0,l))} = O(\varepsilon)$ holds. Hence we derive

$$\|w\|_{C([0,l])} = O(\varepsilon),\quad \|u - u_0\|_{C([0,l])} = O(\varepsilon).$$

Notice that the above type of averaged equation (7) may be interpreted as follows. Consider equation (3) with coefficients "frozen with respect to x," i.e., we replace (3) by a parametric family of equations with periodic coefficients:

$$P_{x_0}u \equiv \frac{d}{dx}\left(\left(\sum_{j=0}^{\infty}\varepsilon^jK^j\left(x_0,\frac{x}{\varepsilon}\right)\right)\frac{du}{dx}\right) - \\ - \sum_{j=0}^{\infty}\varepsilon^jq^j\left(x_0,\frac{x}{\varepsilon}\right)u - \sum_{j=0}^{\infty}\varepsilon^jf^j\left(x_0,\frac{x}{\varepsilon}\right) = 0, \tag{8}$$

where x_0 is a parameter, and we average (8), by maintaining the divergent form. We thus obtain a family of averaged equations

$$\frac{d}{dx}\left(\widehat{K}(x_0)\frac{du}{dx}\right) - \widehat{q}(x_0) - \widehat{f}(x_0) = 0. \tag{9}$$

By "defreezing" x_0, i.e., by substituting x for x_0 in (9), we obtain (7).

We shall not dwell upon investigations of spatial many-dimensional media with slowly variable geometric characteristics. We shall just mention that for them the averaged equations justified rigorously by mathematical methods are the same as those obtained by the principle of frozen coefficients which involves freezing of the slow variable x in the equation's coefficients, followed by averaging and defreezing of the slow variable in the coefficients of the averaged equation (the same principle is valid also for (1)).

An asymptotic expansion of infinite order for the case of the coefficient's equations depending smoothly on slow independent variables (for a many-dimensional, nonlinear case) is given in §5.1.

§6 Heat Transfer Through a System of Screens

When considering a number of applied problems, it is interesting to investigate the process of heat transition in periodic media containing vacuum interlayers or voids where the heat is transferred only by radiation. In other cases, the voids are filled with a gas having infinitesimal coefficients of thermal conductivity and absorption. Then the presence of the gas can often be disregarded, and these voids can be considered as vacuum. The structures and materials containing such interlayers are widely adopted in many quite different technologies to solve the problem of thermal insulation.

Consider the process of heat transfer through a system of screens separated with interlayers of gas. Let a segment [0,l] be cut into sections of length $\varepsilon = \varepsilon_1 + \varepsilon_2$ with temperature T on the intervals $(n\varepsilon,\ n+\varepsilon_1)$, which correspond to the internal points of the screens, satisfying the heat equation

$$C_1\frac{\partial T}{\partial t} = \frac{\partial}{\partial x}\left(K_1\frac{\partial T}{\partial x}\right),$$

and on the intervals $(n\varepsilon + \varepsilon_1,\ (n+1)\varepsilon)$, which correspond to the gas interlayers, satisfying the equation

$$C_2\frac{\partial T}{\partial t} = \frac{\partial}{\partial x}\left(K_2\frac{\partial T}{\partial x}\right),\quad K_i,\ C_i = \text{const}.$$

The radiation heat transfer is taken into account in the following fashion. We neglect it in the solid phase due to a high absorption coefficient, whereas in the gas we neglect the heat absorption because the absorption coefficient is too low.

We get the following boundary values on the interface of the two media: T is a continuous function on [0,l],

$$K_1 \frac{\partial T}{\partial x}\bigg|_{n\varepsilon+\varepsilon_1-0} + g(T)\big|_{n\varepsilon+\varepsilon_1} = K_2 \frac{\partial T}{\partial x}\bigg|_{n\varepsilon+\varepsilon_1+0} + g(T)\big|_{n\varepsilon+\varepsilon},$$

$$K_2 \frac{\partial T}{\partial x}\bigg|_{n\varepsilon+\varepsilon-0} + g(T)\big|_{n\varepsilon+\varepsilon} = K_1 \frac{\partial T}{\partial x}\bigg|_{n\varepsilon+\varepsilon+0} + g(T)\big|_{n\varepsilon+\varepsilon_1},$$

usually $g(T)\theta\sigma T^4$, σ is the Boltzmann constant, and θ is the opacity of the boundary of the solid phase.

The averaged equation for this problem is constructed in heat engineering by the following reasoning. The specific heat-transfer resistance of the solid phase is K_1^{-1}. Heat is transferred in gas in two ways: by conduction and by radiation. Since

$$g(T)\big|_{n\varepsilon+\varepsilon} - g(T)\big|_{n\varepsilon+\varepsilon_1} \approx g'\partial T/\partial x \varepsilon_2,$$

the conductivity coefficient $g'\varepsilon_2$ corresponds to the radiation heat transfer. The total thermal conductivity coefficient in this region is $g'\varepsilon_2 + K_2$, and the specific heat-transfer resistance of the region is $1/(g'\varepsilon_2 + K_2)$. Then the mean heat-transfer resistance is given by

$$\left(\varepsilon_1 \frac{1}{K_1} + \varepsilon_2 \frac{1}{g'\varepsilon_2 + K_2}\right)\varepsilon^{-1}.$$

The mean heat capacity coefficient is naturally found from the formula $(C_1\varepsilon_1 + C_2\varepsilon_2)/\varepsilon$.

Thus, the heat transfer process is described on the average by the equation

$$\frac{C_1\varepsilon_1 + C_2\varepsilon_2}{\varepsilon}\frac{\partial v_0}{\partial t} + \frac{\partial}{\partial x}\left(\varepsilon\left(\frac{\varepsilon_1}{K_1} + \frac{1}{g'(v_0) + K_2/\varepsilon_2}\right)^{-1}\frac{\partial v_0}{\partial x}\right) = 0. \quad (1)$$

In many actual cases, the heat capacity and heat conductivity of the gas interlayer are negligible: $C_2\varepsilon_2 \ll C_1\varepsilon_1$, $K_2 \ll g'\varepsilon_2$. Then the latter equation yields:

$$C_1\frac{\varepsilon_1}{\varepsilon}\frac{\partial v_0}{\partial t} + \frac{\partial}{\partial x}\left(\varepsilon\left(\frac{\varepsilon_1}{K_1} + \frac{1}{g'}\right)^{-1}\frac{\partial v_0}{\partial x}\right) = 0. \quad (2)$$

Equation (2) can describe the heat transfer through a system of screens with vacuum interlayers and is well known in the literature on thermal shields.

Let us consider formally the latter case and obtain an approximation to (1) by the averaging method.

We place the interval $((n-1)\varepsilon, n\varepsilon)$ in correspondence with the nth screen, with the temperature $u(x,t)$ satisfying on each interval the heat equation

$$P_\varepsilon u \equiv \frac{\partial(C(x,t)u)}{\partial t} - \frac{\partial}{\partial x}\left(K(x,t)\frac{\partial u}{\partial x}\right) = f(x,t), \quad C>0, \quad K>0; \quad (3)$$

and the following condition of the radiation heat transfer between screens standing on the boundaries:

$$\left(K\frac{\partial u}{\partial x}\right)\Big|_{x=n\varepsilon-0}=\left(K\frac{\partial u}{\partial x}\right)\Big|_{x=n\varepsilon+0}=g(x,u)\Big|_{x=n\varepsilon-0}^{x=n\varepsilon+0},\quad n=1,\ldots,N-1. \tag{4}$$

Let the total of these sections be N, and $N\varepsilon = l$ be finite and independent of ε. Further, it is assumed that $C = C(x,t,\xi)$, $K = K(x,t,\xi)$, $f = f(x,t,\xi)$, $\xi = x/\varepsilon - 1/2$, with these functions being 1-periodic in ξ and finite, and $g(x,u)$ is continuous with respect to x and u. Substitute the asymptotic series

$$u^{(\infty)} \sim u_0(x,t,\xi)+\varepsilon u_1(x,t,\xi)+\ldots,\quad \xi = x/\varepsilon - 1/2,$$

into (3) and (4), with the functions $u_i(x,t,\xi)$ assumed to be 1-periodic in ξ and continuous everywhere except for the points $\xi - \pm 1/2$. After simplifying, we get

$$P_\varepsilon u^{(\infty)} - f \sim \sum_{i=-2}^{\infty}\varepsilon^i H_i,$$

$$K\frac{\partial u^{(\infty)}}{\partial x}\Big|_{n\varepsilon-0}^{n\varepsilon+0} \sim \sum_{i=-1}^{\infty}\varepsilon^i m_i^1,$$

$$K\frac{\partial u^{(\infty)}}{\partial x}\Big|_{n\varepsilon+0} - g(x,u^{(\infty)})\Big|_{n\varepsilon-0}^{n\varepsilon+0} \sim \sum_{i=-1}^{\infty}\varepsilon^i m_i^2.$$

We require that the coefficients of negative powers of ε are zero. The equality $H_{-2} = -\partial(K\partial u_0/\partial\xi)/\partial\xi = 0$ is satisfied for $u_0 = v(x,t)$; then automatically, $m_{-1}^1 = m_{-1}^2 = 0$. Since $\partial u_0/\partial\xi = 0$,

$$H_{-1} = -\frac{\partial}{\partial x}\left(K\frac{\partial u_0}{\partial\xi}\right)-\frac{\partial K}{\partial\xi}\frac{\partial u_0}{\partial x}-K\frac{\partial^2 u_0}{\partial x\,\partial\xi}-\frac{\partial}{\partial\xi}\left(K\frac{\partial u_1}{\partial\xi}\right)=$$
$$=-\frac{\partial}{\partial\xi}\left(K\left(\frac{\partial v}{\partial x}+\frac{\partial u_1}{\partial\xi}\right)\right);$$

$$H_{-1}=0,\quad\text{if}\quad \frac{\partial v}{\partial x}+\frac{\partial u_1}{\partial\xi}=0,$$

for example, for $u_1(x,t,\xi) = -v/x$.

Next, for each i, we make the coefficients m_i^1 and m^2 to be equal to zero, and H_i independent of ξ. The result is an equation of infinite order of accuracy in the function v, which does not contain the fast variable ξ.

Notice that equation (3) being satisfied up to terms of order ε^q corresponds to the presence of unaccounted-for sources of order ε^q on the right side. Conditions (4) being satisfied at the points $n\varepsilon$ up to terms of order ε^ρ corresponds to the sources of order $\varepsilon^{\rho-1}$ since this mismatch falls on the sections of length ε. Therefore, to obtain an approximation to the averaged equation with accuracy ε^q, the expressions $H_0,\ldots,H_{q-1}$ should be made independent of ξ, and the expressions $m_0^1,\ldots,m_q^1$, $m_0^2,\ldots,m_q^2$ should vanish.

For $i > 1$ we each time take the function u_1 in the form $u_i = w_i + z_i$, where the w_i periodic in ξ satisfy the condition $K(\partial w_i/\partial x + \partial w_{i+1}/\partial\xi) = c_i(x,t)$, $c_i(x,t)$ is continuous in x and is independent of the fast variable ξ, the z_i are functions 1-periodic in ξ such that z_i and $D_i = K(\delta z_i/\delta x + \delta z_{i+1}/\delta\xi)$ are continuous in ξ over the entire period, including the point $\xi \pm 1/2$. For $i = 0, 1$ we assume $w_i = u_i, z_i = 0$. Then all the m_i^1 are automatically equal to zero. The algorithm for successively finding the functions u_i and the coefficients H_i involves the following. We have

$$
\begin{aligned}
H_0 &= \frac{\partial C v}{\partial t} - \frac{\partial}{\partial\xi}\left(K\left(\frac{\partial u_1}{\partial x} + \frac{\partial u_2}{\partial\xi}\right)\right) - f,\\
H_i &= \frac{\partial C w_i}{\partial t} + \frac{\partial C z_i}{\partial t} - \frac{\partial c_i}{\partial x} - \sigma_i,\\
\sigma_i &= \frac{\partial D_i}{\partial x} + \frac{\partial D_{i+1}}{\partial\xi},\\
m_i^2 &= c_i(x,t)\big|_{x=n\varepsilon} + D_i(x,t,\xi)\big|_{\substack{x=n\varepsilon\\ \xi=1/2}} - g_i,
\end{aligned}
\tag{5}
$$

with g_i being a coefficient of ε^i in the expansion $g(x, v + \varepsilon u_1 \ldots)|_{n\varepsilon-0}^{n\varepsilon+0}$.

The equation $\partial D_{i+1}/\partial\xi = \psi$ is solvable for z_{i+2} in the class of functions z_j, provided $\langle\psi\rangle = 0$, with its solution being determined up to an arbitrary term independent of the fast variable ξ. The condition for H_i to be independent of ξ is of the form

$$\frac{\partial C u_i}{\partial t} - \sigma_i = -\bar{\sigma}_i(x,t);$$

therefore, (5) leads to the relation

$$\bar{\sigma}_i = \left\langle \frac{\partial D_i}{\partial x} - \frac{\partial C u_i}{\partial t} \right\rangle.$$

The requirement that H_0 be independent of ξ results in an equation for z_2:

$$\frac{\partial (C - \langle C\rangle) v}{\partial t} - (f - \langle f\rangle) - \frac{\partial}{\partial\xi}\left(K\frac{\partial z_2}{\partial\xi}\right) = 0.$$

After finding z_{i+1} on the condition that H_{i-1} does not depend on ξ, we find c_i from the equation $m_i^2 = 0$. This equation defines c_i only at the points $x = n$; we determine c_i for the rest of the x, assuming formally that $n\varepsilon = x$. Then for the equation $K(\partial w_i/\partial x + \partial w_{i+1}/\partial\xi) = c_i$, we find w_{i+1}. In this way, we obtain an infinite sequence $z_{i+1}, c_i, w_{i+1}, \ldots$ and the averaged equation

$$H_0 - \sum_{i=1}^{\infty} \varepsilon^i (c_i + \bar{\sigma}_i) \sim 0, \quad H_0 = \frac{\partial \langle C\rangle v}{\partial t} - \langle f\rangle$$

of infinite order of accuracy with respect to function v.

Let C, K, f be independent of the fast variable ξ. Then we get consecutively

$$z_2 = 0, \quad c_1 = \frac{\partial g}{\partial v}\frac{\partial v}{\partial x},$$

$$u_2 = w_2 = \frac{\xi^2}{2}\frac{\partial^2 v}{\partial x^2} + \frac{1}{K}\left(\xi - \frac{1}{2}\right)\frac{\partial g}{\partial v}\frac{\partial v}{\partial x},$$

$$z_3 = \frac{C}{6K}(\xi - \xi^3), \quad c_2 = -\frac{1}{K}\left(\frac{\partial g}{\partial v}\right)^2\frac{\partial v}{\partial x}, \ldots$$

The equation for v, accurate up to terms of order $O(\varepsilon^3)$, is

$$\frac{\partial Cv}{\partial t} - f - \varepsilon\frac{\partial}{\partial x}\left(\frac{\partial g}{\partial v}\frac{\partial v}{\partial x}\right) + \varepsilon^2\frac{\partial}{\partial x}\left(\frac{1}{K}\left(\frac{\partial g}{\partial v}\right)^2\frac{\partial v}{\partial x}\right) = 0 \tag{6}$$

and coincides with (1) with the same accuracy ($O(\varepsilon^3)$).

Note the following. Equations (1) and (6) have the same order of accuracy with respect to the small parameter ε, and it may seem at first sight that they are equivalent. But in practice, $\varepsilon g'/K$ is not very small or even $\varepsilon g'/K \gg 1$. Then approximation (1) obtained from plausible considerations reflects the reality much more exactly.

An averaging was carried out for the following multi-dimensional analogues of this problem in [13]:

1) Heat transition in a body with vacuum voids.

2) Heat transition in a system of bodies arranged periodically in vacuum (a system of dust particles).

Note that it was assumed in averaging the problem in 2) that some series converge. This assumption does not hold for the bodies which are not in contact, and so it is possible that the approximations thus obtained do not describe a real situation.

§7 Averaging a Nonlinear Problem of the Elasticity Theory in an Inhomogeneous Rod

Consider a thin rod C composed of $2N$ homogeneous rods

$$C_i = \begin{cases} ((j-1)\,l/N,\, (j-1+\theta_1)\,l/N) & \text{for } i = 2j-1, \\ ((j-1+\theta_1)\,l/N,\, jl/N) & \text{for } i = 2j, \end{cases}$$

$$j = 1, 2, \ldots, N, \quad \theta_1 \in (0, 1) \quad (\text{c f. § 1}).$$

Let the rod be in a field of forces that is aligned with the axis of the rod and is distributed according to the law $f_\varepsilon(x)$. Thus, if the rod is vertical, it is under the action of the gravity field with the distribution function $\rho_\varepsilon(x)$, where $\rho_\varepsilon(x)$ is the linear density of the rod. The stress $\sigma(x) = \sigma_{11}(x)$ in the rod satisfies the equilibrium conditions, i.e., d'Alembert's equation, on each of the rods:

$$\frac{d\sigma}{dx} = f_\varepsilon(x), \quad x \in C_i, \quad i = 1, \ldots, 2N, \tag{1}$$

and the conjugation condition in joints of the rods C_i and C_{i+1}:

$$[\sigma]|_{x^i} = 0. \tag{2}$$

Here $x^i = \xi^i \varepsilon$,

$$\xi^i = \begin{cases} (j - 1 + \theta_1)\varepsilon & \text{for } i = 2j - 1, \\ j\varepsilon & \text{for } i = 2j, \end{cases} \qquad i = 1, \ldots, 2N - 1.$$

In the nonlinear case, the relation between the stress and elongation du/dx in each of the rods C_i is given by the law

$$\sigma(x) = \begin{cases} a^1(du/dx), & x \in C_i, \quad i \text{ odd}, \\ a^2(du/dx), & x \in C_i, \quad i \text{ even}. \end{cases}$$

By substituting the expression for σ into (1) and (2), we get

$$\begin{aligned} \frac{d}{dx} a^1\left(\frac{du}{dx}\right) &= f_\varepsilon(x), \quad x \in C_i, \quad i \text{ odd}, \\ \frac{d}{dx} a^2\left(\frac{du}{dx}\right) &= f_\varepsilon(x), \quad x \in C_i, \quad i \text{ even}, \end{aligned} \tag{3}$$

$$a^1\left(\frac{du}{dx}\right)\Big|_{x=x^i} - a^2\left(\frac{du}{dx}\right)\Big|_{x=x^i} = 0, \quad i = 1, \ldots, 2N - 1. \tag{4}$$

Since translations are continuous due to the continuity of the rod,

$$[u]|_{x=x^i} = 0. \tag{5}$$

Equations (3) and the conjugation conditions (4) and (5) define the problem on longitudinal strain of the rod C. They can be written as

$$\frac{d}{dx} a\left(\frac{x}{\varepsilon}, \frac{du}{dx}\right) - f_\varepsilon(x) = 0, \quad x \neq x^i, \tag{6}$$

$$\left[a\left(\frac{x}{\varepsilon}, \frac{du}{dx}\right)\right]\Big|_{x=x^i} = 0, \tag{7}$$

$$[u]|_{x=x^i} = 0. \tag{8}$$

Here $a(\xi, y)$ is a 1-periodic function in ξ

$$a(\xi, y) = \begin{cases} a^1(y) & \text{for } \xi \in (0, \theta_1), \\ a^2(y) & \text{for } \xi \in (\theta_1, 1). \end{cases}$$

Note that the equations written as (6)–(8) correspond to all types of nonlinearity (geometric and physical).

Let the displacement at the ends of the rod be known as:

$$u(0)=g_1, \quad u(l)=g_2, \tag{9}$$

and let $f_\varepsilon(x)$ equal $f(x, x/\varepsilon)$, where $f(x,\xi)$ is a periodic function with period 1 in ξ. Let the functions $f(x,\xi)$ and $a(\xi, y)$ be infinitely differentiable for $\xi \neq \xi^i (i=1,\dots,2N-1)$, while for $\xi=\xi^i$ these functions and their derivatives may have first-order discontinuities.

We shall seek asymptotics of a solution of problems (6)–(9) in the same way as in §1, as

$$u^{(2)}=v_0(x)+\varepsilon u_1\left(x, \frac{x}{\varepsilon}\right)+\varepsilon^2 u_2\left(x, \frac{x}{\varepsilon}\right), \tag{10}$$

where $u_i(x,\xi)$, $u_2(x,\xi)$ are 1-periodic functions in ξ, infinitely differentiable in x, $v_0(x)\in C^\infty$.

Let us substitute (10) into the left side of equation (6) for $x\neq x^i$ and use the chain rule. We get

$$\begin{aligned}
&\frac{d}{dx} a\left(\frac{x}{\varepsilon}, \frac{dv_0}{dx}+\frac{\partial u_1}{\partial \xi}+\varepsilon\left(\frac{\partial u_1}{\partial x}+\frac{\partial u_2}{\partial \xi}\right)+\varepsilon^2\frac{\partial u_2}{\partial x}\right)-f\left(x, \frac{x}{\varepsilon}\right)=\\
&=\Bigg(\varepsilon^{-1}\frac{\partial}{\partial \xi} a\left(\xi, \frac{dv_0}{dx}+\frac{\partial u_1(x,\xi)}{\partial \xi}+\varepsilon\left(\frac{\partial u_1}{\partial x}(x,\xi)+\frac{\partial u_2}{\partial \xi}(x,\xi)\right)+\right.\\
&\left.+\varepsilon^2\frac{\partial u_2}{\partial x}(x,\xi)\right)+\frac{\partial}{\partial x} a\left(\xi, \frac{dv_0}{dx}+\frac{\partial u_1}{\partial \xi}(x,\xi)+\varepsilon\left(\frac{\partial u_1}{\partial x}(x,\xi)+\right.\right.\\
&\left.\left.+\frac{\partial u_2}{\partial \xi}(x,\xi)\right)+\varepsilon^2\frac{\partial u_2}{\partial x}(x,\xi)\right)-f(x,\xi)\Bigg)\Bigg|_{\xi=x/\varepsilon}.
\end{aligned}$$

We expand $\partial a(\xi,y)/\partial\xi$ in the first term and $a(\xi,y)$ in the second term by Taylor's formula with $(\xi, \partial v_0/\partial x+\partial u_1/\partial\xi)$ as the centre. Then the latter expression is

$$\begin{aligned}
&\varepsilon^{-1}\frac{\partial}{\partial\xi} a\left(\xi, \frac{dv_0}{dx}+\frac{\partial u_1}{\partial\xi}\right)+\frac{\partial}{\partial\xi}\left(\frac{\partial a}{\partial y}\left(\xi, \frac{dv_0}{dx}+\frac{\partial u_1}{\partial\xi}\right)\left(\frac{\partial u_1}{\partial x}+\frac{\partial u_2}{\partial\xi}\right)\right)+\\
&\qquad+\frac{\partial a}{\partial x}\left(\xi, \frac{dv_0}{dx}+\frac{\partial u_1}{\partial\xi}\right)-f(x,\xi)+O(\varepsilon).
\end{aligned}$$

We require that for $\xi\neq\xi^i$ the terms of orders ε^{-1} and ε^0 are zero:

$$\frac{\partial}{\partial\xi} a\left(\xi, \frac{dv_0}{dx}+\frac{\partial u_1}{\partial\xi}\right)=0, \tag{11}$$

$$\begin{aligned}
&\frac{\partial}{\partial\xi}\left(\frac{\partial a}{\partial y}\left(\xi, \frac{dv_0}{dx}+\frac{\partial u_1}{\partial\xi}\right)\left(\frac{\partial u_1}{\partial x}+\frac{\partial u_2}{\partial\xi}\right)\right)+\\
&\qquad+\frac{\partial}{\partial x} a\left(\xi, \frac{dv_0}{dx}+\frac{\partial u_1}{\partial\xi}\right)-f(x,\xi)=0.
\end{aligned} \tag{12}$$

By substituting (10) into the left sides of (7) and (8), we get

$$\left[a\left(\frac{x}{\varepsilon}, \frac{du^{(2)}}{dx}\right)\right]\Big|_{x=x^i} = \left(\left[a\left(\xi, \frac{dv_0}{dx} + \frac{\partial u_1}{\partial \xi}\right)\right]_{\xi=\xi^i} + \right.$$

$$\left. + \varepsilon\left[\frac{\partial a}{\partial y}\left(\xi, \frac{dv_0}{dx} + \frac{\partial u_1}{\partial \xi}\right)\left(\frac{\partial u_1}{\partial x} + \frac{\partial u_2}{\partial \xi}\right) + O(\varepsilon^2)\Big|_{\xi=\xi^i}\right)\Big|_{x=x^i}, \quad (13)$$

$$\left[u^{(2)}\right]_{x=x^i} = \left([v_0]_{x=x^i} + \varepsilon\,[u_1(x, \xi)]_{\xi=\xi^i} + \varepsilon^2\,[u_2(x, \xi)]_{\xi=\xi^i}\right)\big|_{x=x^i}. \quad (14)$$

Let us require that the terms of orders ε^0 and ε^1 and all the summands in (14) vanish (v_0 is continuous):

$$\left[a\left(\xi, \frac{dv_0}{dx} + \frac{\partial u_1}{\partial \xi}\right)\right]_{\xi=\xi^i} = 0, \quad (15)$$

$$\left[\frac{\partial a}{\partial y}\left(\xi, \frac{dv_0}{dx} + \frac{\partial u_1}{\partial \xi}\right)\left(\frac{\partial u_1}{\partial x} + \frac{\partial u_2}{\partial \xi}\right)\right]_{\xi=\xi^i} = 0, \quad (16)$$

$$[u_1(x, \xi)]_{\xi=\xi^i} = 0, \quad (17)$$

$$[u_2(x, \xi)]_{\xi=\xi^i} = 0. \quad (18)$$

(Here we require that (15)–(18) are satisfied for all $x \in [0, l]$ and not only for $x = x^i$.) Finally, by substituting (10) in (9), we have

$$v_0(0) + \varepsilon u_1(0, 0) + \varepsilon^2 u_2(0, 0) = g_1,$$

$$v_0(l) + \varepsilon u_1\left(l, \frac{l}{\varepsilon}\right) + \varepsilon^2 u_2\left(l, \frac{l}{\varepsilon}\right) = v_0(l) + \varepsilon u_1(l, 0) + \varepsilon^2 u_2(l, 0) = g_2,$$

whence

$$v_0(0) = g_1, \quad v_0(l) = g_2, \quad (19)$$

$$u_1(x, 0) = 0, \quad (20)$$

$$u_2(x, 0) = 0. \quad (21)$$

Let there exist a 1-periodic solution $u_1(x, \xi)$ of problems (11), (15), (17), and (20), where dv_0/dx is a parameter independent of ξ. To be more exact, let there be a solution $N_1(\xi, y)$, 1-periodic in ξ, of a parametric family of problems (y is a parameter)

$$\frac{\partial}{\partial \xi}\, a\left(\xi, y + \frac{\partial N_1}{\partial \xi}(\xi, y)\right) = 0, \quad \xi \neq \xi_i,$$

$$\left[a\left(\xi, y + \frac{\partial N_1}{\partial \xi}(\xi, y)\right)\right]_{\xi=\xi^i} = 0,$$

$$[N_1(\xi, y)]_{\xi=\xi^i} = 0, \quad N_1(0, y) = 0, \quad (22)$$

and let

$$u_1(x, \xi) = N_1\left(\xi, \frac{dv_0}{dx}\right). \quad (23)$$

For problems (12), (16), (18), and (21) to have the solution $u_2(x,\xi)$ 1-periodic in ξ, the following relation should hold:

$$\frac{d}{dx}\int_0^1 \left(a\left(\xi,\ \frac{dv_0}{dx}+\frac{\partial u_1}{\partial \xi}\right)-f(x,\ \xi)\right)d\xi=0. \tag{24}$$

For, by integrating (12) over ξ between 0 and 1 and taking into account (16) and the periodicity of $a(\xi,y)$, $u_1(x,\xi)$, $u_2(x,\xi)$ in ξ with period 1, we get

$$\frac{\partial a}{\partial y}\left(\xi,\ \frac{dv_0}{dx}+\frac{\partial u_1}{\partial \xi}\right)\left(\frac{\partial u_1}{\partial x}+\frac{\partial u_2}{\partial \xi}\right)\Bigg|_{\xi=0}^{\xi=1}+$$
$$+\frac{d}{dx}\int_0^1 \left(a\left(\xi,\ \frac{dv_0}{dx}+\frac{\partial u_1}{\partial \xi}\right)-f(x,\ \xi)\right)d\xi=$$
$$=\frac{d}{dx}\int_0^1 \left(a\left(\xi,\ \frac{dv_0}{dx}+\frac{\partial u_1}{\partial \xi}\right)-f(x,\ \xi)\right)d\xi=0.$$

Equation (24) given for $x\in(0,1)$ with the boundary conditions (19) will be called the *averaged problem*. Let a solution of this problem exist, be unique and infinitely differentiable, and let the functions $u_1(x,\xi)$, $u_2(x,\xi)$ be also infinitely smooth for $\xi\neq\xi^i$ (their derivatives may have first-order discontinuities at these points).

Then (11)–(24) imply

$$\max_{x\in[0,l]}\left|\frac{d}{dx}\,a\left(\frac{x}{\varepsilon},\ \frac{du^{(2)}}{dx}\right)-f\left(x,\ \frac{x}{\varepsilon}\right)\right|=O(\varepsilon),\quad x\neq x^i,$$
$$\left[a\left(\frac{x}{\varepsilon},\ \frac{du^{(2)}}{dx}\right)\right]_{x=x^i}=O(\varepsilon^2),$$
$$[u^{(2)}]_{x=x^i}=0,\quad u^{(2)}(0)=g_1,\quad u^{(2)}(l)=g_2.$$

Thus, the approximation of $u^{(2)}$ satisfies (6)–(9) up to a sufficient order of accuracy.

Taking (23) into account, we rewrite the averaged equation as

$$\frac{d}{dx}\,\hat{a}\left(\frac{dv_0}{dx}\right)-\hat{f}(x)=0, \tag{25}$$

where

$$\hat{a}(y)=\left\langle a\left(\xi,\ y+\frac{\partial N_1}{\partial \xi}(\xi,\ y)\right)\right\rangle, \tag{26}$$

$\hat{f}(x)=\langle f(x,\xi)\rangle$, $N_1(\xi,y)$ is a solution of (22). It follows from (22) that

$$a\left(\xi,\ y+\frac{\partial N_1}{\partial \xi}(\xi,\ y)\right)=\hat{a}(y). \tag{27}$$

Thus, what is needed to construct the averaged equation (25) is to solve problem (22) and to define $\hat{a}(y)$ from (26) or, which is the same, from (27).

To illustrate the above-described scheme of constructing the averaged equation, let

$$a(\xi, y) = e^{g(\xi)y} - 1, \quad g(\xi) > 0,$$

where $g(\xi)$ is a piecewise smooth function:

$$g(\xi) \in C^\infty([\xi^i, \xi^{i+1}]), \quad i = 0, \ldots, 2N-1, \quad \xi^0 = 0, \quad \xi^{2N} = l/\varepsilon.$$

1. From (22) we get

(general case)	(example)
$a\left(\xi, y + \frac{\partial N_1}{\partial \xi}\right) = c$	$\exp\left\{g(\xi)\left(y + \frac{\partial N_1}{\partial \xi}\right)\right\} - 1 = c \quad (28)$

2. Solve (28) with respect to $y + \partial N_1/\partial\xi$:

$y + \frac{\partial N_1}{\partial \xi} = \varphi(\xi, c)$	$y + \frac{\partial N_1}{\partial \xi} = \frac{\ln(1+c)}{g(\xi)} \quad (29)$

Here $\varphi(\xi, c)$ is a function inverse to $a(\xi, y)$ with ξ being considered as a parameter:

$$a(\xi, \varphi(\xi, c)) \equiv c.$$

3. We apply the averaging operator $\langle\cdot\rangle$ to (29) and obtain the following, due to (25) and the periodicity of the solution of the problem in ξ:

$y = \langle \varphi(\xi, c)\rangle$	$y = \left\langle \frac{1}{g(\xi)} \right\rangle \ln(1+c) \quad (30)$

4. After solving (30) with respect to c, we see that

$c = \hat{a}(y)$	$c = e^{\hat{g}y} - 1$

where

$$\hat{g} = \langle g^{-1}(\xi)\rangle^{-1}.$$

5. From (29) we obtain:

$\frac{\partial N_1}{\partial \xi} = \varphi(\xi, \hat{a}(y)) - y$	$\frac{\partial N_1}{\partial \xi} = \left(\frac{\hat{g}}{g(\xi)} - 1\right) y \quad (31)$

for $\xi \neq \xi^i$.

The above-described algorithm for constructing the function of $\hat{a}$ implies that

$$\langle \varphi(\xi, \hat{a}(y)) - y \rangle = 0.$$

Then equation (31) with the continuity conditions N_1 and $N_1(0, y) = 0$ is solvable in the class of periodic functions with period 1 in ξ:

$$N_1(\xi) = \int_0^{\xi} (\varphi(\theta, \hat{a}(y)) - y)\, d\theta \quad \Bigg| \quad N_1(\xi) = \int_0^{\xi} \left(\frac{\hat{g}}{g(\theta)} - 1\right) d\theta\, y \quad (32)$$

It can be verified by direct substitution that $N_1(\xi)$ in (32) satisfies the relations (22).

Thus, the averaged equation has the form:

$$\frac{d}{dx}\hat{a}\left(\frac{dv_0}{dx}\right) - \hat{f}(x) = 0 \quad \Bigg| \quad \frac{d}{dx}\left(e^{\hat{g}\, dv_0/dx} - 1\right) - \hat{f}(x) = 0$$

We see that the process of constructing the averaged equation might have been finished at stage 4, because the function $\partial N_1/\partial\xi$ defined at stage 5 does not affect it. The definition of $\partial N_1/\partial\xi$ and N_1 may be needed for a detailed analysis of the behaviour of the strains and for a rigorous justification of the correctness of the averaging.

For the case of several space variables, the definition of the function u_1 may be necessary to construct the averaged equation itself (see Chapter 4).

Notice the following feature, useful for understanding the whole point of the averaging algorithm. The relation $\sigma = a(\xi, \eta)$ determines the dependence of the stress σ on the elongation $\eta = du/dx$, and the function φ, inverse to a, determines the dependence of η on σ: $\eta = \varphi(\xi, \sigma)$. We see that $\hat{\eta} = \langle \varphi(\xi, \hat{\sigma}) \rangle$, $\hat{\eta} = dv_0/dx$ for the averaged equation, and thus the averaging algorithm consists of the direct calculation of the mean over the period for a function expressing the dependence of η on σ.

Let us give a geometric interpretation of the above-described algorithm for averaging. Let a periodic medium consist of two components with different properties, i.e., the function $a(\xi, y)$ is piecewise constant in ξ:

$$a(\xi, y) = \begin{cases} a^1(y), & 0 \leqslant \xi < \theta_1, \\ a^2(y), & \theta_1 \leqslant \xi \leqslant 1. \end{cases}$$

Let $\theta_2 = 1 - \theta_1$. Figure 9 shows plots of the functions $a^1(y)$ and $a^2(y)$.

The function $\varphi(\xi, c)$ is piecewise constant in ξ:

$$\varphi(\xi, c) = \begin{cases} \varphi_1(c) & \text{for } 0 \leqslant \xi < \theta_1, \\ \varphi_2(c) & \text{for } \theta_1 \leqslant \xi < 1, \end{cases}$$

where φ_1 and φ_2 are functions inverse to a^1 and a^2, respectively.

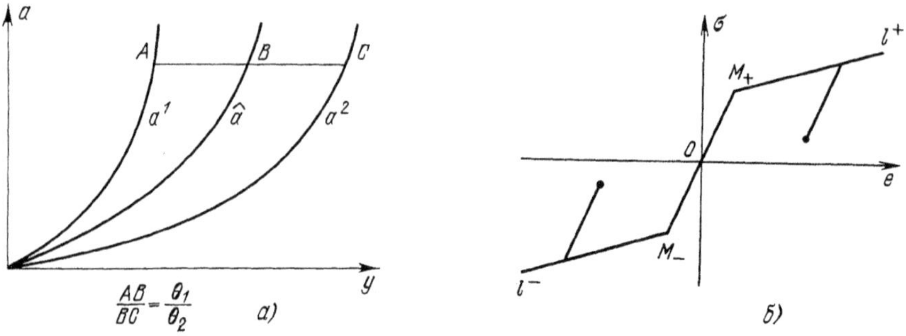

Figure 9

The plot in Fig. 9a can be regarded as a representation of the relationships $\varphi_i(c)$ if φ_i is laid off along the y-axis and c along the x-axis.

The definition of $\langle\varphi(\xi,c)\rangle$ yields $\langle\varphi(\xi,c)\rangle = \theta_1\varphi_1(c)+\theta_2\varphi_2(c)$; for each value of c, the point $(c,\langle\varphi(\xi,c)\rangle)$ divides the interval $((c,\varphi_1(c)),(c,\varphi_2(c)))$ in the ratio θ_2/θ_1. The plot $\hat{a}(y)$ is inverse to this plot. Thus, for any value of $\hat{a}$, the point $(y,\hat{a})$ of the plot of the function $\hat{a}(y)$ divides the interval $((y_1,\hat{a}),(y_2,\hat{a})$ in the ratio θ_2/θ_1, where y_i is the root of the equation $a^i(y_i)=\hat{a}(AB/BC)=\theta_2/\theta_1$ in Fig. 9).

Consider the case of several components:

$$a(\xi,y)=\begin{cases} a^1(y) & \text{for} \quad 0<\xi\leqslant\xi_1, \\ a^i(y) & \text{for} \quad \xi_{i-1}<\xi\leqslant\xi_i, \quad 1<i<p, \\ a^p(y) & \text{for} \quad \xi_{p-1}<\xi\leqslant 1. \end{cases}$$

We set: $\xi_0=0$, $\xi_p=1$, $\xi_i-\xi_{i-1}=0_i$. We plot the functions a^i as above. For each value of c find a point on the straight line $a=c$ whose coordinate $y(c)$ is the weighted mean with the weights θ_i of the y_i-coordinates for the points where the plots of the functions $a^i(y)$ cross the straight line $a=c$.

§8 The System of Equations of Elasticity Theory in a Layered Medium

Consider a layered periodic medium, its layers perpendicular to the axis x_1, under the assumption that the translations depend only on the variable x_1, to be called x below.

The constructions in §7 can be regarded as a particular case of the constructions of the present section if it is assumed that the translations take place only along the axis x_1, that is,

$$u=(u_1,u_2,u_3)=(u_1(x_1),0,0).$$

Let the medium consist of layers of two materials as above. Then we have a system of equations

$$\frac{d}{dx}a^j\left(\frac{x}{\varepsilon}, \frac{du}{dx}\right) = f^j\left(x, \frac{x}{\varepsilon}\right), \quad j = 1, 2, 3, \tag{1}$$
$$\text{for } x^i < x < x^{i+1}, \qquad i = 0, \ldots, 2N-1, \quad x^0 = 0,$$
$$x^{2N} = l, \left[a^j\left(\frac{x}{\varepsilon}, \frac{du}{dx}\right)\right]\Big|_{x=x^i} = 0, \quad i = 1, \ldots, 2N-1.$$

Here

$$a^j\left(\xi, \frac{du}{dx}\right) = \begin{cases} a^{j,1}\left(\dfrac{du}{dx}\right) & \text{for } \xi \in (\xi^{2k}, \xi^{2k+1}), \\ a^{j,2}\left(\dfrac{du}{dx}\right) & \text{for } \xi \in (\xi^{2k+1}, \xi^{2k+2}), \end{cases} \qquad k = 0, \ldots, N-1,$$
$$f^j(x, \xi) = \begin{cases} f^{j,1}(x) & \text{for } \xi^{2k} < \xi < \xi^{2k+1}, \\ f^{j,2}(x) & \text{for } \xi^{2k+1} < \xi < \xi^{2k+2}. \end{cases}$$

The set of these relations can be considered as the equation

$$\frac{d}{dx}a^j\left(\frac{x}{\varepsilon}, \frac{du}{dx}\right) = f^j\left(x, \frac{x}{\varepsilon}\right) \tag{2}$$

for all x in the sense of satisfying the integral identity

$$\int_0^l a^j\left(\frac{x}{\varepsilon}, \frac{du}{dx}\right)\frac{d\psi^j(x)}{dx}\,dx + \int_0^l \psi^j(x)\, f^j\left(x, \frac{x}{\varepsilon}\right)dx = 0 \tag{3}$$
$$\forall \psi = (\psi^1, \psi^2, \psi^3) \in \overset{\circ}{W}{}_2^1((0, l)).$$

The problems considered in §§7 and 8 are in general solvable by quadratures. However, a direct analysis of these explicit solutions is much more complicated than the use of the formalism which we consider here.

We discuss these problems in detail in order to:

1) illustrate the general technique of averaging nonlinear multi-dimensional problems by using a one-dimensional case as an example;

2) give a rigorous justification for the convergence of the solution to that of the averaged problem in the case of discontinuous coefficients starting from the definition of the solution by means of an integral identity (we do not give such proof in the multi-dimensional case).

We average the system of equations

$$\Lambda^j(u) \equiv \frac{d}{dx}a^j\left(x, \frac{x}{\varepsilon}, \frac{du}{dx}\right) = 0, \quad j = 1, \ldots, n, \quad u = (u^1, \ldots, u^n)^T, \tag{4}$$

assuming the a^j are smooth in x and ξ. An asymptotic expansion of the solution is sought in the form

$$u \sim v_0(x) + \varepsilon u_1(x, \xi) + \dots, \quad u_k = (u_k^1, \dots, u_k^n)^T.$$

By substituting the expansion of the solution as in §7 above, we get

$$\Lambda^j(x) \sim h_{-1}^j \varepsilon^{-1} + h_0^j \varepsilon^0 + \dots,$$

$$h_{-1}^j = \frac{\partial}{\partial \xi} a^j\left(x, \xi, \frac{dv_0}{dx} + \frac{\partial u_1}{\partial \xi}\right),$$

$$h_0^j = \frac{\partial}{\partial x} a^j\left(x, \xi, \frac{dv_0}{dx} + \frac{\partial u_1}{\partial \xi}\right) + \frac{\partial}{\partial \xi} a_1^j.$$

Here

$$a_1^j = \frac{\partial a^j}{\partial y^k}\left(x, \xi, \frac{dv_0}{dx} + \frac{\partial u_1}{\partial \xi}\right)\left(\frac{\partial u_1^k}{\partial x} + \frac{\partial u_2^k}{\partial \xi}\right),$$

with $\partial a^j/\partial y^k$ being a partial derivative of the function a^j in the kth component of the vector, we denote, for brevity, by $\partial a^j/\partial^k(x, \xi, dv_0/dx + \partial u_1/\partial \xi)$ the result of the substitution of $dv_0/dx + \partial u_1/\partial \xi$ for y in the expression $\partial a^j(x, \xi, y)/\partial y_k$.

We obtain the following system of equations on the condition that all the h_{-1}^j are zero:

$$a^j\left(x, \xi, \frac{dv_0^1}{dx} + \frac{\partial u_1^1}{\partial \xi}, \dots, \frac{dv_0^n}{dx} + \frac{\partial u_1^n}{\partial \xi}\right) = c^j(x).$$

After solving this system of nonlinear equations with respect to $dv_0^k/dx + \partial u_1^k/\partial \xi$, we get the system of equations

$$dv_0^k/dx + \partial u_1^k/\partial \xi = \varphi^k(x, \xi, c^1, \dots, c^n).$$

We apply the operator $\langle \cdot \rangle$ to both sides; then $\frac{dv_0^k}{dx} = \langle \varphi^k(x, \xi, c^1, \dots, c^n) \rangle$. We solve the set of these relations with respect to $c^1, \dots, c^n$ and get $c^k(x) = \hat{a}^k(x, dv_0/dx)$. We require that the h_0^j be zero; then $\langle h_0^j \rangle = 0$.

We have

$$\langle h_0^j \rangle = \left\langle \frac{\partial}{\partial x} a^j\left(x, \xi, \frac{dv_0}{dx} + \frac{\partial u_1}{\partial \xi}\right) \right\rangle.$$

Since

$$a^j\left(x, \xi, \frac{dv_0}{dx} + \frac{\partial u_1}{\partial \xi}\right) = c^j(x) = \hat{a}^j\left(x, \frac{dv_0}{dx}\right),$$

then the condition $\langle h_0^j \rangle = 0$ means that

$$\frac{d}{dx} \hat{a}^j\left(x, \frac{dv_0}{dx}\right) = 0.$$

We obtain the system of equations which must be satisfied by the function

$v_0(x)$. The question of solvability of the auxiliary system of nonlinear equations that occurred above and the question of how close the solution of the original problem is to that of the averaged problem remain to be discussed.

Let us make an accurate investigation of these questions in the case when equations (4) are Euler equations for a functional. We seek

$$\inf_{u \in U} I(u), \quad I(u) = \int_0^l \Phi\left(x, \xi, \frac{du}{dx}\right) dx; \tag{5}$$

with U being the set of functions $u \in W_2^1((0,l))$ which satisfy the conditions $u(0) = g_1, u(l) = g_2$. We assume the following conditions hold: there exist $\xi_1, \dots, \xi_p$ such that $\xi_0 = 0 < \xi_1 < \cdots < \xi_p = 1$, and in the domains $R_k = \{(x, \xi, y_1, \dots, y_n) | \xi \in (\xi_{k-1}, \xi_k), -\infty < x, y_1, \dots, y_n < \infty\}$, the functions

$$\frac{\partial^{i_0+i_1+i_2}\Phi(x, \xi, y)}{\partial x^{i_0} \partial y_{j_1}^{i_1} \partial y_{j_2}^{i_2}}, \quad i_0 = 0, 1, \quad 0 \leqslant i_1 + i_2 \leqslant 2, \quad j_1, j_2 = 1, \dots, n,$$

are continuous and their absolute values are uniformly bounded; there exist $\kappa_1, \kappa_2, 0 < \kappa_1 \leqslant \kappa_2 < \infty$ such that

$$0 < \varkappa_1 \eta_j \eta_j \leqslant \frac{\partial^2 \Phi}{\partial y_{j_1} \partial y_{j_2}} \eta_{j_1} \eta_{j_2} \leqslant \varkappa_2 \eta_j \eta_j \tag{6}$$

(summation over j is between 1 and n) for all the values of the arguments and arbitrary vectors $\eta = (\eta_i, \dots, \eta_n) \neq (0, \dots, 0)$; that is Φ is strongly convex.

It is well known from the calculus of variations that a solution of the extreme value problem (5) exists, and for any $\psi = (\psi^1, \dots, \psi^n) \in \overset{\circ}{W}{}_2^1((0,l))$ satisfies the relation

$$\lambda(u, \psi) = \int_0^l a^j \frac{d\psi^j}{dx} dx = 0; \tag{7}$$

here $a^j = \partial\Phi\partial y_j$. Relation (7) coincides with the integral identity in the definition of a generalized solution.

Let us justify all the stages of the averaging considered above.

Lemma 1. *With condition* (6), *the system of equations*

$$a^j(x, \xi, y) = c^j$$

is solvable in the vector y for all values of c^j.

Proof: Consider the function $\psi(x, \xi, y) = \Phi(x, \xi, y) - c^j y_j$. The function ψ also satisfies the condition of strong convexity and therefore has an extreme point. At this point, $\partial\psi/\partial y_j = 0$, that is, $a^j(x,\xi,y) - c^j = 0$. Thus, the system $a^j(x,\xi,y_1,\ldots,y_n) = c^j$ is solvable for the values of y_j. The lemma is proved. (The convexity of Φ also implies that the solution of y is unique.)

By definition of the functions φ^j, we have $a^i(x,\xi,c)) = c^i$, $i = 1,\ldots,n$, $\varphi = (\varphi^1,\ldots,\varphi^n)$, $c = (c^1,\ldots,c^n)$. After differentiating each of the equations with respect to each variable c^j, we obtain the set of relations

$$\frac{\partial a^i(x,\ \xi,\ \varphi(x,\ \xi,\ c))}{\partial y_k}\,\frac{\partial\varphi^k}{\partial c^j} = \delta_{ij}, \qquad i,\ j = 1,\ \ldots,\ n,$$

or $\|\partial a^i/\partial y_k\|\|\partial\varphi^k/\partial c^j\| = E$.

Thus, $\|\partial\varphi^k/\partial c^j\| = \|\partial a^i/\partial y_k\|^{-1}$. We have

$$\frac{\partial a^i}{\partial y_k} = \frac{\partial^2\Phi}{\partial y_i \partial y_k}; \tag{8}$$

it follows from (8) that the matrix $\partial a^i/\partial y_k$ is symmetric and satisfies the relations (6).

By the norm of a matrix, we shall mean henceforth the so-called *third norm*:

$$\| A \|_3 = \sup_{\|y\|_3 \neq 0} \frac{\| Ay \|_3}{\| y \|_3},$$

with $\|y\|_3 = \sqrt{y_j y_j}$ being the Euclidean norm.

It is well known from matrix theory that if A is symmetric and

$$\varkappa_1(\eta,\ \eta) \leqslant (A\eta,\ \eta) \leqslant \varkappa_2(\eta,\ \eta), \quad \varkappa_1 > 0, \tag{9}$$

then A^{-1} is symmetric:

$$\varkappa_2^{-1}(\eta,\ \eta) \leqslant (A^{-1}\eta,\ \eta) \leqslant \varkappa_1^{-1}(\eta,\ \eta), \quad \| A \|_3 \leqslant \varkappa_2, \quad \| A^{-1} \|_3 \leqslant \varkappa_1^{-1}.$$

Since for $A = \|\partial a^i/\partial y\|$ relation (9) holds due to (6),

$$\varkappa_2^{-1}\eta_j\eta_j \leqslant \frac{\partial\varphi^k}{\partial c^j}\,\eta_k\eta_j \leqslant \varkappa_1^{-1}\eta_j\eta_j$$

and the matrix $\|\partial\varphi k/\partial c^j\|$ is symmetric.

This relation holds for any η. Therefore, after taking the mean over the period, we get

$$\varkappa_2^{-1}\eta_j\eta_j \leqslant \left\langle \frac{\partial\varphi^k}{\partial c^j} \right\rangle \eta_k\eta_j \leqslant \varkappa_1^{-1}\eta_j\eta_j. \tag{10}$$

The matrix $\|\delta\overline{\varphi}^k/\partial c^j\|$, $\overline{\varphi}^k = \langle\varphi^k\rangle$ is symmetric, too. Since $\|\partial\overline{\varphi}^k/\partial c^j\|$ is symmetric, $\operatorname{rot}\varphi = 0$. Consequently, the vector $\overline{\varphi}$ is a gradient, and there exists a function $\widehat{F}(x,c)$ such that $\overline{\varphi}^k = \partial\hat{F}/\partial c^k$.

Due to (10), the function F satisfies inequality (6) with the constants κ_2^{-1} and κ_1^{-1}. Therefore, Lemma 1 implies the solvability of the system of equations

$$\overline{\varphi}^k(x, c) = \frac{dv_0^k}{dx}, \qquad k = 1, \ldots, n,$$

in the set of variables $c^1, \ldots, c^n$. By expressing c from this system, we see that

$$c^j = \widehat{a}^j\left(x, \frac{dv_0}{dx}\right).$$

In just the same way as (10) was obtained from (6), we derive that the matrix $\|\partial\hat{a}^k/\partial y_j\|$ satisfied the conditions

$$\varkappa_1\eta_j\eta_j \leqslant \frac{\partial\widehat{a}^k}{\partial y_j}\eta_k\eta_j \leqslant \varkappa_2\eta_j\eta_j \tag{11}$$

and is symmetric.

This matrix being symmetric implies that there exists a function $\widehat{\Phi}$ such that $\partial\widehat{\Phi}/\partial y_k = \hat{a}^k$. For the function Φ, the relations (11) will be written as

$$\varkappa_1\eta_j\eta_j \leqslant \frac{\partial^2\widehat{\Phi}}{\partial y_k\partial y_j}\eta_k\eta_j \leqslant \varkappa_2\eta_j\eta_j. \tag{12}$$

It turns out (we shall justify this assertion in §6.3 for the case of several space variables) that the function Φ can be expressed as

$$\widehat{\Phi} = \left\langle \Phi\left(x, \xi, \frac{dv_0}{dx} + \frac{\partial u_1}{\partial\xi}\right)\right\rangle.$$

Lemma 2. *Let a function $U(y_1, \ldots, y_n)$ satisfy the condition*

$$\varkappa_1\eta_j\eta_j \leqslant \frac{\partial^2 U}{\partial y_j\partial y_k}\eta_j\eta_k, \quad \varkappa_1 > 0.$$

Then for any $y_1, \ldots, y_n$, $z_1, \ldots, z_n$ the relation

$$\varkappa_1(z_j - y_j)(z_j - y_j) \leqslant \left(\frac{\partial U(z_1, \ldots, z_n)}{\partial z_j} - \frac{\partial U(y_1, \ldots, y_n)}{\partial y_j}\right)(z_j - y_j). \tag{13}$$

holds.

We do not give proof of the lemma, because its assertion follows from the criterion of strong convexity (see [38]).

Let us demonstrate that the solution of the original problem (5) is close to that of the extreme value problem of the functional

$$\inf_{v_0 \in U} \hat{I}(v_0) = \int_0^l \hat{\Phi}\left(x, \frac{dv_0}{dx}\right) dx \tag{14}$$

(for the same boundary conditions $v_0(0) = g_1$, $v_0(l) = g_2$).

With condition (13), it follows from the calculus of variations that a solution of the extreme value problem (14) exists. The extremum condition of the functional $\hat{I}$ implies that the equality

$$\int_0^l \hat{a}^j\left(x, \frac{dv_0}{dx}\right) \frac{d\psi^j}{dx}\, dx = 0 \quad \forall \psi \in \overset{\circ}{W}{}_2^1((0, l)) \tag{15}$$

holds.

When constructing the averaged system, we had the relations

$$dv_0^j/dx + \partial u_1^j/\partial\xi = \varphi^j(x, \xi, c^1, \ldots, c^n), \quad c^j = \hat{a}^j(x, dv_0/dx).$$

Thus,

$$\frac{\partial u_1^j}{\partial\xi} = \varphi^j\left(x, \xi, \hat{a}^1\left(x, \frac{dv_0}{dx}\right), \ldots, \hat{a}^n\left(x, \frac{dv_0}{dx}\right)\right) - \frac{dv_0^j}{dx}; \tag{16}$$

with the right side satisfying the condition $\langle \varphi^j - dv_0^j/dx \rangle = 0$, according to the construction of the functions $\hat{a}^j$. Let us find a solution $\overline{u}_1^j(x, \xi, dv_0/dx)$, 1-periodic in ξ, of this equation. We select an arbitrary constant $c_j(x, dv_0/dx)$ in $\overline{u}_1^j$ such that $\overline{u}_1^j|\xi = 0$ for all x and all dv_0/dx.

Let $w^j = v_0^j(x) e \overline{u}_1^j$. The functions w^j satisfy the same boundary conditions as u and v_0. It follows from the definition of $\overline{u}_1^j$ that $\partial\overline{u}_1^j/\partial\xi - \partial u_1^j/\partial\xi$; therefore,

$$\lambda(w, \psi) = \int_0^l a^j\left(x, \xi, \frac{dv_0}{dx} + \frac{\partial u_1}{\partial\xi} + \varepsilon \frac{\partial \overline{u}_1}{\partial x}\right) \frac{d\psi^j}{dx}\, dx. \tag{17}$$

Due to the differentiability of $a^j(x, \xi, y)$ with respect to the variables y_k, we have

$$\left| a^j\left(x, \xi, \frac{dv_0}{dx} + \frac{\partial u_1}{\partial\xi} + \varepsilon \frac{\partial \overline{u}_1}{\partial x}\right) - a^j\left(x, \xi, \frac{dv_0}{dx} + \frac{\partial u_1}{\partial\xi}\right) \right| \leqslant \varepsilon M_{jk} \left| \frac{\partial \overline{u}_1^k}{\partial x} \right|. \tag{18}$$

Under the earlier assumption that Φ is smooth, it can be shown by standard methods of mathematical analysis that in $\bigcup_k R_k$ the functions a^j, φ^j, $\overline{\varphi}^j$, a^j have uniformly bounded derivatives

$$\frac{\partial^{i_0+i_1}}{\partial x^{i_0} \partial y_k^{i_1}}, \quad i_0, \ i_1 = 0, 1, \quad 0 \leqslant k \leqslant n,$$

and the function (x, y) has the same derivatives in x and y_k as Φ. Then it follows from the differential equation theory that the solution of an averaged variational problem is twice infinitely differentiable with respect to x. By now taking the equation in u_1, we see that its right side is differentiable with respect to x. It implies that u_i is also infinitely differentiable in x; we obtain from (16) that u_i is piecewise infinitely differentiable in ξ, and the derivative $\partial \overline{u}_1/\partial x$ is uniformly bounded.

Thus, the right side of (18) does not exceed $c_1\varepsilon$. Then (17) implies the estimate

$$\left| \lambda(w, \psi) - \int_0^l a^j\left(x, \xi, \frac{dv_0}{dx} + \frac{\partial u_1}{\partial \xi}\right) \frac{d\psi^j}{dx}\, dx \right| \leqslant c_2 \varepsilon \sqrt{\int_0^l \frac{d\psi^j}{dx}\frac{d\psi^j}{dx}\, dx}.$$

From the algorithm for constructing the functions $\hat{a}^j$, we get

$$a^j\left(x, \xi, \frac{dv_0}{dx} + \frac{\partial u_1}{\partial \xi}\right) = c^j(x) = \widehat{a}^j\left(x, \frac{dv_0}{dx}\right). \tag{19}$$

Consequently,

$$\int_0^l a^j\left(x, \xi, \frac{dv_0}{dx} + \frac{\partial u_1}{\partial \xi}\right)\frac{d\psi^j}{dx}\, dx = \int_0^l \widehat{a}^j\left(x, \frac{dv_0}{dx}\right)\frac{d\psi^j}{dx} dx = 0;$$

the latter equality being valid according to (15), since $\hat{a}^j(x,y) = \partial\widehat{\Phi}(x,y)/\partial y_j$. It follows that

$$|\lambda(w, \psi)| \leqslant c_2\varepsilon \sqrt{\int_0^l \frac{d\psi^j}{dx}\frac{d\psi^j}{dx}\, dx}.$$

Since $\lambda(u, \psi) = 0$,

$$|\lambda(w, \psi) - \lambda(u, \psi)| \leqslant c_2 \varepsilon \|\psi\|_{\overset{\circ}{W}{}_2^1((0,l))}.$$

Assuming that $\psi = w - u$ and by using the inequality (13), we obtain

$$\begin{aligned}
&\lambda(w, w-u) - \lambda(u, w-u) = \\
&\quad = \int_0^l \left(a^j\left(x, \xi, \frac{dw}{dx}\right) - a^j\left(x, \xi, \frac{du}{dx}\right)\right)\left(\frac{dw^j}{dx} - \frac{du^j}{dx}\right) dx \geqslant \\
&\quad \geqslant \varkappa_1 \int_0^l \left(\frac{dw^j}{dx} - \frac{du^j}{dx}\right)\left(\frac{dw^j}{dx} - \frac{du^j}{dx}\right) dx = \varkappa_1 \|w - u\|^2_{\overset{\circ}{W}{}_2^1((0,l))}.
\end{aligned}$$

Thus,

$$\varkappa_1 \| w - u \|^2_{\overset{\circ}{W}{}^1_2((0,l))} \leqslant c_2 \varepsilon \| w - u \|_{\overset{\circ}{W}{}^1_2((0,l))},$$

and therefore,

$$\| w - u \|_{\overset{\circ}{W}{}^1_2((0,l))} \leqslant \frac{c_2 \varepsilon}{\varkappa_1} = c_3 \varepsilon. \tag{20}$$

The embedding theorem implies $\|w - u\|_{c([0,l])} \leqslant c_4\varepsilon$, and since $\overline{u}_1$ is bounded, $v_0\|_{c([0,l])}$, $\|u - v_0\|_{C([0,l])} \leqslant c_4\varepsilon$.

We have the inequality

$$\left\| \frac{dw}{dx} - \left(\frac{dv_0}{dx} + \frac{\partial u_1}{\partial \xi} \right) \right\|_{C([0,l])} \leqslant c_6 \varepsilon.$$

From the inequality and from (20), we get

$$\left\| \frac{du}{dx} - \left(\frac{dv_0}{dx} + \frac{\partial u_1}{\partial \xi} \right) \right\|_{L_2((0,l))} \leqslant c_7 \varepsilon. \tag{21}$$

It follows from (21) that

$$\left\| a^j \left(x, \xi, \frac{du}{dx} \right) - a^j \left(x, \xi, \frac{dv_0}{dx} + \frac{\partial u_1}{\partial x} \right) \right\|_{L_2((0,l))} \leqslant c_8 \varepsilon, \quad j = 1, \ldots, n. \tag{22}$$

Applying (19), we obtain

$$\left\| a^j \left(x, \xi, \frac{du}{dx} \right) - \widehat{a}^j \left(x, \frac{dv_0}{dx} \right) \right\|_{L_2((0,l))} \leqslant c_9 \varepsilon, \quad j = 1, \ldots, n.$$

In reality, the L_2-norm can be replaced by the uniform one in (21) and (22). The two values under the norm sign are constant; therefore,

$$\begin{gathered} \sqrt{l} \left| a^j \left(x, \xi, \frac{du}{dx} \right) - \widehat{a}^j \left(x, \frac{dv_0}{dx} \right) \right| \leqslant c_9 \varepsilon, \\ \left| a^j \left(x, \xi, \frac{du}{dx} \right) - \widehat{a}^j \left(x, \frac{dv_0}{dx} \right) \right| \leqslant c_{10} \varepsilon, \\ \left| a^j \left(x, \xi, \frac{du}{dx} \right) - a^j \left(x, \xi, \frac{dv_0}{dx} + \frac{\partial u_1}{\partial \xi} \right) \right| \leqslant c_{11} \varepsilon. \end{gathered} \tag{23}$$

Let us use relation (13) for $y_j = du^j/dx$, $z_j = dv_0^j/dx + \partial u_1^j/\partial\xi$. We get

$$\left(a^j \left(x, \xi, \frac{du}{dx} \right) - a^j \left(x, \xi, \frac{dv_0}{dx} + \frac{\partial u_1}{\partial \xi} \right) \right) \left(\frac{du^j}{dx} - \left(\frac{dv_0^j}{dx} + \frac{\partial u_1^j}{\partial \xi} \right) \right) \geqslant$$

$$\geqslant \varkappa \sum_{j=1}^{n} \left(\frac{du^j}{dx} - \left(\frac{dv_0^j}{dx} + \frac{\partial u_1^j}{\partial \xi} \right) \right)^2.$$

On the other hand, (23) implies that the left side of the inequality does not exceed

$$c_{11}\varepsilon\sum_{j=1}^{n}\left|\frac{du^j}{dx}-\left(\frac{dv_0^j}{dx}+\frac{\partial u_1^j}{\partial\xi}\right)\right|\leqslant$$

$$\leqslant c_{11}\sqrt{n}\varepsilon\sqrt{\sum_{j=1}^{n}\left(\frac{du^j}{dx}-\left(\frac{dv_0^j}{dx}+\frac{\partial u_1^j}{\partial\xi}\right)\right)^2}.$$

We conclude from these estimates that

$$\sqrt{\sum_{j=1}^{n}\left(\frac{du^j}{dx}-\left(\frac{dv_0^j}{dx}+\frac{\partial u_1^j}{\partial\xi}\right)\right)^2}\leqslant\frac{c_{11}\sqrt{n}\varepsilon}{\varkappa_1}.$$

§9 Considerations Permitting Reduction of Calculations in Constructing Averaged Equations

In the case of averaging the nonlinear system of equations (8.1), the averaging algorithm described in §3 needs approximate solutions of many systems of nonlinear equations to be realized numerically. If we are interested in small deformations, the approximation to the functions $\hat{a}^j$ can be obtained in a more constructive fashion.

Let the a^j be expandable in the asymptotic series:

$$a^j\sim\sum_{l=1}^{\infty}a^j_{i_1\ldots i_l}(x,\xi)\frac{du^{i_1}}{dx}\cdots\frac{du^{i_l}}{dx} \tag{1}$$

(in this section, summation over repeated indices $i_1,\ldots,i_l$ is between 1 and n).

Then the conditions $h^j_{-1}=0$ from §8 are equivalent asymptotically to the following equations:

$$\frac{\partial}{\partial\xi}\left(\sum_{l=1}^{\infty}a^j_{i_1\ldots i_l}(x,\xi)\left(\frac{dv_0^{i_1}}{dx}+\frac{\partial u_1^{i_1}}{\partial\xi}\right)\cdots\left(\frac{dv_0^{i_l}}{dx}+\frac{\partial u_1^{i_l}}{\partial\xi}\right)\right)\sim 0. \tag{2}$$

We shall seek the functions u_1^k in the form

$$u_1^k\sim\sum_{l=1}^{\infty}A^k_{i_1\ldots i_l}(x,\xi)\frac{dv_0^{i_1}}{dx}\cdots\frac{dv_0^{i_l}}{dx}. \tag{3}$$

We substitute this series into (2) and simplify. We obtain the system of equations

$$\sum_{l=0}^{\infty} h^j_{i_1 \ldots i_l}(x, \xi) \frac{dv_0^{i_1}}{dx} \cdots \frac{dv_0^{i_l}}{dx} \sim 0$$

with the coefficients $h^j_{i_1 \ldots i_l}(x, \xi)$ expressed in terms of the known functions $a^j_{j_1 \cdots j_l}$ and the unknown functions $A^k_{j_1 \cdots j_l}$.

By requiring that the coefficients $h^j_{i_1 \ldots i_l}$ vanish, we get the asymptotic equality

$$\frac{\partial}{\partial \xi} a^j \left(x, \xi, \frac{dv_0}{dx} + \frac{\partial u_1}{\partial \xi} \right) \sim 0.$$

For $l = 1$, we have

$$h^j_{i_1} = \frac{\partial}{\partial \xi} \left(a^j_q(x, \xi) \left(\delta_{q i_1} + \frac{\partial A^q_{i_1}}{\partial \xi} \right) \right) = 0.$$

These equations have the same structure as the one in $N_1(\xi)$ for the linear system (see §3):

$$\frac{d}{dx} \left(K \left(\frac{x}{\varepsilon} \right) \frac{du}{dx} \right) = 0.$$

Namely, we get

$$a^j_q(x, \xi) \left(\delta_{q i_1} + \frac{\partial A^q_{i_1}}{\partial \xi} \right) = c^j_{i_1}(x),$$

or

$$a \left(E + \frac{\partial A}{\partial \xi} \right) = C, \tag{4}$$

where $a = \|a^j_q\|$, $A = \|A^q_{i_1}\|$, $C = \|c^j_{i_1}\|$ are matrices.

Hence, $E + \frac{\partial A}{\partial \xi} = a^{-1} C$. Let us apply the operation $\langle \cdot \rangle$ to both sides: $E = \langle a^{-1} \rangle C$, that is $C = \langle a^{-1} \rangle^{-1}$. Then we find the matrix A from (4).

The subsequent (for $l > 1$) equations are of the form:

$$h^j_{i_1 \cdots i_l} = \frac{\partial}{\partial \xi} \left(a^j_q(x, \xi) \frac{\partial}{\partial \xi} A^q_{i_1 \ldots i_l} + T^j_{i_1 \ldots i_l}(x, \xi) \right) = 0. \tag{5}$$

Functions $T^j_{i_1 \ldots i_l}(x, \xi)$ are defined by $A^q_{j_1 \cdots j_p}$ for $p < l$.

Subsequently, we thus consider the set of equations (5) for $l = 2, \ldots,$ and from them we find the sets of values $A^q_{i_1\ldots i_l}$ corresponding to these l. Relations (5) are combined into sets of vector equations

$$\frac{\partial}{\partial\xi}\left(a\frac{\partial}{\partial\xi}A_{i_1\ldots i_l}+T_{i_1\ldots i_l}\right)=0,$$
$$A_{i_1\ldots i_l}=\left(A^1_{i_1\ldots i_l},\ldots,A^n_{i_1\ldots i_l}\right)^T,$$
$$T_{i_1\ldots i_l}=\left(T^1_{i_1\ldots i_l},\ldots,T^n_{i_1\ldots i_l}\right)^T.$$

Hence

$$a\frac{\partial}{\partial\xi}A_{i_1\ldots i_l}+T_{i_1\ldots i_l}=C_{i_1\ldots i_l},$$
$$\frac{\partial}{\partial\xi}A_{i_1\ldots i_l}+a^{-1}T_{i_1\ldots i_l}=a^{-1}C_{i_1\ldots i_l}. \tag{6}$$

After applying the operation $\langle\cdot\rangle$, we get the equality $\langle a^{-1}T_{i_1\ldots i_l}\rangle = \langle a^{-1}\rangle C_{i_1\ldots i_l}$, whence we define $C_{i_1\ldots i_l}$. Then we find $A_{i_1\ldots i_l}$ form (6).

With series (3) found, we substitute them in expansions (2) and define $\hat{a}^j = \langle a^j(x,\xi,dv_0/dx+\partial u_1/\partial\xi)$.

A similar procedure can be employed also in the case of several space variables $x=(x_1,\ldots,x_s)$. The difference is that in order to determine each of the coefficients A, a numerical solution of the problem over the period will be needed for the same linear elliptic system with different (for every coefficient) right sides [2].

We wrote down series for the functions u^j_1 in the form such that the terms corresponding to various rearrangements of the set of indices $(i_1,\ldots,i_l)$ cannot be simplified. Just as in averaging up to high orders of accuracy, this notation may yield more information about the averaged system (see Chapter 9).

Note another possibility of making calculations less laborious in the course of averaging. Suppose the functions a^j can be represented as $a^j = a^j(x,\xi,e_1,\ldots,e_r$, $e_i(u) = b_{ik}\frac{du^k}{dx}$, where the b_{ik} are constants. The system of equations in the functions u^q_1 has the form

$$\frac{\partial}{\partial\xi}a^j\left(x,\ \xi,\ e_1(v_0)+b_{1q}\frac{\partial u^q_1}{\partial\xi},\ \ldots,\ e_r(v_0)+b_{rq}\frac{\partial u^q_1}{\partial\xi}\right)=0.$$

The functions $c^j(x)$ and u^j_1 had to be sought for all collections $(dv^1_0/dx,\ldots,dv^n_0/dx)$ before; it can be seen from the case in question that the unknown functions u^j_1 may be sought in the form $u^j_1(x,\xi,e_1(v_0),\ldots,e_r(v_0))$. If $r<n$, we get a smaller dimension of the range of parameters for which the functions c_i, u^j_1 are sought.

For instance, in the multi-dimensional case of averaging problems of geometrically linear theory of elasticity, all the functions a^j depend only on the strains $e_{ik} = \frac{1}{2}(\frac{\partial u^i}{\partial x_k}+\frac{\partial u^k}{\partial x_i})$. Then the reasoning similar to the above leads to the conclusion that the functions u^j_1 and $\hat{a}^j_q$ depend only on the mean strains

$e_{ik}(v_0) = (\partial v_0^i/\partial x_k + \partial v_0^k/\partial x_i)/2$; the dimension of the number of parameters is $n(n+1)/2$, whereas in the general case, we have n^2 independent parameters $\partial v_0^i/\partial x_k$ (see Chapter 6).

§10 Nonstationary Nonlinear Problems

This section will deal briefly with the results of averaging nonstationary problems. We shall not discuss in detail the formalism of averaging for this case because of the following circumstance which we first encountered in §1.

As compared with the stationary case $P_x u = f$, another term $-R(x/\varepsilon)\partial^k u/\partial t^k$, which does not contain differentiation with respect to x, is added to the equation. When a system of asymptotic series is substituted into the equation, the presence of such terms does not change the expansion terms of orders ε^{-2} and ε^{-1}. Therefore, the equations to define u_1, u_2 are of the same form. The result is that the term $\widehat{P}_x v$ in the averaged equation

$$-\widehat{R}\frac{\partial^2 v}{\partial t^2} + \widehat{P}_x v = \widehat{f}$$

is the same as in averaging the equation $P_x u = l$, with $\widehat{R}$ and $\hat{f}$ being given directly by the formulas $\widehat{R} = \langle R\rangle$, $\hat{f} = \langle f\rangle$.

The nonstationary analogue of equation (7.6) is

$$R\left(\frac{x}{\varepsilon}\right)\frac{\partial^2 u}{\partial t^2} - \frac{\partial}{\partial x}a\left(\frac{x}{\varepsilon}, \frac{\partial u}{\partial x}\right) = f\left(x, t, \frac{x}{\varepsilon}\right), \tag{1}$$

where $R(\xi)$ is a function of the linear density of the rod, periodic in ξ with period 1, $f(x,t,\xi)$ is the density of a longitudinal external force, and $u(x,t)$ is a displacement of a point from equilibrium. Contact continuity conditions of the displacement and stress (7.7) and (7.8) are set on the joints of the rods.

Let the boundary and initial values be

$$u(0, t) = g_1(t), \qquad u(l, t) = g_2(t),$$
$$u(x, 0) = \varphi(x), \quad \frac{\partial u}{\partial t}(x, 0) = \psi(x),$$

where f, g_i, φ, ψ are infinitely differentiable functions.

Asymptotics of the solution of the problem can be found in the form

$$u^{(2)} = v_0(x, t) + \varepsilon u_1\left(x, t, \frac{x}{\varepsilon}\right) + \varepsilon^2 u_2\left(x, t, \frac{x}{\varepsilon}\right), \tag{2}$$

where $u_1(x,t,\xi)$, $u_2(x,t,\xi)$ are functions 1-periodic in ξ, infinitely differentiable with respect to x and $v_0 \in C^\infty$.

By substituting (2) into (1) and by combining terms of the same order in ε, we obtain problems for defining u_1, u_2, and v_0 in the same way as in §7. Thus,

$$u_1(x,\ t,\ \xi) = N_1(\xi,\ \partial v_0/\partial x),$$

where $N_1(\xi, y)$ is a solution 1-periodic in ξ of the parametric family of problems (7.22), $v_0(x,t)$ is a solution of the averaged equation

$$\widehat{R}\,\frac{\partial^2 v_0}{\partial t^2} - \frac{\partial}{\partial x}\,\widehat{a}\left(\frac{\partial v_0}{\partial x}\right) - \widehat{f}\,(x, t) = 0$$

with the boundary and initial values

$$v_0\,(0,\,t) = g_1\,(t), \qquad v_0\,(l,\,t) = g_2\,(t),$$

$$v_0\,(x,\,0) = \varphi\,(x), \quad \frac{\partial v_0}{\partial t}\,(x,\,0) = \psi\,(x),$$

where

$$\widehat{a}\,(y) = \int_0^1 a\left(\xi,\, y + \frac{\partial N_1}{\partial \xi}\,(\xi,\, y)\right) d\xi,$$

$$\widehat{R} = \langle R\,(\xi)\rangle, \quad \widehat{f}\,(x,\,t) = \langle f\,(x,\,t,\,\xi)\rangle.$$

The nonlinear heat equation

$$S\left(\frac{x}{\varepsilon}\right)\frac{\partial u}{\partial t} - \frac{\partial}{\partial x}\left(K\left(\frac{x}{\varepsilon},\, u\right)\frac{\partial u}{\partial x}\right) = f\left(x,\,t,\,\frac{x}{\varepsilon}\right),$$

where $S(\xi)$, $K(\xi, u)$ are functions 1-periodic in ε, can be averaged by the same method. The averaged equation is

$$\widehat{S}\,\frac{\partial v_0}{\partial t} - \frac{\partial}{\partial x}\left(\widehat{K}\,(v_0)\,\frac{\partial v_0}{\partial x}\right) = \widehat{f}\,(x,\,t),$$

where $\widehat{S} = \langle S(\xi)\rangle$, $\widehat{K}(y) = \langle K^{-1}(\xi, y)\rangle^{-1}$.

§11 Averaging Equations with Rapidly Oscillating Nonperiodic Coefficients

The problem of averaging processes in media with random properties is of great interest for the applications. Most of the results in this area concern the one-dimensional case.

Consider an elementary model of a random coefficient. Let the segment $[0, l]$ be split into the parts $C_1 = [0, \varepsilon]$, $C_2 = (\varepsilon, 2\varepsilon], \ldots, C_N = ((N-1)\varepsilon, N\varepsilon]$,where $N\varepsilon = l$. Let $\xi_\alpha = (\alpha = 1, 2, 3)$ be a random variable with expectation 1, which satisfies the conditions $0 < \kappa_1 \leqslant \xi_\alpha < \infty$ for all its realizations.

Let $\xi_\alpha^i, \ldots, \xi_\alpha^N$ be independent realizations of this random variable. Assume $\xi_\alpha(x) = \xi_\alpha^i$ for $x \in C_i$ $(i = 1, \ldots, N)$. Let $K_\alpha(x)$ $(\alpha = 1, 2, 3)$ be continuously differentiable on $[0, l]$. By the random coefficient, we mean the function $k_\alpha(x) = K_\alpha(x)\xi_\alpha(x)$. The following inequality (Kolmogorov inequality) was proved in [33]:

$$\mathsf{M} \sup_{[0,l]} \left| \int_0^x (\xi_\alpha(t) - 1)\, dt \right| \leqslant c_0 \sqrt{\varepsilon}. \tag{1}$$

Let us estimate $\mathsf{M} \sup\limits_{[0,l]} | \int_0^x (k_\alpha(t) - K_\alpha(t))dt|$. After integration by parts, we get

$$g(x) = \int_0^x (k_\alpha(t) - K_\alpha(t))\, dt = \int_0^x K_\alpha(t)\, d\left(\int_0^t (\xi_\alpha(\tau) - 1)\, d\tau \right) =$$

$$= K_\alpha(x) \int_0^x (\xi_\alpha(\tau) - 1)\, d\tau - \int_0^x \left(\int_0^t (\xi_\alpha(\tau) - 1)\, d\tau \right) K'_\alpha(t)\, dt\,.$$

Hence,

$$|g(x)| \leqslant \left(\max_x |K_\alpha(x)| + \int_0^x |K'_\alpha(t)|\, dt \right) \max_x \left| \int_0^x (\xi_\alpha(t) - 1)\, dt \right|.$$

This and (1) imply

$$\mathsf{M} \sup_{[0,l]} \left| \int_0^x (k_\alpha(t) - K_\alpha(t))\, dt \right| \leqslant \left(\max_x |K_\alpha(x)| + \operatorname*{Var}_{[0,l]} |K_\alpha(x)| \right) c_0 \sqrt{\varepsilon} = c_1 \sqrt{\varepsilon}. \tag{2}$$

Consider the case $K_\alpha \geqslant K > 0$; we set

$$\widehat{k}_\alpha(x) \equiv (\mathsf{M}(1/k_\alpha))^{-1}.$$

If the random variable ξ_α does not equal identically 1, then

$$\frac{1}{\widehat{k}_\alpha(x)} K_\alpha(x) > 1.$$

According to (2), we have

$$\mathsf{M} \sup_x \left| \int_0^x \left(\frac{1}{k_\alpha(t)} - \frac{1}{\widehat{k}_\alpha(t)} \right) dt \right| \leqslant c_2 \sqrt{\varepsilon}. \tag{3}$$

For $F(x) = \int_0^z f(t)dt$ we denote $\|F\|_p = (\int_0^l |F(x)|^p dz)^{1/p}$ by $\|f\|_p^{(-1)}$, supposing for $p = \infty$ that $\|f\|_\infty^{(-1)} = \|F\|_\infty = \sup_{z\in[0,l]} |F(z)|$.

The well-known inequality

$$\|F\|_{p_1} \leqslant \|F\|_{p_2} l^{1/p_1 - 1/p_2}, \quad p_1 \leqslant p_2,$$

implies

$$\|f\|_{p_1}^{(-1)} \leqslant \|f\|_{p_2}^{(-1)} l^{1/p_1 - 1/p_2}, \quad p_1 \leqslant p_2.$$

Thus, inequalities (2) and (3) mean that

$$\mathsf{M}\| k_\alpha - K_\alpha \|_p^{(-1)} \leqslant c_1' \sqrt{\varepsilon}, \quad \mathsf{M}\| k_\alpha^{-1} - \widehat{k}_\alpha^{-1} \|_p^{(-1)} \leqslant c_2' \sqrt{\varepsilon}$$

for any $p \in [1, \infty]$.

Consider the boundary value problem

$$Lu \equiv \frac{d}{dx}\left(k_1(x)\frac{du}{dx}\right) - k_2(x)\,u = k_3(x), \quad u(0) = u(l) = 0, \tag{4}$$

where $k_\alpha(x)$ are some random coefficients constructed by the above-described method on the basis of some given, continuously differentiable functions $K_\alpha(x)$ and random functions $\xi_\alpha (0 < K_1(x) < \infty,\ 0 \leqslant K_2(x))$.

Consider also the boundary value problem

$$\frac{d}{dx}\left(\widehat{k}_1(x)\frac{dv}{dx}\right) - K_2(x)\,v = K_3(x) \tag{5}$$

for the boundary values $v(0) = v(l) = 0$.

THEOREM 1. *The following inequalities hold for the above-formulated assumptions:*

$$\mathsf{M}\| u - v \|_{C([0,l])} \leqslant c_3 \sqrt{\varepsilon},$$

$$\mathsf{M}\left\| k_1 \frac{du}{dx} - \widehat{k}_1 \frac{dv}{dx} \right\|_{C([0,l])} \leqslant c_3 \sqrt{\varepsilon}.$$

The expectation is taken with respect to the measure defined by the product of measures in the spaces of the random variables ξ_1, ξ_2, ξ_3.

When solutions of equations with close coefficients are compared, the estimates of the closeness of the solutions are obtained via estimates of the closeness of the coefficients in the norm L_p. The coefficients of equations (1) and (5) are close in the so-called negative norms $\|\cdot\|_p^{(-1)}$. Thus, in order to estimate how close solutions of the equations with random and smooth coefficients are, the closeness estimates are to be obtained via estimates of the closeness of the coefficients in some negative norms.

The leading considerations that serve to construct an approximation to a solution of equation (4) with rapidly oscillating coefficients are as follows.

Let k_1 be variables of order 1 with the characteristic size of variation of order ε, their derivatives being of order ε^{-1}. By analogy with the case of periodic coefficients, we seek the approximation in the form

$$u(x) \sim v(x) + N(x)\frac{dv}{dx} = u^{(1)}(x),$$

where $v(x)$ is a smooth function, $N(x)$ is a function of order ε, and $d^q N(x)/dx^q$ is a function of order ε^{1-q}. Under this assumption, the expression $Lu^{(1)}$ contains a term of order ε^{-1} and does not contain terms of order ε^{-2}:

$$Lu^{(1)} = \frac{d}{dx}\left(k_1\left(1 + \frac{dN}{dx}\right)\right)\frac{dv}{dx} + O(1),$$

in general, $\frac{d}{dx}(k_1(1 + \frac{dN}{dx}))\frac{dv}{dx}$ may turn out to be a variable of order ε^{-1}. We require that it have order $O(1)$.

Then $k_1(1 + dN/dx) = c(x)$ is a smooth function. Hence,

$$N(x) = \int_0^x \left(\frac{c(t)}{k_1(t)} - 1\right) dt = \int_0^x c(t)\left(\frac{1}{k_1(t)} - \frac{1}{c(t)}\right) dt =$$

$$= \int_0^x c(t)\, d\left(\int_0^t \left(\frac{1}{k_1(\tau)} - \frac{1}{c(\tau)}\right) d\tau\right);$$

we see that the smallness of $\int_0^t (1/k_1(\tau) - 1/c(\tau))d\tau$ is important for that of $N(x)$.

So it is natural to suppose that some norm $\|1/k_1 - 1/c\|_p^{(-1)}$ must be small. By analogy with the above-discussed case of the equation with rapidly oscillating coefficients, it is natural to assume that $c(x) = \hat{k}_1(x)$.

Let us estimate the closeness between solutions of equations (4) and (5) by

$$d = d_1 + d_2 + d_3,\ d_1 = \| k_1^{-1} - \widehat{k}_1^{-1} \|_\infty^{(-1)},$$
$$d_2 = \| k_2 - K_2 \|_2^{(-1)},\ d_3 = \| k_3 - K_3 \|_2^{(-1)}.$$

The solutions of equations (4) and (5) satisfy the relations

$$\lambda_1(u, \varphi) = \int_0^l \left(k_1 \frac{du}{dx}\frac{d\varphi}{dx} + k_2 u\varphi + k_3\varphi\right) dx = 0,$$

$$\lambda_2(v, \varphi) = \int_0^l \left(\widehat{k}_1 \frac{dv}{dx}\frac{d\varphi}{dx} + K_2 v\varphi + K_3\varphi\right) dx = 0 \quad \forall \varphi \in \overset{\circ}{W}{}_2^1((0, l)). \tag{6}$$

Note that it suffices for our presentation that this equality hold for any continuous, piecewise continuously differentiable function.

We set $w(x) = v(x) + (N(x) - N(l)x/l)dv/dx$. We have

$$|w(x) - v(x)| \leqslant 2\|N(x)\|_{C([0,l])}\left\|\frac{dv}{dx}\right\|_{C([0,l])},$$

$$k_1(x)\frac{dw}{dx} = k_1(x)\left(\frac{dv}{dx} + \left(\frac{dN}{dx} - \frac{N(l)}{l}\right)\frac{dv}{dx}\right) + \qquad (7)$$

$$+ k_1(x)\left(N(x) - N(l)\frac{x}{l}\right)\frac{d^2v}{dx^2} = \widehat{k}_1(x)\frac{dv}{dx} + r(x);$$

and for $r(x) = -k_1(x)\frac{N(l)}{l}\frac{dv}{dx} + k_1(x)\left(N(x) - N(l)\frac{x}{l}\right)\frac{d^2v}{dx^2}$, the estimate is

$$|r(x)| \leqslant \|k_1\|_{C([0,l])}\|N\|_{C([0,l])}\left(\frac{1}{l}\left\|\frac{dv}{dx}\right\|_{C([0,l])} + 2\left\|\frac{d^2v}{dx^2}\right\|_{C([0,l])}\right). \qquad (8)$$

The chain of relations

$$N(x) = \int_0^x \widehat{k}_1(t)\left(\frac{1}{k_1(t)} - \frac{1}{\widehat{k}_1(t)}\right)dt = \int_0^x \widehat{k}_1 d\left(\int_0^t\left(\frac{1}{k_1(\tau)} - \frac{1}{\widehat{k}_1(\tau)}\right)d\tau\right) =$$

$$= \widehat{k}_1(x)\int_0^x\left(\frac{1}{k_1(t)} - \frac{1}{\widehat{k}_1(t)}\right)dt - \int_0^x\left(\int_0^t\left(\frac{1}{k_1(\tau)} - \frac{1}{\widehat{k}_1(\tau)}\right)d\tau\right)\widehat{k}_1'(t)\,dt$$

holds. It implies

$$\|N(x)\|_{C([0,l])} \leqslant \left(\|\widehat{k}_1\|_{C([0,l])} + \int_0^l |\widehat{k}_1'(t)|\,dt\right)d_1. \qquad (9)$$

The estimate derived from (8) and (9) is

$$\left|\int_0^l\left(k_1(x)\frac{dw}{dx} - \widehat{k}_1(x)\frac{dv}{dx}\right)\frac{d\psi}{dx}dx\right| \leqslant c_4 d_1\|\psi\|_{\overset{\circ}{W}{}_2^1((0,l))}. \qquad (10)$$

Here and elsewhere, the c_i are constants depending on $||k_1||_{C([0,l])}$,

$$\int_0^l |\widehat{k}_1'|\,dt,\ \inf_{[0,l]} k_1,\ \|v\|_{C([0,l])} + \|dv/dx\|_{C([0,l])} + \|d^2v/dx^2\|_{C([0,l])},$$

which are finite for problems (4) and (5); in this section,

$$\|\psi\|_{\overset{\circ}{W}{}_2^1((0,l))} = \sqrt{\int_0^l\left(\frac{d\psi}{dx}\right)^2 dx}.$$

The following equality holds:

$$\int_0^l k_2 w\psi\, dx - \int_0^l K_2 v\psi\, dx = I_1 + I_2,$$

where

$$I_1 = \int_0^l (k_2 - K_2)\, v\psi\, dx, \quad I_2 = \int_0^l k_2\, (w - v)\, \psi\, dx.$$

Using inequality (7), we get

$$|I_2| \leqslant \| k_2(x) \|_{C([0,l])} 2 \| N(x) \|_{C([0,l])} \left\| \frac{dv}{dx} \right\|_{C([0,l])} \| \psi \|_{L_1((0,l))} \leqslant$$
$$\leqslant \frac{2}{3} \| k_2 \|_{C([0,l])} \| N \|_{C([0,l])} \left\| \frac{dv}{dx} \right\|_{C([0,l])} \| \psi \|_{\overset{\circ}{W}{}_2^1((0,l))} l^{3/2};$$

here we have also used the well-known estimate

$$\| \psi \|_{L_1((0,l))} \leqslant \frac{1}{\pi} l^{3/2} \| \psi \|_{\overset{\circ}{W}{}_2^1((0,l))}.$$

By integrating by parts, we get

$$I_1 = \int_0^l (k_2 - K_2)\, v\psi\, dx = - \int_0^l \left(\int_0^x (k_2(t) - K_2(t))\, dt \right) \left(\frac{dv}{dx} \psi + v \frac{d\psi}{dx} \right) dx.$$

Then, by using the estimate $\|\psi\|_{L_2((0,l))} \leqslant l/\pi \|\psi\|_{\overset{\circ}{W}{}_2^1((0,l))}$, we have

$$|I_1| \leqslant \left\| \int_0^x (k_2(t) - K_2(t))\, dt \right\|_{L_2((0,l))} \left(\left\| \frac{dv}{dx} \right\|_{C([0,l])} \| \psi \|_{L_2((0,l))} + \right.$$
$$\left. + \| v \|_{C([0,l])} \| \psi \|_{\overset{\circ}{W}{}_2^1((0,l))} \right) \leqslant \left\| \int_0^x (k_2(t) - K_2(t))\, dt \right\|_{L_2((0,l))} \left(\frac{l}{\pi} \left\| \frac{dv}{dx} \right\|_{C([0,l])} + \right.$$
$$\left. + \| v \|_{C([0,l])} \right) \| \psi \|_{\overset{\circ}{W}{}_2^1((0,l))}.$$

Thus,

$$\left| \int_0^l k_2 w\psi\, dx - \int_0^l K_2 v\psi\, dx \right| \leqslant c_5 d_2 \| \psi \|_{\overset{\circ}{W}{}_2^1((0,l))}. \tag{11}$$

Next we have

$$\int_0^l k_3\psi\,dx - \int_0^l K_3\psi\,dx = -\int_0^l \left(\int_0^x (k_3(t) - K_3(t))dt\right)\frac{d\psi}{dx}\,dx \leqslant$$
$$\leqslant \left\| \int_0^x (k_3(t) - K_3(t))dt \right\|_{L_2((0,l))} \|\psi\|_{\overset{\circ}{W}{}_2^1((0,l))}.$$

Consequently,

$$\left|\int_0^l (k_3\psi - K_3\psi)\,dx\right| \leqslant c_6 d_3 \|\psi\|_{\overset{\circ}{W}{}_2^1((0,l))}, \quad c_6 = 1. \tag{12}$$

(10)–(12) imply

$$\left|\int_0^l \left(k_1\frac{dw}{dx}\frac{d\psi}{dx} + k_2 w\psi + k_3\psi\right)dx - \int_0^l \left(\hat{k}_1\frac{dv}{dx}\frac{d\psi}{dx} + K_2 v\psi + K_3\psi\right)dx\right| \leqslant$$
$$\leqslant \left(\sum_{i=1}^{3} c_{i+3} d_i\right)\|\psi\|_{\overset{\circ}{W}{}_2^1((0,l))} \leqslant c_7 d \|\psi\|_{\overset{\circ}{W}{}_2^1((0,l))}, \quad c_7 = \max_{i=4,5,6} c_i,$$

i. e. $|\lambda_1(w,\psi) - \lambda_2(v,\psi)| \leqslant c_7 d \|\psi\|_{\overset{\circ}{W}{}_2^1((0,l))}$.

Since $\lambda_2(v,\psi) = \lambda_1(u,\psi)$ from (6),

$$\left|\int_0^l \left(k_1\frac{d(w-u)}{dx}\frac{d\psi}{dx} + k_2(w-u)\psi\right)dx\right| \leqslant c_7 d \|\psi\|_{\overset{\circ}{W}{}_2^1((0,l))}.$$

We have $(w-u)|_{x=0} = (w-u)|_{x=l} = 0$; it follows from the theory of differential equations that $w - u \in \overset{\circ}{W}{}_2^1$. Therefore, this function can replace ψ in the latter inequality.

Since $k_1 \geqslant K\kappa_1 > 0$, $k_2 \geqslant 0$, we get

$$k_1\int_0^l \left(\frac{d(u-w)}{dx}\right)^2 dx \leqslant c_7 d \sqrt{\int_0^l \left(\frac{d(u-w)}{dx}\right)^2 dx};$$

and so

$$\|u - w\|_{\overset{\circ}{W}{}_2^1((0,l))} \leqslant \frac{c_7 d}{K\varkappa_1}. \tag{13}$$

By using the estimate $\|u\|_{C([0,l])} \leqslant \sqrt{l/2}\|u\|_{\overset{\circ}{W}{}_2^1((0,l))}$, we obtain $|u - w| \leqslant c_8 d$, and therefore,

$$|u - v| \leqslant c_9 d. \tag{14}$$

From (8) we derive

$$\left| \frac{dw}{dx} - \frac{1}{k_1(x)} \hat{k}_1(x) \frac{dv}{dx} \right| \leqslant c_9 d,$$

and, consequently,

$$\left| k_1 \frac{dw}{dx} - \hat{k}_1(x) \frac{dv}{dx} \right| \leqslant c_{10} d.$$

The above and (13) imply

$$\left\| k_1 \frac{du}{dx} - \hat{k}_1(x) \frac{dv}{dx} \right\|_{L_2((0,l))} \leqslant c_{11} d. \tag{15}$$

Let us estimate the same value in uniform metric. In the case under consideration, when k_1 has a finite number of the discontinuity points ($x = j\varepsilon$), and $\hat{k}_1(x)$ is continuous, the theory of differential equations implies that du/dx is continuous everywhere except at these points, and u, v, $k_1 du/dx$, dv/dx, $\hat{k}_1 dv/dx$ are continuous everywhere on $[0, l]$.

By subtracting (5) from (4) and by integrating the obtained expression over $[x_1, x_2]$, we get

$$\left(k_1 \frac{du}{dx} - \hat{k}_1 \frac{dv}{dx} \right) \Big|_{x_1}^{x_2} - \int_{x_1}^{x_2} (k_2 u - K_2 v)\, dt = \int_{x_1}^{x_2} (k_3 - K_3)\, dt.$$

We have

$$\int_{x_1}^{x_2} (k_2 u - K_2 v)\, dt = \int_{x_1}^{x_2} (k_2 - K_2) v\, dx + \int_{x_1}^{x_2} k_2 (u - v)\, dt;$$

the first integral is transformed in much the same way as I_1:

$$\int_{x_1}^{x_2} (k_2 - K_2) v\, dt =$$

$$= v(x) \int_0^x (k_2(\tau) - K_2(\tau))\, d\tau \Big|_{x_1}^{x_2} - \int_{x_1}^{x_2} \left(\int_0^t (k_2(\tau) - K_2(\tau))\, d\tau \right) v' dt.$$

It implies

$$\left| \int_{x_1}^{x_2} (k_2 - K_2) v\, dt \right| \leqslant c_{12} \| k_2 - K_2 \|_\infty^{(-1)}, \quad \left| \int_{x_1}^{x_2} k_2 (u - v)\, dt \right| \leqslant c_{13} d$$

from estimate (14). Thus,

$$\left| \left(k_1 \frac{du}{dx} - \hat{k}_1 \frac{dv}{dx} \right) \Big|_{x_1}^{x_2} \right| \leqslant c_{12} \| k_2 - K_2 \|_\infty^{(-1)} + c_{13} d. \tag{16}$$

Lemma 1. *Let for all $x_1, x_2 \in [0, l]$ the inequalities*

$$|z(x_2) - z(x_1)| \leqslant \beta, \quad \sqrt{\int_0^l z^2(x)\,dx} \leqslant \gamma$$

hold. Then $|z(x)| \leqslant \omega = \beta + \gamma/\sqrt{l}$ for all x.

Proof: Suppose the opposite is true, i.e., we have $|z(\overline{x})| > \omega$ at some point $\overline{x}$. Then from the first condition for all z, we have $|z(x)| \geqslant |z(\overline{x})| - \beta > \gamma/\sqrt{l}$. Consequently,

$$\sqrt{\int_0^l z^2(x)\,dx} > \sqrt{\int_0^l \frac{\gamma^2}{l}\,dx} = \gamma,$$

a contradiction. The lemma is proved.

Estimates (15), (16), and Lemma 1 applied to the function $z(x) = k_1 \frac{du}{dx} - \hat{k}_1 \frac{dv}{dx}$ imply that

$$\left| k_1 \frac{du}{dx} - \hat{k}_1 \frac{dv}{dx} \right| \leqslant c_{12} \| k_2 - K_2 \|_\infty^{(-1)} + c_{13} d + c_{11} \frac{d}{\sqrt{l}}.$$

By substituting the expression for d there, we get

$$\left| k_1 \frac{du}{dx} - \hat{k}_1 \frac{dv}{dx} \right| \leqslant c_{14} \tilde{\alpha},$$

$$\tilde{\alpha} = \left(\| k_1^{-1} - \hat{k}_1^{-1} \|_\infty^{(-1)} + \| k_2 - K_2 \|_\infty^{(-1)} + \| k_3 - K_3 \|_2^{(-1)} \right).$$

Since (2) and (3) imply that $\mathbf{M}d$, $\mathbf{M}\tilde{\alpha} = O(\sqrt{\varepsilon})$, it follows from (14) that $\mathbf{M}|u - v| = O(\sqrt{\varepsilon})$ and from (17) that

$$\mathbf{M} \left| k_1 \frac{du}{dx} - \hat{k}_1 \frac{dv}{dx} \right| = O(\sqrt{\varepsilon}).$$

The theorem is proved.

A similar result might have been obtained on the basis of [153]. The case of equations of higher orders was discussed in [183, 224].

Just as in the foregoing sections, the above scheme of averaging extends to the case of parabolic and hyperbolic equations ([17]). See the multi-dimensional case in [63, 161].

§12 Problems of Plasticity and Dynamics of Viscous Fluid as Described by Functions Depending on Fast Variables[1]

The formalism elaborated upon above results eventually in the following procedure. While solving auxiliary problems on a cell, we determine the effective

[1]This section was written jointly with M. E. Eglit.

characteristics of the medium, i.e., we find state equations which relate the average values of stresses to elongations. Then we solve the system of "averaged" equations corresponding to the "averaged" medium.

For the problems considered below, this approach is either inapplicable (due to failure to split the original problem into separate problems for fast and for slow variables) or is not effective enough (the amount of computations required to construct and solve averaged equations is excessive). It may be expedient to describe such processes which would depend both on fast and slow variables with the characteristic size of variation for the two groups of variables not tending to zero as $\varepsilon \to 0$. This expansion of the technique is convenient for describing processes in a periodic viscous, heat conducting, compressible fluid (gas) and problems like those of plasticity theory. The formal constructions carried out below are not supported by rigorous mathematical justifications; this justification is unlikely at present, because the correctness of the problems is yet to be proved.

The problems of plasticity theory differ from those discussed earlier in that there is no fixed connection between strains and stresses. A characteristic feature of the proposed description of processes in such media is that it does not seek a relationship between strains and stresses at any moment of time, but these quantities are considered as independent functions. We shall illustrate this approach with reference to the equation already investigated:

$$R(\xi)\frac{\partial^2 u}{\partial t^2} - \frac{\partial}{\partial x}\left(K(\xi)\frac{\partial u}{\partial x}\right) = 0, \quad \xi = \frac{x}{\varepsilon}.$$

Let us rewrite it as a system of equations:

$$R(\xi)\frac{\partial^2 u}{\partial t^2} - \frac{\partial \sigma}{\partial x} = 0, \quad e - \frac{1}{K(\xi)}\sigma = 0, \quad e = \frac{\partial u}{\partial x}. \tag{1}$$

Suppose a solution of the problem is written as

$$u \sim u_0(x,\ t,\ \xi) + \varepsilon u_1(x,\ t,\ \xi) + \ldots,$$
$$\sigma \sim \sigma_0(x,\ t,\ \xi) + \varepsilon \sigma_1(x,\ t,\ \xi) + \ldots$$

By substituting these expansions into (1), we get

$$-\frac{1}{\varepsilon}\frac{\partial \sigma_0}{\partial \xi} + \left(R(\xi)\frac{\partial^2 u_0}{\partial t^2} - \frac{\partial \sigma_0}{\partial x} - \frac{\partial \sigma_1}{\partial \xi}\right) + \ldots \sim 0,$$
$$\frac{1}{\varepsilon}\frac{\partial u_0}{\partial \xi} + \left(\frac{\partial u_0}{\partial x} + \frac{\partial u_1}{\partial \xi} - \frac{1}{K(\xi)}\sigma_0\right) + \ldots \sim 0;$$

the terms of order ε^{-1} being equal to 0 imply that $\sigma_0 = \sigma_0(x,t)$, $u_0 = u_0(x,t)$. Further, we have

$$R\frac{\partial^2 u_0}{\partial t^2} - \frac{\partial \sigma_0}{\partial x} - \frac{\partial \sigma_1}{\partial \xi} = 0, \quad \frac{\partial u_0}{\partial x} + \frac{\partial u_1}{\partial \xi} - \frac{1}{K(\xi)}\sigma_0 = 0.$$

After applying the operation $\langle\cdot\rangle$, we get

$$\langle R\rangle \frac{\partial^2 u_0}{\partial t^2} - \frac{\partial \sigma_0}{\partial x} = 0, \quad e_0 - \left\langle \frac{1}{K} \right\rangle \sigma_0 = 0, \quad e_0 = \frac{\partial u_0}{\partial x}. \tag{2}$$

By eliminating σ_0, we can derive the averaged equation

$$\langle R\rangle \frac{\partial^2 u_0}{\partial t^2} - \frac{\partial}{\partial x}\left(\left\langle \frac{1}{K} \right\rangle^{-1} \frac{\partial u_0}{\partial x}\right) = 0. \tag{3}$$

Another approach to the investigation of (2) involves u_0 and σ_0 being considered as independent functions and system (2) being solved, for example, numerically. Let F^{mn} be approximations to the values of function $f(mh, n\tau)$; with h, τ being step sizes of the mesh for x and t, respectively, while m is not necessarily an integer; the values of u_0^{mn} and $\sigma_0^{m+1/2,n}$ are found in the process of calculations. The finite-difference approximation of system (2) can be taken as

$$\langle R\rangle \frac{u_0^{m,n+1} - 2u_0^{m,n} + u_0^{m,n-1}}{\tau^2} - \frac{\sigma_0^{m+1/2,n} - \sigma_0^{m-1/2,n}}{h} = 0, \tag{4}$$

$$\frac{u_0^{m+1,n} - u_0^{m,n}}{h} - \left\langle \frac{1}{K} \right\rangle \sigma_0^{m+1/2,n} = 0. \tag{5}$$

Let all the $u_0^{m,n}$ and $\sigma_0^{m+1/2,n}$ be known for $n \leqslant k$, $(1 \leqslant k)$. We define $u_0^{m,k+1}$ from (4) and then $\sigma_0^{m+1/2,k+1}$ from (5). We have moved one step forward in time.

Equations (4) and (5) are equivalent to the following mesh equation approximating (3):

$$\langle R\rangle \frac{(u_0^{m,n+1} - 2u_0^{mn} + u_0^{m,n-1})}{\tau^2 - \langle 1/K\rangle^{-1}} \times \frac{(u_0^{m+1,n} - 2u_0^{m,n} + u_0^{m-1,n})}{h^2} = 0;$$

a solution of this equation converges to that of (3) for $\tau^2 \leqslant h^2\langle 1/K\rangle\langle R\rangle$.

Consider an elementary one-dimensional model for the problems of plasticity in a one-dimensional periodic medium; the translation vector and the first column of the stress tensor are related by

$$R(\xi) \frac{\partial^2 u}{\partial t^2} - \frac{\partial \sigma}{\partial x} = 0; \tag{6}$$

the first column of the elongation tensor $e = \partial u/\partial x$ and the vector σ are related by

$$\frac{\partial e}{\partial t} - G\left(\xi, e, \chi, \sigma, \frac{\partial \sigma}{\partial t}\right) = 0, \quad \frac{\partial \chi}{\partial t} - H\left(\xi, e, \chi, \sigma, \frac{\partial \sigma}{\partial t}\right) = 0; \tag{7}$$

χ is a vector with components called strain hardening parameters, and R, G, H are 1-periodic in the variable $\xi = x/\varepsilon$.

REMARK 1: In theories like those of flow neglecting temperature effects, the defining relations for plastic bodies with strain hardening are often written in the form (called associated law)

$$\left.\begin{aligned} de^{p}_{ij} &= \mathscr{H}\frac{\partial f}{\partial\sigma^{ij}}d'f \\ d\chi_s &= A^s d'f \end{aligned}\right\} \quad \text{for } f\left(\|\sigma^{kl}\|, \|e^{p}_{ij}\|, \chi\right) = 0, \quad d'f = \frac{\partial f}{\partial\sigma^{kl}}\,d\sigma^{kl} > 0,$$

$$\left.\begin{aligned} de^{p}_{ij} &= 0 \\ d\chi_s &= 0 \end{aligned}\right\} \quad \text{for } f\left(\|\sigma^{kl}\|, \|e^{p}_{ij}\|, \chi\right) < 0, \quad \text{and also } f = 0, \quad d'f < 0. \tag{7'}$$

Here $e^{p}_{ij} = e_{ij} - e^{e}_{ij} = e_{ij} - \beta_{ijkl}\sigma_{kl}$, e^{p}_{ij}, e^{e}_{ij}, e_{ij} are components of the tensors of plastic, elastic, and total strains, σ_{ij} are components of the stress tensor, $f(\|\sigma_{kl}\|, \|e^{p}_{ij}\|, \chi)$ is a function of loading, and κ, A^s are the functions of σ_{kl}, e^{μ}_{ij}, χ_s. Equations such as $d\chi = \sigma_{ij}de^{P}_{ij}$ (i.e., $A = \mathcal{H}(\partial f/\partial\sigma^{ij})\theta^{i,j}$), or $d\chi = \sqrt{\partial e^{p}_{ij}\partial e^{p}_{ij}}$ (i.e., $A = \sqrt{\kappa^2(\partial f/\partial\sigma_{ij})(\partial f/\partial\sigma_{ij})}$) are taken for the parameters χ.

By keeping in mind that e coincides in our notation with a column of the strain tensor e_{ij}, one can see that in the one-dimensional case relations (7′) are of the same form as (7).

Figure 9b shows a part of the stress-strain diagram (7) for simple tension-compression of a linearly hardening material, with the plastic strains themselves and additional parameters acting as hardening parameters, and the vectors of the components of the vector χ being absent. The function G is defined by the relations

$$G\left(e, \sigma, \frac{\partial\sigma}{\partial t}\right) = \begin{cases} \beta\dfrac{\partial\sigma}{\partial t} & \text{for } |\sigma - \beta e| = \gamma, \quad (\sigma - \beta e)\partial\sigma/\partial t > 0, \\ \alpha\dfrac{\partial\sigma}{\partial t} & \text{for the remaining } e, \sigma\text{: } |\sigma - \beta e| \leqslant \gamma \quad (\alpha > \beta); \end{cases}$$

of the pair of values (e, σ) for which $|\sigma - \beta e| > \gamma$ are inadmissible.

When loaded, the points (e, σ) determining strains and stresses in a material move from the state $(e, \sigma) = (0, 0)$ to the right along OM_+ and then along l^+ if $\partial\sigma/\partial t > 0$, or to the left along OM_- and then along l^- if $\partial\sigma/\partial t < 0$, with the rays $l^\pm$ lying on the straight lines $\sigma - \beta e = \pm\gamma$. Consider the case when $\partial\sigma/\partial t > 0$. Further, if $\partial\sigma/\partial t < 0$, the point (e, σ) moves along the straight line with inclination α and parallel to the segment M_-M_+ (we are describing the case when $\sigma - \beta e > -\gamma$, that is, (e, σ) lies above the continuation of the ray l^-). If, further, $\sigma\partial\sigma/\partial t > 0$, then there is first a movement along the straight line with inclination α until crossing with l^+ and then along l^+.

In a periodic medium consisting of components with such properties, β, α, γ are 1-periodic functions of the variable $\zeta = x/\varepsilon$.

As above, we seek a solution in the form

$$\begin{aligned} u &\sim u_0(x, t, \xi) + \varepsilon u_1(x, t, \xi) + \ldots, \\ \sigma &\sim \sigma_0(x, t, \xi) + \varepsilon\sigma_1(x, t, \xi) + \ldots, \\ \chi &\sim \chi_0(x, t, \xi) + \varepsilon\chi_1(x, t, \xi) + \ldots \end{aligned}$$

We immediately assume that $\partial u_0/\partial\xi = 0$. Otherwise, infinitely large strains $e \sim 1/\varepsilon(\partial u_0/\partial\xi)$ occur as $\varepsilon \to 0$, which is incompatible with some models of continuum mechanics. By substituting the asymptotic expansion into (6) and (7), we get

$$-\frac{1}{\varepsilon}\frac{\partial\sigma_0}{\partial\xi} + \left(R\frac{\partial^2 u_0}{\partial t^2} - \frac{\partial\sigma_0}{\partial x} - \frac{\partial\sigma_1}{\partial\xi}\right) + \ldots \sim 0,$$
$$\frac{\partial(e_0+e_1)}{\partial t} - G\left(\xi,\, e_0+e_1,\, \chi_0,\, \sigma_0,\, \frac{\partial\sigma_0}{\partial t}\right) + \ldots \sim 0,$$
$$\frac{\partial\chi_0}{\partial t} - H\left(\xi,\, e_0+e_1,\, \chi_0,\, \sigma_0,\, \frac{\partial\sigma_0}{\partial t}\right) + \ldots \sim 0;$$

here $e_0 = \partial u_0/\partial x$, $e_1 = \partial u_1/\partial\xi$.

It follows from the first, second, and third equations that

$$\frac{\partial\sigma_0}{\partial\xi} = 0, \quad \langle R\rangle\frac{\partial^2 u_0}{\partial t^2} - \frac{\partial\sigma_0}{\partial x} = 0, \tag{8}$$

$$\begin{aligned} \frac{\partial(e_0+e_1)}{\partial t} - G\left(\xi,\, e_0+e_1,\, \chi_0,\, \sigma_0,\, \frac{\partial\sigma_0}{\partial t}\right) = 0,\\ \frac{\partial\chi_0}{\partial t} - H\left(\xi,\, e_0+e_1,\, \chi_0,\, \sigma_0,\, \frac{\partial\sigma_0}{\partial t}\right) = 0. \end{aligned} \tag{9}$$

The system of equations in which we are interested is

$$\begin{aligned} \langle R\rangle\frac{\partial^2 u_0}{\partial t^2} - \frac{\partial\sigma_0}{\partial x} = 0,\\ \frac{\partial(e_0+e_1)}{\partial t} - G\left(\xi,\, e_0+e_1,\, \chi_0,\, \sigma_0,\, \frac{\partial\sigma_0}{\partial t}\right) = 0,\\ \frac{\partial\chi_0}{\partial t} - H\left(\xi,\, e_0+e_1,\, \chi_0,\, \sigma_0,\, \frac{\partial\sigma_0}{\partial t}\right) = 0,\\ e_0 = \frac{\partial u_0}{\partial x}, \quad e_1 = e_1(x,\, t,\, \xi); \end{aligned} \tag{10}$$

here $e_1(x,t,\xi)$ is an unknown function satisfying the condition $\langle e_1\rangle = 0$. If u_0, σ_0, e_0, e_1 is a solution of this system, then for

$$\sigma_1 = \left(\int^{\xi} (\langle R\rangle - R(\eta))\, d\eta\right)\frac{\partial^2 u_0}{\partial t^2}, \quad u_1(t,\, x,\, \xi) = \int^{\xi} e_1(t,\, x,\, \eta)\, d\eta$$

the original relations (6) and (7) are satisfied up to $O(\varepsilon)$. Note that system (10) contains the functions e_1 and χ_0 that depend on the fast variable ξ.

The following scheme of calculations can be proposed to solve the Cauchy (initial value) problem for system (10). The values of $u_0^{m,n}$, $\sigma_0^{m+1/2,n}$, $e_0^{m+1/2,n}$, $e_1^{m+1/2,n}(\xi)$, $\chi_0^{m+1/2,n}$ are sought. Let all the values be known for $n \leqslant k$. From the first equation of (10) approximated in (4), we find $u_0^{m,k+1}$ and then

$e_0^{m+1/2,k+1} = (u_0^{m+1,k+1} - u_0^{m,k+1})/h$. Approximation of the third equation of (10) is taken as

$$\frac{\chi_0^{m+1/2,k+1}(\xi) - \chi_0^{m+1/2,k}(\xi)}{\tau} - H\left(\xi,\ e_0^{m+1/2,k} + \right.$$
$$\left. + e_1^{m+1/2,k}(\xi),\ \chi_0^{m+1/2,k}(\xi), \frac{\sigma_0^{m+1/2,k+1} - \sigma_0^{m+1/2,k}}{\tau}\right) = 0, \quad (11)$$

and of the second equation as

$$\frac{\left(e_0^{m+1/2,k+1} + e_1^{m+1/2,k+1}(\xi)\right) - \left(e_0^{m+1/2,k} + e_1^{m+1/2,k}(\xi)\right)}{\tau} -$$
$$-G\left(\xi,\ e_0^{m+1/2,k} + e_1^{m+1/2,k}(\xi),\ \sigma_0^{m+1/2,k},\ \frac{\sigma_0^{m+1/2,k+1} - \sigma_0^{m+1/2,k}}{\tau}\right) = 0,$$

The latter relation can be written as $\langle e_1(x,t,\xi)\rangle = 0$,

$$e_1^{m+1/2,k+1} = \psi_{m,k}\left(\xi,\ \chi_0^{m+1/2,k},\ \sigma_0^{m+1/2,k+1}\right), \quad (12)$$

where $\psi_{m,k}$ is some known function. Since $\langle e_1\rangle = 0$,

$$0 = \langle \psi_{m,k}\left(\xi,\ \chi_0^{m+1/2,k},\ \sigma_0^{m+1/2,k+1}\right)\rangle.$$

By solving this equation together with (1), we find $\sigma_0^{m+1/2,k+1}$ and $\chi_0^{m+1/2,k+1}(\xi)$. Then we calculate $e_1^{m+1/2,k+1}(\xi)$ from formula (12). The trouble with this solution algorithm is that an additional variable ξ is introduced into the consideration, and the problem must be quantized in this variable (meaning the need to store and compute the functions depending on the variable ξ).

The medium in question often consists of several homogeneous layers, that is, $G(\xi, e, \sigma, \partial\sigma/\partial t) = g_k(e, \sigma, \partial\sigma/\partial t)$, $H(\xi, e, \chi, \sigma, \partial\sigma/\partial t) = h_k(e, \chi, \sigma, \partial\sigma/\partial t)$ for $\omega_{k-1} < \xi < \omega_k, \omega_0 = 0, \ldots, \omega_k = 1$; we set $\theta_k = \omega_k - \omega_{k-1}$.

Let it be at some moment of time t_0 that

$$\left.\begin{aligned} e_1(x,t,\xi) &= e_1^k(x,t) \\ \chi(x,t,\xi) &= \chi^k(x,t) \end{aligned}\right\} \text{ for } \omega_{k-1} < \xi \leqslant \omega_k, \quad (13)$$

We conclude from (10) that $\partial e_1/\partial t$, $\partial \chi_0/\partial t$ do not depend on ξ in the interval $\omega_{k-1} < \xi \leqslant \omega_k$. This naturally implies that

$$\left.\begin{aligned} e_1(x,t,\xi) &= e_1^k(x,t) \\ \chi(x,t,\xi) &= \chi^k(x,t) \end{aligned}\right\} \text{ for } \omega_{k-1} < \xi \leqslant \omega_k$$

for $t > t_0$. Then the system of equations (10) becomes a system of equations in the functions $u_0(x,t)$, $\sigma_0(x,t)$, $e_0(x,t)$, $e_1^k(x,t)$, $\chi_0^k(x,t)$, $k = 1, \ldots, l$, where the condition $\langle e_1(x,t,\xi)\rangle = 0$ is replaced by $\sum_{k=1}^{l} \theta_k e_1^k(x,t) = 0$.

We shall describe the averaging of the problem of plasticity in a quasi-stationary case by means of an elementary example, where the vector χ is absent and we have the following system of equations instead of (6) and (7):

$$\frac{\partial e}{\partial t} - g\left(\xi,\ e,\ \sigma,\ \frac{\partial \sigma}{\partial t}\right) = 0, \quad \sigma = \sigma^0(t), \quad e = \frac{\partial u}{\partial x}, \tag{14}$$

with g being 1-periodic in ξ, $0 \leqslant x \leqslant 1$.

Two types of boundary values are the most common:

$$u(0,t) = u_0(t), \quad \sigma^0(t) = \sigma_l(t) \text{ — a given function}; \tag{15}$$

$$u(0,t) = u_0(t), \quad u(l,t) = u_l(t). \tag{16}$$

In case (14) and (15) are replaced by (9),

$$\frac{\partial (e_0 + e_1)}{\partial t} - g\left(\xi,\ e_0 + e_1,\ \sigma_l^{\cdot}(t),\ \frac{\partial \sigma_l^{\cdot}}{\partial t}\right) = 0,$$

$$e_0 = e_0(x,t), \quad e_1 = e_1(x,t,\xi), \quad \langle e_1\rangle = 0.$$

The finite-difference approximation of these equations has the form

$$\frac{\left(e_0^{m+1/2,k+1} + e_1^{m+1/2,k+1}(\xi)\right) - \left(e_0^{m+1/2,k} + e_1^{m+1/2,k}(\xi)\right)}{\tau} =$$

$$= g\left(\xi,\ e_0^{m+1/2,k} + e_1^{m+1/2,k}(\xi),\ \sigma_l(k\tau),\ \left.\frac{\partial \sigma_l}{\partial t}\right|_{k\tau}\right); \tag{17}$$

the application of $\langle\cdot\rangle$ yields the explicit formula for defining $e_0^{m+1/2,k+1}$:

$$e_0^{m+1/2,k+1} = e_0^{m+1/2,k} + \tau\left\langle g\left(\xi,\ e_0^{m+1/2,k} + e_1^{m+1/2,k},\ \sigma_l(k\tau),\ \left.\frac{\partial \sigma_l}{\partial t}\right|_{k\tau}\right)\right\rangle;$$

knowing $e_0^{m+1/2,k+1}$, we find $e_1^{m+1/2,k+1}$ from (17).

When necessary, at the required moments of time, we also find

$$u_0(x,t) = \int_0^x e_0(y,t)\,dy + u_0(t).$$

We take the right side of (17) for problems (14) and (16) in the form

$$g\left(\xi_1,\ e_0^{m+1/2,k} + e_1^{m+1/2,k},\ \sigma_k^0,\ \frac{\sigma_{k+1}^0 - \sigma_k^0}{\tau}\right).$$

The boundary values (16) imply

$$\int_0^l e_0(y,t)\,dy = u_0(l,t) - u_0(0,t) \approx u_l(t) - u_0(t);$$

and the following equation is taken as an approximation of this approximate equality,

$$h \sum_{m=0}^{M-1} e_0^{m+1/2,k+1} = u_l((k+1)\tau) - u_0((k+1)\tau); \tag{18}$$

After applying the operation $\langle\cdot\rangle$ to (17), we get

$$e_0^{m+1/2,k+1} = e_0^{m+1/2,k} + \tau \left\langle g\left(\xi,\ e_0^{m+1/2,k} + e_1^{m+1/2,k},\ \sigma_k^0,\ \frac{\sigma_{k+1}^0 - \sigma_k^0}{\tau}\right)\right\rangle.$$

By substituting this expression into (18), we obtain an equation in σ_{k+y}^0, and once it is found, we proceed as for problem (14) and (15).

REMARK 2: Consider the case when u, σ are scalars and for each fixed ξ,

$$g\left(\xi,\ e,\ \sigma,\ \frac{\partial\sigma}{\partial t}\right) \equiv g_0(\xi,\ e,\ \sigma)\frac{\partial\sigma}{\partial t}$$

except for some set G_ξ of the points (e,σ) of zero measure; at the points of this set,

$$g\left(\xi,\ e,\ \sigma,\ \frac{\partial\sigma}{\partial t}\right) = g^0\left(\xi,\ e,\ \sigma,\ \operatorname{sign}\frac{\partial\sigma}{\partial t}\right)\frac{\partial\sigma}{\partial t}.$$

Some additional assumptions made about the structure of G_ξ enable us to get rid of the function e_1 and to obtain an equation $\partial e_0/\partial t = \hat{g}(e_0, \sigma_0, \partial\sigma_0/\partial t)$ instead of (9). The former does not contain fast variables.

REMARK 3: The above-described procedure of constructing equations of type (8) and (9) that do not contain functions with the characteristic size of variation of order ε can be extended to equations of a more general nature,

$$\frac{\partial\alpha}{\partial t} + \frac{\partial\sigma}{\partial x} = \beta, \quad e = \frac{\partial u}{\partial x}, \quad g = 0; \tag{19}$$

here α, β, g are functions of the arguments ξ, x, t, u, $u_i, \ldots, u_{in}$, σ, $\sigma_t, \ldots, \sigma_{tn}$, e, $e_t, \ldots, e_{tn}$ and of some expressions

$$A_i u = \int_0^t a_i(\xi, x, t, \tau)\, u(t-\tau)\, d\tau,$$

$$B_i \sigma = \int_0^t b_i(\xi, x, t, \tau)\, \sigma(t-\tau)\, d\tau,$$

$$C_i e = \int_0^t c_i(\xi, x, t, \tau)\, e(t-\tau)\, d\tau, \qquad i = 1, \ldots, m,$$

1-periodic in ξ. The system of type (8) and (9) assumes the form

$$\frac{\partial \langle\alpha_0\rangle}{\partial t} + \frac{\partial \sigma_0}{\partial x} = \langle\beta_0\rangle, \quad g_0 = 0,$$
$$e_0 = \frac{\partial u_0}{\partial x}, \quad \sigma_0 = \sigma_0(x, t), \quad \langle e_1\rangle = 0,$$

where α_0, β_0, g_0 is the result of the substitution of $u = u_0(x,t)$, $e = e_0(x,t) + e_1(x,t,\xi)$, $\sigma = \sigma_0(x,t)$ into appropriate functions.

REMARK 4: In many typical cases, the vector u can be split naturally into two vectors u^1, u^2, $u = (u^1, u^2)$ such that the relation $g = 0$ can be written more specifically as $g_1 = 0$, $g_2 = g_3$, with g_1, g_2, g_3 being functions of the arguments x, t, $u, \ldots, u_{t^n}$, $\sigma, \ldots, \sigma_{t^n}$, $e^1, \ldots, e^1_{t^n}$, $A_1 u, \ldots, A_m u$, $B_1\sigma, \ldots, B_m\sigma$, $C_1^1 e^1, \ldots, C_m^1 e^1$. In addition, g_1, g_2 are 1-periodic in ξ, and g_3 depends linearly on $e^2, \ldots, e^2_{t^n}$, $C_1^2 e^2, \ldots, C_m^2 e^2$; $e^1 = \partial u^1/\partial x$, $e^2 = \partial u^2/\partial x$, g_3 is independent of e_1 as well as of the variable ξ. Then, instead of (9), we get the relations $g_1^0 = 0$, $\langle g_2^0\rangle = g_3^0$, $\langle e_1^1\rangle = 0$; here g_1^0 is the result of the substitution of $u^1 = u_0^1$, $u^2 = u_0^2$, $\sigma = \sigma_0$, $e_0^1 = \frac{\partial u_0^1}{\partial x}$, $e_0^2 = \frac{\partial u_0^2}{\partial x}$, $e^1 = e_0^1 + e_1^1$, $e^2 = e_0^2$ into appropriate functions, and the u_i^k are coefficients of the asymptotic expansion

$$u^k \sim u_0^k + \varepsilon u_1^k + \ldots, \quad e_1^k = \partial u_1^k/\partial\xi.$$

This concrete form of g made it possible to reduce the number of unknown functions depending on the fast variable, namely, not to introduce e_1^2.

Consider a system of equations describing one-dimensional motion of a viscous, heat-conducting, compressible medium:

$$L_1 = \frac{\partial W}{\partial t} - \frac{\partial v}{\partial s} = 0, \quad L_2 = \frac{\partial v}{\partial t} + \frac{\partial (p - \lambda(\partial v/\partial s))}{\partial s} = 0,$$
$$L_3 = \frac{\partial (E + v^2/2)}{\partial t} + \frac{\partial}{\partial s}\left(\left(p - \lambda\frac{\partial v}{\partial s}\right)v\right) - \frac{\partial}{\partial s}\left(\varkappa\frac{\partial T}{\partial s}\right) = 0. \tag{19}$$

Here s is a Lebesque coordinate, $ds = \rho dx$, ρ is the density, $W = 1/\rho$ a specific volume, v the speed, p the pressure; T is the temperature, and $\lambda = (\lambda_1 + 2\mu_1)\rho$, $\kappa = \kappa_1\rho$, λ_1, μ_1, κ_1 are the coefficients of viscosity and thermal conductivity of the medium. It is assumed that λ, κ and $W(p,T)$, $E(p,T)$ defining state equations are 1-periodic functions of the variable $\xi = \frac{s}{\varepsilon}$: $W = W(\xi, p, T)$, $E = E(\xi, p, T)$.

Let us first do formal averaging of the problem, assuming that the coefficients are differentiable with respect to ξ. If we represent a solution of system (1) as series in powers of ξ,

$$\begin{aligned} v(s, t) &\sim v_0(s, t, \xi) + \varepsilon v_1(s, t, \xi) + \varepsilon^2 v_2(s, t, \xi) + \ldots \\ p(s, t) &\sim p_0(s, t, \xi) + \varepsilon p_1(s, t, \xi) + \varepsilon^2 p_2(s, t, \xi) + \ldots \\ T(s, t) &\sim T_0(s, t, \xi) + \varepsilon T_1(s, t, \xi) + \varepsilon^2 T_2(s, t, \xi) + \ldots \end{aligned} \tag{20}$$

and consider λ, κ to be finite, it can be shown by means of (19) that v_0 and T_0 are independent of the fast variable ξ, but p_0 does depend on ξ in the general cases. Indeed, let us substitute expansion (20) into (19). The results can be written as

$$L_i \sim \sum_{k=-2}^{\infty} \varepsilon^k l_{i,k} \sim 0, \qquad i = 1, 2, 3,$$

or $l_{i,k} = 0$; the equations $l_{i,-2} = 0$ are satisfied for $\partial v_0/\partial\xi = 0$, $\partial T_0/\partial\xi = 0$, that is,

$$v_0 = v_0(s,\ t), \quad T_0 = T_0(s,\ t). \tag{21}$$

The equations $l_{i,-1} = 0$, using (21), can be written as

$$l_{1,-1} \equiv 0, \quad l_{2,-1} \equiv \frac{\partial}{\partial\xi}\left(p_0 - \lambda_0\left(\frac{\partial v_0}{\partial s} + \frac{\partial v_1}{\partial\xi}\right)\right) = 0,$$

$$l_{3,-1} \equiv \frac{\partial}{\partial\xi}\left(v_0\left(p_0 - \lambda_0\left(\frac{\partial v_0}{\partial s} + \frac{\partial v_1}{\partial\xi}\right)\right)\right) - \frac{\partial}{\partial\xi}\left(\varkappa_0\left(\frac{\partial T_0}{\partial s} + \frac{\partial T_1}{\partial\xi}\right)\right) = 0;$$

here $\lambda_0 = \lambda(\xi, T_0(s,t), p_0(s,t,\xi))$, $\kappa_0 = \kappa(\xi, T_0(s,t,\xi))$. Therefore,

$$p_0 - \lambda_0\left(\frac{\partial v_0}{\partial s} + \frac{\partial v_1}{\partial\xi}\right) = C_1(s,t), \quad \varkappa_0\left(\frac{\partial T_0}{\partial s} + \frac{\partial T_1}{\partial\xi}\right) = C_2(s,t), \tag{22}$$

where $C_1(s,t)$, $C_2(s,t)$ are yet unknown functions independent of the fast variable.

Consider the conditions $l_{i,0} = 0$. We have

$$l_{1,0} \equiv \frac{\partial W_0}{\partial t} - \frac{\partial v_0}{\partial s} - \frac{\partial v_1}{\partial\xi} = 0, \quad l_{2,0} \equiv \frac{\partial v_0}{\partial t} + \frac{\partial C_1}{\partial s} + \frac{\partial H_{2,0}}{\partial\xi} = 0,$$
$$l_{3,0} \equiv \frac{\partial\left(E_0 + v_0^2/2\right)}{\partial t} + \frac{\partial v_0 C_1}{\partial s} - \frac{\partial C_2}{\partial s} + \frac{\partial H_{3,0}}{\partial\xi} = 0. \tag{23}$$

Here $H_{2,0}$, $H_{3,0}$ are some known expressions periodic in ξ, $W_0 = W(\xi, T_0(s,t), p_0(s,t,\xi))$, $E_0 = E(\xi, T_0(s,t), p_0(s,t,\xi))$.

By applying the operator $\langle\cdot\rangle$ to equations (23), we get:

$$\frac{\partial v_0}{\partial t} + \frac{\partial C_1}{\partial s} = 0, \quad \frac{\partial\langle W_0\rangle}{\partial t} - \frac{\partial v_0}{\partial s} = 0,$$

$$\frac{\partial}{\partial t}\left(\langle E_0\rangle + \frac{v_0^2}{2}\right) + \frac{\partial v_0 C_1}{\partial s} - \frac{\partial C_2}{\partial s} = 0,$$

The following relations are derived from (22):

$$C_1 = \left\langle\frac{p_0}{\lambda_0}\right\rangle \cdot \frac{1}{\langle\lambda_0^{-1}\rangle} - \frac{\partial v_0}{\partial s} \cdot \frac{1}{\langle\lambda_0^{-1}\rangle}, \quad C_2 = \frac{\partial T_0}{\partial s} \cdot \frac{1}{\langle\varkappa_0^{-1}\rangle}.$$

As a result, the equations $l_{i,-1} = 0$, $l_{i,0} = 0$ yield the following system containing only functions v_0, p_0, T_0:

$$\frac{\partial W_0}{\partial t} - \frac{\bar{\lambda}_0}{\lambda_0}\frac{\partial v_0}{\partial s} + \frac{\bar{\lambda}_0}{\lambda_0}\left\langle\frac{p_0}{\lambda_0}\right\rangle - \frac{p_0}{\lambda_0} = 0,$$

$$\frac{\partial v_0}{\partial t} + \frac{\partial}{\partial s}\left(\bar{\lambda}_0\left(\left\langle\frac{p_0}{\lambda_0}\right\rangle - \frac{\partial v_0}{\partial s}\right)\right) = 0,$$

$$\frac{\partial}{\partial t}\left(\langle E_0\rangle + \frac{v_0^2}{2}\right) + \frac{\partial}{\partial s}\left(\bar{\lambda}_0\left(\left\langle\frac{p_0}{\lambda_0}\right\rangle - \frac{\partial v_0}{\partial s}\right)v_0\right) - \frac{\partial}{\partial s}\left(\bar{\varkappa}_0\frac{\partial T_0}{\partial s}\right) = 0. \tag{24}$$

We used the notation

$$\frac{1}{\bar{\lambda}_0} = \left\langle\frac{1}{\lambda_0}\right\rangle, \quad \frac{1}{\bar{\varkappa}_0} = \left\langle\frac{1}{\varkappa_0}\right\rangle.$$

System (24) is a set of integro-differential equations, $\langle p_0/\lambda_0\rangle$, $\overline{\lambda}_0$, $\overline{\kappa}_0$, $\langle E_0\rangle$ contain an unknown function p_0 under the integration sign. If a solution of the system is found, then v_1, T_1 can be selected so that the equations (22) are satisfied and p_1, v_2, T_2 (which appear in the expressions for $H_{2,0}$, $H_{3,0}$) so that the equations (23) are also satisfied. Then the original equations (19) will be satisfied up to terms of order $O(\varepsilon)$.

Just as in problems of plasticity, the variable ξ cannot be gotten rid of even in the zeroth approximation for the medium: p_0 depends on ξ and this dependence is important. Note that this effect occurs when both the viscosity and compressibility of the medium are taken into consideration simultaneously.

Consider a periodic medium consisting of a recurrent system of layers with distinct properties. If at some moment of time t, the pressure p_k inside of each layer is uniform along it, $p_k = p_k(s,t)$, then due to (24) $\partial p_k/\partial t$ does not depend on the fast variable ξ either, and therefore it can be assumed that always $p_k = p_k(s,t)$. Then the integrals in (24) become sums and (24) becomes a system of differential equations in the functions $v_0(s,t)$, $T_0(s,t)$, $p_k(s,t)$. The above-described method of averaging can be used as an effective tool in this case. The resultant system implies the possibility of quite different pressures arising in adjacent layers when a wave travels through the medium.

REMARK 5: In the case of discontinuous coefficients and solutions, the equations (19) are understood as a set of conservation laws:

$$\int_{s_1}^{s_2}\varphi_i(y,t)\,dy\bigg|_{t_1}^{t_2} + \int_{t_1}^{t_2}\psi_i(s,\tau)\,d\tau\bigg|_{s_1}^{s_2} = 0, \tag{25}$$

$$\varphi_1 = w, \quad \psi_1 = -v, \qquad \varphi_2 = v, \quad \psi_2 = \left(p - \lambda\frac{\partial v}{\partial s}\right),$$

$$\varphi_3 = E + \frac{v^2}{2}, \quad \psi_3 = \psi_2 v - \varkappa\frac{\partial T}{\partial s}.$$

For $\lambda > 0$, $\kappa > 0$, and natural assumptions that the solution is smooth, it can be demonstrated that the formally constructed approximations to the solution, $p_0(s,t,\xi)$, $v_0(s,t,\xi)+\varepsilon v_1(s,t,\xi)$, $T_0(s,t)+\varepsilon T_1(s,t,\xi)$ satisfy the conservation laws (25) up to $O(\varepsilon)$. This circumstance is evidence of the fact that the averaged equations (24) hold also in the case of discontinuous coefficients.

REMARK 6: Another description of (19) by functions depending both on fast and slow variables can be obtained in the same way as in Remark 4. Let u be a translation, that is, $v=\partial u/\partial t$; the first of the equations (19) is written as $\frac{\partial}{\partial t}(w-\frac{\partial u}{\partial s})=0$, and therefore,

$$W-\frac{\partial u}{\partial s}=\left(W-\frac{\partial u}{\partial s}\right)\bigg|_{t=0}. \tag{26}$$

Suppose $(W-\partial u/\partial s)|_{t=0}$ is some function $C(s,\xi)$ with the characteristic size of variation in s and ξ of order $l\gg\varepsilon$ and 1, respectively.

Then (26) implies $W=\partial u/\partial s+C(s,\xi)$; by transforming the equation $W=W(\xi,p,T)$ with respect to p, we get $p=p(\xi,W,T)=p(\xi,\partial u/\partial s+C(s,\xi),T)=P(s,\xi,\partial u/\partial s,T)$; after substituting into the expressions for E,κ,λ, we see that

$$E=\mathscr{E}(s,\ \xi,\ \partial u/\partial s,\ T),\quad \varkappa=\mathscr{K}(s,\ \xi,\ \partial u/\partial s,\ T),\quad \lambda=\Lambda(s,\ \xi,\ \partial u/\partial s,\ T).$$

The second two equations of (19) are rewritten as a system

$$\begin{gathered}\frac{\partial^2 u}{\partial t^2}+\frac{\partial\sigma^1}{\partial s}=0,\quad \frac{\partial\left(\mathscr{E}(s,\xi,e^1,T)+(\partial u/\partial t)^2/2\right)}{\partial t}+\frac{\partial\sigma^2}{\partial s}=0,\\ \frac{\partial e^1}{\partial t}=\frac{P(s,\xi,e^1,T)-\sigma^1}{\Lambda(s,\xi,e^1,T)},\quad e^2=\frac{\sigma^1(\partial u/\partial t)-\sigma^2}{\mathscr{K}(s,\xi,e^1,T)};\end{gathered} \tag{27}$$

here $e^1=\partial u/\partial s$, $e^2=\partial T/\partial s$.

The above system belongs to the class of systems described in Remark 4. We seek an expansion of the solution in the form

$$\begin{aligned}&u\sim u_0(s,t)+\varepsilon u_1(s,t,\xi)+\dots,\\ &\sigma^1\sim\sigma_0^1(s,t)+\varepsilon\sigma_1^1(t,s,\xi)+\dots,\\ &e^1\sim e_0^1(s,t)+e_1^1(s,t,\xi)+\dots,\quad \langle e_1^1\rangle=0,\\ &e^2\sim e_0^2(s,t)+e_1^2(s,t,\xi)+\dots,\quad \langle e_1^2\rangle=0,\\ &T\sim T_0(s,t)+\varepsilon T_1(s,t,\xi)+\dots\end{aligned}$$

After substituting into (27) and following the construction scheme of Remark 4, we get the system of equations

$$\frac{\partial^2 u_0}{\partial t^2}+\frac{\partial\sigma_0^1}{\partial s}=0,$$

$$\frac{\partial\left(\langle\mathscr{E}(s,\xi,e_0^1+e_1^1,T_0)\rangle+(\partial u_0/\partial t)^2/2\right)}{\partial t}+\frac{\partial\sigma^2}{\partial s}=0,$$

$$\frac{\partial\left(e_0^1+e_1^1\right)}{\partial t}=\frac{P\left(s,\xi,e_0^1+e_1^1,T_0\right)-\sigma^1}{\Lambda\left(s,\xi,e_0^1+e_1^1,T_0\right)},$$

$$e_0^2=\left(\sigma^1\frac{\partial u_0}{\partial t}-\sigma^2\right)\left\langle\frac{1}{\mathscr{X}\left(s,\xi,e_0^1+e_1^1,T_0\right)}\right\rangle$$

$$e_0^1=\frac{\partial u_0}{\partial s},\quad \langle e_1^1\rangle=0,\quad e_0^2=\frac{\partial T_0}{\partial s}.$$

REMARK 7: By slightly complicating the method of averaging the integral-differential equation in §4, one can average system (19) in the form of a system of nonlinear integral-differential equations with a delay, which does not contain fast variables.

CHAPTER 4

Averaging Basic Equations of Mathematical Physics

In this chapter we will consider partial differential equations which describe physical processes in composite materials. Section 1 of the Introduction cited an example of thermal conductivity problems (1)–(3) for a fibrous composite (Fig. 1 shows its section by a plane perpendicular to the fibers, ε is the side of a periodic cell, $\varepsilon \ll 1$). By analogy with §3 of the Introduction let us introduce the "fast" variables $\xi_q = x_q/\varepsilon$. Then the elementary cell—the square $(0,\varepsilon) \times (0,\varepsilon)$ in the variables x_1, x_2—turns into the unit square in the variables ξ_1, ξ_2 (see Fig. 2). Again we denote by $\mathcal{B}$ the domain in the unit square which corresponds to the fiber cross-section and by $\mathcal{M}$ the domain occupied by the matrix. Let

$$K(\xi_1, \xi_2) = \begin{cases} K_{\mathrm{B}}, & (\xi_1, \xi_2) \in \mathcal{B}, \\ K_{\mathrm{M}}, & (\xi_1, \xi_2) \in \mathcal{M}, \end{cases}$$

where $K(\xi_1, \xi_3)$ is a periodic function with period 1 in ξ_1 and ξ_2. Then the conductivity coefficient $K_\varepsilon(x_1, x_2)$ in §1 of the Introduction can be represented as $K_\varepsilon(x_1, x_2) = K(x_1/\varepsilon, x_2/\varepsilon)$. Note that $K(x_1, x_2)$ s a periodic function of x_1 and x_2 with period ε.

Equation (1.1) with contact conditions (1.2), (1.3) of the Introduction describes a stationary thermal field in a composite material consisting of isotropic components. In the case of anisotropic components, the heat conduction properties are described by a positive-definite symmetric matrix $\|A_{ij}(x_1/\varepsilon, x_2/\varepsilon)\|$, where

$$A_{kj}(\xi) = A_{kj}(\xi_1, \xi_2) = \begin{cases} A_{\mathrm{B}kj}, & \xi \in \mathcal{B}, \\ A_{\mathrm{M}kj}, & \xi \in \mathcal{M}, \end{cases}$$

the $A_{kj}(\xi)$ are functions 1-periodic in each of the variables ξ_1 and ξ_2. The matrix $\|A_{kj}\|$ is called the *thermal conductivity tensor*. The temperature $u(x_1, x_2)$ satisfies the elliptic equation

$$\sum_{k,j=1}^{2} \frac{\partial}{\partial x_k} \left(A_{kj}\left(\frac{x_1}{\varepsilon}, \frac{x_2}{\varepsilon}\right) \frac{\partial u}{\partial x_j} \right) = f(x_1, x_2) \tag{1}$$

everywhere off the lines Σ separating the fiber and the matrix. The following conditions for the continuity of the temperature and density of the heat flow are satisfied on the discontinuity lines of coefficients

$$[u]_\Sigma = 0, \tag{2}$$

$$\left[\frac{\partial u}{\partial \nu}\right]_\Sigma \equiv \left[\sum_{k,j=1}^{2} A_{kj}\left(\frac{x_1}{\varepsilon}, \frac{x_2}{\varepsilon}\right)\frac{\partial u}{\partial x_j}\, n_k\right]_\Sigma = 0, \tag{3}$$

where n_k is the cosine of the angle between the normal to the discontinuity line and the kth coordinate direction.

Let G be a bounded domain in $\mathbf{R}^2$, independent of ε, on whose boundary ∂G the constant temperature $u(x_1, x_2) = 0$ is maintained. Then equation (1) for $x = (x_1, x_2) \in G\backslash\Sigma$, contact conditions (2) and (3), and the boundary condition

$$u(x) = 0, \quad x \in \partial G, \tag{4}$$

constitute a problem having a unique solution (see §1.3).

In §4.1, the asymptotics, as $\varepsilon \downarrow 0$, of a solution of problems (1)–(4) is constructed, first for smooth coefficients, and then by analogy with §§2.4 and 3.1, the procedure is extended to discontinuous (piecewise smooth) coefficients corresponding to actual composites. It turns out that the solution of problems (1)–(4) coincides up to terms or order ε with a smooth function $v_0(x_1, x_2)$ independent of ε. This function is a solution of the averaged problem with the constant coefficients $\hat{A}_{kj}$:

$$\sum_{k,j=1}^{2} \widehat{A}_{kj} \frac{\partial^2 v_0}{\partial x_k \partial x_j} = f(x_1, x_2), \quad x \in G,$$

$$v_0|_{\partial G} = 0.$$

Thus, the coefficients $\hat{A}_{kj}$ define the thermal conductivity tensor of such a hypothetical homogeneous material whose properties are close to the macroscopic (effective) properties of the original inhomogeneous composite material.

Definition of the coefficients $\hat{A}_{kj}$ is related to the numerical solution of a "problem on a cell" similar to the one-dimensional problem (2.3.9), namely,

$$\widehat{A}_{kj} = \sum_{l=1}^{2} \int_0^1\int_0^1 A_{kl}(\xi_1, \xi_2)\frac{\partial}{\partial \xi_l}(N_j + \xi_j)\, d\xi_1 d\xi_2,$$

where $N_j(\xi_1, \xi_2)$ is a solution, periodic in ξ_1 and ξ_2, of the equation

$$\sum_{k,l=1}^{2} \frac{\partial}{\partial \xi_k}\left(A_{kl}(\xi_1, \xi_2)\frac{\partial}{\partial \xi_l}(N_j + \xi_j)\right) = 0 \tag{5}$$

with the contact conditions on the discontinuity lines of coefficients

$$[N_j] = 0, \tag{6}$$

$$\left[\sum_{k,l=1}^{2} A_{kl}(\xi_1, \xi_2) \frac{\partial}{\partial \xi_l} (N_j + \xi_j) n_k \right] = 0, \tag{7}$$

where n_k is the cosine of the angle between the normal to the discontinuity line and the kth coordinate direction.

Microscopic properties, such as local heat flows, can also be described exactly when higher asymptotic approximations are taken into account.

For one space variable x_1, the formalism of averaging can yield by quadrature an asymptotic expansion of any order of accuracy. Problems on a cell (5)–(7) cannot be solved analytically in the multi-dimensional case. Nevertheless, these constructions make it possible:

– to obtain an averaged equation and to calculate approximately the effective and local characteristics of simulated processes;

– to construct an asymptotic expansion of a solution of an equation with rapidly oscillating coefficients following an algorithm whose steps do not contain the small parameter ε and can be realized by a computer;

– to justify enlargement of ε in the numerical count of the equation with rapidly oscillating coefficients to a value which would make the original problem solvable by a modern computer.

It means the following. In solving multi-dimensional (especially nonlinear) problems, the amount of calculations needed to construct averaged equations is often excessive. The following approach proposed by Yanenko may help in this case: We solve the problem taking an ε much greater than it is in reality, analyze the solution thus obtained by extracting the principal and first terms of the expansion of the solution, and then extrapolate the result to the case of small ε, if necessary. Numerical experiments that confirm the soundness of this approach are described in [190]. We shall discuss them in detail at the end of §1.

§1 Averaging Stationary Thermal Fields in a Composite

1. Let the thermal conductivity tensor be defined at each point $x = (x_1, \dots, x_s)$ of a medium by the matrix-valued function $\|A_{kj}(x/\varepsilon)\|(k, j = 1, \dots, s)$, $A_{kj}(\xi)$ being 1-periodic functions, $A_{kj}(\xi) \in C^\infty(R^s)$, $\xi = (\xi_1, \dots, \xi_s)$ (so the functions $A_{kj}(x/\varepsilon)$ are ε-periodic in all x). Let G be a bounded s-dimensional domain with an infinitely smooth boundary. In G, we consider the equation

$$Pu \equiv \frac{\partial}{\partial x_k}\left(A_{kj}\left(\frac{x}{\varepsilon}\right)\frac{\partial u}{\partial x_j}\right) - f(x) = 0, \quad x \in G, \tag{1}$$

with the boundary condition

$$u|_{\partial G} = 0, \tag{2}$$

where $f(x) \in C^\infty(\overline{G})$. The summation sign over repeated indices from 1 to s (s is the dimension of the space) is omitted.

It is assumed that the matrix $\|A_{kj}(\xi)\|$ is symmetric and positive-definite at each point of ξ:

$$A_{kj}(\xi) = A_{jk}(\xi),$$
$$A_{kj}(\xi)\eta_k\eta_j \geqslant \varkappa_1\eta_k\eta_k \quad \forall\eta = (\eta_1, \ldots, \eta_s) \in \mathbf{R}^s, \tag{3}$$

where the constant $\kappa_1 > 0$ is independent of ε and ξ.

The main objective of this Section is to construct asymptotics for a solution of the boundary value problem (1), (2) as $\varepsilon \downarrow 0$ and to generalize the procedure of constructing asymptotics for the case of piecewise smooth coefficients $A_{jk}(\xi)$.

By analogy with §3.1, we seek the asymptotics of the solution of (1) and (2) in the form

$$u^{(2)}(x) = u_0\left(x, \frac{x}{\varepsilon}\right) + \varepsilon u_1\left(x, \frac{x}{\varepsilon}\right) + \varepsilon^2 u_2\left(x, \frac{x}{\varepsilon}\right) = \sum_{i=0}^{2} \varepsilon^i u_i\left(x, \frac{x}{\varepsilon}\right), \tag{4}$$

where the $u_i(x,\xi)$ are 1-periodic, infinitely differentiable functions.

Let us substitute the sum (4) into the left side of equation (1) and apply the chain rule $F(x, x/\varepsilon)$:

$$\frac{d}{dx_h} F\left(x, \frac{x}{\varepsilon}\right) = \left(\frac{\partial}{\partial x_h} F(x, \xi) + \varepsilon^{-1} \frac{\partial}{\partial \xi_h} F(x, \xi)\right)\Big|_{\xi = x/\varepsilon}. \tag{5}$$

We set $L_{\alpha\beta}u(x,\xi) = \sum\limits_{k,j=1}^{s} \frac{\partial}{\partial\alpha_k}\left(A_{kj}(\xi)\frac{\partial u(x,\xi)}{\partial\beta_j}\right)$. Thus

$$\begin{aligned} Pu^{(2)} = \Big(\sum_{i=0}^{2} (\varepsilon^i L_{xx}u_i + \varepsilon^{i-1}L_{x\xi}u_i + \varepsilon^{i-1}L_{\xi x}u_i + \varepsilon^{i-2}L_{\xi\xi}u_i) - f\Big)\Big|_{\xi=x/\varepsilon} = \\ = \big(\varepsilon^{-2}L_{\xi\xi}u_0 + \varepsilon^{-1}(L_{\xi\xi}u_1 + L_{\xi x}u_0 + L_{x\xi}u_0) + \\ + \varepsilon^0(L_{\xi\xi}u_2 + L_{\xi x}u_1 + L_{x\xi}u_1 + L_{xx}u_0 - f) + \\ + \varepsilon^1(L_{\xi x}u_2 + L_{x\xi}u_2 + L_{xx}u_1) + \varepsilon^2 L_{xx}u_2\big)\big|_{\xi=x/\varepsilon}. \end{aligned} \tag{6}$$

Equate the terms of orders ε^{-2}, ε^{-1}, ε^0 to zero. The terms of orders ε and ε^2 then make up the discrepancy of equation (1). So $u^{(2)}$ will satisfy equation (1) up to terms of order ε.

Thus we define the functions $u_0(x,\xi)$, $u_1(x,\xi)$, $u_2(x,\xi)$ from the conditions

$$L_{\xi\xi}u_0 = 0, \tag{7}$$
$$L_{\xi\xi}u_1 = -L_{\xi x}u_0 - L_{x\xi}u_0, \tag{8}$$
$$L_{\xi\xi}u_2 = -L_{\xi x}u_1 - L_{x\xi}u_1 - L_{xx}u_0 + f, \tag{9}$$

where x and ξ are regarded as independent variables. Let us formulate an analogue of Lemma 2.3.1. It is a corollary of Theorem 1 in the Supplement (see also [43]).

Lemma 1. *Let $A_{kj}(\xi)$, $F(\xi)$ be 1-periodic differentiable functions, $A_{kj}(\xi)$ satisfy conditions (3). Then a necessary and sufficient condition for a 1-periodic solution of the equation*

$$L_{\xi\xi}N \equiv \frac{\partial}{\partial \xi_h}\left(A_{hj}(\xi)\frac{\partial}{\partial \xi_j}N(\xi)\right) = F(\xi) \tag{10}$$

to exist is

$$\langle F\rangle \equiv \int_0^1 \dots \int_0^1 F(\xi)\, d\xi_1 \dots d\xi_s = 0.$$

Here the general 1-periodic solution of equation (10) is written as $N(\xi) = \overline{N}(\xi) + C$, where $\overline{N}(\xi)$ is a solution of (10) with zero mean over period: $\langle \overline{N}\rangle = 0$, C is an arbitrary constant.

Lemma 1 implies that the 1-periodic solution $u_0(x,\xi)$ of (7) is independent of ξ, that is,

$$u_0(x, \xi) = v_0(x). \tag{11}$$

Then (8) can be rewritten as

$$L_{\xi\xi}u_1 + \frac{\partial}{\partial \xi_h}\left(A_{hj}(\xi)\frac{\partial v_0}{\partial x_j}\right) = 0, \tag{12}$$

where x is a parameter.

The necessary and sufficient condition for the solvability of (12) in the 1-periodic function $u_1(x,\xi)$ holds:

$$\langle -L_{\xi x}v_0\rangle = \int_0^1 \dots \int_0^1 \left(-\frac{\partial}{\partial \xi_h}\left(A_{hj}(\xi)\frac{\partial v_0}{\partial x_j}\right)\right) d\xi_1 \dots d\xi_s =$$

$$= -\frac{\partial v_0}{\partial x_j}\left\langle \frac{\partial}{\partial \xi_h} A_{hj}\right\rangle = 0. \tag{13}$$

Equality (13) results from the 1-periodicity of the functions A_{kj}.

The solution of problem (12) can be represented as

$$u_1(x, \xi) = N_{i_1}(\xi)\frac{\partial v_0}{\partial x_{i_1}}, \tag{14}$$

where $N_{i_1}(\xi)(i_1 = 1, \dots, s)$ is a 1-periodic solution of the equation

$$L_{\xi\xi}N_{i_1} + \frac{\partial}{\partial \xi_k} A_{ki_1}(\xi) = 0,$$

or

$$L_{\xi\xi}(N_{i_1} + \xi_{i_1}) = 0. \tag{15}$$

Equation (15) is called the *equation on a periodic cell.* It must be solved to define the coefficients of the averaged equation. By Lemma 1 a solution of equation (15) is found up to an arbitrary constant. The choice of the constant is fixed by the condition $\langle N_{i_1} \rangle = 0$.

Further, considering relation (11) and (14), we rewrite (9) as

$$L_{\xi\xi}u_2(x, \xi) + \frac{\partial}{\partial \xi_k}(A_{kj}(\xi) N_{i_1}(\xi)) \frac{\partial^2 v_0}{\partial x_j \partial x_{i_1}} + \\ + \left(A_{kj}(\xi) \frac{\partial N_{i_1}(\xi)}{\partial \xi_j} \right) \frac{\partial^2 v_0}{\partial x_k \partial x_{i_1}} + A_{kj}(\xi) \frac{\partial^2 v_0}{\partial x_k \partial x_j} = f(x),$$

from which, by changing symbols of the indices, we get

$$L_{\xi\xi}u_2 + \left(\frac{\partial}{\partial \xi_k}(A_{ki_1}(\xi) N_{i_2}(\xi)) + A_{i_1 j}(\xi) \frac{\partial N_{i_2}}{\partial \xi_j} + A_{i_1 i_2}(\xi) \right) \frac{\partial^2 v_0}{\partial x_{i_1} \partial x_{i_2}} = f(x). \tag{16}$$

The necessary and sufficient condition for the solvability of (16) in the class of 1-periodic functions determines the averaged equation

$$\left\langle \frac{\partial}{\partial \xi_k}(A_{ki_1}(\xi) N_{i_2}(\xi)) + A_{i_1 j}(\xi) \frac{\partial N_{i_2}(\xi)}{\partial \xi_j} + A_{i_1 i_2}(\xi) \right\rangle \frac{\partial^2 v_0}{\partial x_{i_1} \partial x_{i_2}} = f(x). \tag{17}$$

Since $\left\langle \frac{\partial}{\partial \xi_k}(A_{ki_1}(\xi) N_{i_2}(\xi)) \right\rangle = 0$, because $A_{ki_1}(\xi) N_{i_2}(\xi)$ is 1-periodic, equation (17) can be rewritten as

$$\widehat{A}_{i_1 i_2} \frac{\partial^2 v_0}{\partial x_{i_1} \partial x_{i_2}} = f(x), \tag{18}$$

where

$$\widehat{A}_{i_1 i_2} = \left\langle A_{i_1 j}(\xi) \frac{\partial N_{i_2}(\xi)}{\partial \xi_j} + A_{i_1 i_2}(\xi) \right\rangle = \left\langle A_{i_1 j} \frac{\partial M_{i_2}}{\partial \xi_j} \right\rangle, \tag{19}$$

$M_i \equiv N_i + \xi$; the coefficients $\hat{A}_{i_1 i_2}$ of (18) are constant.

THEOREM 1. *The Coefficients $\hat{A}_{i_1 i_2}$ satisfy the conditions of symmetry $\hat{A}_{i_1 i_2} = \hat{A}_{i_2 i_1}$ and ellipticity $\hat{A}_{i_1 i_2 i_1 i_2} \geqslant \kappa_1 \eta_{i_1} \eta_{i_1}$.*

This Theorem follows from a more general statement that will be formulated and proved in §6.1, but we give the proof in the particular case under consideration for the sake of clearness.

PROOF: It follows from (15) that for any 1-periodic differentiable function $\varphi(\xi)$ the equality

$$\left\langle A_{kj}(\xi) \frac{\partial M_{i_2}}{\partial \xi_j} \frac{\partial \varphi}{\partial \xi_k} \right\rangle = 0, \qquad i_2 = 1, \ldots, s \tag{20}$$

holds.

By rewriting (19) as

$$\hat{A}_{i_1 i_2} = \left\langle A_{kj}(\xi) \frac{\partial M_{i_2}}{\partial \xi_j} \frac{\partial \xi_{i_1}}{\partial \xi_k} \right\rangle$$

and by adding it to (20) for $\varphi = N_{i_1}(\xi)$, we get

$$\hat{A}_{i_1 i_2} = \left\langle A_{kj}(\xi) \frac{\partial M_{i_2}}{\partial \xi_j} \frac{\partial M_{i_1}}{\partial \xi_k} \right\rangle. \tag{21}$$

Representation (21) and the symmetry of A_{kj} imply the symmetry of the coefficients $\hat{A}_{i_1 i_2}$. Let us prove the positive-definiteness of the matrix $\|\hat{A}_{i_1 i_2}\|$:

$$\hat{A}_{i_1 i_2} \eta_{i_1} \eta_{i_2} = \left\langle A_{kj}(\xi) \frac{\partial}{\partial \xi_j} (\eta_{i_2} M_{i_2}) \frac{\partial}{\partial \xi_k} (\eta_{i_1} M_{i_1}) \right\rangle \geqslant \varkappa_1 \eta_k \eta_k.$$

Due to the positive-definiteness of the matrix of $A_{kj}(\xi)$, we see that

$$A_{kj}(\xi) \frac{\partial}{\partial \xi_j} (\eta_{i_2} M_{i_2}) \frac{\partial}{\partial \xi_k} (\eta_{i_1} M_{i_1}) \geqslant \varkappa_1 \sum_{k=1}^{s} \left(\frac{\partial}{\partial \xi_k} (\eta_{i_1} M_{i_1}) \right)^2,$$

whence, by using the Cauchy–Buniakowski–Schwarz inequality, we get

$$\hat{A}_{i_1 i_2} \eta_{i_1} \eta_{i_2} \geqslant \varkappa_1 \sum_{k=1}^{s} \left\langle \left(\frac{\partial}{\partial \xi_k} (\eta_{i_1} M_{i_1}) \right)^2 \right\rangle \geqslant$$

$$\geqslant \varkappa_1 \sum_{k=1}^{s} \left\langle \frac{\partial}{\partial \xi_k} (\eta_{i_1} M_{i_1}) \right\rangle^2. \tag{22}$$

Since $N_{i_1}(\xi)$ is a periodic function with period 1, $\left\langle \frac{\partial}{\partial \xi_k} \eta_{i_1} N_{i_1} \right\rangle = 0$. Therefore the right side of (22) is

$$\varkappa_1 \sum_{k=1}^{s} \left\langle \eta_{i_1} \frac{\partial \xi_{i_1}}{\partial \xi_k} \right\rangle^2 = \varkappa_1 \sum_{k=1}^{s} \eta_k^2.$$

Thus the averaged equation (18) maintains the type of the original because it is elliptic.

Let v_0 be a solution of (18); as was mentioned above, equation (16) is solvable. Let us represent $f(x)$ in the form (18). Then $u_2(x,\xi)$ is a solution of the equation

$$L_{\xi\xi} u_2 + T_{i_1 i_2}(\xi) \frac{\partial^2 v_0}{\partial x_{i_1} \partial x_{i_2}} = \widehat{A}_{i_1 i_2} \frac{\partial^2 v_0}{\partial x_{i_1} \partial x_{i_2}},$$

where

$$T_{i_1 i_2} = \frac{\partial}{\partial \xi_k} \left(A_{k i_1}(\xi) N_{i_2} \right) + A_{i_1 j}(\xi) \frac{\partial N_{i_2}}{\partial \xi_j} + A_{i_1 i_2}(\xi).$$

By the same reasoning as in constructing $u_1(x,\xi)$, we get

$$u_2(x, \xi) = N_{i_1 i_2}(\xi) \frac{\partial^2 v_0}{\partial x_{i_1} \partial x_{i_2}}, \tag{23}$$

where the $N_{i_1 i_2}(\xi)$ are 1-periodic solutions of the equations

$$L_{\xi\xi} N_{i_1 i_2} + T_{i_1 i_2}(\xi) = \widehat{A}_{i_1 i_2}. \tag{24}$$

Therefore, if the $N_{i_1}(\xi)$ are solutions of equations (15), the $N_{i_1 i_2}(\xi)$ are solutions of equations (24), $v_0(x)$ is a solution of the averaged equation, and $v_0(x) \in C^\infty(\overline{G})$, then the function

$$u^{(2)} = v_0 + \varepsilon N_{i_1}\left(\frac{x}{\varepsilon}\right) \frac{\partial v_0}{\partial x_{i_1}} + \varepsilon^2 N_{i_1 i_2}\left(\frac{x}{\varepsilon}\right) \frac{\partial^2 v_0}{\partial x_{i_1} \partial x_{i_2}} \tag{25}$$

is a solution of the equation $Pu^{(2)} = \varepsilon r^{(2)}(x,\varepsilon)$; here $\varepsilon r^{(2)}(x,\varepsilon)$ is the sum of terms in (6) of orders ε^1 and ε^2. For $v_0(x) \in C^4(\overline{G})$, there exists a constant c, independent of ε, such that

$$|r^{(2)}(x, \varepsilon)| \leqslant c, \quad x \in G. \tag{26}$$

Substituting $u^{(2)}$ into the boundary condition (2) we see that the averaged equation (18) should be supplemented by the boundary condition

$$v_0|_{\partial G} = 0. \tag{27}$$

Since $\partial G \in C^\infty$, $f(x) \in C^\infty(\overline{G})$, then the solution of problems (18), (27) is infinitely differentiable in the domain G up to its boundary. Thus, estimate (26) is proved completely and

$$u^{(2)}|_{\partial G} = \varepsilon u_1(x, x/\varepsilon)|_{\partial G} + \varepsilon^2 u_2(x, x/\varepsilon)|_{\partial G}.$$

Let $\chi(x)$ be an infinitely differentiable function whose support (the domain where the function is nonzero) belongs to the interior of the ε-neighbourhood of the boundary of the domain $G, \chi|_{\partial G} = 1, |\chi| \leqslant 1$, $\|\varepsilon\partial\chi/\partial x_j\|_{C(\overline{G})} \leqslant c_1$, where c_1 is independent of ε. Then the function

$$a_k(x, \varepsilon) = A_{kj}\left(\frac{x}{\varepsilon}\right)\left(\varepsilon \frac{\partial}{\partial x_j}(\chi u_1) + \varepsilon^2 \frac{\partial}{\partial x_j}(\chi u_2)\right)$$

is bounded in absolute value by a constant independent of ε, and its support belongs to the interior of the ε-neighbourhood of the boundary of G, and so $\|a_k(x, \varepsilon)\|^2_{L_2(G)} = O(\varepsilon)$.

Consider the function

$$\tilde{u}^{(2)} = u^{(2)} - \chi(x)\left(\varepsilon u_1\left(x, \frac{x}{\varepsilon}\right) + \varepsilon^2 u_2\left(x, \frac{x}{\varepsilon}\right)\right). \tag{28}$$

It satisfies the boundary condition (2) exactly, and on the other hand,

$$P\tilde{u}^{(2)} = \varepsilon r^{(2)}(x, \varepsilon) - \frac{\partial}{\partial x_k} a_k(x, \varepsilon). \tag{29}$$

By subtracting equation (1) from (29), we get

$$\frac{\partial}{\partial x_k}\left(A_{kj}\left(\frac{x}{\varepsilon}\right)\frac{\partial(\tilde{u}^{(2)} - u)}{\partial x_j}\right) = \varepsilon r^{(2)}(x, \varepsilon) - \frac{\partial}{\partial x_k} a_k(x, \varepsilon),$$

whence, considering that $(\tilde{u}^{(2)} - u)|_{\partial G} = 0$, we obtain the estimate (Theorem 1.3.1)

$$\|\tilde{u}^{(2)} - u\|_{W_2^1(G)} \leqslant c_2\left(\varepsilon\|r^{(2)}(x, \varepsilon)\|_{L_2(G)} + \|a_k(x, \varepsilon)\|_{L_2(G)}\right) \leqslant c_3\sqrt{\varepsilon}, \tag{30}$$

where c_2, c_3 are constants independent of ε.

Representation (28) and estimate (30) imply that

$$\|u^{(2)} - u\|_{W_2^1(G)} \leqslant c_4 \sqrt{\varepsilon}, \tag{31}$$

$$\|u^{(1)} - u\|_{W_2^1(G)} \leqslant c_5 \sqrt{\varepsilon}, \tag{32}$$

$$\|v_0 - u\|_{L_2(G)} \leqslant c_6 \sqrt{\varepsilon}, \tag{33}$$

where

$$u^{(1)} = v_0 + \varepsilon N_{i_1}\left(\frac{x}{\varepsilon}\right)\frac{\partial v_0}{\partial x_{i_1}}, \tag{34}$$

with the constants c_4, c_5, c_6 being independent of ε.

Estimate (32) implies the closeness of the flows:

$$\left\| A_{kj}\left(\frac{x}{\varepsilon}\right)\frac{\partial u}{\partial x_j} - A_{kj}\left(\frac{x}{\varepsilon}\right)\frac{d}{dx_j}\left(v_0 + \varepsilon N_{i_1}\left(\frac{x}{\varepsilon}\right)\frac{\partial v_0}{\partial x_{i_1}}\right)\right\|_{L_2(G)} = O(\sqrt{\varepsilon}); \tag{35}$$

here d/dx_j is the total derivative with respect to x_j.

Thus, the following theorem was proved:

THEOREM 2. *Let the $N_{i_1}(\xi)$ be 1-periodic solutions of equations* (15), *the $N_{i_1 i_2}(\xi)$ 1-periodic solutions of equations* (24), *$v_0(x)$ a solution of the averaged problem* (18), (27), *let $u^{(1)}$, $u^{(2)}$ be defined by* (34), (25), *and u an exact solution of problem* (1), (2). *Then the estimates* (31)–(33), *and* (35) *hold.*

Note that the generalized maximum principle for an elliptic equation (see [82]) implies $\|u^{(2)} - u\|_{C(\overline{G})} = O(\varepsilon)$, from which $\|v_0 - u\|_{C(\overline{G})} = O(\varepsilon)$.

2. Let the $A_{kj}(\xi)$ be 1-periodic piecewise smooth functions, that is the $A_{kj}(\xi)$ are infinitely differentiable except on some smooth nonintersecting surfaces Σ_1, everywhere up to these surfaces (only a finite number of the Σ_1 intersect the unit cube), and on the Σ_1, the coefficients $A_{kj}(\xi)$ and all their derivatives may suffer first-order discontinuities. It is assumed that the $A_{kj}(\xi)$ still satisfy conditions (3). Problem (1), (2) is considered in the bounded domain G with a smooth boundary, where $f(x) \in C^\infty(\overline{G})$. In this case, either the solution is understood in the generalized sense, or equation (1) is regarded as defined only off the surfaces of discontinuity of the coefficients, whereas on the surfaces, the following contact conditions hold:

$$[u] = 0, \tag{36}$$

$$\left[\frac{\partial u}{\partial \nu}\right] = 0, \tag{37}$$

where $\partial u/\partial \nu$ is a derivative with respect to the conormal:

$$\frac{\partial u}{\partial \nu} = \sum_{k,j=1}^{s} A_{kj} \frac{\partial u}{\partial x_j} n_k.$$

Again we will try to find asymptotics in the form of (4), where $u_i(x, \xi)$ are 1-periodic functions smooth in x. By substituting (4) into equation (1) everywhere off the coefficients' discontinuity surfaces, we get relations (7)–(9) for $\xi \notin \Sigma_l$.

Let us substitute (4) into the left side of the relations on discontinuities (36), (37). We set

$$\frac{\partial}{\partial \nu_\alpha} u(x, \xi) = \sum_{k,j=1}^{s} A_{kj}(\xi) \frac{\partial u(x, \xi)}{\partial \alpha_j} n_k.$$

We have

$$[u^{(2)}] = \big([u_0(x, \xi)]\,|_{\xi\in\Sigma_l} + \varepsilon\,[u_1(x, \xi)]\,|_{\xi\in\Sigma_l} + \varepsilon^2\,[u_2(x, \xi)]\,|_{\xi\in\Sigma_l}\big)\big|_{x=\varepsilon\xi}, \tag{38}$$

$$\left[\frac{\partial u^{(2)}}{\partial \nu}\right] = \left(\varepsilon^{-1}\left[\frac{\partial u_0}{\partial \nu_\xi}\right]\Big|_{\xi\in\Sigma_l} + \left[\frac{\partial u_1}{\partial \nu_\xi} + \frac{\partial u_0}{\partial \nu_x}\right]\Big|_{\xi\in\Sigma_l} + \varepsilon\left[\frac{\partial u_2}{\partial \nu_\xi} + \frac{\partial u_1}{\partial \nu_x}\right]\Big|_{\xi\in\Sigma_l} + \varepsilon^2\left[\frac{\partial u_2}{\partial \nu_x}\right]\Big|_{\xi\in\Sigma_l}\right)\Big|_{x=\varepsilon\xi}. \tag{39}$$

We require that three highest terms in each of the relations (38), (39) vanish, with ξ and x being considered as independent variables. Then

$$[u_0(x, \xi)]\,|_{\xi\in\Sigma_l} = 0, \tag{40}$$

$$\left[\frac{\partial u_0}{\partial \nu_\xi}\right]\Big|_{\xi\in\Sigma_l} = 0, \tag{41}$$

$$[u_1(x, \xi)]\,|_{\xi\in\Sigma_l} = 0, \tag{42}$$

$$\left[\frac{\partial u_1}{\partial \nu_\xi} + \frac{\partial u_0}{\partial \nu_x}\right]\Big|_{\xi\in\Sigma_l} = 0, \tag{43}$$

$$[u_2(x, \xi)]\,|_{\xi\in\Sigma_l} = 0, \tag{44}$$

$$\left[\frac{\partial u_2}{\partial \nu_\xi} + \frac{\partial u_1}{\partial \nu_x}\right]\Big|_{\xi\in\Sigma_l} = 0. \tag{45}$$

We shall need the following analogue of Lemma 1.

LEMMA 2. *Let the $A_{kj}(\xi)$, $F_0(\xi)$, $F_k(\xi)$ be 1-periodic piecewise smooth functions, and let the $A_{kj}(\xi)$ satisfy condition (3). Then a necessary and sufficient condition for a 1-periodic solution of the problem*

$$L_{\xi\xi}N = F_0(\xi) + \frac{\partial}{\partial \xi_k} F_k(\xi) \tag{46}$$

to exist for $\xi \notin \Sigma_l$,

$$[N]\,|_{\xi\in\Sigma_l} = 0, \tag{47}$$

$$\left[\left(A_{kj}(\xi)\frac{\partial N}{\partial \xi_j} - F_k(\xi)\right)n_k\right]\Big|_{\xi\in\Sigma_l} = 0 \tag{48}$$

is $\langle F_0\rangle = 0$.

The general 1-periodic solution of problems (46)–(48) is written as $N(\xi) = \overline{N}(\xi) + C$, where $\overline{N}(\xi)$ is a solution with the zero mean over the period and C is an arbitrary constant. Lemma 2 follows from Theorem 1 of the Supplement.

The conditions of Lemma 2 are satisfied for problem (7) when $\xi \notin \Sigma_1$ with (40), (41) and so only functions independent of ξ, that is $u_0(x,\xi) = v_0(x)$, are its solutions.

Lemma 2 implies that equation (12) ($\xi \notin \Sigma_1$) with the contact conditions (42), (43) has the 1-periodic solution $u_i(x,\xi)$, and $u_1(x,\xi)$ can be represented as (14), where the $N_{i_1}(\xi)$ ($i_1 = 1,\dots,s$) are 1-periodic generalized solutions of equations (15), that is

$$L_{\xi\xi}(N_{i_1} + \xi_{i_1}) = 0, \quad \xi \notin \Sigma_l, \tag{49}$$

$$[N_{i_1}]\,|_{\xi\in\Sigma_l} = 0, \tag{50}$$

$$\left[\frac{\partial (N_{i_1} + \xi_{i_1})}{\partial \nu_\xi}\right]\Big|_{\xi\in\Sigma_l} = 0. \tag{51}$$

By representing (9) as (16), and relation (45) as

$$\left[\left(A_{kj}(\xi)\frac{\partial u_2}{\partial \xi_j} + A_{ki_1}N_{i_2}(\xi)\frac{\partial^2 v_0}{\partial x_{i_1}\partial x_{i_2}}\right)n_k\right]\Big|_{\xi\in\Sigma_l} = 0,$$

we see that for a solution of problem (9), (44), (45) to exist and be unique, it is necessary and sufficient that condition (18) hold, with the $\hat{A}_{i_1 i_2}$ being defined by (19). Theorem 1 remains valid for the coefficients $\hat{A}_{i_1 i_2}$.

The function $u_2(x,\xi)$ can be represented as (23), where the $N_{i_1 i_2}(\xi)$ are 1-periodic generalized solutions of equation (24).

By substituting $u^{(2)}$ into the boundary condition (2), we obtain the boundary condition for v_0 (27). Since the $\hat{A}_{i_1 i_2}$ are constant, $f(x) \in C^\infty(\overline{G})$, $\partial G \in C^\infty$, it follows that $v_0(x) \in C^\infty(\overline{G})$, and so $u^{(2)}$ is a solution of the problem

$$Pu^{(2)} = \varepsilon r^{(2)}(x, \varepsilon), \quad x \in G, \quad x/\varepsilon \notin \Sigma_l,$$
$$[u^{(2)}]\,|_{x/\varepsilon\in\Sigma_l} = 0,$$
$$\left[\frac{\partial u^{(2)}}{\partial \nu} + \varepsilon^2 A_{kj}\left(\frac{x}{\varepsilon}\right)\frac{\partial u_2}{\partial x_j} n_k\right]\Bigg|_{x/\varepsilon\in\Sigma_l} = 0,$$
$$u^{(2)}\,|_{\partial G} = \varepsilon u_1\,|_{\partial G} + \varepsilon^2 u_2\,|_{\partial G},$$

where

$$r^{(2)}(x, \varepsilon) = r_0^{(2)}(x, \varepsilon) + \varepsilon\partial r_k^{(2)}(x, \varepsilon)/\partial x_k,$$
$$r_0^{(2)}(x, \varepsilon) = (L_{x\xi}u_2 + L_{xx}u_1)\,|_{\xi=x/\varepsilon},$$
$$r_k^{(2)}(x, \varepsilon) = A_{kj}\left(\frac{x}{\varepsilon}\right)\frac{\partial u_2}{\partial x_j}\left(x, \frac{x}{\varepsilon}\right),$$
$$|r_0^{(2)}| \leqslant c, \quad |r_k^{(2)}| \leqslant c, \quad x \in G,$$

c is independent of ε.

Let us define $u^{(2)}$ from formula (28); the function $w(x) = u^{(2)} - u$ is a solution of the problem

$$\frac{\partial}{\partial x_k}\left(A_{kj}\left(\frac{x}{\varepsilon}\right)\frac{\partial w}{\partial x_j}\right) =$$
$$= \varepsilon r_0^{(2)}(x, \varepsilon) + \frac{\partial}{\partial x_k}\left(\varepsilon^2 r_k^{(2)}(x, \varepsilon) - a_k(x, \varepsilon)\right), \quad \frac{x}{\varepsilon} \notin \Sigma_l, \tag{52}$$

$$[w]\,|_{x/\varepsilon\in\Sigma_l} = 0,$$
$$\left[\frac{\partial w}{\partial \nu} + \left(\varepsilon^2 r_k^{(2)} - a_k\right)n_k\right]\Bigg|_{x/\varepsilon\in\Sigma_l} = 0, \tag{53}$$
$$w\,|_{\partial G} = 0,$$

where

$$\|a_k(x, \varepsilon)\|^2_{L_2(G)} = O(\varepsilon).$$

As in the one-dimensional case (§3.1), the derivative's discontinuity on the conormal agrees with the divergent term on the right side:

$$\left[\frac{\partial \widetilde{u}^{(2)}}{\partial \nu}\right]_{x/\varepsilon\in\Sigma_l} = \left[\left(\varepsilon^2 r_k^{(2)} - a_k\right)n_k\right]\,|_{x/\varepsilon\in\Sigma_l}.$$

This means that w is the generalized solution of problem (52), (53) and Theorem 1.3.1 implies estimate (30); consequently, all the estimates of Theorem 2 (the N_{i_1} and $N_{i_1 i_2}$ are the generalized solutions of equations (15), (24)).

The results of the present section are extended to domains with piecewise smooth boundaries on the following conditions.

The solution $v_0(x)$ of problem (18), (27) has all the derivatives up to the 4th order including, continuous up to the boundary of G. It suffices to require just the continuity of the third-order derivatives [8] to justify asymptotics by the integral identity technique (see §8.1).

The requirement that the function χ be infinitely differentiable is unimportant; it suffices that it be continuous and piecewise smooth. Such a function χ can be constructed for any domain with a piecewise smooth boundary, for example,

$$\chi = \omega\left(\frac{\rho(x, \partial G)}{\varepsilon}\right),$$

where $\rho(x, \partial G)$ is the distance from x to ∂G,

$$\omega(t) = \begin{cases} 1 - t & \text{for } 0 \leqslant t \leqslant 1, \\ 0 & \text{for } t \geqslant 1. \end{cases}$$

Therefore, to construct the asymptotics v_0 or $u^{(1)}$ of the solution of problems (1), (36), (37), and (2), one must:

– solve cell problems (49)–(51) in the class of 1-periodic functions;

– calculate the effective heat conductivity tensor by formula (19);

– solve equation (18) with constant coefficients and boundary condition (27).

Notice that the asymptotics of v_0 traces only macroeffects, whereas the asymptotics of $u^{(1)}$ reflects also the microscopic variation of the field, which is imposed on v_0.

Numerical methods such as finite-difference or variate difference techniques (e.g., the finite element method) are used to solve problems (49)–(51) and (18), (27).

The periodicity conditions in (49)–(51) are nonstandard: as a rule, the sets of programs available are not designed for such conditions. However, in the case when the periodic cell is symmetric (see §6.3), the periodicity conditions can be replaced by more common boundary conditions. If a composite material consists of isotropic components and the periodic cell resembles the one shown in Figure 10, the problems (49)–(51) reduce to a problem in a quarter of the cell $\{(\xi_1, \xi_2) | 0 < \xi_1 < 1/2,\ 0 < \xi_2 < 1/2\}$:

$$\begin{gathered} L_{\xi\xi} M_1 = 0, \quad \xi \notin C, \\ [M_1]|_C = \left[K(\xi)\frac{\partial M_1}{\partial n}\right]\bigg|_C = 0, \\ M_1(0, \xi_2) = 0, \quad M_1(1/2, \xi_2) = 1/2, \\ \frac{\partial M_1}{\partial \xi_2}(\xi_1, 0) = \frac{\partial M_1}{\partial \xi_2}\left(\xi_1, \frac{1}{2}\right) = 0. \end{gathered} \tag{54}$$

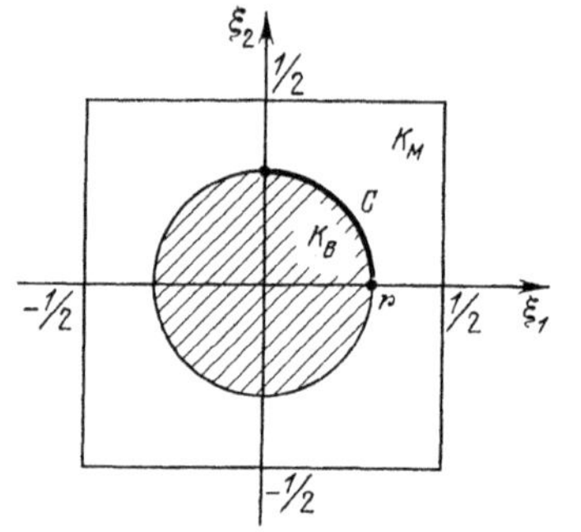

Figure 10

Here $K(x/\varepsilon)$ is the heat conductivity coefficient at x,

$$K(\xi) = \begin{cases} K_B & \text{for } \sqrt{\xi_1^2+\xi_2^2} < r, \\ K_M & \text{for } \sqrt{\xi_1^2+\xi_2^2} > r, \end{cases}$$

where C is the arc of the circle with radius r. Due to the symmetry of problems (49)–(51), $M_1(\xi_1,\xi_2) = M_2(\xi_2,\xi_1)$; odd continuation of M_1 in ξ_1 and M_2 in ξ_2 and even continuation of M_1 in ξ_2 and of M_2 in ξ_1 over the whole periodic cell $\{(\xi_1,\xi_2)| -1/2 < \xi_1 < 1/2, -1/2 < \xi_2 < 1/2\}$ yield the functions $N_1+\xi_1$ and $N_2+\xi$, respectively, that is, $N_1(\xi) = M_1(\xi) - \xi_1$, $N_2(\xi) = M_2(\xi) - \xi_2$.

Note that it is not the functions N_1 and N_2 themselves but their derivatives, or rather $K(\xi)\partial M_{i_2/\partial\xi_{i_1}}$, are needed to calculate effective coefficients:

$$\widehat{A}_{i_1 i_2} = \int_{-1/2}^{1/2}\int_{-1/2}^{1/2} K(\xi_1, \xi_2)\frac{\partial M_{i_2}}{\partial \xi_{i_1}}\, d\xi_1 d\xi_2,$$

and in our case, due to the symmetry of problems (49)–(51),

$$\widehat{A}_{i_1 i_2} = \delta_{i_1 i_2}\widehat{K},$$

where

$$\widehat{K} = 4\int_0^{1/2}\int_0^{1/2} K(\xi_1, \xi_2)\frac{\partial M_1}{\partial \xi_1}\, d\xi_1 d\xi_2 \tag{55}$$

is the effective coefficient of heat conductivity. So the medium shown in Fig. 1 is isotropic if its components are isotropic and it suffices to solve only problem (54) to find $\widehat{K}$.

The author of [190] investigated the relationship between the effective heat conductivity coefficient $\widehat{K}$ and the shape of the fiber cross-section, with the constant area $\mathcal{B}$ (concentration of the fibers) being 0.25, $K_B = 10$, $K_M = 1$. Some of the results are given in Fig. 11.

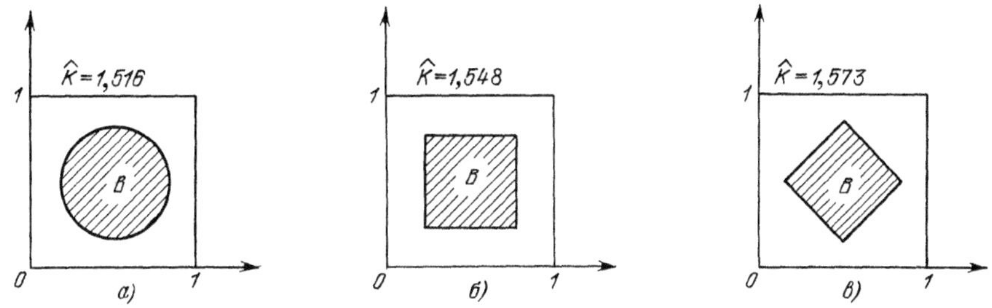

Figure 11

In the same work, a numerical solution of problems (1) and (2) was compared with the asymptotic solutions v_0, $u^{(1)}$, and $u^{(2)}$ for various values of ε ($\varepsilon = 0.5$, $\varepsilon = 0.25$, $\varepsilon = 0.125$) on the structure of the type shown in Fig. 11b (square), and the domain $G = (0,1) \times (0,1)$ (Figs. 12 and 13).

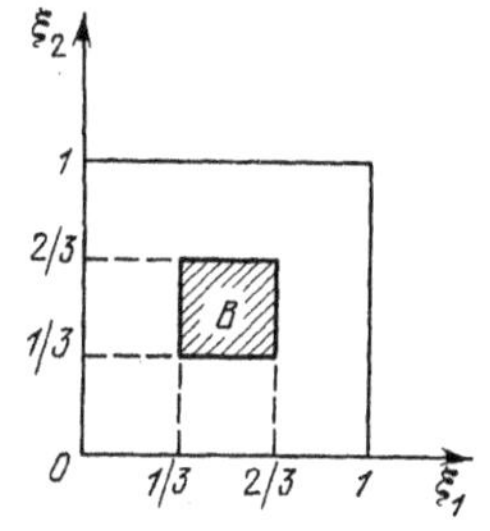

Figure 12

Problem (1), (2) was considered for $f \equiv 10$, $A_{kj}(\xi) = \delta_{kj} K(\xi)$, where

$$K(\xi) = \begin{cases} \lambda & \text{for } \xi \in \mathcal{B}, \\ 1 & \text{for } \xi \in Q \setminus \mathcal{B}, \end{cases}$$

$$\mathcal{B} = (1/4, 3/4) \times (1/4, 3/4), \quad Q = (0,1) \times (0,1).$$

The finite element method was used for numerical solution of (1), (2), which involved splitting the domain G into several thousands of triangles (from 1152 to 18432, depending on the parameters of a problem; see [190] for detail). The numerical solution was compared with the asymptotic v_0, $u^{(1)}$, $u^{(2)}$ in the norms C, L_2, W_2^1. The results of the comparison for various values of ε and λ are listed in Tables 2–4. Figures 14–16 show graphs of numerical asymptotic solutions on the section $x_1 = x_2$, see Fig. 13. (It can be seen from the results of the calculations that only the approximation $u^{(2)}$ appears to be acceptable for $\lambda = 1/114$.)

REMARK 1: Due to the periodicity of all the data for problem (49)–(51), it can be considered on the cube $Q = (0,1)^s$. In this case, the periodic solution of problem (49)–(51) given for Q is understood as a function satisfying the conditions of (49)–(51) for $\xi \in Q$ and the conditions

$$N_{i_1}\big|_{\xi_i=0} = N_{i_1}\big|_{\xi_i=1}, \quad \frac{\partial N_{i_1}}{\partial \xi_j}\bigg|_{\xi_i=0} = \frac{\partial N_{i_1}}{\partial \xi_j}\bigg|_{\xi_i=1}$$

for $i, j = 1, \ldots, s$ (the values of the derivatives for $\xi_i = 0$ are assumed to have been found).

REMARK 2: The following estimate was proved in [217]: Let $A(\xi) = \|A_{kj}(\xi)\|$ be an $s \times s$ matrix of the coefficients of equation (1), and let $\hat{A} = \|\hat{A}_{kj}\|$ be an $s \times s$ matrix of the effective coefficients, then

$$\langle A^{-1}\rangle^{-1} \leqslant \widehat{A} \leqslant \langle A\rangle. \tag{56}$$

Table 2

Convergence of the difference $u - v_0$ for $\lambda = 10$, $\lambda = 1/114$ [190]

	$\lambda=10$			$\lambda=1/114$		
ε	$\sup\lvert u - v_0\rvert$	$\dfrac{\Vert u - v_0\Vert_{L_2}}{\Vert v_0\Vert_{L_2}}$	$\dfrac{\Vert u - v_0\Vert_{W_2^1}}{\Vert v_0\Vert_{W_2^1}}$	$\sup\lvert u - v_0\rvert$	$\dfrac{\Vert u - v_0\Vert_{L_2}}{\Vert v_0\Vert_{L_2}}$	$\dfrac{\Vert u - v_0\Vert_{W_2^1}}{\Vert v_0\Vert_{W_2^1}}$
1/2	0.128	0,1	0.310	2.24	0.56	0.973
1/4	0.072	0.054	0,357	0,554	0.187	0.92
1/8	0.040	0.028	0.376	0.133	0,049	0,77

Table 3

Convergence of the difference $u - u^{(1)}$ for $\lambda = 10$, $\lambda = 1/114$ [190]

	$\lambda=10$			$\lambda=1/114$		
ε	$\sup\lvert u - u^{(1)}\rvert$	$\dfrac{\Vert u - u^{(1)}\Vert_{L_2}}{\Vert u^{(1)}\Vert_{L_2}}$	$\dfrac{\Vert u - u^{(1)}\Vert_{W_2^1}}{\Vert u^{(1)}\Vert_{W_2^1}}$	$\sup\lvert u - u^{(1)}\rvert$	$\dfrac{\Vert u - u^{(1)}\Vert_{L_2}}{\Vert u^{(1)}\Vert_{L_2}}$	$\dfrac{\Vert u - u^{(1)}\Vert_{W_2^1}}{\Vert u^{(1)}\Vert_{W_2^1}}$
1/2	0.020	0.032	0.085	2,24	0.56	0.952
1/4	0.0097	0.012	0.050	0.554	0.183	0.889
1/8	0.0033	0.004	0.033	0.133	0.042	0.707

Table 4

Convergence of the difference $u - u^{(2)}$ for $\lambda = 1/114$ [190]

ε	$\sup\lvert u - u^{(2)}\rvert$	$\dfrac{\Vert u - u^{(2)}\Vert_{L_2}}{\Vert u^{(2)}\Vert_{L_2}}$	$\dfrac{\Vert u - u^{(2)}\Vert_{W_2^1}}{\Vert u^{(2)}\Vert_{W_2^1}}$
1/2	0.181	0.031	0.181
1/4	0.011	0.012	0.012
1/8	0.013	0.012	0.034

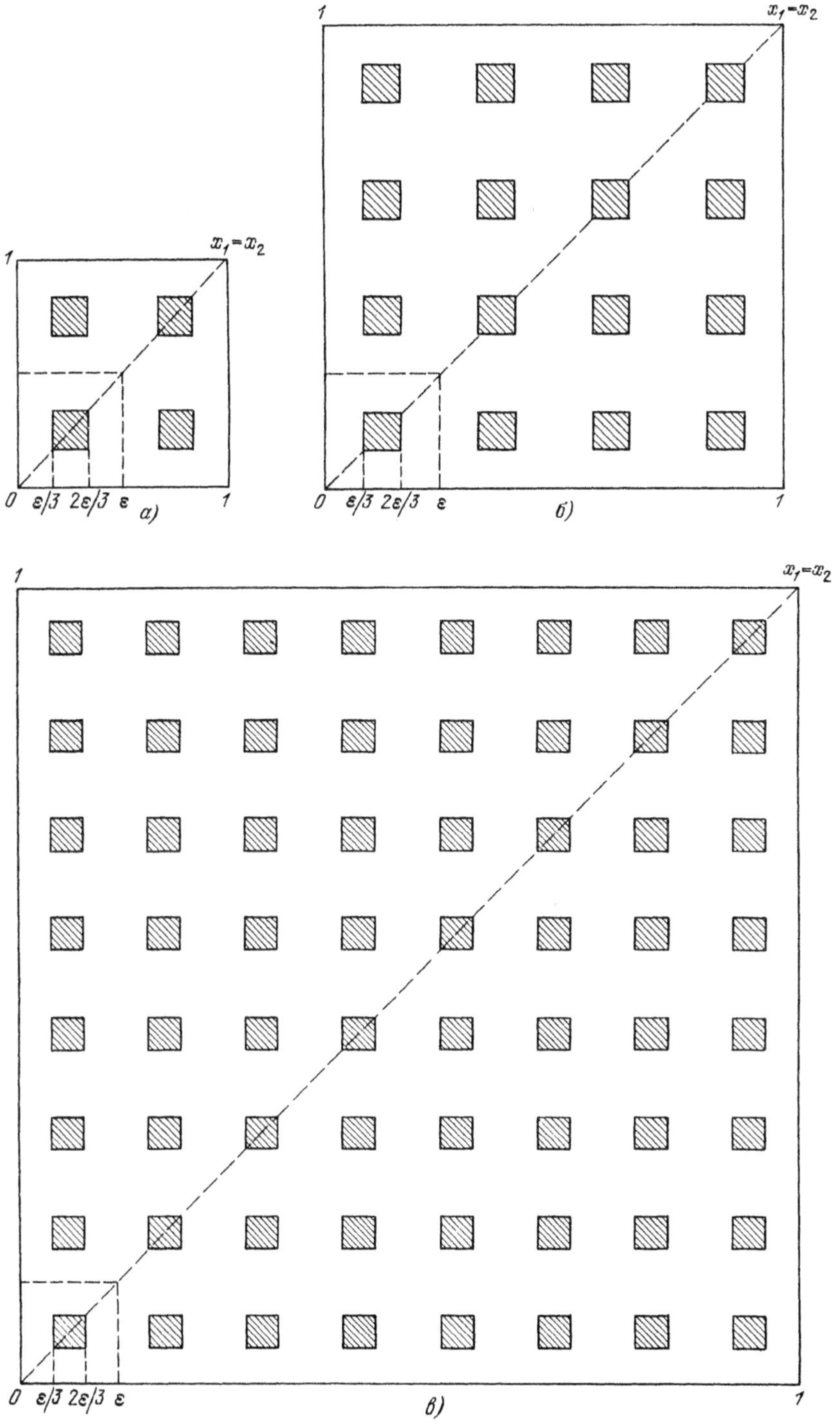

Figure 13

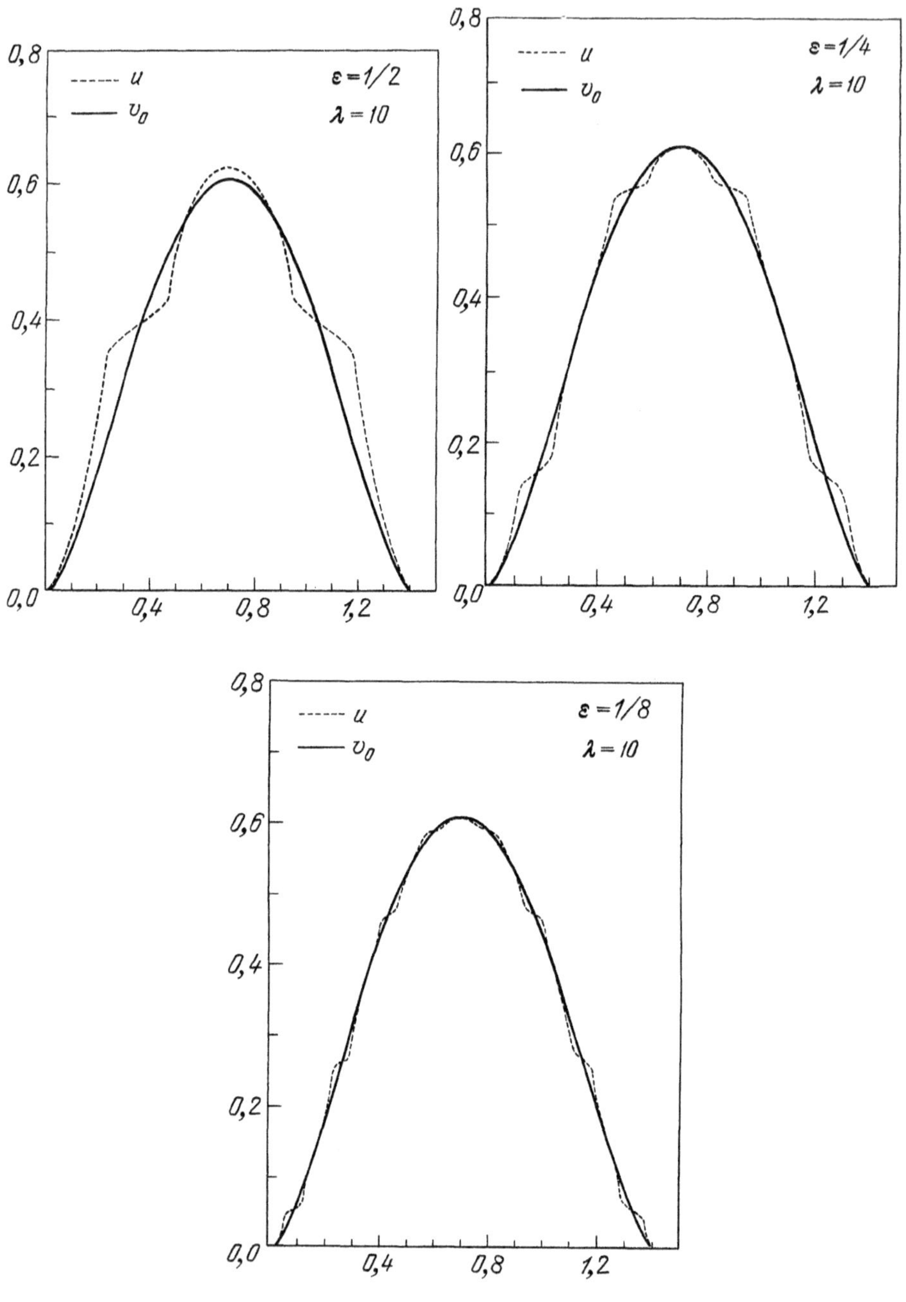

Figure 14a

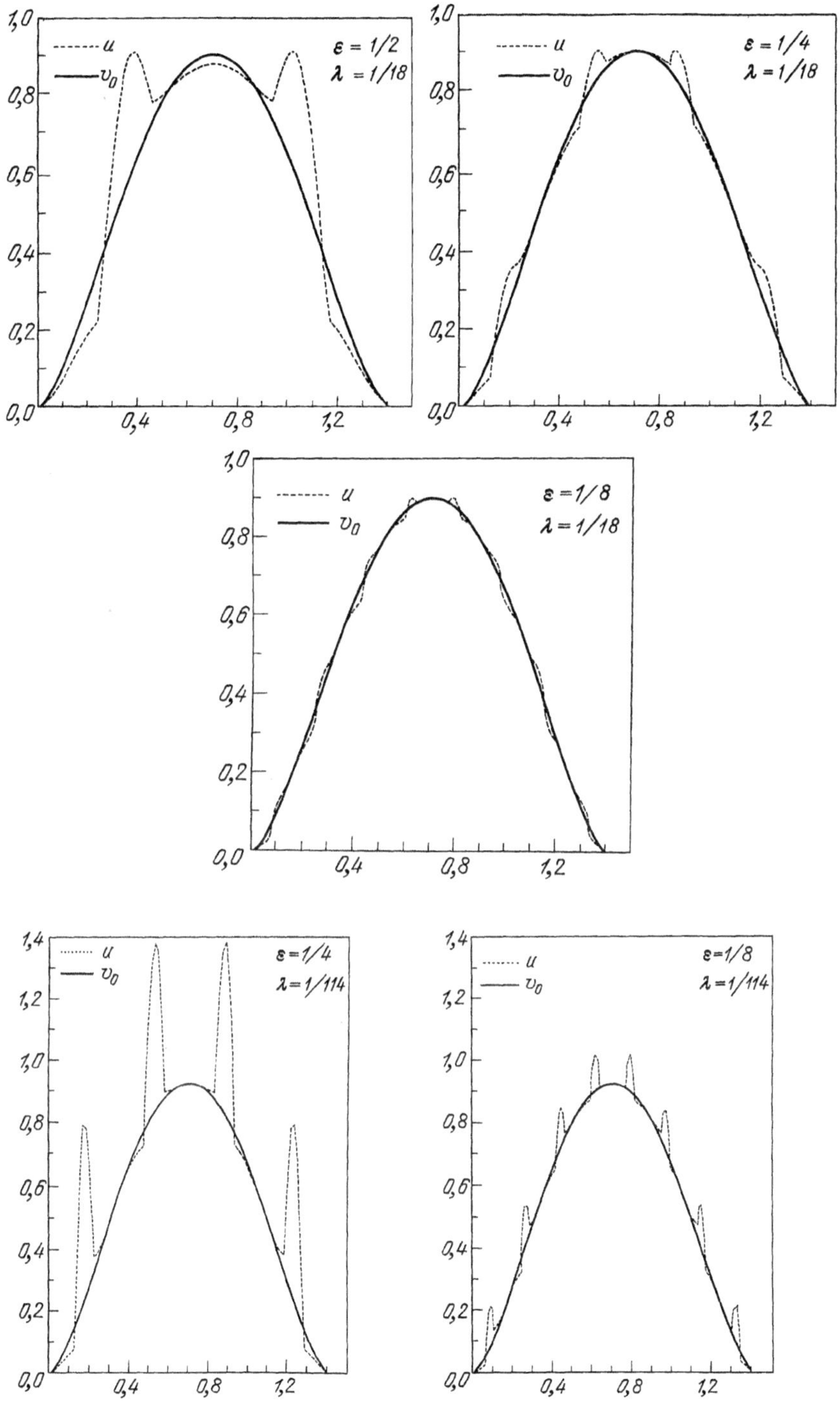
u
v0
ε = 1/2
λ = 1/18
ε = 1/4
λ = 1/18
ε = 1/8
λ = 1/18
ε = 1/4
λ = 1/114
ε = 1/8
λ = 1/114
0,0
0,2
0,4
0,6
0,8
1,0
1,2
1,4
0,4
0,8
1,2

Figure 14b

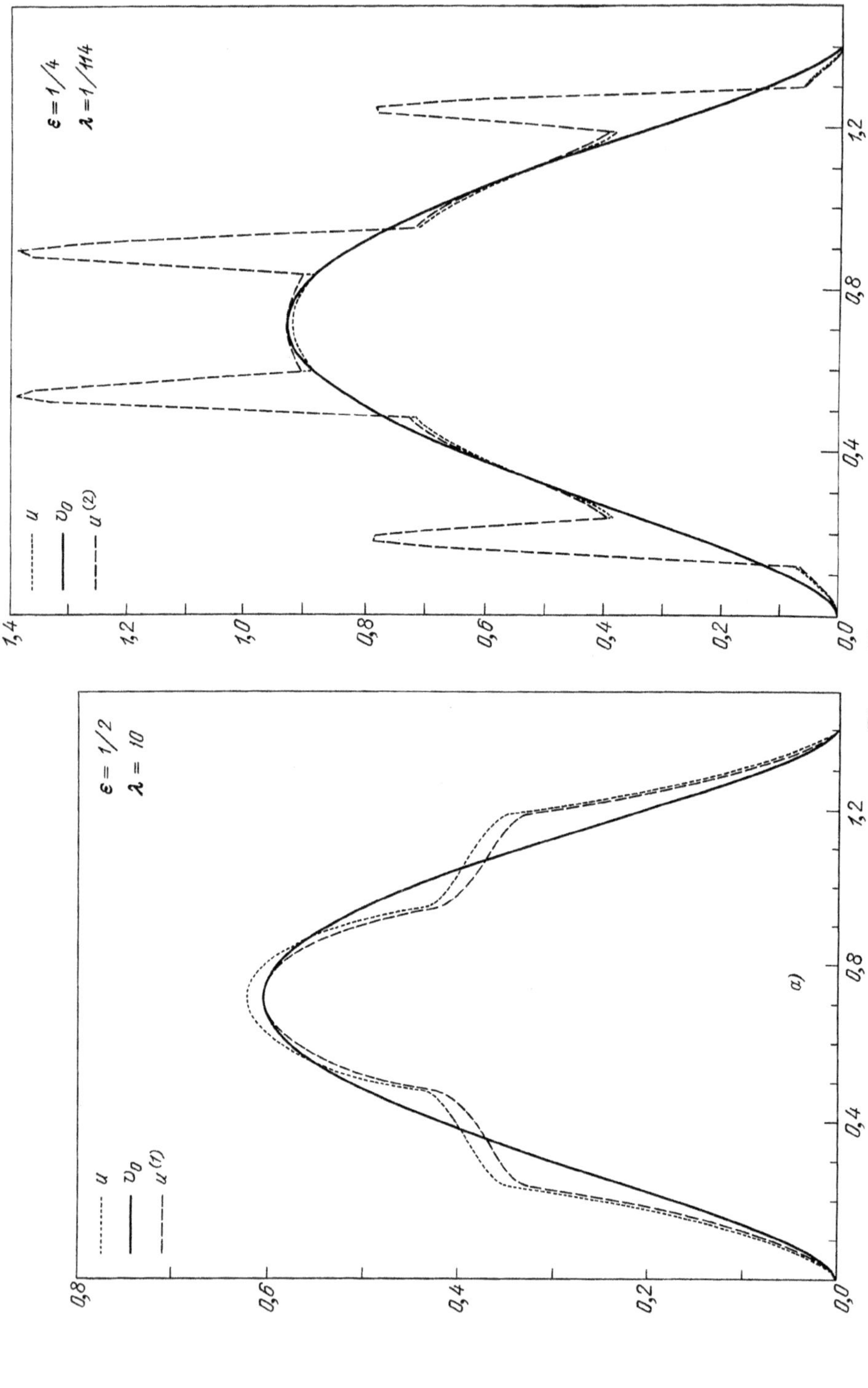

u
v_0
$u^{(2)}$
ε = 1/4
λ = 1/114
0,0
0,2
0,4
0,6
0,8
1,0
1,2
1,4
0,4
0,8
1,2
u
v_0
$u^{(1)}$
ε = 1/2
λ = 10
a)
0,0
0,2
0,4
0,6
0,8
0,4
0,8
1,2

Figure 15

Remark 3: We can give an estimate for the coefficients $\hat{A}_{qq}$ in the case of $A_{kj}(\xi) = \delta_{kj}K(\xi)$, $s = 3$, which is more exact than (56). (Other estimates, more exact than (56), can be found in [89, 90].) We denote $\int_0^1 \cdot d\xi_l$ by $\langle\cdot\rangle$ and let $W_2^{1(j)}$ be the class of functions $M(\xi)$ satisfying the conditions $M(\xi) \in W_2^1(G)$ for any $G \subset \mathbf{R}^3$, while $M(\xi)$ is periodic in all the ξ_l except ξ_j, $M|_{\xi_j=1} - M|_{\xi_j=0} = 1$ (with the values of the other ξ_l fixed).

Formula (21) means that $\hat{A}_{qq} = \left\langle K(\xi)\frac{\partial M_q}{\partial \xi_k}\frac{\partial M_q}{\partial \xi_k}\right\rangle$, where M_q belongs to $W_2^{1(q)}$ and is a generalized solution of the equation $L_{\xi\xi}M_q = 0$. This is the Euler equation for the functional $\left\langle K(\xi)\frac{\partial M}{\xi_k}\frac{\partial M}{\partial \xi_k}\right\rangle$ and consequently

$$\hat{A}_{qq} = \inf_{M \in W_2^{1(q)}} \left\langle K(\xi)\frac{\partial M}{\partial \xi_k}\frac{\partial M}{\partial \xi_k}\right\rangle.$$

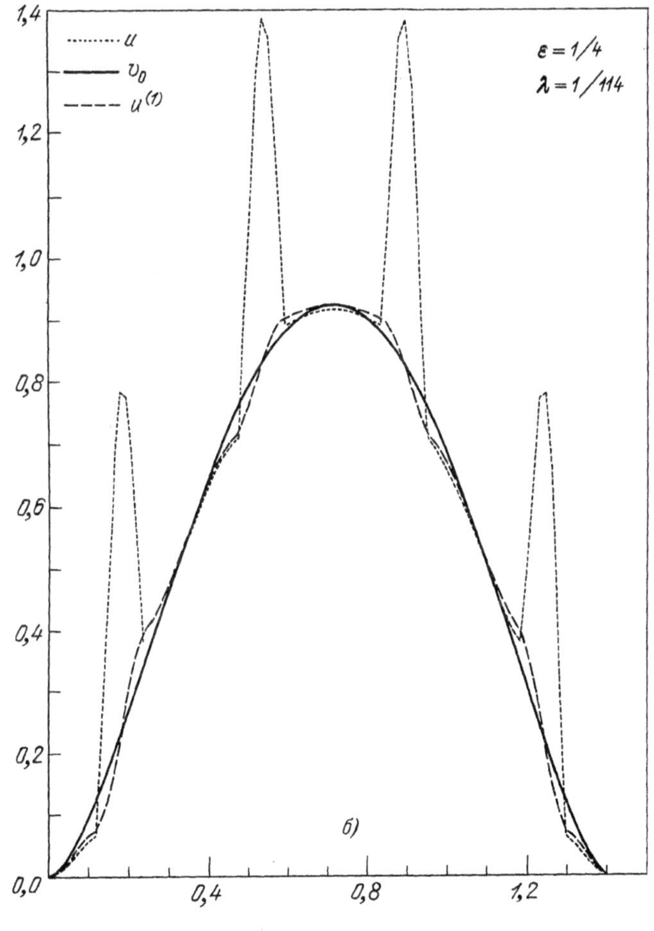

Figure 16

For definiteness let us estimate $\hat{A}_{11}$:

$$\hat{A}_{11}=\inf_{M\in W_2^{1(1)}}\left\langle K(\xi)\frac{\partial M}{\partial\xi_k}\frac{\partial M}{\partial\xi_k}\right\rangle\geqslant A_{11}^{-}=\inf_{M\in W_2^{1(1)}}\left\langle K(\xi)\left(\frac{\partial M}{\partial\xi_1}\right)^2\right\rangle.$$

By solving the Euler equation, we see that the lower bound of A_{11}^{-} is achieved on M, such that

$$\frac{\partial M}{\partial\xi_1}=K^{-1}(\xi)\langle K^{-1}(\xi)\rangle_1^{-1}. \tag{57}$$

By substituting (57) into the expression of A_{11}^{-}, we get

$$A_{11}^{-}=\langle\langle\langle K^{-1}(\xi)\rangle_1^{-1}\rangle_2\rangle_3.$$

The upper estimate is produced by minimizing

$$\left\langle K(\xi)\frac{\partial M}{\partial\xi_k}\frac{\partial M}{\partial\xi_k}\right\rangle$$

for the functions which depend only on ξ_1: $M=M(\xi_1)$, rather than for all the $M\in W_2^{1(1)}$.

We have

$$\hat{A}_{11}\leqslant\hat{A}_{11}^{+}=\inf_{M(\xi_1)\in W_2^{1(1)}}\left\langle K(\xi)\left(\frac{\partial M}{\partial\xi_1}\right)^2\right\rangle=$$

$$=\inf_{M(\xi_1)\in W_2^{1(1)}}\left\langle\langle\langle K(\xi)\rangle_2\rangle_3\left(\frac{\partial M}{\partial\xi_1}\right)^2\right\rangle_1.$$

Thanks to Euler's equation, we obtain $A_{11}^{+}=\langle\langle\langle K(\xi)\rangle_2\rangle_3^{-1}\rangle_1^{-1}$. Thus,

$$\langle\langle\langle K^{-1}(\xi)\rangle_1^{-1}\rangle_2\rangle_3\leqslant\hat{A}_{11}\leqslant\langle\langle\langle K(\xi)\rangle_2\rangle_3^{-1}\rangle_1^{-1}.$$

It follows from the Cauchy–Buniakowski–Schwarz inequality that for $K(\xi)\not\equiv K(\xi_1)$, always $\langle K^{-1}(\xi)\rangle^{-1}<A_{11}^{-}$, and for $K(\xi)\not\equiv K(\xi_2,\xi_3)$ always $A_{11}^{+}<K(\xi)$.

REMARK 4: Consider the fibrous composite whose cross-section is shown in Fig. 1. Let the axis $0x_3$ be colinear to the fibers. Following the averaging procedure described above, we can see that the effective heat conductivity tensor is of the form $\hat{A}_{i_1i_2}=\delta_{i_1i_2}\widehat{K}_{i_1}$, where $\widehat{K}_{i_1}$ coincides with the coefficient $\widehat{K}$ calculated above (see (55)) for $i_1=1,2$ (i.e., in the directions orthogonal to the fibers, the heat conductivity coefficient equals $\widehat{K}$), and $\widehat{K}_3=\langle K(\xi)\rangle$. The latter equality follows from $N_3(\xi)\equiv 0$. Thus, the conductivity coefficient along the fibers equals the usual mean of $K(\xi)$.

§2 Asymptotic Expansion of Solution of the Stationary Heat Conduction Problem

The technique described in Chapter 2 made it possible to obtain not only zero approximation to a solution but also its asymptotic expansion. In this section, we shall illustrate the method used to carry out this expansion by means of an elementary multi-dimensional example.

The main objective of the section is to construct and justify an asymptotic expansion of a solution of the equation

$$L_\varepsilon u \equiv \frac{\partial}{\partial x_k}\left(A_{kj}\left(\frac{x}{\varepsilon}\right)\frac{\partial u}{\partial x_j}\right) = f(x), \quad x \in \mathbf{R}^s, \tag{1}$$

where the coefficients $A_{kj}(\xi)$ are 1-periodic functions satisfying the condition

$$\begin{gathered} A_{kj}(\xi) = A_{jk}(\xi), \\ A_{kj}(\xi)\,\eta_k\eta_j \geqslant \varkappa_1\eta_k\eta_k \quad \forall(\eta_1, \ldots, \eta_s) \in \mathbf{R}^s, \end{gathered} \tag{2}$$

κ_1 is independent of ε. It is assumed that $f(x) \in C^\infty(\mathbf{R}^s)$, $f(x)$ is a function T-periodic in x_q, T a number of order 1, a multiple of ε.

The coefficients $A_{kj}(\xi)$ are piecewise smooth functions in the sense of §1, item 2. The contact relations (1.36), (1.37) hold on the surface of discontinuity of the coefficients Σ_1.

Let

$$\langle f\rangle_T \equiv \int_0^T \ldots \int_0^T f(x)\,dx_1 \ldots dx_s = 0. \tag{3}$$

Then, according to Theorem 1 of the Supplement, there exists a generalized solution of equation (1), T-periodic in $x_1, \ldots, x_s$, and having the zero mean $\langle u\rangle_T = 0$. It satisfies the inequality

$$\|u\|_{W_2^1(Q_T)} \leqslant c\|f\|_{L_2(Q_T)},$$

where $c = \text{const} > 0$, c is independent of ε, and Q_T is the periodicity cube $\{(x \in \mathbf{R}^s | x_i \in (0, T),\ i = 1, \ldots, s)\}$.

Let us first construct an asymptotic expansion for the infinitely differentiable coefficients $A_{kj}(\xi)$ and then consider the case of piecewise smooth coefficients.

By generalizing the asymptotics (1.25), we shall seek the f.a.s. of equation (1) as the asymptotic series

$$u^{(\infty)} \sim \left(\sum_{l=0}^{\infty} \varepsilon^l \sum_{|i|=l} N_i(\xi)\,D^i v(x)\right)\Bigg|_{\xi=x/\varepsilon}, \tag{4}$$

where $i = (i_1, \ldots, i_l)$, $i_j \in 1, \ldots, s$ is a multi-index (an analogue of the index i), the number $|i| = l$ of the components of the multi-index we call its *length*,

$D_v^i = \partial^l v/\partial x_{i_1} \ldots \partial x_{i_l}$, the $N_i(\xi)$ are functions periodic in $\xi_1, \ldots, \xi_2$ with period 1, and v/x is a function T-periodic in $x_1, \ldots, x_s$, independent of the fast variables x_j/ε. (Formally, N_i loses its index when $l = 0$, and we denote it by $N_\emptyset$.)

As in §2.3, the form of series (4) is arrived at automatically if f.a.s. is sought as

$$u^{(\infty)} \sim \sum_{l=0}^{\infty} \varepsilon^l u_l\left(x, \frac{x}{\varepsilon}\right), \tag{5}$$

where the $u_l(x, \xi)$ are 1-periodic functions.

By substituting the formal series (4) into (1) and by taking (1.5) into account, we get

$$\begin{aligned} f(x) \sim L_\varepsilon u^{(\infty)} \sim \Bigg(&\sum_{l=0}^{\infty} \varepsilon^l \sum_{|i|=l} A_{kj}(\xi) N_i(\xi) \frac{\partial^2 D^i v}{\partial x_k \partial x_j} + \\ &+ \varepsilon^{-1} \sum_{l=0}^{\infty} \varepsilon^l \sum_{|i|=l} A_{kj}(\xi) \frac{\partial N_i(\xi)}{\partial \xi_j} \frac{\partial D^i v}{\partial x_k} + \\ &+ \varepsilon^{-1} \sum_{l=0}^{\infty} \varepsilon^l \sum_{|i|=l} \frac{\partial}{\partial \xi_k} (A_{kj}(\xi) N_i(\xi)) \frac{\partial D^i v}{\partial x_j} + \\ &+ \varepsilon^{-2} \sum_{l=0}^{\infty} \varepsilon^l \sum_{|i|=l} \frac{\partial}{\partial \xi_k} \left(A_{kj}(\xi) \frac{\partial}{\partial \xi_j} N_i(\xi) \right) D^i v \Bigg)\Bigg|_{\xi = x/\varepsilon}. \end{aligned}$$

By making the change of indices $r = l + 2$, $i_1 = k$, $i_2 = j$ in the first sum, $q = l + 1$, $i_1 = k$ in the second sum, and $q = l + 1$, $i_1 = j$ in the third sum, we get

$$\begin{aligned} L_\varepsilon u^{(\infty)} \sim \Bigg(&\sum_{r=2}^{\infty} \varepsilon^{r-2} \sum_{|i|=r} A_{i_1 i_2}(\xi) N_{i_3 \ldots i_r}(\xi) D^i v + \\ &+ \sum_{q=1}^{\infty} \varepsilon^{q-2} \sum_{|i|=q} A_{i_1 j}(\xi) \frac{\partial N_{i_2 \ldots i_q}}{\partial \xi_j} D^i v + \\ &+ \sum_{q=1}^{\infty} \varepsilon^{q-2} \sum_{|i|=q} \frac{\partial}{\partial \xi_k} (A_{k i_1}(\xi) N_{i_2 \ldots i_q}) D^i v + \\ &+ \sum_{l=0}^{\infty} \varepsilon^{l-2} \sum_{|i|=l} \frac{\partial}{\partial \xi_k} \left(A_{kj}(\xi) \frac{\partial}{\partial \xi_j} N_i \right) D^i v \Bigg)\Bigg|_{\xi = x/\varepsilon}. \end{aligned}$$

Identifying the indices r, q, l yields

$$f(x) \sim \sum_{l=0}^{\infty} \varepsilon^{l-2} \sum_{|i|=l} H_i(\xi) D^i v \Big|_{\xi = x/\varepsilon}, \tag{6}$$

where

$$H_i(\xi) = L_{\xi\xi}N_{i_1\ldots i_l} + \frac{\partial}{\partial\xi_k}(A_{ki_1}(\xi)N_{i_2\ldots i_l}) + \\ + A_{i_1 j}(\xi)\frac{\partial N_{i_2\ldots i_l}}{\partial \xi_j} + A_{i_1 i_2}(\xi)N_{i_3\ldots i_l}, \quad |i| \geqslant 2, \tag{7}$$

$$H_{i_1}(\xi) = L_{\xi\xi}N_{i_1} + \frac{\partial}{\partial\xi_k}(A_{ki_1}(\xi)N_{\varnothing}) + A_{i_1 j}(\xi)\frac{\partial N_{\varnothing}}{\partial\xi_j}, \quad |i| = 1, \tag{8}$$

$$H_{\varnothing}(\xi) = L_{\xi\xi}N_{\varnothing}, \quad |i| = 0. \tag{9}$$

Note that formula (6) can be extended also to the cases $|i| = 1$, $|i| = 0$ if we consider $N_i = 0$ for $|i| < 0$.

We assume $N_\emptyset = 1$, i.e., series (4) starts with v. Then

$$H_{i_1}(\xi) = \frac{\partial}{\partial\xi_k}\left(A_{kj}(\xi)\frac{\partial N_{i_1}}{\partial\xi_j} + A_{ki_1}(\xi)\right), \quad H_{\varnothing}(\xi) \equiv 0.$$

By setting

$$T_i(\xi) = \frac{\partial}{\partial\xi_k}(A_{ki_1}(\xi)N_{i_2\ldots i_l}) + A_{i_1 j}(\xi)\frac{\partial N_{i_2\ldots i_l}}{\partial\xi_j} + A_{i_1 i_2}(\xi)N_{i_3\ldots i_l}, \tag{10}$$

we can rewrite relations (7)–(9) as $H_i(\xi) = L_{\xi\xi}N_i + T_i(\xi)$.

Let us order N_i so as to have $N_j < N_i$ (the priority of N_i is higher than that of N_j) if $|j| < |i|$ (i is "longer" than j), with the functions N_i with the same length of $|i|$ being ordered arbitrarily. Note that T_i is defined by the functions N_j with the priority $|j| < |i|$. This feature can be used for the recurrent definition of N_i by means of the functions N_j, of lower priority, if some requirement is imposed on $H_i(\xi)$.

Let us require that

$$H_i(\xi) = h_i, \tag{11}$$

where $h_i = \text{const}$ and

$$h_{\varnothing} = h_{i_1} = 0. \tag{12}$$

If we can find N_i and h_i satisfying (11) and (12), then (6) reduces to the problem with constant coefficients for v (see (14)). This problem can be solved by the standard method of expanding v in powers of ε (see below).

We shall construct N_i and h_i be induction. Let all the N_i of the priority $|i| < k$ be found. We define $T_i(\xi)$ for all the $i : |i| = k$ from formula (10) (the N_i with the negative $|i|$ are equal to zero).

We set $h_i = \langle T_i(\xi)\rangle$. Then, according to Theorem 1 of the Supplement, there exists a unique 1-periodic solution of the problem

$$LN_i + T_i(\xi) = h_i, \quad \langle N_i\rangle = 0. \tag{13}$$

By solving this problem, we find N_i for each multi-index i of length $|i| = k$.

The step of induction is completed. The basis of the induction: $k = 0$, $N_\emptyset = 1$.

Notice that

$$h_\varnothing = 0, \quad h_{i_1} = \left\langle \frac{\partial}{\partial \xi_k} A_{ki_1}(\xi) \right\rangle = 0.$$

Taking into account (11), (12), we rewrite (6) as

$$f(x) \sim L_\varepsilon u^{(\infty)} \sim$$

$$\sim \sum_{l=2}^{\infty} \varepsilon^{l-2} \sum_{|i|=l} h_i D^i v \sim \sum_{k,j=1}^{s} h_{kj} \frac{\partial^2 v}{\partial x_k \partial x_j} + \sum_{l=3}^{\infty} \varepsilon^{l-2} \sum_{|i|=l} h_i D^i v. \tag{14}$$

Relation (14) is called the *averaged equation* of infinite order. By Theorem 1.1, $h_{kj}\partial^2 v/\partial x_k \partial x_j$, its leading term with respect to ε, is elliptic.

The f.a.s. v of (14) is sought in the form of the series

$$v(x) \sim \sum_{j=0}^{\infty} \varepsilon^j v_j(x), \tag{15}$$

where the $v_j(x)$ are T-periodic infinitely differentiable functions. By substituting (15) into (14) and by equating terms of the same order in ε, we obtain a recurrent chain of problems for v_j:

$$\hat{L} v_j \equiv h_{kr} \frac{\partial^2 v_j}{\partial x_k \partial x_r} = f_j(x); \tag{16}$$

the $v_j(x)$ are T-periodic functions. Here

$$f_0 = f, \quad f_j = -\sum_{l=3}^{j+2} \sum_{|i|=l} h_i D^i v_{j+2-l}. \tag{17}$$

In addition, we require that the following relation hold:

$$\langle v_j \rangle_T = 0. \tag{18}$$

THEOREM 1. *For any $j = 0, 1, \ldots$ there exists a unique infinitely differentiable function v_j, T-periodic in $x_1, \ldots, x_s$, and satisfying the relations* (16)–(18).

The proof of Theorem 1 is based on the fact that for any j, we can obtain by induction,

$$\langle f_j \rangle_T = -\sum_{l=3}^{j+2} \sum_{|i|=l} h_i \langle D^i v_{j+2-l} \rangle_T = 0, \qquad j = 0, 1, \ldots$$

Then Theorem 1 of the Supplement is applied. This is the way the f.a.s. of equation (1) was constructed. Let us prove that $u^{(\infty)}$ is an asymptotic expansion of the solution.

Let

$$v^{(K)}(x) = \sum_{j=0}^{K} \varepsilon^j v_j(x), \tag{19}$$

$$u^{(K)}(x) = \sum_{l=0}^{K+1} \varepsilon^l \sum_{|i|=l} N_i(\xi) D^i v^{(K)} \Big|_{\xi = x/\varepsilon}. \tag{20}$$

Theorem 2. *The function $u^{(K)}$ is a solution of the equation*

$$L_\varepsilon u^{(K)} = f + \varepsilon^K \theta(x, \varepsilon)$$

in the class of periodic functions $x_1, \dots, x_s$ with period T, where $|\theta(x,\varepsilon)| \leqslant c_1, c_1 = \text{const} > 0$, and c_1 is independent of ε. The estimates

$$|\langle u^{(K)} \rangle_T| \leqslant c_2 \varepsilon^\alpha \quad \forall \alpha > 0,$$

$$\| u^{(K)} - u \|_{W_2^1(Q_T)} \leqslant c_3 \varepsilon^{K+1}, \quad c_2, c_3 > 0$$

hold.

Proof: By substituting $u^{(K)}$ into the left side of equation (1) and by making transformations similar to those described above, we see that $L_\varepsilon u^{(K)} = f + \varepsilon^K 0(x,\varepsilon)$, where $|\theta(x,\varepsilon)| \leqslant c_1$. Theorem 1 of the Supplement implies

$$\left\| \frac{\partial (u^{(K)} - u)}{\partial x_j} \right\|_{L_2(Q_T)} \leqslant c_4 \varepsilon^K.$$

In the same way as in §2.3, taking into account the form of $u^{(K)}$, we get

$$\left\| \frac{\partial (u^{(K)} - u)}{\partial x_j} \right\|_{L_2(Q_T)} \leqslant c_5 \varepsilon^{K+1}.$$

Let us prove that $\langle u^{(\infty)} \rangle_T \sim 0$. By $C_r(A)$ we denote the class of functions $f(x)$ satisfying the conditions

$$\left| \frac{\partial^{j_1 + \dots + j_l} f(x)}{\partial x_1^{j_1} \dots \partial x_l^{j_l}} \right| \leqslant A, \quad 0 \leqslant l \leqslant r.$$

Lemma 1. *Let $f(x) \in C_r(A)$, $f(x)$ be a periodic function with period T in x and $T = n\varepsilon$, n being a natural number, and let $N(\varepsilon)$ be a periodic function with period 1, integrable on Q. Then*

$$\left| \left\langle f(x) N\left(\frac{x}{\varepsilon}\right) \right\rangle_T - \langle f(x) \rangle_T \langle N \rangle \right| \leqslant c(r, s) A T^s \varepsilon^r \langle |N| \rangle.$$

PROOF: We represent the integral $\langle f(x)N(x/\varepsilon)\rangle_T$ as

$$\left\langle N(\xi) \sum_{m_1,\dots,m_s=0}^{n-1} f(\xi + m\varepsilon)\,\varepsilon^s \right\rangle.$$

According to the estimate of the remainder in the multi-dimensional formula of rectangles for the periodic case,

$$\left| \sum_{m_1,\dots,m_s=0}^{n-1} f(\xi + m\varepsilon)\,\varepsilon^s - \langle f(x)\rangle_T \right| \leqslant c(r, s)\, A\varepsilon^r T^s.$$

This implies the required estimate.

COROLLARY. *Let, under the conditions of Lemma 1, $f \in C^\infty$, $\langle N\rangle = 0$. Then for any r,*

$$|\langle f(x)N(x/\varepsilon)\rangle_T| \leqslant c(r, s) A_r \varepsilon^r T^s \langle |N|\rangle,$$

and therefore, $\langle f(x)N(x/\varepsilon)\rangle_T \sim 0$.

We have

$$\langle u^{(K)}\rangle_T = \sum_{l=0}^{K+1} \sum_{|i|=l} \sum_{j=0}^{K} \varepsilon^{l+j} \left\langle N_i\left(\frac{x}{\varepsilon}\right) D^i v_j \right\rangle_T$$

Since $f(x) \in C^\infty$, then $v(x) \in C^\infty$. If $|i| > 0$, then $\langle N_i\rangle = 0$, and from Lemma 1, $\langle N_i(x/\varepsilon)D^i v_j\rangle_T \sim 0$. Since $N_\emptyset = 1$ for $|i| = 0$, then $\langle N_\emptyset(x/\varepsilon)v_j\rangle_T = \langle v_j\rangle_T$. Thus, $\langle u^{(K)} - u\rangle_T = \langle u_T^{(K)}\rangle \sim 0$, and therefore, $\langle u^{(\infty)}\rangle_T \sim 0$.

According to the Poincaré–Friedrichs inequality

$$\|u^{(K)} - u\|_{W_2^1(Q_T)} \leqslant c_6 \left(|\langle u^{(K)} - u\rangle| + \sum_{j=1}^{s} \left\| \frac{\partial}{\partial x_j}(u^{(K)} - u)\right\|_{L_2(Q_T)} \right) \leqslant c_3 \varepsilon^{K+1}.$$

The theorem is proved.

REMARK 1: In the case of the piecewise smooth $A_{kj}(\xi)$ and by complete analogy with §2.5, we substitute series (4) into the contact relations on Σ_1, and in view of expansion (6) and requirement (11), we get the following problems for $N_i(\xi)$:

$$\begin{gathered} L_{\xi\xi}N_i + T_i(\xi) = h_i, \quad \xi \notin \bigcup_{l=1}^{\infty} \Sigma_l, \\ [N_i] = 0, \quad \xi \in \Sigma_l, \\ \left[\left(A_{kj}\frac{\partial N_i}{\partial \xi_j} + A_{ki_1}N_{i_2\dots i_l}\right) n_k\right] = 0, \quad \xi \in \Sigma_l, \\ \langle N_i(\xi)\rangle = 0, \end{gathered} \tag{21}$$

where the $N_i(\xi)$ are 1-periodic functions,

$$h_i = \left\langle A_{i_1 j} \frac{\partial N_{i_2 \ldots i_l}}{\partial \xi_j} + A_{i_1 i_2} N_{i_3 \ldots i_l} \right\rangle.$$

In other words, in the case if piecewise smooth coefficients, the $N_i(\xi)$ are generalized solutions of problem (13). By Theorem 1 of the Supplement, this problem is uniquely solvable. Once all the solutions N_i of problem (21) are defined, (6) transforms into the averaged equation of infinite order (14) which is solved in the same way as in the case of the smooth coefficients $A_{kj}(\xi)$.

THEOREM 3. *The function $u^{(K)}$ of the form (20), constructed from N_i of (21), is a solution of the problem*

$$L_\varepsilon u^{(K)} = f + \varepsilon^K \theta_1(x, \varepsilon), \quad \frac{x}{\varepsilon} \notin \bigcup_{l=1}^{\infty} \Sigma_l, \tag{22}$$

$$\left[u^{(K)}\right] = 0, \quad \frac{x}{\varepsilon} \in \bigcup_{l=1}^{\infty} \Sigma_l, \tag{23}$$

$$\left[\frac{\partial u^{(K)}}{\partial \nu}\right] = \varepsilon^{K+1} \theta_2(x, \varepsilon), \quad \frac{x}{\varepsilon} \in \bigcup_{l=1}^{\infty} \Sigma_l, \tag{24}$$

$$\langle u^{(K)} \rangle_T = \theta_3(\varepsilon),$$

$u^{(K)}$ is a T-periodic function of the variables x_q, where $|\theta_1(x,\varepsilon)| \leqslant c_1$, $|\theta_2(x,\varepsilon)| \leqslant c_2$, $|\theta_3(\varepsilon)| \leqslant c_3(\alpha)\varepsilon^\alpha$ for any α, c_1, c_2, $c_3(\alpha)$ are positive constants independent of ε, and $n_k(\xi)$ are the direction cosines of the normal to Σ_1 at the point ξ.

The estimate

$$\| u^{(K)} - u \|_{W_2^1(Q_T)} \leqslant c\varepsilon^{K+1}, \quad c = \text{const} > 0 \tag{25}$$

holds.

PROOF: Equation (22) is obtained as in Theorem 2. Formula (23) follows from the continuity of the $N_i(\xi)$ and $v^{(K)}(x)$. We prove (24):

$$\left[\frac{\partial u^{(K)}}{\partial \nu}\right] = \left[\sum_{l=0}^{K+1} \varepsilon^{l-1} \sum_{|i|=l} A_{kj}(\xi) \frac{\partial N_{i_1 \ldots i_l}}{\partial \xi_j} n_k(\xi) D^i v^{(K)} + \right.$$
$$\left. + \sum_{l=1}^{K+2} \varepsilon^{l-1} \sum_{|i|=l} A_{k i_1}(\xi) N_{i_2 \ldots i_l} n_k(\xi) D^i v^{(K)} \right] \Bigg|_{\xi = x/\varepsilon \in \Sigma_l} =$$
$$= \left[\sum_{l=0}^{K+1} \varepsilon^{l-1} \sum_{|i|=l} \left(A_{kj}(\xi) \frac{\partial N_{i_1 \ldots i_l}}{\partial \xi_j} + A_{k i_1} N_{i_2 \ldots i_l} \right) n_k(\xi) D^i v^{(K)} \right] \Bigg|_{\xi = x/\varepsilon \in \Sigma_l} +$$
$$+ \varepsilon^{K+1} \left[\sum_{|i|=K+2} A_{k i_1}(\xi) N_{i_2 \ldots i_{K+2}} n_k(\xi) D^i v^{(K)} \right] \Bigg|_{\xi = x/\varepsilon \in \Sigma_l}.$$

It follows from the contact relations for N_i (see (21)) that

$$[\partial u^{(K)}/\partial \nu] = \varepsilon^{K+1}\theta_2(x, \varepsilon),$$

where

$$\theta_2(x, \varepsilon) = \left[\sum_{|i|=K+2} A_{ki_1}(\xi) N_{i_2 \dots i_{K+2}}{}^{n_h}(\xi) D^i v^{(K)}\right]\Bigg|_{\xi = x/\varepsilon \in \Sigma_l},$$

$$|\theta_2(x, \varepsilon)| \leqslant c_2, \quad c_2 = \text{const} > 0.$$

The inequality $|\theta_3(\varepsilon)| \leqslant c_3(\alpha)\varepsilon^\alpha$ follows from Lemma 1. Estimate (22) follows from Theorem 1 of the Supplement.

REMARK 2: By using the generalized maximum principle, we can get

$$\| u^{(K)} - u \|_{C(Q_T)} \leqslant c\varepsilon^{K+2}. \tag{26}$$

REMARK 3: The estimates (22)–(26) remain true when $u^{(K)}$ is replaced by

$$\tilde{u}^{(K)} = \sum_{l=0}^{K+1} \varepsilon^l \sum_{|i|=l} N_i\left(\frac{x}{\varepsilon}\right) D^i v^{(K-l+1)}.$$

§3 Stationary Thermal Field in a Porous Medium

Porous media, including felt, wood, foam plastics, powders, and materials with systems of cracks, have long been used in engineering. They can be modelled by a continuous medium with periodic or random voids. Thus in percolation theory, soil is sometimes modelled by various periodic arrangements of balls (see [128], p. 21) or a system of capillaries with their radii distributed in a predetermined manner. Chapter 8 will deal with skeletal structures that are also "porous media."

Let Q be the cube $\{\xi \in \mathbf{R}^s | 0 < \xi_j < 1, j = 1, \dots, s\}$. Let us put a set of a finite number of domains into Q and continue it periodically (with period 1) over the whole space $\mathbf{R}^s$. We denote by $\mathcal{A}$ the union of all resultant sets. Suppose the boundary $\partial\mathcal{A}$ is a smooth manifold, and $\mathbf{R}^s \setminus \mathcal{A}$ is a connected domain. We denote by $\mathcal{A}_\varepsilon$ the set obtained from $\mathcal{A}$ by homothetic contraction in ε^{-1} times, that is, $\mathcal{A}_\varepsilon = \{x \in \mathbf{R}^s | x/\varepsilon \in \mathcal{A}\}$ (Fig. 17).

Let G be a bounded domain in $\mathbf{R}^s$ with an infinitely smooth boundary. Consider the equation

$$L_\varepsilon u = f(x), \quad x \in G \setminus \mathcal{A}_\varepsilon, \tag{1}$$

with the boundary condition on $\partial G \setminus \partial \mathcal{A}_\varepsilon$:

$$u|_{\partial G \setminus \partial \mathcal{A}_\varepsilon} = 0; \tag{2}$$

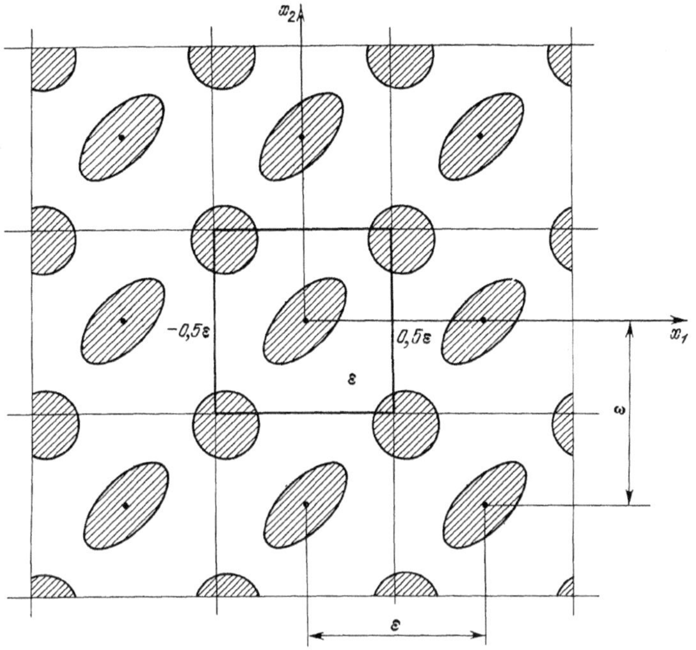

Figure 17

here the $A_{kj}(\xi)$ are 1-periodic, infinitely differentiable functions satisfying the conditions (1.3) $(k, j = 1, \dots, s)$, $f(x) \in C^\infty(\overline{G})$.

The Neumann boundary condition

$$\left.\frac{\partial u}{\partial \nu}\right|_{\partial \mathcal{A}_\varepsilon \cap G} = 0, \tag{3_1}$$

is set on the boundary $\partial \mathcal{A}_\varepsilon \cap G$, where $\partial u/\partial \nu = A_{kj} \partial u/\partial x_j n_k$, and n_k is the cosine of the angle between the normal to $\partial \mathcal{A}_\varepsilon$ and the kth coordinate direction. Condition (3_1) means that there is no heat exchange with the medium inside the voids $\mathcal{A}_\varepsilon$.

In reality, there is a heat exchange by radiation between the parts of the body's boundary within the sight range even if the voids are filled with vacuum. This circumstance affects only the terms of order $O(\varepsilon)$ in the averaged equation, as is shown by [13].

The objective of this section is to construct asymptotics for a solution of problem (1), (2), (3). We assume as usual

$$u^{(2)} = u_0(x, \xi) + \varepsilon u_1(x, \xi) + \varepsilon^2 u_2(x, \xi), \tag{4}$$

where $\xi = x/\varepsilon$, $u_l(x, \xi)$ are 1-periodic, infinitely differentiable.

By substituting (4) into (1) and by using the indirect differentiation formula, we combine terms of the same order for ε. We obtain expression (1.6), which implies the relations (1.7)–(1.9), where x acts as a parameter.

By substituting (4) into the left side of (3_1), we get

$$\left.\frac{\partial u^{(2)}}{\partial \nu}\right|_{\partial \mathcal{A}_\varepsilon} = \left(\varepsilon^{-1}\frac{\partial u_0}{\partial \nu_\xi} + \left(\frac{\partial u_1}{\partial \nu_\xi} + \frac{\partial u_0}{\partial \nu_x}\right) + \right.$$
$$\left. + \varepsilon\left(\frac{\partial u_2}{\partial \nu_\xi} + \frac{\partial u_1}{\partial \nu_x}\right) + \varepsilon^2 \frac{\partial u_2}{\partial \nu_x}\right)\Bigg|_{\xi = x/\varepsilon}. \tag{5}$$

We require the relations

$$\left.\frac{\partial u_0}{\partial \nu_\xi}\right|_{\xi \in \partial \mathcal{A}} = 0, \tag{6}$$

$$\left.\left(\frac{\partial u_1}{\partial \nu_\xi} + \frac{\partial u_0}{\partial \nu_x}\right)\right|_{\xi \in \partial \mathcal{A}} = 0, \tag{7}$$

$$\left.\left(\frac{\partial u_2}{\partial \nu_\xi} + \frac{\partial u_1}{\partial \nu_x}\right)\right|_{\xi \in \partial \mathcal{A}} = 0. \tag{8}$$

Theorem 2 of the Supplement implies the following lemma.

Lemma 1. *Let the $A_{kj}(\xi)$, $F_0(\xi)$, $F_k(\xi)$ be 1-periodic differentiable functions, with the $A_{kj}(\xi)$ satisfying condition (1.3). Then a necessary and sufficient condition for a 1-periodic solution of the equation*

$$L_{\xi\xi} N = F_0(\xi) + \frac{\partial F_h(\xi)}{\partial \xi_h}, \quad \xi \in \mathbf{R}^s \setminus \mathcal{A}, \tag{9}$$

with the boundary condition

$$\left.\left(A_{kj}(\xi)\frac{\partial}{\partial \xi_j} N(\xi) - F_k(\xi)\right) n_k\right|_{\xi \in \partial \mathcal{A}} = 0 \tag{10}$$

to exist is

$$\langle F_0(\xi)\rangle^{\mathcal{A}} \equiv \int_{Q \setminus \mathcal{A}} F_0(\xi)\, d\xi = 0. \tag{11}$$

Consider pairs of the relations (1.7), (6); (1.8), (7); and (1.9), (8) as problems of type (9), (10) in $u_0(x,\xi)$, $u_1(x,\xi)$, and $u_2(x,\xi)$, respectively (x is a parameter).

Lemma 1 implies that the first two problems (1.7), (6) and (1.8), (7) are solvable, while by analogy with §1

$$u_0(x, \xi) = v_0(x), \quad u_1(x, \xi) = N_{i_1}(\xi)\frac{\partial v_0}{\partial x_{i_1}},$$

where the $N_{i_1}(\xi)$ are the 1-periodic solution of the problems

$$L_{\xi\xi}(N_{i_1} + \xi_{i_1}) = 0, \quad \xi \in \mathbf{R}^s \setminus \mathscr{A}, \tag{12}$$

$$\frac{\partial}{\partial \nu_\xi}(N_{i_1} + \xi_{i_1})|_{\xi \in \partial\mathscr{A}} = 0. \tag{13}$$

By rewriting (1.9) as (1.16) and condition (8) as

$$\left(A_{kj}(\xi) \frac{\partial u_2(x,\xi)}{\partial \xi_j} + A_{ki_1}(\xi) N_{i_2}(\xi) \frac{\partial^2 v_0}{\partial x_{i_1} \partial x_{i_2}} \right) n_k \Bigg|_{\xi \in \partial\mathscr{A}} = 0, \tag{14}$$

we see that condition (11) is equivalent to

$$\left\langle \left(A_{i_1 j}(\xi) \frac{\partial N_{i_2}}{\partial \xi_j} + A_{i_1 i_2}(\xi) \right) \frac{\partial^2 v_0}{\partial x_{i_1} \partial x_{i_2}} \right\rangle^{\mathscr{A}} = \langle f(x) \rangle^{\mathscr{A}},$$

that is,

$$\hat{A}_{i_1 i_2} \frac{\partial^2 v_0}{\partial x_{i_1} \partial x_{i_2}} = f(x), \tag{15}$$

where

$$\hat{A}_{i_1 i_2} = \left\langle A_{i_1 j}(\xi) \frac{\partial N_{i_2}}{\partial \xi_j} + A_{i_1 i_2}(\xi) \right\rangle^{\mathscr{A}} \operatorname{mes}(Q \setminus A). \tag{16}$$

THEOREM 1. *(It can be proved in the same way as Theorem 1.1). The matrix of the coefficients $\hat{A}_{i_1 i_2}$ is symmetric and positive-definite.*

Thus, as in §1, the necessary and sufficient condition of the solvability of the problem for $u_2(x,\xi)$ has led to the averaged equation in $v_0(x)$.

By substituting $u^{(2)}$ into the boundary condition (2), we see that the averaged equation (15) must be supplemented by the boundary condition

$$v_0|_{\partial G} = 0. \tag{17}$$

Note that $v_0 \in C^\infty(\overline{G})$.

By analogy with §1, it can be proved that $u_2(x,\xi)$ is represented as in (1.23), where $N_{i_1 i_2}(\xi)$ is a periodic solution of (1.24) with period 1 in $\mathbf{R}^s \setminus A$ with the boundary condition

$$\left(A_{kj}(\xi) \frac{\partial N_{i_1 i_2}}{\partial \xi_j} + A_{ki_1}(\xi) N_{i_2} \right) n_k \Bigg|_{\partial\mathscr{A}} = 0. \tag{18}$$

Thus, $u^{(2)}$ is a solution of the problem

$$L_\varepsilon u^{(2)} = f(x) + \varepsilon r^{(2)}(x, \varepsilon), \quad x \in G\backslash\mathcal{A}_\varepsilon,$$

$$\left.\frac{\partial u^{(2)}}{\partial \nu}\right|_{\partial\mathcal{A}_\varepsilon\cap G} = \varepsilon^2 A_{kj} N_{i_1 i_2} n_k \left.\frac{\partial^3 v_0}{\partial x_j \partial x_{i_1} \partial x_{i_2}}\right|_{\partial\mathcal{A}_\varepsilon\cap G},$$

$$u^{(2)}|_{\partial G} = \varepsilon u_1\left(x, \frac{x}{\varepsilon}\right)\Big|_{\partial G} + \varepsilon^2 u_2\left(x, \frac{x}{\varepsilon}\right)\Big|_{\partial G}.$$

As in §1, let us introduce the function

$$\widetilde{u}^{(2)} = u^{(2)} - \chi(x)\left(\varepsilon u_1\left(x, \frac{x}{\varepsilon}\right) + \varepsilon^2 u_2\left(x, \frac{x}{\varepsilon}\right)\right).$$

It satisfies the relations

$$L_\varepsilon\left(\widetilde{u}^{(2)} - u\right) = \varepsilon r^{(2)}(x, \varepsilon) - \frac{\partial}{\partial x_k} a_k(x, \varepsilon), \quad x \in G\backslash\mathcal{A}_\varepsilon,$$

$$\left.\left(\frac{\partial(\widetilde{u}^{(2)} - u)}{\partial \nu} + a_k n_k\right)\right|_{\partial\mathcal{A}_\varepsilon\cap G} = \varepsilon^2 A_{kj} N_{i_1 i_2}\left(\frac{x}{\varepsilon}\right) \left.\frac{\partial^3 v_0}{\partial x_j \partial x_{i_1} \partial x_{i_2}} n_k\right|_{\partial\mathcal{A}_\varepsilon\cap G},$$

$$(\widetilde{u}^{(2)} - u)|_{\partial G} = 0.$$

By using the estimates from [76], similar to the a priori estimates from Theorem 2, we get

$$\|\widetilde{u}^{(2)} - u\|_{W_2^1(G\backslash\mathcal{A}_\varepsilon)} = O(\sqrt{\varepsilon}).$$

This implies the following theorem.

THEOREM 2. *Let the $N_{i_1}(\xi)$ be 1-periodic solutions of problems (12) and (13), let the $N_{i_1 i_2}$ be 1-periodic solutions of the equations (1.2) with boundary conditions (18), let $v_0(x)$ be a solution of the averaged problem (15), (17), let $u^{(1)}$, $u^{(2)}$ be defined by (1.34), (1.15), and let u be an exact solution of problem (1), (2), (3_1). Then the estimates*

$$\|u^{(2)} - u\|_{W_2^1(G\backslash\mathcal{A}_\varepsilon)} \leqslant c_1\sqrt{\varepsilon},$$

$$\|u^{(1)} - u\|_{W_2^1(G\backslash\mathcal{A}_\varepsilon)} \leqslant c_2\sqrt{\varepsilon},$$

$$\|v_0 - u\|_{L_2(G\backslash\mathcal{A}_\varepsilon)} \leqslant c_3\sqrt{\varepsilon},$$

hold, where the c_i are independent of ε.

REMARK 1: Problem (1), (2) may be considered with the Dirichlet conditions

$$u|_{\partial\mathcal{A}_\varepsilon\cap G} = 0 \qquad (3_2)$$

or with the third-type conditions

$$\left(\frac{\partial u}{\partial \nu}+b\left(\frac{x}{\varepsilon}\right)u\right)\Big|_{\partial\mathscr{A}_\varepsilon\cap G}=0, \tag{3_3}$$

where $b(\xi)$ is a 1-periodic, infinitely differentiable function, $\int_{\partial\mathcal{A}\cap Q} b(\xi)d\xi = 0$. Condition ($3_2$) corresponds to the maintenance of zero temperature on the boundary $\mathcal{A}_\varepsilon$. Condition (3_3) describes heat exchange with the medium inside the voids $\mathcal{A}_\varepsilon$ according to Newton's law.

In addition, we require for (3_2) that $f(x)$ vanish in a neighbourhood of ∂G.

The asymptotics of the solution of problem (1), (2), (3_3) is constructed in the same manner as that of (1), (2), (3_1).

The Dirichlet problem (1), (2), (3_2) has a solution of order ε^2, and therefore it is found as

$$u^{(2)}=\left(\varepsilon^2 u_0(x,\xi)+\varepsilon^3 u_1(x,\xi)+\varepsilon^4 u_2(x,\xi)\right)\big|_{\xi=x/\varepsilon}. \tag{19}$$

Substitution of (19) into (1) and (3_2) yields the following problems which define $u_l(x,\xi)$:

$$L_{\xi\xi}u_0=f(x),\quad \xi\in\mathbf{R}^s\setminus\mathscr{A},\quad u_0(x,\xi)\,|_{\partial\mathscr{A}}=0; \tag{20}$$

$$L_{\xi\xi}u_1+L_{\xi x}u_0+L_{x\xi}u_0=0,\quad \xi\in\mathbf{R}^s\setminus\mathscr{A},\quad u_1(x,\xi)\,|_{\partial\mathscr{A}}=0; \tag{21}$$

$$L_{\xi\xi}u_2+L_{\xi x}u_1+L_{x\xi}u_1+L_{xx}u_0=0,\quad u_2(x,\xi)\,|_{\partial\mathscr{A}}=0. \tag{22}$$

It can be proved that each of the problems (20), (21), and (22) is solvable in the class of the functions $u_0(x,\xi)$, $u_1(x,\xi)$, and $u_2(x,\xi)$, respectively, periodic in $\xi_1,\dots\xi_s$ with period 1.

Note that $u_0(x,\xi)$ depends on ξ. Since $f(x)$ vanishes in a neighbourhood O of the boundary of the domain G, then the $u_l(x,\xi)$ also vanish for $x\in O$, and thus $u^{(2)}$ satisfies the boundary conditions (2) and (3_2) exactly. Using the a priori estimate from Theorem 1.3.1, we get $\|u^{(2)}-u\|_{W_2^1(G\setminus\mathcal{A}_\varepsilon)}=O(\varepsilon^4)$. See more detailed investigations of problems in porous media in [39, 109, 203].

Remark 2: As in §1, the results of the present section are generalized for the case of equation (1) with piecewise smooth coefficients, which simulates a composite material with a periodic system of cracks or pores.

§4 Averaging a Stationary System of Equations of Elasticity Theory in Composite and Porous Materials

Let the $A_{kj}(\xi)$ be 1-periodic $s\times s$ matrix-valued functions with piecewise smooth elements (see §1).

Let the elements of the matrices $A_{kj}(\xi)$ satisfy the conditions

$$a_{kj}^{ih}(\xi)=a_{ij}^{kh}(\xi)=a_{kh}^{ij}(\xi)=a_{jk}^{hi}(\xi),\qquad a_{kj}^{ih}(\xi)\,\eta_{ik}\eta_{hj}\geqslant\varkappa_1\eta_{ik}\eta_{ik}, \tag{1}$$

for a symmetric matrix $\|\eta_{ik}\|$, where $\kappa_1 > 0$ is a constant.

The linear boundary value problem for the system of equations of elasticity theory is written as

$$L_\varepsilon u \equiv \frac{\partial}{\partial x_k}\left(A_{kj}\left(\frac{x}{\varepsilon}\right)\frac{\partial u}{\partial x_j}\right) = f(x), \quad x \in G, \tag{2}$$

and the matrices A_{kj} satisfy (1). Consider this problem for the boundary value

$$u|_{\partial G} = 0, \tag{3}$$

where $u(x)$, $f(x)$ are s-dimensional vector-valued functions, $f(x) \in C^\infty(\overline{G})$, G is a bounded domain in $\mathbf{R}^s$, and $\partial G \in C^\infty$. The solution of problem (2), (3) is understood in the generalized sense (see Chapter 1). Problem (2), (3) describes the strained state of an elastic specimen made of a composite material that occupies the domain G and is subjected to external forces distributed with density $f(x)$.

The asymptotics of the solution of (2), (3) is constructed as (1.4), where the $u_l(x\xi)$ are 1-periodic s-dimensional vector-valued functions. The construction of the vector-valued function $u^{(2)}$ follows the appropriate constructing procedure in §1, with the following corrections:

1) The $A_{kj}(\xi)$ are $s \times s$ matrices everywhere, and u, u_l, and v_0 are vector-valued functions.

2) $N_{i_1}(\xi)$ in (1.14) and $N_{i_1 i_2}(\xi)$ in (1.23) are $s \times s$ matrix-valued functions, respectively, satisfying the systems of equations

$$L_{\xi\xi} M_{i_1} = 0, \quad M_{i_1} \equiv N_{i_1} + \xi_{i_1} E \tag{4}$$

(analogue of (1.15)),

$$L_{\xi\xi} N_{i_1 i_2} + T_{i_1 i_2}(\xi) = \hat{A}_{i_1 i_2}$$

(cf. (1.24)) where E is the unit matrix, and the $\hat{A}_{i_1 i_2}$ are constant $s \times s$ matrices defined by

$$\hat{A}_{i_1 i_2} = \left\langle A_{i_1 j}(\xi)\frac{\partial M_{i_2}}{\partial \xi_j}\right\rangle. \tag{5}$$

The matrix-valued functions $T_{i_1 i_2}(\xi)$ are found from

$$T_{i_1 i_2}(\xi) = \frac{\partial}{\partial \xi_k}\left(A_{k i_1} N_{i_2}\right) + A_{i_1 j}\frac{\partial N_{i_2}}{\partial \xi_j} + A_{i_1 i_2}.$$

3) F, f_0, f_k in Lemmas 1.1 and 1.2 are $s \times s$ matrices, and condition (1.3) is replaced by (1). The proof of Lemmas 1.1 and 1.2 for the system of elasticity

theory is based on the first Korn inequality, which is valid for periodic functions with the zero mean over the periodicity cube. This inequality implies

$$\left\langle \left(A_{kj}(\xi) \frac{\partial \varphi}{\partial \xi_j}, \frac{\partial \varphi}{\partial \xi_k} \right) \right\rangle \geqslant c \left\langle \left(\frac{\partial \varphi}{\partial \xi_k}, \frac{\partial \varphi}{\partial \xi_k} \right) \right\rangle,$$

where $\varphi(\xi)$ is a 1-periodic differentiable vector-valued function, and $c > 0$ is a constant independent of φ.

4) Theorem 1.1 is replaced by the following analogue (see §6.1): The matrices $\hat{A}_{i_1 i_2}$ satisfy the conditions of type (1).

5) Finally, the estimate from Theorem 1.3.4 is used to prove Theorem 1.2.

Thus, to construct the asymptotics of a solution of problem (2), (3), one must:

– solve the following problems on a cell in the class of 1-periodic $s \times s$ matrix-valued functions $N_{i_1}(\xi)$ $(i_1 = 1, \ldots, 5)$:

$$\begin{gathered} L_{\xi\xi}(N_{i_1} + \xi_{i_1} E) = 0, \quad \xi \in \mathbf{R}^s \setminus \Sigma_l^1, \\ [N_{i_1}(\xi)]|_{\Sigma_l} = 0, \quad \left[\frac{\partial}{\partial \nu_\xi} (N_{i_1} + \xi_{i_1} E) \right] \Big|_{\Sigma_l} = 0; \end{gathered} \tag{6}$$

– calculate the effective elasticity tensor by formula (5);

– solve the averaged system of equations of elasticity theory with the constant elasticity tensor:

$$\widehat{A}_{i_1 i_2} \frac{\partial^2 v_0}{\partial x_{i_1} \partial x_{i_2}} = f(x), \quad x \in G,$$

with the boundary value $v_0|_{\partial G} = 0$.

The vector-valued function $v_0(x)$ approximates the displacement field $u(x)$ with accuracy of order ε and describes macroscopic effects. The approximation $u^{(1)} = v_0 + \varepsilon N_{i_1}(\xi) \frac{\partial v_0}{\partial x_{i_1}}$ characterizes the microscopic variation of the field; the stresses $\sigma_k(x) = A_{kj}\left(\frac{x}{\varepsilon}\right) \frac{\partial u}{\partial x_j}$ can be calculated by the approximate formulas

$$\sigma_k \approx A_{kj}\left(\frac{x}{\varepsilon}\right) \frac{\partial u^{(1)}}{\partial x_j} = A_{kj}\left(\frac{x}{\varepsilon}\right) \frac{\partial M_{i_1}}{\partial \xi_j} \frac{\partial v_0}{\partial x_{i_1}} + \varepsilon A_{kj}\left(\frac{x}{\varepsilon}\right) N_{i_1}\left(\frac{x}{\varepsilon}\right) \frac{\partial^2 v_0}{\partial x_j \partial x_{i_1}},$$

that is,

$$\sigma_k \approx \widetilde{A}_{k i_1} \frac{\partial v_0}{\partial x_{i_1}},$$

where $A_{kj} = A_{kj} M_i / \partial \xi_j$ or for the components

$$\sigma_k^j \approx \widetilde{a}_{kp}^{jq}\left(\frac{x}{\varepsilon}\right) \frac{\partial v_0^q}{\partial x_p} = \widetilde{a}_{kp}^{jq}\left(\frac{x}{\varepsilon}\right) \widehat{e}_p^q$$

(summation over repeated indices from 1 to s is understood). Here $\hat{e}_p^q$ is the tensor of mean strains:

$$\hat{e}_p^q = \frac{1}{2}\left(\frac{\partial v_0^q}{\partial x_p} + \frac{\partial v_0^p}{\partial x_q}\right)$$

and as will be shown in the proof of Theorem 6.1.3,

$$\tilde{a}_{mj}^{ki} = \tilde{a}_{kj}^{mi} = \tilde{a}_{mi}^{kj}.$$

We have introduced the notation $\tilde{A}_{kj}$ from the following considerations:

1. The effective coefficients $\hat{A}_{kj}$ are calculated via the explicit formulas $\hat{A}_{kj} = \langle \tilde{A}_{kl} \rangle$.

2. If the periodicity domain satisfies symmetry conditions (see Introduction), the system of equations (4) in the matrices M_j falls into systems of standard boundary-value problems in the column vectors of the M_j. When these problems are solved by using routine program packs, stress tensors are generated for the problems of elasticity theory, which just coincide with the elements of the $\tilde{A}_{kj}$.

REMARK 1: Estimates of the elements $\hat{a}_{kk}^{ii}$ from variational considerations similar to those set forth in Remark 3 in §1 have been obtained in [16]. They may be of interest for use in the case of low values of the ratio of the Lamé coefficients μ/λ.

REMARK 2: The procedure of constructing an asymptotic expansion is also extended from §2 to equations of elasticity theory wherein all the $N_i(\xi) \equiv N_{i_1 \ldots i_l}(\xi)$ are $s \times s$ matrix-valued functions, and v is an s-dimensional vector.

REMARK 3: Asymptotics of a solution of a problem of elasticity theory in a porous medium can be constructed if the $A_{kj}(\xi)$ in §3 are considered $s \times s$ matrix-valued functions. The "problems on a cell" take the form

$$L_{\xi\xi}\left(N_{i_1} + \xi_{i_1} E\right) = 0, \quad \xi \in \mathbf{R}^s \setminus \mathcal{A},$$

$$\frac{\partial}{\partial \nu_\xi}\left(N_{i_1} + \xi_{i_1} E\right)\Big|_{\partial \mathcal{A}} = 0;$$

the $N_{i_1}(\xi)$ are sought in the class of 1-periodic functions, the effective elasticity tensor is calculated by formula (3.16), and the averaged problem takes the form of (3.15), (3.17).

The Corn inequality in $G \setminus \mathcal{A}_\varepsilon$ with the constant independent of ε should be used in addition to justifying the asymptotics. Proof of this inequality is given in [26].

V. I. Gorbachev has considered a problem on effective characteristics of a porous medium made of a homogeneous isotropic material [227].

The domain $\mathcal{A}$ is of the form:

$$\text{a)} \ \left\{\left|\xi_1 - \frac{1}{2}\right|, \left|\xi_2 - \frac{1}{2}\right| \leqslant r, \ \ 0 \leqslant \xi_3 \leqslant 1\right\},$$

or

$$\text{b)}\ \left\{\left(\xi_1 - \frac{1}{2}\right)^2 + \left(\xi_2 - \frac{1}{2}\right)^2 \leqslant r^2,\ \ 0 \leqslant \xi_3 \leqslant 1\right\}.$$

Thus, the pores penetrate the medium. Calculations by this method produced the following results.

Let E be Young's modulus and ν be Poisson's ratio of the initial material. The equivalent homogeneous medium is anisotropic and is characterized by six independent effective constants. They are $\widehat{E}_{11}$, Young's modulus of tension along the x_1-axis; $\hat{\nu}_{12}$, Poisson's ratio characterizing contraction in the direction of the x_2-axis when tension is in the direction of the x_1-axis; $\widehat{G}_{12}$, the modulus of elasticity in shear in the plane $0x_1x_2$; $\widehat{E}_{33}$, Young's modulus in tension in the direction of the x_3-axis (along the pores); $\hat{\nu}_{13}$, Poisson's ratio characterizing contraction in the direction of the x_3-axis when tension is in the direction of the x_1-axis; and $\widehat{G}_{13}$, the modulus of elasticity in shear in the plane $0x_1x_3$. The remaining three constants are expressed in terms of those listed above: $\widehat{E}_{22} = \widehat{E}_{11}$, $\widehat{G}_{23} = \widehat{G}_{13}$, $\hat{\nu}_{23} = \hat{\nu}_{13}$. In addition, there are equalities: $\hat{\nu}_{21} = \hat{\nu}_{12}$, $\hat{\nu}_{31} = \hat{\nu}_{32} = \hat{\nu}_{13}\widehat{E}_{33}/\widehat{E}_{11}$. The results of appropriate calculations for $E = 1$, $\nu = 0.05$, $\nu = 0.25$, and $\nu = 0.45$ are given as graphs in Fig. 18a, depending on the volume fraction of the pore V_a. Solid lines correspond to the square pore and dash lines to the round pore. Crosses mark values of $\widehat{E}_{11}$, $\widehat{G}_{12}$, $\hat{\nu}_{12}$ for the round pore found in [45] by means of the theory of doubly periodic functions. There is a good agreement of the results produced by different methods.

The value of Poisson's ratio for the material of a cell and the shape of a pore affect most of all the value of $\hat{\nu}_{12}$. The curves in Fig. 18b are marked by Figs. 1, 2, 3. They correspond to $\nu = 0.05$, $\nu = 0.25$, and $\nu = 0.45$, respectively. It can be seen from the curves that for $\nu < 0.25$ and a small porosity (less than 20 percent), the situation may arise when $\hat{\nu}_{12} > \nu$.

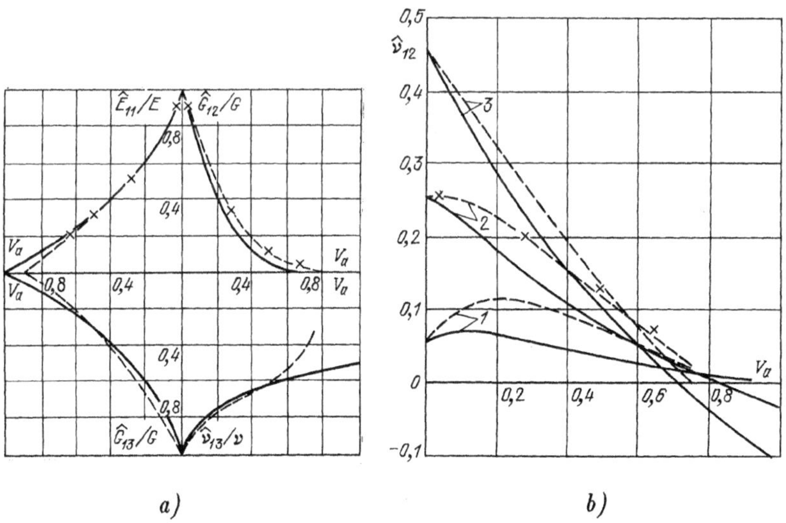

Figure 18

This effect is more pronounced for the round pore (dash lines). In this case, $\max \hat{\nu}_{12} \approx 0.111$ for $\nu = 0.5$, with $V_a \approx 0.2$, that is, $\hat{\nu}_{12}$ is more than twice as high as ν. In the case of the square pore, $\max \hat{\nu}_{12} \approx 0.072$ for $\nu = 0.05$ with $V_a \approx 0.1$. These data point to the substantially nonlinear dependence of $\hat{\nu}_{12}$ on ν and V_a.

§5 Nonstationary Systems of Equations of Elasticity and Diffusion Theory

1. Consider the nonstationary problem

$$Pu \equiv R\left(\frac{x}{\varepsilon}\right)\frac{\partial^2 u}{\partial t^2} + S\left(\frac{x}{\varepsilon}\right)\frac{\partial u}{\partial t} - \frac{\partial}{\partial x_m}\left(A_{mk}\left(\frac{x}{\varepsilon}\right)\frac{\partial u}{\partial x_k}\right) - f(x, t) = 0, \quad x \in G, \quad t > 0, \qquad (1)$$

$$u\,|_{\partial G} = \varphi(x, t)\,|_{\partial G}, \quad t > 0, \qquad (2)$$

$$u\,|_{t=0} = \left.\frac{\partial u}{\partial t}\right|_{t=0} = 0, \qquad (3)$$

where the $R(\xi)$, $S(\xi)$, $A_{mk}(\xi)$ are 1-periodic $n \times n$ matrix-valued functions piecewise smooth in the sense of §1, f and φ are n-dimensional vector-valued functions, and $f(x,t)$, $\varphi(x,t) \in C^\infty(\overline{G} \times [0,+\infty))$. For the system of elasticity equations, the matrix $R(\xi)$ takes the form $\rho(\xi)E$, $S(\xi) \equiv 0$; here $\rho(\xi)$ is the density at the point ξ, and $s \leqslant n$.

The following assumptions are also made:

1) $(R(\xi)\eta, \eta) \geqslant \kappa_0(\eta, \eta)$ for any real-valued, n-dimensional vector η.
2) One of the following conditions is satisfied:
   a) $n = s$, and the collection of the matrices $A_{mk}(\xi)$ satisfies the conditions (4.1);
   b) For any collection of real vectors $\eta_i, \dots, \eta_s$,

$$(A_{mk}(\xi)\eta_k, \ \eta_m) \geqslant \varkappa_1(\eta_k, \ \eta_k). \qquad (4)$$

Here κ_0, $\kappa_1 > 0$ are independent of x, ξ, ε.

We assume that all the initial data agree, i.e., for example, f, φ, and their derivatives of all orders vanish at $t = 0$ on ∂G.

The asymptotics of the solution of problem (1)–(3) is constructed in much the same way as in §3.1:

$$u^{(2)}(x, t) = u_0\left(x, t, \frac{x}{\varepsilon}\right) + \varepsilon u_1\left(x, t, \frac{x}{\varepsilon}\right) + \varepsilon^2 u_2\left(x, t, \frac{x}{\varepsilon}\right), \qquad (5)$$

where the $u_i(x,t,\xi)$ are functions periodic in $\xi_1, \dots, \xi_s$ with period 1, infinitely differentiable with respect to x, t, ξ everywhere except on the discontinuity surfaces $\overline{G} \times [0, \infty) \times \Sigma$.

Then, as we did in §§3.1 and 1, we substitute (5) into (1), simplify and equate the terms of orders ε^{-2}, ε^{-1}, and ε^{-0} to zero. We get relations of type (1.7)–(1.9). Similar substitution is made also in the contact conditions, which yield the equalities (1.40)–(1.45). We prove in the same way as in §§3.1 and 1 that $u_0(x,t,\xi)$ is independent of ξ, while the vector-valued function $u_1(x,t,\xi)$ can be represented as (1.14), where the $N_{i_1}(\xi)$ are 1-periodic solutions of problem (4.6). Next, given the solvability of the problem for $u_2(x,t,\xi)$, we get the averaged equation

$$\widehat{R}\frac{\partial^2 v_0}{\partial t^2}+\widehat{S}\frac{\partial v_0}{\partial t}-\widehat{A}_{i_1 i_2}\frac{\partial^2 v_0}{\partial x_{i_1}\partial x_{i_2}}=f(x,t),\quad x\in G,\quad t>0, \tag{6}$$

where

$$\widehat{R}=\langle R\rangle\equiv\int_Q R(\xi)\,d\xi>0,\quad \widehat{S}=\langle S\rangle,$$

$\widehat{A}_{i_1 i_2}$ are defined from (4.5), and Q is the unit cube.

By substituting $u^{(2)}$ into the boundary and initial conditions, we obtain those for v_0:

$$v_0|_{\partial G}=\varphi(x,t)|_{\partial G},\quad v_0|_{t=0}=\left.\frac{\partial v_0}{\partial t}\right|_{t=0}=0. \tag{7}$$

Problem (6), (7) has the unique solution $v_0(x,t)\in C^\infty(\overline{G}\times[0,+\infty))$.

The asymptotics $u^{(2)}$ thus constructed is justified by repeating the procedure described in §1.

We introduce the function $u^{(2)}$ by formula (1.28). It can be proved that $u^{(2)}$ is a generalized solution of the problem

$$R\left(\frac{x}{\varepsilon}\right)\frac{\partial^2\widetilde{u}^{(2)}}{\partial t^2}+S\left(\frac{x}{\varepsilon}\right)\frac{\partial\widetilde{u}^{(2)}}{\partial t}-\frac{\partial}{\partial x_m}\left(A_{mk}\left(\frac{x}{\varepsilon}\right)\frac{\partial\widetilde{u}^{(2)}}{\partial x_k}\right)=$$
$$=f(x,t)+\varepsilon r_0^{(2)}(x,t,\varepsilon)-\frac{\partial}{\partial x_k}A_m(x,t,\varepsilon), \tag{8}$$

$$\widetilde{u}^{(2)}|_{\partial G}=\varphi(x,t)|_{\partial G},\quad \widetilde{u}^{(2)}|_{t=0}=\varepsilon^2u_2\left(x,0,\frac{x}{\varepsilon}\right)(1-\chi(x)), \tag{9}$$
$$\left.\frac{\partial\widetilde{u}^{(2)}}{\partial t}\right|_{t=0}=\varepsilon^2\frac{\partial u_2}{\partial t}\left(x,0,\frac{x}{\varepsilon}\right)(1-\chi(x)),$$

where

$$\max_{t\in[0,t_0]}\|A_m(x,t,\varepsilon)\|_{L_2(G)},\quad \left\|\frac{\partial}{\partial t}A_m(x,t,\varepsilon)\right\|_{L_2(G\times(0,t_0))}\leqslant c_1(t_0)\sqrt{\varepsilon},$$
$$\|r_0^{(2)}(x,t,\varepsilon)\|_{L_2(G\times(0,t_0))}\leqslant c_2(t_0)\,\varepsilon,$$
$$\left\|u_2\left(x,0,\frac{x}{\varepsilon}\right)(1-\chi(x))\right\|_{W_2^1(G)}\leqslant c_3(t_0)\,\varepsilon^{-1},$$
$$\left\|\frac{\partial u_2}{\partial t}\left(x,0,\frac{x}{\varepsilon}\right)(1-\chi(x))\right\|_{W_2^1(G)}\leqslant c_4(t_0)\,\varepsilon^{-1};$$

here $c_1(t_0), \dots, c_4(t_0)$ are constants independent of ε.

By subtracting (1), (2) from (8), (9) and by using the a priori estimates of Theorem 1.3.5, we get

$$\| \tilde{u}^{(2)} - u \|_{W_2^1(G\times(0,t_0))} \leqslant c(t_0)\sqrt{\varepsilon},$$

and consequently the estimates

$$\left\| v_0 + \varepsilon u_1\left(x, t, \frac{x}{\varepsilon}\right) - u \right\|_{W_2^1(G\times(0,t_0))} = O(\sqrt{\varepsilon}),$$
$$\| v_0 - u \|_{L_2(G\times(0,t_0))} = O(\sqrt{\varepsilon}), \tag{10}$$

which justify the asymptotics constructed above. (This estimate can be found in [8] for somewhat different conditions.)

Remark 1: The scheme for the construction of $u^{(2)}$ extends also to the case of the system

$$S\left(\frac{x}{\varepsilon}\right)\frac{\partial u}{\partial t} - \frac{\partial}{\partial x_m}\left(A_{mk}\left(\frac{x}{\varepsilon}\right)\frac{\partial u}{\partial x_k}\right) = f(x, t), \quad x \in G, \quad t > 0,$$

with the boundary and initial conditions

$$u|_{\partial G} = \varphi(x, t)|_{\partial G}, \quad t > 0, \quad u|_{t=0} = 0;$$

estimates of closeness are obtained assuming $(S\eta, \eta) \geqslant \kappa_2(\eta, \eta)$ for any real vector η.

2. Let $f(x,t)$ be a function T-periodic in $x_1, \dots, x_s$ (T is a number of order 1, a multiple of ε), $f(x,t) \in C^\infty(R^s \times [0, +\infty))$.

$$f(x, 0) = \frac{\partial^l f}{\partial t^l}(x, 0) = 0, \qquad l = 1, 2, \dots \tag{11}$$

Let matrix-valued coefficients $R(\xi)$, $S(\xi)$, $A_{mk}(\xi)$ satisfy the conditions formulated at the beginning of the section.

Using the method of [155], we can demonstrate that there exists a unique solution $u(x,t)$, T-periodic in $x_1, \dots, x_s$, of equation (1) defined in $\mathbf{R}^s \times (0, +\infty)$ for homogeneous initial values.

The f.a.s. of problem (1), (3) is found as the series

$$u^{(\infty)} \sim \sum_{p,q,l=0}^{\infty} \varepsilon^{p+q+l} \sum_{|i|=l} N_{qi}^p D^{q,i} v(x, t)\Big|_{\xi=x/\varepsilon} \equiv$$
$$\equiv \sum_{r=0}^{\infty} \varepsilon^r \sum_{q=0}^{r} \sum_{p=0}^{q} \sum_{|i|=r-q-p} N_{qi}^p(\xi) D^{q,i} v(x, t)\Big|_{\xi=x/\varepsilon}, \tag{12}$$

where i is a multi-index $(i_1, \dots, i_l)$, $|i| = l$ is its length,

$$\sum_{|i|=l} = \sum_{i_1=1}^{s} \cdots \sum_{i_l=1}^{s}, \qquad D^{q,i}v = \frac{\partial^{q+|i|}v}{\partial t^q \partial x_{i_1} \dots \partial x_{i_l}},$$

the $N^p_{qi}(\xi)$ are 1-periodic $n \times n$ matrices, $v(x,t)$ is a T-periodic vector function independent of the fast variables, and $v \sim \sum_{j=0}^{\infty} \varepsilon^j v_j(x,t)$.

By substituting (12) into (1) and by analogy with §2, we get

$$Pu^{(\infty)} \sim \sum_{p,q,l=0}^{\infty} \varepsilon^{p+q+l-2} \sum_{|i|=l} H^p_{qi}(\xi) D^{q,i}v \Big|_{\xi=x/\varepsilon} - f, \tag{13}$$

where

$$H^p_{qi}(\xi) = -L_{\xi\xi}N^p_{qi} + T^p_{qi}(\xi), \tag{14}$$

$$T^p_{qi_1\dots i_l}(\xi) = -A_{i_1 j}\frac{\partial N^p_{qi_2\dots i_l}}{\partial \xi_j} - \frac{\partial}{\partial \xi_j}\left(A_{ji_1}N^p_{qi_2\dots i_l}\right) - \\ - A_{i_1 i_2}N^p_{qi_3\dots i_l} + RN^p_{q-2 i_1\dots i_l} + SN^{p-1}_{q-1 i_1\dots i_l}. \tag{15}$$

(N^0_0 is the unit matrix, and the matrices with negative indices are zero matrices.) Note that $H^0_0 = 0$.

A similar substitution into the contact conditions (1.36), (1.37) gives

$$[u^{(\infty)}] \sim \left(\sum_{p,q,l=0}^{\infty} \varepsilon^{p+q+l} \sum_{|i|=l} [N^p_{qi}(\xi)] D^{q,i}v \right)\Big|_{\xi=x/\varepsilon} = 0,$$

$$\left[\frac{\partial u^{(\infty)}}{\partial \nu}\right] \sim \left(\sum_{p,q,l=0}^{\infty} \varepsilon^{p+q+l-1} \sum_{|i|=l} \left[\left(A_{rm}\frac{\partial N^p_{qi}}{\partial \xi_m} + \right.\right.\right. \\ \left.\left.\left. + A_{ri_1}N^p_{qi_2\dots i_l} \right) n_r(\xi) \right] D^{q,i}v \right)\Big|_{\xi=x/\varepsilon} \sim 0. \tag{16}$$

We require that each of the expressions H^p_{qi} be a constant matrix h^p_{qi}, with the coefficients of terms of order ε^{-1} in (13) being zeroes:

$$H^p_{qi} = h^p_{qi}, \quad h^0_1 = h^0_{01} = \dots = h^0_{0s} = h^1_0 = 0, \tag{17}$$

and that we have the relations (16) due to the equalities

$$[N^p_{qi}] = 0, \quad \xi \in \Sigma_l; \\ \left[\left(A_{rm}\frac{\partial N^p_{qi}}{\partial \xi_m} + A_{ri_1}N^p_{qi_2\dots i_l} \right) n_r(\xi) \right] = 0, \quad \xi \in \Sigma_l. \tag{18}$$

If we can select N^p_{qi} and h^p_{qi} in order to get (17), (18), then the system

$$\sum_{\substack{p,q,l\geqslant 0\\ p+q+l\geqslant 2}} \varepsilon^{p+q+l-2} \sum_{|i|=l} h^p_{qi} D^{q,i} v - f(x,t) \sim 0 \tag{19}$$

can be considered as an averaged one: If v is the f.a.s. (19), then $u^{(\infty)}$ from (12) is the f.a.s. of problem (1), (2).

The construction of the N^p_{qi} and h^p_{qi} can be carried out by increasing $p+q+|i|$; consequently: We set $N^0_0 = E$; let all the N^p_{qi} and h^p_{ql} be calculated for $p+q+|i| \leqslant K_0 - 1$; we calculate the T^p_{qi} for $p+q+|i| = K$ by (15); we assume

$$h^p_{qi} = \Big\langle -A_{i_1 j}\frac{\partial N^p_{qi_2\ldots i_l}}{\partial \xi_j} - A_{i_1 i_2} N^p_{qi_3\ldots i_l} + RN^p_{q-2i_1\ldots i_l} + \\ + SN^{p-1}_{q-1\,i_1\ldots i_l}\Big\rangle \tag{20}$$

and find the 1-periodic functions N^p_{qi} for $p+q+|i| = K_0$ from the equations

$$-L_{\xi\xi} N^p_{qi} + T^p_{qi} = h^p_{qi} \tag{21}$$

with the contact conditions (18).

Theorem 1 of the Supplement implies that (20) is a necessary and sufficient condition for the solvability of problem (21), (18). It turns out with this algorithm for constructing the N^p_{qi} and h^p_{qi} that by isolating the leading term of (19), we get

$$\hat{L}v - f + \sum_{\substack{p,\,q,\,l\geqslant 0\\ p+q+l\geqslant 3}} \varepsilon^{p+q+l-2} \sum_{|i|=l} h^p_{qi} D^{q,i} v \sim 0,$$

where $\hat{L} = \hat{R}\frac{\partial^2}{\partial t^2} + \hat{S}\frac{\partial}{\partial t} - \frac{\partial}{\partial x_k}\left(\hat{A}_{kr}\frac{\partial}{\partial x_r}\right)$, $\hat{R} = \langle R\rangle$, $\hat{S} = \langle S\rangle$, $\hat{A}_{kr}$ were defined in §4.

We find the f.a.s. of (19) as the series

$$v(x,\ t) \sim \sum_{j=0}^{\infty} \varepsilon^j v_j(x,\ t), \tag{22}$$

where the $v_j(x,t)$ are infinitely differentiable functions periodic in x_q with period T.

By substituting (22) into (19) as in §2, we get the recurrent chain of problems for v_j:

$$\hat{L}v_j = f_j(x,\ t),$$

$$f_0 = f, \quad f_j = -\sum_{r=3}^{j+2}\sum_{p=0}^{r}\sum_{q=0}^{r-p}\sum_{|i|=r-q-p} h^p_{qi} D^{q,i} v_{j-r+2}, \tag{23}$$

where the f_j are defined by $v_0, \ldots, v_{j-1}$. Let us add the initial values

$$v_j(x, 0) = 0, \quad \partial v_j/\partial t\,(x, 0) = 0.$$

to the systems (23).

By carrying out the induction with respect to j and taking (11) into account, we can demonstrate that $\frac{\partial^l}{\partial t^l} D^i v_j(x, 0)$, $l = 0, 1, \ldots$ This implies

$$u^{(\infty)}(x, 0) \sim 0, \qquad \partial u^{(\infty)}/\partial t\,(x, 0) \sim 0.$$

By analogy with §2, we may consider the truncated sums

$$u^{(K)}(x, t) = \sum_{r=0}^{K+1} \varepsilon^r \sum_{q=0}^{r} \sum_{p=0}^{q} \sum_{|i|=r-q-p} N_{qi}^{p}(\xi)\, D^{q,i} v^{(K)}(x, t)\Big|_{\xi=x/\varepsilon},$$

$$v^{(K)} = \sum_{j=0}^{K} \varepsilon^j v_j(x, t)$$

and obtain the estimates of the closeness of $u^{(K)}$ and u.

The above-described formalism of averaging fits in the framework of a more general scheme in §§5.1 and 5.2.

§6 Averaging Nonstationary Nonlinear System of Equations of Elasticity Theory

Consider elastic vibrations of a bounded domain $G \subset \mathbf{R}^s$ containing a medium with an ε-periodic structure. The equations of motion are of the form

$$\rho\left(\frac{x}{\varepsilon}\right)\frac{\partial^2 u^i}{\partial t^2} - \frac{\partial}{\partial x_m}\sigma_m^i = f^i\left(x, t, \frac{x}{\varepsilon}\right), \quad x \in G; \tag{1}$$

here the u^i are the components of the translation vector u, the σ_m^i are the components of the stress tensor: $\sigma = (\sigma_1, \ldots, \sigma)$, $\sigma_m = (\sigma_m^1, \ldots, \sigma_m^s)$, $\rho(x/\varepsilon)$ is the density of the medium at the point x, the f^i are the components of the vector f and of the density of the external field of forces, and $\rho(\xi)$, $f^i(x, t, \xi)$ are 1-periodic functions. The continuity conditions for the components of the translations and for the vector of normal stresses are satisfied on the interfaces Σ_ε of different materials:

$$[u]\,|_{\Sigma_\varepsilon} = 0, \quad [\sigma n]\,|_{\Sigma_\varepsilon} = [\sigma_m n_m]\,|_{\Sigma_\varepsilon} = 0. \tag{2}$$

The stress tensor is related to the tensor of relative displacements

$$\nabla_x u = \left\| \partial u^i/\partial x_j \right\| = \begin{bmatrix} \partial u^1/\partial x_1 & \ldots & \partial u^1/\partial x_s \\ \cdot & \cdots & \cdot \\ \partial u^s/\partial x_1 & \ldots & \partial u^s/\partial x_s \end{bmatrix},$$

defined by the law

$$\sigma_l^k = a_l^k(x/\varepsilon,\ \nabla_x u), \tag{3}$$

where the $a_l^h(\xi, \nabla_x u)$ are 1-periodic functions. This form of dependence encompasses both geometric and physical type of nonlinearity.

By substituting (3) into (1) and (2), we get

$$\rho\left(\frac{x}{\varepsilon}\right)\frac{\partial^2 u}{\partial t^2} - \frac{\partial}{\partial x_m} A_m\left(\frac{x}{\varepsilon},\ \nabla_x u\right) - f\left(x,\ t,\ \frac{x}{\varepsilon}\right) = 0,\quad x \in G,\quad t > 0, \tag{4}$$

$$[u]\,|_{\Sigma_\varepsilon} = 0,\quad \left[A_m\left(\frac{x}{\varepsilon},\ \nabla_x u\right) n_m\right] = 0, \tag{5}$$

where $A_m = (a_m^1, \dots, a_m^s)^T$.

We will assume that $f(x,t,\xi)$, $a_l^k(\xi, y)$ are infinitely differentiable functions everywhere out of the interfaces Σ (up to these interfaces), that they may have a first-order discontinuity on the Σ, and the interfaces are assumed to be sufficiently smooth (see §1).

We find the asymptotics of the solution of problem (4), (5) as

$$u^{(2)} = v_0(x,\ t) + \varepsilon u_1\left(x,\ t,\ \frac{x}{\varepsilon}\right) + \varepsilon^2 u_2\left(x,\ t,\ \frac{x}{\varepsilon}\right), \tag{6}$$

where $u_1(x,t,\xi)$, $u_2(x,t,\xi)$ are 1-periodic functions.

By substituting (6) into the left side of equation (4) and by applying the chain rule, we get

$$\begin{aligned}
&\rho\left(\frac{x}{\varepsilon}\right)\frac{\partial^2 u^{(2)}\left(x,\ t,\frac{x}{\varepsilon}\right)}{\partial t^2} - \\
&\qquad - \frac{\partial}{\partial x_m}\left(A_m\left(\xi,\ \varepsilon^{-1}\nabla_\xi u^{(2)}(x,\ t,\ \xi) + \nabla_x u^{(2)}(x,\ t,\ \xi)\right)\Big|_{\xi=x/\varepsilon}\right) - \\
&\quad - f\left(x,\ t,\ \frac{x}{\varepsilon}\right) = \Big(-\varepsilon^{-1}\frac{\partial}{\partial \xi_m} A_m\big(\xi,\ \nabla_\xi u_1(x,\ t,\ \xi) + \nabla_x v_0(x,\ t) + \\
&\quad + \varepsilon\left(\nabla_\xi u_2(x,\ t,\ \xi) + \nabla_x u_1(x,\ t,\ \xi)\right) + \varepsilon^2 \nabla_x u_2(x,\ t,\ \xi)\big) + \\
&\quad + \rho(\xi)\frac{\partial^2 v_0}{\partial t^2} - \frac{\partial}{\partial x_m} A_m\big(\xi,\ \nabla_\xi u_1(x,\ t,\ \xi) + \nabla_x v_0(x,\ t) + \\
&+ \varepsilon\left(\nabla_\xi u_2(x,\ t,\ \xi) + \nabla_x u_1(x,\ t,\ \xi)\right) + \varepsilon^2 \nabla_x u_2(x,\ t,\ \xi)\big)\Big)\Big|_{\xi=x/\varepsilon} + O(\varepsilon);
\end{aligned} \tag{7}$$

here

$$\nabla_\xi u = \begin{bmatrix} \partial u^1/\partial \xi_1 & \dots & \partial u^1/\partial \xi_s \\ \cdot & \cdots & \cdot \\ \partial u^s/\partial \xi_1 & \dots & \partial u^s/\partial \xi_s \end{bmatrix}.$$

For fixed ξ, we expand $\partial A_m(\xi,y)/\partial \xi_m$ and $\partial A_m(\xi,y)/\partial x_m$ in (7) by using Taylor's formula, with the point $(\xi, \nabla_\xi u_1(x,t,\xi)+\nabla_x v_0(x,t))$ as centre. We get

$$\Big(-\varepsilon^{-1}\frac{\partial}{\partial \xi_m} A_m(\xi, \nabla_\xi u_1(x,t,\xi)+\nabla_x v_0(x,t)) -$$
$$-\frac{\partial}{\partial \xi_m}\big(\overline{\nabla}_y A_m(\xi, \nabla_\xi u_1(x,t,\xi)+\nabla_x v_0(x,t))\otimes$$
$$\otimes(\nabla_\xi u_2(x,t,\xi)+\nabla_x u_1(x,t,\xi))\big)+\rho(\xi)\frac{\partial^2 v_0}{\partial t^2}-$$
$$-\frac{\partial}{\partial x_m}A_m(\xi,\nabla_\xi u_1(x,t,\xi)+\nabla_x v_0(x,t))-f(x,t,\xi)+O(\varepsilon)\Big)\Big|_{\xi=x/\varepsilon};$$

here

$$\overline{\nabla}_y A_m(\xi,w)\otimes z=\sum_{i,j=1}^{s}\frac{\partial}{\partial y_j^i}A_m(\xi,y)\,z_j^i\Big|_{y=w},$$
$$y=\|y_j^i\|,\quad z=\|z_j^i\|,\quad w=\|w_j^i\|.$$

We require that the terms of orders ε^{-1}, ε^0 vanish everywhere except on the interfaces Σ, that is

$$\frac{\partial}{\partial \xi_m}A_m(\xi,\nabla_\xi u_1(x,t,\xi)+\nabla_x v_0(x,t))=0, \tag{8}$$

$$-\frac{\partial}{\partial \xi_m}\big(\overline{\nabla}_y A_m(\xi,\nabla_\xi u_1(x,t,\xi)+\nabla_x v_0(x,t))\otimes$$
$$\otimes(\nabla_\xi u_2(x,t,\xi)+\nabla_x u_1(x,t,\xi))\big)+\rho(\xi)\frac{\partial^2 v_0}{\partial t^2}-$$
$$-\frac{\partial}{\partial x_m}A_m(\xi,\nabla_\xi u_1(x,t,\xi)+\nabla_x v_0(x,t))-f(x,t,\xi)=0. \tag{9}$$

By substituting then (6) into the contact relations (2), we obtain contact relations on Σ for the functions u_1, u_2:

$$[u_1(x,t,\xi)]\,|_{\xi\in\Sigma}=0, \tag{10}$$
$$\big[\big(A_m(\xi,\nabla_\xi u^1(x,t,\xi)+\nabla_x v(x,t))\big)n_m\big]\,|_{\xi\in\Sigma}=0, \tag{11}$$
$$[u_2(x,t,\xi)]\,|_{\xi\in\Sigma}=0, \tag{12}$$
$$\big[\overline{\nabla}_y A_m(\xi,\nabla_\xi u_1(x,t,\xi)+\nabla_x v_0(x,t))\otimes$$
$$\otimes(\nabla_\xi u_2(x,t,\xi)+\nabla_x u_1(x,t,\xi))\,n_m\big]\,|_{\xi\in\Sigma}=0. \tag{13}$$

We shall seek the 1-periodic functions u_1, u_2 from the relations (8), (10), (11) and (9), (12), (13), respectively.

We seek a solution of problem (8), (10), (11) as $u_1(x,t,\xi)=N_1(\xi,\nabla_x v_0)$, where $N_1(\xi,y)$ is a 1-periodic solution of the parametric family of problems (y

is a matrix-valued parameter)

$$\frac{\partial}{\partial \xi_m} A_m(\xi, \nabla_\xi N_1(\xi, y) + y) = 0, \quad \xi \notin \Sigma, \tag{14}$$

$$[N_1(\xi, y)]|_\Sigma = 0, \tag{15}$$

$$[A_m(\xi, \nabla_\xi N_1(\xi, y) + y) n_m]|_\Sigma = 0. \tag{16}$$

We shall assume that the solution of (14)–(16) exists.

If $F(\xi)$ is a 1-periodic functions, then $\langle \partial F(\xi)/\partial \xi_m \rangle = 0$. So a necessary condition for the 1-periodic solution $u_2(x,t,\xi)$ of problem (9), (12), (13) to exist is the identity

$$\left\langle \rho(\xi) \frac{\partial^2 v_0}{\partial t^2} - \frac{\partial}{\partial x_m} A_m(\xi, \nabla_\xi u_1(x, t, \xi) + \nabla_x v_0(x, t)) - f(x, t, \xi) \right\rangle = 0,$$

which can be rewritten as

$$\hat{\rho} \frac{\partial^2 v_0}{\partial t^2} - \frac{\partial}{\partial x_m} \hat{A}_m(\nabla_x v_0) - \hat{f}(x, t) = 0, \tag{17}$$

where $\hat{\rho} = \langle \rho(\xi) \rangle$, $\hat{A}_m(y) = \langle A_m(\xi, \nabla_\xi N_1(\xi, y) + y) \rangle$, $\hat{f}(x,t) = \langle f(x,t,\xi) \rangle$.

It is natural that the system of equations (17) be called the averaged system: if $v_0(x,t) \in C^\infty(\mathbf{R}^s \times [0, +\infty))$ is a solution of the system of equations (17) and $u_1(x,t,\xi)$, $u_2(x,t,\xi)$ are 1-periodic solution of problems (8), (10), (11) and (9), (12), (13), infinitely differentiable everywhere except on the discontinuity surfaces A_m (up to these surfaces), then the system of equations (4) is satisfied up to $O(\varepsilon)$, the continuity condition u is satisfied exactly, and the continuity condition σn in (5) up to $O(\varepsilon^2)$.

If, in addition to the system of equations (4) and conditions (5), the boundary and initial values

$$u|_{\partial G} = \varphi(x, t)|_{\partial G}, \tag{18}$$

$$u|_{t=0} = \psi_1(x), \quad \left.\frac{\partial u}{\partial t}\right|_{t=0} = \psi_2(x), \tag{19}$$

are set, where φ, ψ_1, $\psi_2 \in C^\infty$, then we put additionally the boundary and initial conditions for v_0:

$$v_0|_{\partial G} = \varphi(x, t)|_{\partial G}, \quad v_0|_{t=0} = \psi_1(x), \quad \left.\frac{\partial v_0}{\partial t}\right|_{t=0} = \psi_2(x).$$

Then $u^{(2)}$ satisfies the conditions (18), (19) up to $O(\varepsilon)$.

The closeness of $u^{(2)}$ and u, v_0 and u can be obtained in some norm (see §5) in the linear case from a priori estimates and the smallness of the discrepancy of problem (4), (5), (18), (19).

By analogy with problem (4), (5), we may consider the nonlinear equation of elasticity theory (4) in the porous medium $\mathbf{R}^s \setminus \mathcal{A}_\varepsilon$ with the Neumann condition on $\partial \mathcal{A}_\varepsilon$:

$$A_m(x/\varepsilon, \nabla_x u)\, n_m = 0. \tag{20}$$

Below we shall give the form of the averaged system for problem (1), (20). Let $N_1(\xi, y)$ be a 1-periodic solution of the parametric family of problems

$$\frac{\partial}{\partial \xi_m} A_m(\xi, \nabla_\xi N_1(\xi, y) + y) = 0, \quad \xi \in \mathbf{R}^s \setminus \mathcal{A},$$
$$A_m(\xi, \nabla_\xi N_1(\xi, y) + y)\, n_m = 0, \quad \xi \in \partial \mathcal{A},$$

$$\widehat{\rho} = \langle \rho(\xi) \rangle^{\mathcal{A}}, \quad \widehat{A}_m(y) = \langle A_m(\xi, N_1(\xi, y) + y) \rangle^{\mathcal{A}},$$
$$\widehat{f}(x, t) = \langle f(x, t, \xi) \rangle^{\mathcal{A}}.$$

Then the averaged system is of the form (17), where $x \in \mathbb{R}^s$, $t > 0$.

Remark 1: The method described in this section can produce the asymptotics of a solution of a nonlinear system of equations in a more general form, for example

$$R\left(\frac{x}{\varepsilon}, u, x, t\right) \frac{\partial^2 u}{\partial t^2} + S\left(\frac{x}{\varepsilon}, u, x, t\right) \frac{\partial u}{\partial t} -$$
$$- \frac{\partial}{\partial x_m} A_m\left(\frac{x}{\varepsilon}, \nabla_x u, u, x, t\right) - f\left(\frac{x}{\varepsilon}, u, x, t\right) = 0,$$

where R, S are $n \times n$ matrices, A_m, f, u are n-dimensional vectors, and $x \in \mathbb{R}^s$.

General formalism of averaging nonlinear partial differential equations will be considered in Chapter 5.

§7 Averaging Stokes and Navier–Stokes Equations. The Derivation of the Percolation Law for a Porous Medium (Darcy's Law)[1]

Consider the stationary Stokes equations

$$-\nu \Delta u + \nabla_x p = f(x), \quad x \in \mathbf{R}^s \setminus \mathcal{A}_\varepsilon,$$
$$\operatorname{div} u = 0, \tag{1}$$

and the Navier–Stokes equations

[1]This section was written by D. B. Volkov.

$$-\nu\,\Delta u + u^k \frac{\partial u}{\partial x_k} + \nabla_x p = f(x), \quad x \in \mathbf{R}^s \setminus \mathscr{A}_\varepsilon, \tag{2}$$
$$\operatorname{div} u = 0,$$

with boundary values

$$u\,|_{\partial\mathscr{A}_\varepsilon} = 0, \tag{3}$$

where $\mathbf{R}^s \backslash \mathcal{A}_\varepsilon$ is a domain from §3, which simulates a porous medium.

The boundary value problems (1), (3) and (2), (3) describe a steady motion of an incompressible viscous fluid in a porous medium in linear and nonlinear approximation, respectively. Here ν is the coefficient of kinematic viscosity, u is the velocity vector, p is the ratio of the pressure head to the density, and $f(x)$ is the specific density of external mass forces. Let $f(x)$ be a T-periodic vector-valued function of the variables $x_1, \ldots, x_s$ and let it belong to the class $C^\infty(\mathbf{R}^s)$, $u(x)$ being assumed also to be a T-periodic function.

The f.a.s. $u^{(\infty)}$ and $p^{(\infty)}$ of problems (1), (3) and (2), (3) will be found as the asymptotic series

$$u^{(\infty)} \sim v(x) + \sum_{l=0}^{\infty} \varepsilon^l u_l(x, \xi),$$
$$p^{(\infty)} \sim \varepsilon^{-2} q(x) + \sum_{l=0}^{\infty} \varepsilon^{l-1} p_l(x, \xi), \tag{4}$$
$$\xi = x/\varepsilon, \quad \xi \in \mathbf{R}^s \setminus \mathscr{A}, \quad x \in \mathbf{R}^s.$$

In the linear case,

$$u_l(x, \xi) = \sum_{|i|=l} N_i(\xi)\, D^i v,$$
$$p_l(x, \xi) = \sum_{|i|=l} m_i(\xi)\, D^i v, \quad i = (i_1, \ldots, i_l), \tag{5}$$

where the $N_i \equiv N_{i_1 \ldots i_l}$ are 1-periodic $s \times s$ matrices and the $m_i \equiv m_{i_1 \ldots i_l}$ are 1-periodic vector-valued functions of dimension s.

In this section we denote by $\langle \cdot \rangle_0^{\mathcal{A}}$ the operator of averaging over period:

$$\langle f(x, \xi) \rangle_0^{\mathscr{A}} = \frac{1}{\operatorname{mes}(Q \setminus \mathscr{A})} \int_{Q \setminus \mathscr{A}} f(x, \xi)\, d\xi.$$

We require that the functions u_l satisfy the relations

$$\langle u_l(x,\xi)\rangle_0^{\mathscr{A}} = 0,$$
$$u_l(x,\xi) + \delta_{l0}v(x)\,|_{\xi\in\partial\mathscr{A}} = 0, \quad x \in \mathbf{R}^s. \tag{6}$$

In the linear case, (6) are rewritten as

$$\langle N_i(\xi)\rangle_0^{\mathscr{A}} = 0, \quad N_i + \delta_{l0}E\,|_{\partial\mathscr{A}} = 0. \tag{7}$$

Substitution of (4) into (1), considering (5), yields an asymptotic expansion for the equation of motion and incompressibility:

$$\sum_{l=0}^{\infty} \varepsilon^{l-2} \sum_{|i|=l} H_i(\xi)\, D^i v + \varepsilon^{-2}\nabla_x q - f(x) \sim 0,$$
$$\sum_{l=0}^{\infty} \varepsilon^{l-1} \sum_{|i|=l} g_i(\xi)\, D^i v + \operatorname{div} v \sim 0, \tag{8}$$

where

$$H_{i_1\ldots i_l} = -\nu\Delta_\xi N_{i_1\ldots i_l} + \left(\nabla_\xi m_{i_1\ldots i_l}\right)^T + T_{i_1\ldots i_l}, \quad \Delta_\xi = \sum_{j=1}^{s}\frac{\partial^2}{\partial\xi_j^2},$$
$$g_{i_1\ldots i_l} = \operatorname{div}_\xi N_{i_1\ldots i_l} + N^{i_1}_{i_2\ldots i_l},$$

$$T_{i_1\ldots i_l} = -2\nu\frac{\partial N_{i_2\ldots i_l}}{\partial \xi_{i_1}} - \delta_{i_1 i_2}\nu N_{i_3\ldots i_l} + e_{i_1} * m_{i_2\ldots i_l} - \nu\delta_{l2}\delta_{i_1 i_2}E.$$

Here e_{i_1} is the vector $(\delta_{i_1 1}, \delta_{i_1 2}, \ldots, \delta_{i_1 s})^T$, the divergence of a matrix is taken column-wise, the gradient of a vector is taken component-wise (see §6), $N^{i_1}_{i_2\ldots i_l}$ is the i_1th row-vector of the matrix $N_{i_2\ldots i_l}$, and the tensor product $a\otimes b$ of two vectors of dimension s form the $s\times s$ matrix c with the components $c_{ij} = a_i b_j$.

Let us select $N_{i_1\ldots i_l}$ and $m_{i_1\ldots i_l}$ so as to have (7) for each l, with $H_{i_1\ldots i_l} \equiv h_{i_1\ldots i_l}$ being some constant matrices, and $g_{i_1\ldots i_l} \equiv 0$ (cf. §2). We get the boundary value problem

$$-\nu\Delta_\xi N_{i_1\ldots i_l} + \left(\nabla_\xi m_{i_1\ldots i_l}\right)^T + T_{i_1\ldots i_l} = h_{i_1\ldots i_l},$$
$$\operatorname{div}_\xi N_{i_1\ldots i_l} + N^{i_1}_{i_2\ldots i_l} = 0, \quad \xi \in \mathbf{R}^s\setminus\mathscr{A},$$
$$N_{i_1\ldots i_l} + \delta_{l0}E\,|_{\partial\mathscr{A}} = 0, \quad \langle N_{i_1\ldots i_l}\rangle_0^{\mathscr{A}} = 0. \tag{9}$$

We represent N_i, m_i as

$$N_i(\xi) = \overline{N}_i(\xi) + R(\xi)h_i, \quad m_i(\xi) = \overline{m}_i(\xi) + S(\xi)h_i,$$
$$h_i = -(\langle R(\xi)\rangle_0^{\mathscr{A}})^{-1}\langle \overline{N}_i(\xi)\rangle_0^{\mathscr{A}}, \quad \overline{N}_\varnothing = -E, \quad \overline{m}_\varnothing = 0, \tag{10}$$

where $\overline{N}_i$, $\overline{m}_i(|i| \geqslant 1)$ are solutions of the boundary value problem (9) with $h_i = 0$, and R and S satisfy the system of equations

$$-\nu\Delta_\xi R + (\nabla_\xi S)^T = E, \quad \xi \in \mathbf{R}^s\backslash\mathscr{A},$$
$$\operatorname{div}_\xi R = 0, \quad R|_{\partial\mathscr{A}} = 0. \tag{11}$$

It can be verified directly that the substitution of (10) into (9) yields the identity. The functions $\overline{m}_i$ and S are defined up to an arbitrary constant. We can fix it by requiring that $\langle\overline{m}_i\rangle_0^{\mathcal{A}} = 0$, $\langle S\rangle_0^{\mathcal{A}} = 0$.

Now we shall prove several statements which imply the solvability of the boundary value problem (9) for any i.

Lemma 1. *$\langle R\rangle_0^{\mathcal{A}}$ is a symmetric, positive-definite matrix (see [214, p. 139]).*

Lemma 2. *Let $F(\xi) \in L_2(Q\backslash\mathcal{A})$ be a 1-periodic function $u\langle F\rangle_0^{\mathcal{A}} = 0$ and the boundary $\partial\mathcal{A}$ be sufficiently smooth. Then there exists a 1-periodic vector-valued function $a(\xi) \in W_2^{\text{loc}}$ such that $\operatorname{div}_\xi a = F$, $\xi \in \mathbf{R}^s\backslash\mathcal{A}$, $a|_{\partial\mathcal{A}} = 0$. (The notation $f(\xi) \in W_2^{loc}$ here means that $f(\xi) \in W_2^1(G)$ for any compact set $G \subset R^s\backslash\mathcal{A}$.)*

Proof: The vector-valued function a is constructed in a manner similar to that described in [75], Chapter 1, §2. We represent a as $a(\xi) = b(\xi) + c(\xi)$, where $b = \nabla_\xi\varphi$, $(b,n)|_{\partial\mathcal{A}} = 0$, $c - \operatorname{rot}_\xi d$, $c|_{\partial\mathcal{A}} = -b|_{\partial\mathcal{A}} \equiv \beta$, $(\beta, n)|_{\partial\mathcal{A}} = 0$.

The potential φ satisfies the Neumann boundary value problem

$$\Delta\varphi = F, \quad \xi \in \mathbf{R}^s\backslash\mathscr{A}, \quad \frac{\partial\varphi}{\partial n}\bigg|_{\partial\mathscr{A}} = 0,$$

which is solvable in the class of periodic functions if and only if $\langle F\rangle_0^{\mathcal{A}} = 0$. If the boundary $\partial\mathcal{A}$ is sufficiently smooth, for example $\partial\mathcal{A} \subset C^2$, then $\varphi \in W_2^2(Q\backslash\mathcal{A})$, $\beta \in W_2^{1/2}(\partial\mathcal{A})$ (see [76]).

The vector c was constructed in [75], while d can be chosen to be equal to zero in $Q\backslash B$, where B is a domain satisfying the condition $\overline{\mathcal{A}} \subset \overline{B}$, $\overline{B} \subset Q$; thus permitting to extend $d(\xi)$ periodically to the whole space $\mathbf{R}^s$. For $\beta \in W_2^{1/2}(\partial\mathcal{A})$ (the boundary is twice differentiable with bounded derivative) $d \in W_2^2(Q\backslash\mathcal{A})$ (see [75]). Since $\varphi \in W_2^2(Q\backslash\mathcal{A})$, then $\beta \in W_2^{1/2}(\partial\mathcal{A})$, and, consequently, $a \in W_2^1(Q\backslash\mathcal{A})$. The lemma is proved.

Theorem 1. *Let 1-periodic vector-valued functions $F_k(k = 0, 1, \ldots, s)$ and a function φ belong to $L_2(Q\backslash\mathcal{A})$ and let the boundary $\partial\mathcal{A}$ be infinitely smooth. Then the system of equations*

$$-\nu\Delta u+\nabla p+F_0+\partial F_h/\partial\xi_h=0,\quad \xi\in\mathbf{R}^s\setminus\mathcal{A},\qquad(12)$$
$$\operatorname{div}u+\varphi=0,\quad u|_{\partial\mathcal{A}}=0,$$

has a unique generalized solution in the class of periodic functions if and only if $\langle\varphi\rangle_0^{\mathcal{A}}=0$, *while* $u\in W_2^1(Q\backslash\mathcal{A})$, $p\in L_2(Q\backslash\mathcal{A})$.

Proof: Necessity. We integrate the second equation in (12) over $Q\backslash\mathcal{A}$, use the formula of integration by parts for periodic functions from $W_2^1(Q\backslash\mathcal{A})$, and take into account the boundary value $u|_{\partial\mathcal{A}}=0$. We get

$$0=\langle\operatorname{div}u\rangle_0^{\mathcal{A}}+\langle\varphi(\xi)\rangle_0^{\mathcal{A}}=\frac{1}{\operatorname{mes}(Q\setminus\mathcal{A})}\int\limits_{\partial\mathcal{A}\cap Q}(u,n)\,dS+\langle\varphi\rangle_0^{\mathcal{A}}=\langle\varphi\rangle_0^{\mathcal{A}}.$$

Sufficiency. We seek the solution as $u=a+w$, $a\in W_2^1(Q\backslash\mathcal{A})$, where $\operatorname{div}a+\varphi=0$, $\xi\in\mathbf{R}^s\backslash\mathcal{A}$, $a|_{\partial\mathcal{A}}=0$. The existence of a is proved by Lemma 2; the function w satisfies the system of equations

$$-\nu\Delta w+\nabla p+F_0+\frac{\partial(F_k-\nu\varphi_k)}{\partial\xi_k}=0,\quad \operatorname{div}w=0,\quad w|_{\partial\mathcal{A}}=0,$$

where $\varphi_i=\partial a/\partial\xi_i\in L_2(Q\backslash\mathcal{A})$. The solvability of this system in the class of the 1-periodic functions $w\in W_2^1(Q\backslash\mathcal{A})$ and $p\in L_2(Q\backslash\mathcal{A})$ is investigated in a standard fashion [75]. The theorem is proved.

Thus, for any i, problem (9) is solvable. The unknown v, q are given by the averaged system

$$\varepsilon^{-2}(hv+\nabla_x q)+\sum_{l=1}^{\infty}\varepsilon^{l-2}\sum_{|i|=l}h_iD^iv\sim f(x),\qquad \operatorname{div}v\sim 0,\ x\in\mathbf{R}^s.\qquad(13)$$

Solution (13) is sought as the series

$$v\sim\varepsilon^2\sum_{j=0}^{\infty}\varepsilon^jv_j(x),\quad q\sim\varepsilon^2\sum_{j=0}^{\infty}\varepsilon^jq_j.$$

After substituting v and q into (13), we see that v_j satisfies the Darcy equations with recurrently defined right sides:

$$hv_j+\nabla_xq_j=\Phi_j(x),\qquad \operatorname{div}v_j=0,$$

where $\Phi_j=\delta_{j0}f(x)-\sum_{k=1}^{j}\sum_{|i|=k}h_iD^iv_{j-k}$. The f.a.s. of problem (1), (3) is constructed.

In particular, for v_0 we have

$$hv_0+\nabla_xq_0=f(x),\quad \operatorname{div}v_0=0.$$

For $f(x) = 0$, Darcy's law states that the mean velocity is directly proportional to the mean pressure gradient with the proportionality factor: $h^{-1} = \langle R(\xi)\rangle_0^A$, where $R(\xi)$ is a solution of "cell problem" (11). The matrix h^{-1} is called the percolation tensor of a porous medium.

Note that the mean over period of $u^{(\infty)}$ and $p^{(\infty)}$ coincide with the solution of the averaged system: $\langle u^{(\infty)}(x,\xi)\rangle_0^A \sim v(x)$, $\langle p^{(\infty)}(x,\xi)\rangle_0^A \sim \varepsilon^{-2}q(x)$.

The Navier–Stokes equations (2) are investigated in a similar manner. After substituting (4) into (2), we get

$$\sum_{l=0}^{\infty} \varepsilon^{l-2} H_l(x,\xi) + \varepsilon^{-2}\nabla_x q - f(x) \sim 0,$$

$$\sum_{l=0}^{\infty} \varepsilon^{l-1} g_l(x,\xi) + \operatorname{div} v \sim 0,$$

where

$$H_l = -\nu\Delta_\xi u_l + \nabla_\xi p_l - \nu\left(\frac{\partial^2 u_{l-1}}{\partial\xi_j\partial x_j} + \frac{\partial^2 u_{l-1}}{\partial x_j\partial\xi_j}\right) -$$

$$-\nu\Delta_x u_{l-2} + \nabla_x p_{l-1} + \sum_{k=0}^{l-1} u_k^r \frac{\partial u_{l-1-k}}{\partial\xi_r} + v^r\frac{\partial u_{l-1}}{\partial\xi_r} +$$

$$+\sum_{k=0}^{l-2} u_k^r\frac{\partial u_{l-2-k}}{\partial x_r} + v^r\frac{\partial u_{l-2}}{\partial x_r} + u_{l-2}^r\frac{\partial v}{\partial x_r} + \delta_{l2}\left(-\nu\Delta_x v + v_r\frac{\partial v}{\partial x_r}\right),$$

$$g_l = \operatorname{div}_\xi u_l + \operatorname{div}_x u_{l-1}.$$

We choose u_l, p_l as before so as to have (6) and

$$H_l(x,\xi) = h_l(x),$$
$$g_l(x,\xi) = 0 \quad \forall x \in \mathbf{R}^s. \tag{14}$$

For $l = 0$, the equation does not include nonlinear terms; hence the solution takes the same form as in the linear case: $u_0 = N(\xi)v(x)$, $p_0 = m(\xi)v(x)$, $h_0 = hv$. It means that the Darcy equations are the leading terms of the averaged system, as they are in the linear case. It can be proved by induction that the linear terms in H_l and g_l coincide with $\sum_{|i|=l} H_i D^i v$ and $\sum_{|i|=l} g_i D^i v$ from (8), and the nonlinear terms are of the form

$$\Lambda = \sum_{k=1}^{l}\Lambda_k,$$

$$\Lambda_k = \sum_{|I_1|=0}^{l-k}\sum_{|I_2|=0}^{l-k}\cdots\sum_{|I_{k+1}|=0}^{l-k} P_{\alpha_1\alpha_2\ldots\alpha_{k+1}}^{I_1 I_2\ldots I_{k+1}}(\xi)\, D^{I_1}v^{\alpha_1} D^{I_2}v^{\alpha_2}\ldots D^{I_{k+1}}v^{\alpha_{k+1}}, \tag{15}$$

where the v^α are the components of the vector $v(x)$, the I with indices are multi-

indices ($I = (i_l, \dots, i_r)$, $0 \leqslant r \leqslant l-k$), $D^I = \partial^{|I|}/\partial x_{i_1} \dots \partial x_{i_1} \dots \partial x_{i_r}$, *and the* $P(\xi)$ *with indices are known periodic functions defined unambiguously from the step* $l-1$. *Summation over repeated Greek indices from 1 to s is understood.*

Thus, Λ_k *includes products of the terms that are all possible derivatives of orders* $0 \leqslant r \leqslant l-k$ *of the components of* $v(x)$. *The functions* u_l, p_l, h_l *should be sought for as sums (5) and nonlinear additions with structure (15). By substituting these into (4) and by equating the coefficients of nonlinear terms, we get a series of problems of type (9), with the coefficients of the linear terms giving problem (9) exactly.*

For the Stokes problem, the asymptotic expansions $u \sim u^{(\infty)}$, $p \sim p^{(\infty)}$ *are justified following the standard scheme: We substitute the truncated series* $u^{(K)}$, $p^{(K)}$ *into system (1) or (2), demonstrate the smallness of the discrepancy and estimate the error* $(u - u^{(K)}, p - p^{(K)})$ *by using the a priori estimates from [75, 77] similar to those given in Chapter 1. In the case of the problem in the bounded domain G, the closeness of zero approximation to a solution of problem (1) (in the weak sense) was justified in [214].*

If the domain G is a layer, then the asymptotics $u \sim u^{(\infty)}$, $p \sim p^{(\infty)}$ *is constructed by means of the boundary layer technique set forth in Chapter 9, and it was justified (this time in the strong sense) in [40].*

REMARK 1: If the right side $f(x)$ in system (2) is replaced by $f(x)/\varepsilon$, then the expansion u in (4) starts with the terms of order ε^{-1}, and the averaged equation will look differently.

§8 Averaging in Case of Short-Wave Propagation

The foregoing sections discussed the averaging of problems with the characteristic size of the solution variation being much larger than the period of the structure. Now we consider the propagation of high-frequency waves with the wavelength much smaller than the period of the structure for the case of one unknown function and one space variable, which is described by the equation

$$L_{x,t}u = R(\xi)\frac{\partial^2 u}{\partial t^2} - \frac{\partial}{\partial x}\left(A(\xi)\frac{\partial u}{\partial x}\right) = 0, \tag{1}$$

where $\xi = x/\varepsilon$ and R, A are infinitely differentiable, 1-periodic functions. A strongly oscillating solution of equation (1) is sought by the standard scheme of wave physics as

$$u = Se^{i\varphi/\eta}, \tag{2}$$

where S and φ are real-valued.

Next we assume that $\eta \ll \varepsilon$; we set $\mu = \eta^2/\varepsilon^2$, $b = A/R$, $B = RS^2$. By substituting u into (1), we get

$$L_{x,t}u = \left(-\frac{RS}{\eta^2} L_1(\varphi, B) + \frac{i}{S\eta} L_2(\varphi, B)\right) e^{i\varphi/\eta},$$

$$L_1(\varphi, B) = \left(\frac{\partial\varphi}{\partial t}\right)^2 - b\left(\frac{\partial\varphi}{\partial x}\right)^2 -$$
$$- \varepsilon^2\mu\left(\frac{1}{\sqrt{B}}\frac{\partial^2\sqrt{B}}{\partial t^2} - \frac{1}{\sqrt{BR}}\frac{\partial}{\partial x}\left(bR\frac{\partial}{\partial x}\sqrt{\frac{B}{R}}\right)\right), \qquad (3)$$
$$L_2(\varphi, B) = \frac{\partial}{\partial t}\left(B\frac{\partial\varphi}{\partial t}\right) - \frac{\partial}{\partial x}\left(bB\frac{\partial\varphi}{\partial x}\right).$$

By equating the real and imaginary terms to zero, we get the equations

$$L_1(\varphi, B) = 0, \quad L_2(\varphi, B) = 0. \qquad (4)$$

The solution is sought as the asymptotic series

$$\varphi \sim \psi(x, t) + \sum_{n=1}^{\infty}\sum_{m=0}^{\infty} \varepsilon^n\mu^m N_{nm}(x, t, \xi),$$
$$B \sim \sum_{n=0}^{\infty}\sum_{m=0}^{\infty} \varepsilon^n\mu^m Q_{nm}(x, t, \xi).$$

Let us substitute the series into (4); we expand the resultant expressions into series in powers of ε, μ. We will try to make the expansion coefficients of the negative powers of ε and μ vanish and to make those of the positive powers independent of the fast variable ξ.

We have

$$L_1(\varphi, B) \sim \sum_{k,l=0}^{\infty} \varepsilon^k\mu^l H_{kl}, \quad L_2(\varphi, B) \sim \sum_{k,l=0}^{\infty} \varepsilon^{k-1}\mu^l G_{kl},$$

where, in particular,

$$H_{00} = \left(\frac{\partial\psi}{\partial t}\right)^2 - b\left(\frac{\partial\psi}{\partial x} + \frac{\partial N_{10}}{\partial\xi}\right)^2,$$
$$G_{00} = -\frac{\partial}{\partial\xi}\left(b\left(\frac{\partial\psi}{\partial x} + \frac{\partial N_{10}}{\partial\xi}\right)Q_{00}\right).$$

We require that H_{00} be independent of the fast variable

$$H_{00} = h_{00}(x, t), \quad \left(\frac{\partial\psi}{\partial t}\right)^2 - b\left(\frac{\partial\psi}{\partial x} + \frac{\partial N_{10}}{\partial\xi}\right)^2 = h_{00}(x, t).$$

This implies

$$\sqrt{\frac{(\partial\psi/\partial t)^2 - h_{00}}{b}} = \frac{\partial\psi}{\partial x} + \frac{\partial N_{10}}{\partial\xi}. \tag{5}$$

By taking the mean over period $\langle\cdot\rangle$, we get $\sqrt{\left(\frac{\partial\Psi}{\partial t}\right)^2 - h_{00}}\left\langle\frac{1}{\sqrt{b}}\right\rangle = \frac{\partial\Psi}{\partial x}$. Consequently, $h_{00} = \left(\frac{\partial\Psi}{\partial t}\right)^2 - \hat{c}^2\left(\frac{\partial\Psi}{\partial x}\right)^2$, where

$$\frac{1}{\hat{c}} = \left\langle\frac{1}{\sqrt{b}}\right\rangle. \tag{6}$$

We have obtained the following equation as the zeroth approximation:

$$h_{00} = \left(\frac{\partial\psi}{\partial t}\right)^2 - \hat{c}^2\left(\frac{\partial\psi}{\partial x}\right)^2 = 0,$$

which is an equation for the phase of a high-frequency wave in a homogeneous medium with the perturbation speed $\hat{c}$. The value $\sqrt{b(\xi)} = c(\xi)$ equals the local perturbation velocity, and therefore, formula (6) is quite natural for the average speed: If the leading edge of the signal travels at the speed $c(\xi)$, it will take just the time $\int_0^\varepsilon \frac{dx}{c(\xi)} = \frac{\varepsilon}{\hat{c}}$, for the leading edge to pass over the period ε of the structure, i.e., its average speed is equal to $\hat{c}$.

By substituting h_{00} into (6), we get

$$\frac{\partial N_{10}}{\partial\xi} = \left(\frac{\hat{c}}{\sqrt{b}} - 1\right)\frac{\partial\psi}{\partial x}. \tag{7}$$

The function

$$N_{10} = \left(\int_0^\xi\left(\frac{\hat{c}}{\sqrt{b(\tau)}} - 1\right)d\tau\right)\frac{\partial\psi}{\partial x}$$

is periodic in ξ, since $\langle\hat{c}/\sqrt{b(\xi)}\rangle = 1$. We require that the equality $G_{00} = 0$ hold. Then

$$bQ_{00}\left(\frac{\partial\psi}{\partial x} + \frac{\partial N_{10}}{\partial\xi}\right) = g_{00}(x, t). \tag{8}$$

By substituting $\hat{c}\partial\psi/\partial x$ for $\sqrt{b}(\partial\Psi/\partial x + \partial N_{10}/\partial\xi)$ in (8), we get $\sqrt{b}Q_{00}\hat{c}\frac{\delta\psi}{\partial x} = g_{00}(x,t)$. For $Q_{00} = h(x,t)/\sqrt{b}$, with $h(x,t)$ being arbitrary, the left side of the latter expression is independent of ξ, and (8) holds for $g_{00}(x,t) = h(x,t)\hat{c}\partial\psi/\partial x$.

For definiteness, we take Q_{00} as $Q_{00} = \frac{\hat{c}}{\sqrt{b}}q(x,t)$, where $q(x,t)$ is an unknown function; then $\langle Q_{00}\rangle = q(x,t)$.

The coefficients H_{kl}, G_{kl} for $(k,l) \neq (0,0)$ are of the form

$$H_{kl} = -b\left(\frac{\partial\psi}{\partial x} + \frac{\partial N_{10}}{\partial\xi}\right)\frac{\partial N_{k+1,l}}{\partial\xi} + T_{kl},$$

$$G_{kl} = -\frac{\partial}{\partial\xi}\left(b\left(\frac{\partial\psi}{\partial x} + \frac{\partial N_{10}}{\partial\xi}\right)Q_{kl}\right) + W_{kl};$$

The T_{kl} are defined by the definition of N_{ij} for $i \leqslant k$, $j \leqslant l$, and Q_{ij} for $i \leqslant k$, $j \leqslant l-1$; W_{kl} by the definition of N_{ij} for $i \leqslant k+1$, $j \leqslant 1$, and Q_{ij} for $i \leqslant k-1$, $j \leqslant l-1$. Due to (7) we have

$$H_{kl} = \sqrt{b}\,\hat{c}\,\frac{\partial\psi}{\partial x}\frac{\partial N_{k+1,l}}{\partial\xi} + T_{kl}.$$

We require that the following equality hold:

$$H_{kl} = h_{kl}(x,\ t); \tag{9}$$

then $\hat{c}\frac{\partial\psi}{\partial x}\frac{\partial N_{k+1,l}}{\partial\xi} + \frac{T_{kl}}{\sqrt{b}} = \frac{h_{kl}}{\sqrt{b}}$. By taking the mean over period, we get $\langle T_{kl}/\sqrt{b}\rangle = h_{kl}\langle 1/\sqrt{b}\rangle$, that is, $h_{kl} = \langle T_{kl}/\sqrt{b}\rangle\langle 1/\sqrt{b}\rangle^{-1}$. Next we find $N_{k+1,l}$. We require that the following condition be satisfied: $G_{kl} = g_{kl}(x,t)$ is independent of ξ. By applying $\langle\cdot\rangle$, we get

$$g_{kl} = \langle G_{kl}\rangle = \langle W_{kl}\rangle. \tag{10}$$

From the equation $G_{kl} = W_{kl}$ in the function Q_{kl}, we have

$$\frac{\partial}{\partial\xi}\left(\sqrt{b}\,\hat{c}\,\frac{\partial\psi}{\partial x}Q_{kl}\right) = W_{kl} - \langle W_{kl}\rangle. \tag{11}$$

Hence we find Q_{kl} (periodic in ξ).

We set the following order of seeking the h_{kl}, N_{kl}, g_{kl}, and Q_{kl} (other arrangements are also possible). Suppose all the Q_{kl} with $k+l < p$ and N_{kl} with $k+l \leqslant p$ have already been found. Then all the values of T_{kl} with $k+l = p+1$ are known, and from (9) we find the N_{kl} with $k+l = p+1$. Now we know all the W_{kl} with $k+l = p$, and from (11) we find the Q_{kl} with $k+l = p$. We have passed to $p+1$.

Under condition (10) all the $g_{0l} = 0$ for $l > 0$; let us define g_{10}. We have

$$g_{10} = \langle G_{10}\rangle = \frac{\partial}{\partial t}\left(\left\langle\frac{\hat{c}}{\sqrt{b}}\right\rangle q\frac{\partial\psi}{\partial t}\right) - \frac{\partial}{\partial x}\left(\left\langle\sqrt{b}\,\hat{c}\left(\frac{\partial\psi}{\partial x} + \frac{\partial N_{10}}{\partial\xi}\right)\right\rangle q\right).$$

According to the definition of $\hat{c}$, we have $\langle\hat{c}/\sqrt{b}\rangle = 1$, and according to (7),

$$\sqrt{b}\,\hat{c}\left(\frac{\partial\psi}{\partial x} + \frac{\partial N_{10}}{\partial\xi}\right) = \hat{c}^2\frac{\partial\psi}{\partial x};$$

therefore,

$$\left\langle \sqrt{b}\,\hat{c}\left(\frac{\partial \psi}{\partial x} + \frac{\partial N_{10}}{\partial \xi}\right)\right\rangle = \hat{c}^2 \frac{\partial \psi}{\partial x}.$$

Thus,

$$g_{10} = \frac{\partial}{\partial t}\left(q \frac{\partial \psi}{\partial t}\right) - \frac{\partial}{\partial x}\left(\hat{c}^2 q \frac{\partial \psi}{\partial x}\right). \tag{12}$$

We have obtained the averaging of infinite order of accuracy

$$\begin{aligned} \hat{L}_1(\psi, q) &\sim \sum_{k,l=0}^{\infty} \varepsilon^k \mu^l h_{kl} \sim 0, \\ \hat{L}_2(\psi, q) &\sim \sum_{k,l=0}^{\infty} \varepsilon^k \mu^l g_{k+1,l} \sim 0, \end{aligned} \tag{13}$$

where the leading terms h_{00}, g_{10} were given by (7), (12). The remaining h_{kl} g_{kl} are calculated by this recurrent process, and each of them turns out to be a differential operator of finite order for the functions ψ and q. Note that the N_{kl} and Q_{kl} and, consequently, the subsequent h_{kl}, g_{kl} are defined up to an arbitrary term independent of ξ. Should the arbitrary terms be chosen more elaborately, as dependent of ψ, q, t, x, the result would be an averaged system containing accordingly a more complex dependence on ψ, q, t, x.

Solution of the system can be sought as a regular expansion

$$\psi \sim \sum_{k,l=0}^{\infty} \varepsilon^k \mu^l \psi_{kl}(x, t), \quad q \sim \sum_{k,l=0}^{\infty} \varepsilon^k \mu^l q_{kl}(x, t).$$

Consider the wave equation

$$\frac{\partial^2 w}{\partial t^2} - \frac{\partial}{\partial x}\left(\hat{c}^2 \frac{\partial w}{\partial x}\right) = 0 \tag{14}$$

with the constant coefficient $\hat{c}$. We shall seek a solution of this equation, strongly oscillating in x, as (2); by repeating the calculations given above for $R = R(\xi)$ and $A = A(\xi)$, we see that this solution is of the form $v \sim (q + \eta q_1 + \dots)e^{i(\varphi/\eta + \dots)}$, where q and φ are solutions of the system $h_{00} = 0$, $g_{10} = 0$. So, as far as the short-wave propagation is concerned, the equations (1), (14) are equivalent asymptotically.

Consider the case when the waves are not necessarily orthogonal to the layers. The averaging procedure differs from that described above only in technicalities (more general problems covering the cases of the system of equations of elasticity theory for a medium with the many-dimensional periodicity domain were considered in [12]).

As a model, we can take the equation

$$R(\xi)\frac{\partial^2 u}{\partial t^2} = \frac{\partial}{\partial x_i}\left(A(\xi)\frac{\partial u}{\partial x_i}\right),$$

where summation over repeated indices from 1 to 3 is understood.

Then $H_{00} = c^2(\xi)((\partial\psi/\partial x_1 + \partial N_{10}/\partial\xi_1)^2 + (\partial\psi/\partial x_2)^2 + (\partial\psi/\partial x_3)^2)$. We require that H_{00} be equal to $h_{00}(x,t)$; we get

$$\frac{\partial\psi}{\partial x_1} + \frac{\partial N_{10}}{\partial\xi_1} = \pm\sqrt{\frac{h_{00}}{c^2(\xi)} - \left(\frac{\partial\psi}{\partial x_2}\right)^2 - \left(\frac{\partial\psi}{\partial x_3}\right)^2}. \tag{15}$$

After applying the operation $\langle\cdot\rangle$, we obtain an equation that relates h_{00} and ψ:

$$\frac{\partial\psi}{\partial x_1} = \left\langle\sqrt{\frac{h_{00}}{c^2(\xi_1)} - \left(\frac{\partial\psi}{\partial x_2}\right)^2 - \left(\frac{\partial\psi}{\partial x_3}\right)^2}\right\rangle. \tag{16}$$

For definiteness we have taken + on the right side, which corresponds to the waves travelling at an acute angle to the x_1-axis.

Let $c_1 = \sup c(\xi_1)$, provided

$$\sqrt{\left(\frac{\partial\psi}{\partial x_2}\right)^2 + \left(\frac{\partial\psi}{\partial x_3}\right)^2}\left(\left|\frac{\partial\psi}{\partial x_1}\right|\right)^{-1}\left\langle\sqrt{\frac{c_1^2}{c^2(\xi_1)} - 1}\right\rangle \leqslant 1 \tag{17}$$

equation (16) is solvable for h_{00}.

In fact, consider the function

$$H(y) = \left|\frac{\partial\psi}{\partial x_1}\right| - \left\langle\sqrt{\frac{y}{c^2(\xi_1)} - \left(\frac{\partial\psi}{\partial x_2}\right)^2 - \left(\frac{\partial\psi}{\partial x_3}\right)^2}\right\rangle.$$

We have

$$H\left(c_1^2\left(\left(\frac{\partial\psi}{\partial x_2}\right)^2 + \left(\frac{\partial\psi}{\partial x_3}\right)^2\right)\right) =$$

$$= \left|\frac{\partial\psi}{\partial x_1}\right| - \sqrt{\left(\frac{\partial\psi}{\partial x_2}\right)^2 + \left(\frac{\partial\psi}{\partial x_3}\right)^2}\left\langle\sqrt{\frac{c_1^2}{c^2(\xi_1)} - 1}\right\rangle \geqslant 0;$$

and according to (17),

$$H\left(c_1^2\left(\left(\frac{\partial\psi}{\partial x_1}\right)^2+\left(\frac{\partial\psi}{\partial x_2}\right)^2+\left(\frac{\partial\psi}{\partial x_3}\right)^2\right)\right)=$$

$$=\left|\frac{\partial\psi}{\partial x_1}\right|-\left\langle\left(\frac{(\partial\psi/\partial x_1)^2 c_1^2}{c^2(\xi_1)}+\left(\left(\frac{\partial\psi}{\partial x_2}\right)^2+\left(\frac{\partial\psi}{\partial x_3}\right)^2\right)\left(\frac{c_1^2}{c^2(\xi_1)}-1\right)\right)^{1/2}\right\rangle\leqslant$$

$$\leqslant\left|\frac{\partial\psi}{\partial x_1}\right|-\left|\frac{\partial\psi}{\partial x_1}\right|\left\langle\frac{c_1}{c(\xi_1)}\right\rangle\leqslant 0.$$

Thus, for some intermediate y, the continuous function $H(y)$ vanishes. Having defined $h_{00}(\partial\psi/\partial x_1, \partial\psi/\partial x_2, \partial\psi/\partial x_3)$, we find N_{10} from equation (15). Consider a periodic system composed of layers of substances with different values of the coefficients $c(\xi)$:

$$c(\xi)=\begin{cases} c_1, & 0\leqslant\xi<\theta_1, \\ c_2, & \theta_1\leqslant\xi<1, \end{cases} \qquad c_1>c_2.$$

The solvability condition takes the form

$$\sqrt{\left(\frac{\partial\psi}{\partial x_2}\right)^2+\left(\frac{\partial\psi}{\partial x_3}\right)^2}\left(\left|\frac{\partial\psi}{\partial x_1}\right|\right)^{-1}\left(\theta_2\sqrt{\frac{c_1^2}{c_2^2}-1}\right)\leqslant 1.$$

Thus, the tangent of the angle formed by the normal to the wave front and the x_1-axis does not exceed $\left(\theta_2\sqrt{c_1^2/c_2^2-1}\right)^{-1}$

It may appear at first sight that this circumstance does not make sense; it is well known that the high-frequency perturbations propagate following the laws of geometric optics, and therefore the light beam can go at an angle to the x_1-axis arbitrarily close to π/s. But in fact, there is no inconsistency here. The point is that after averaging, we get an anisotropic medium, where the angle between the direction of the beam and the wave front is other than $\pi/2$.

The problem considered in this section has an important distinction as compared with the ones discussed above and to be discussed below. It was said in Chapter 1 that the investigation of equations with nonsmooth (discontinuous) coefficients may be reduced to investigating those with smooth coefficients by smoothing as we take the sequence of infinitely differentiable coefficients $R_k(\xi)$, $A_k(\xi)$, $\|R_k(\xi)-R(\xi)\|_{L_p}\to 0$, $\|A_k(\xi)-A(\xi)\|_{L_p}\to 0$ as $k\to\infty$ for any $p(1\leqslant p<\infty)$. We average the problems for the coefficients A_k, R_k and pass to the limit as $k\to\infty$. It turns out that in all the other cases the coefficients of averaged equations tend to some finite limits that determine the coefficients of the averaged equation for the original problem. In the case considered here, we fail to justify this statement for all the coefficients h_{nm}, g_{nm} of (13).

It is customary to assume that oscillations with a wavelength η much smaller than the period of a structure ε cannot travel through this structure, except for a discrete set of wavelengths. Let us analyze this statement for the case of the infinitely differentiable $R(\xi)$ and $A(\xi)$.

The system of equations $h_{00} = 0$, $g_{10} = 0$ is satisfied by the pair of functions $\varphi_0 = x - \hat{c}t$, $q_0 = 1$. By using the method of regular expansions, we can construct an asymptotic solution of (13) as the finite series in powers of ε, μ:

$$\psi^{(l,r)} = \varphi_0 + \sum_{n=1}^{l} \sum_{m=0}^{r} \varepsilon^n \mu^m \psi_{nm}(x, t),$$

$$q^{(l,r)} = q_0 + \sum_{n=1}^{l} \sum_{m=0}^{r} \varepsilon^n \mu^m q_{nm}(x, t),$$

which satisfies (13) up to $O(\varepsilon^{l+1} + \mu^{r+1})$. It can serve as a basis for constructing the finite series in powers of ε, μ:

$$\varphi^{(l,r)} = \varphi_0(x, t) + \sum_{n=1}^{l} \sum_{m=0}^{r} \varepsilon^n \mu^m N_{nm},$$

$$B^{(l,r)} = \sum_{n=0}^{l} \sum_{m=0}^{r} \varepsilon^n \mu^m Q_{nm}, \quad Q_{00} = \frac{\hat{c}}{c(\xi)},$$

which satisfy (4) up to $O\left(\frac{\varepsilon^{l+1} + \mu^{r+1}}{\varepsilon}\right)$. Then for $(l, r) = \sqrt{B^{(l,r)}/R} e^{i\varphi^{(l,r)}/\eta}$, and due to (3), we get

$$L_{x,t} u^{(l,r)} = O\left(\frac{\varepsilon^{l+1} + \mu^{r+1}}{\varepsilon \eta^2}\right).$$

Consider the case $\eta = \varepsilon(\varepsilon/X)^\gamma$, $\gamma > 0$, X being the characteristic size of the domain where the process is investigated. Then for any ω with sufficiently large l, r, we have

$$L u^{(l,r)} = O(\varepsilon^\omega).$$

Let u be a solution of (1), with the initial data close to those for $u^{(l,r)}$,

$$\| u(x, 0) - u^{(l,r)}(x, 0) \|_{C^1} = O(\varepsilon^\omega),$$

$$\left\| \frac{\partial u}{\partial t}(x, 0) - \frac{\partial u^{(l,r)}}{\partial t}(x, 0) \right\|_{C^1} = O(\varepsilon^\omega);$$

from the closeness estimate for solutions of (1) and by the theory of differential equations, we obtain $\sup_x |u(x,t) - u^{(l,r)}(x,t)| \leqslant c(t)\varepsilon^\omega < \infty$ for any $t < \infty$.

The solution $u^{(l,r)}$ is of the form

$$u^{(l,r)} \sim \left(\sqrt{\frac{\hat{c}}{cR}} + \dots \right) \exp\left\{ i \frac{x - \hat{c}t}{\eta} + \dots \right\}$$

and it decays a little in passing through the periodic structure under consideration.

It is the impossibility of passing to the limit in the case of the discontinuous coefficients mentioned above that appears to make the difference between the cases of smooth and discontinuous coefficients. The latter was considered in [59].

Many materials have transition zones in the region of interface of different substances, where there is a smooth transition from the properties of the matter in one area to the properties of the matter in the other area.

If we suppose that the characteristic size of variation of coefficients in this zone is of order $\delta \gg \eta(\delta = \varepsilon(\varepsilon/X)^{\beta}, \beta < \gamma)$, then by the same reasoning as above, we can justify the statement that the solution with the wavelength $\eta \ll \varepsilon$ propagates without much decay.

§9 Averaging the Transition Equation for a Periodic Medium

The numerical solution of the transition equation for a periodic medium is very laborious. In view of this fact, an approximation of the problem for a periodic medium by a problem for a homogeneous medium (homogenization) is appropriate.

Consider the multi-speed transition equation

$$L\varphi = \frac{\partial\varphi}{\partial t} + v_j \frac{\partial\varphi}{\partial x_j} + \sigma\varphi - P\varphi = F; \tag{1}$$

here

$$\varphi = \varphi(x, t, v), \quad \sigma = \sigma(\xi, x, t, v), \quad \sigma, \Sigma \geqslant 0,$$

$$F = F(\xi, x, t, v), \quad P\varphi = \int_{\mathbf{R}^3} \Sigma(\xi, x, t, v, v')\, \varphi(x, t, v')\, dv',$$

$$x \in \mathbf{R}^3, \quad \xi = x/\varepsilon \quad v = (v_1, v_2, v_3), \quad v' = (v_1', v_2', v_3');$$

with summation over repeated indices between $1 \leqslant j \leqslant 3$ being understood. The coefficients σ, Σ, F are 1-periodic in the variables ξ and sufficiently smooth in the variables x.

A typical case is when, after introduction of dimensionless scales, equation (1) takes the form

$$\frac{\partial\varphi}{\partial t} + \frac{v_j}{\varepsilon}\frac{\partial\varphi}{\partial x_j} + \frac{1}{\varepsilon^2}\left(\sum_{q=0}^{\infty} \alpha_q(\xi, x, t, v)\,\varepsilon^q\right)\varphi - \frac{1}{\varepsilon^2}\sum_{q=0}^{\infty}\varepsilon^q P_q\varphi =$$

$$= \sum_{q=0}^{\infty} \varepsilon^q f_q(\xi, x, t, v), \tag{2}$$

where the operators P_q are given by the relations

$$P_q\varphi = \int_{\mathbf{R}^3} p_q(\xi, x, t, v, v')\varphi(x, t, v')dv'.$$

We will assume the following conditions to hold:

1) $P_q^1 = \alpha_q(\xi, x, t, v)$ for $q = 0, 1$.

2) The kernel $p_0(\xi, x, t, v, v')$ is nonnegative and is represented as $K(\xi, x, t, v, v')\mu(v')$, where K is a function symmetric in v, v',

$$\int_{\mathbf{R}^3} \mu(v')\,dv' = 1, \quad \int_{\mathbf{R}^3} \mu(v)\,v^j dv = 0 \quad \text{for} \quad j = 1, 2, 3. \tag{3}$$

Usually $\mu(v)$ depends only on $|v|$ and is a distribution. For instance, in the simplest case of the one-speed transition equation, μ is a δ-function: $\mu(v) = \frac{1}{4\pi|v^0|^2}\delta(|v| - v^0)$, $|v| = \sqrt{v_1^2 + v_1^2 + v_3^2}$, and $\int_{\mathbf{R}^3} p_0\varphi dv'$ turns into an integral over the surface of the sphere $|v| = v^0$. It is often the case for the multi-speed transition equation that $\mu(v) = \sum_{j=1}^{l} c_j\delta(|v| - v^j)$, where c_j, v^j are numbers, $l < \infty$, c_j, $v^j > 0$.

Since in the expression defining P_q the integration is carried out only over v', the equality $P_q 1 = \alpha_q$ implies that $P_{q\varphi} = \alpha_q\varphi$ for $\varphi = \varphi(\xi, x, t)$.

In particular, from condition 1) we get

$$P_q\varphi = \alpha_q\varphi \quad \text{for} \quad \varphi = \varphi(x, t), \qquad q = 0, 1. \tag{4}$$

An asymptotic expansion of the solution is sought as

$$\varphi(x, t, v) \sim \sum_{q=0}^{\infty} \varepsilon^q \varphi_q(\xi, x, t, v),$$

where the φ_q are functions 1-periodic in ξ. By substituting this a.e. into (2) and by simplifying, we obtain a chain of relations

$$L_0\varphi_0 \equiv v_j\frac{\partial\varphi}{\partial\xi_j} + \alpha_0\varphi_0 - P_0\varphi_0 = 0,$$

$$v_j\frac{\partial\varphi_0}{\partial x_j} + v_j\frac{\partial\varphi_1}{\partial\xi_j} + \alpha_0\varphi_1 + \alpha_1\varphi_0 - P_0\varphi_1 - P_1\varphi_0 = 0, \tag{5}$$

$$\frac{\partial\varphi_0}{\partial t} + v_j\frac{\partial\varphi_1}{\partial x_j} + v_j\frac{\partial\varphi_2}{\partial\xi_j} + \alpha_0\varphi_2 + \alpha_1\varphi_1 + \alpha_2\varphi_0 - {} - P_0\varphi_2 - P_1\varphi_1 - P_2\varphi_0 = f_0.$$

Due to (4), the first relation holds for an arbitrary $\varphi_0 = \varphi(x, t)$ independent of the fast variables ξ and of v. Each of the subsequent equations takes the form

$L_0\varphi_n = g_n$; the function g_n is defined by $\varphi_0, \dots, \varphi_{n-1}$. Thus, the solvability of the equation

$$L_0\varphi = g. \tag{6}$$

is a decisive factor in making an asymptotic expansion (4) possible. We set $\{f\} = \int_{\mathbf{R}^3} f\mu(v)dv$. If the operator $\{\langle\cdot\rangle\}$ is applied to (6), then we get $0 = \{\langle g\rangle\}$, considering the symmetry of K in v, v'. Thus, the condition

$$\{\langle g\rangle\} = 0 \tag{7}$$

is necessary for (6) to be solvable. If $\varphi(\xi, x, t, v)$ is a solution of (6), then $\varphi + a$ is also a solution of (6) with an arbitrary $a = a(x, t)$.

Further, we assume α_0, p_0 are such that condition (7) makes (6) solvable.

From (4) we have $\alpha_1\varphi_0 - P_1\varphi_0 = 0$, and therefore the second equation in (5) is written as

$$L_0\varphi_1 + v_j\frac{\partial\varphi_0}{\partial x_j} = 0; \tag{8}$$

and the solvability condition (7) is satisfied since

$$\left\{\left\langle v_j\frac{\partial\varphi_0}{\partial x_j}\right\rangle\right\} = \{v_j\}\frac{\partial\varphi_0}{\partial x_j} = 0$$

due to (3).

Since the right side of (8) is linear with respect to $\partial\varphi_0/\partial x_j$, we seek a solution of this equation in the form $\varphi_1 = N_k\partial\varphi_0/\partial x_k$ as in §1. By substituting φ_1 into (8), we obtain equations in the functions N_k: $L_0N_k = -v_k$. By virtue of (3), the solvability condition is satisfied for these equations.

The next equation in (5) is rewritten as

$$L_0\varphi_2 = f_0 - \frac{\partial\varphi_0}{\partial t} - v_j\frac{\partial\varphi_1}{\partial x_j} - \alpha_1\varphi_1 - \alpha_2\varphi_0 + P_1\varphi_1 + P_2\varphi_0.$$

For the equation to be satisfied, the following solvability condition must hold:

$$\{\langle f_0\rangle\} - \frac{\partial\varphi_0}{\partial t} - \left\{\left\langle v_j\frac{\partial\varphi_1}{\partial x_j}\right\rangle\right\} - \{\langle\alpha_1\varphi_1\rangle\} - \\ - \{\langle\alpha_2\rangle\}\varphi_0 + \{\langle P_1\varphi_1\rangle\} + \{\langle P_2 1\rangle\}\varphi_0 = 0. \tag{9}$$

Substituting $\varphi_1 = N_k\partial\varphi_0/\partial x_k$ into (9) yields

$$\frac{\partial\varphi_0}{\partial t} - \frac{\partial}{\partial x_j}\left(A_{jh}\frac{\partial\varphi_0}{\partial x_h}\right) + A_k\frac{\partial\varphi_0}{\partial x_k} + A\varphi_0 = \{\langle f_0\rangle\}; \tag{10}$$

here $A_{kj} = \{\langle v_j N_k \rangle\}$, $A_k = \{\langle \alpha_1 N_k - P_1 N_k \rangle\}$, $A = \{\langle \alpha_2 - P_2 1 \rangle\}$.

The subsequent equations (4) are of the form $L_0 \varphi_{n+1} = g_n(\varphi_0, \dots, \varphi_n)$. Each time, the solvability condition for φ_{n+1} forms an equation in the preceding functions φ. Consider the Cauchy problem for equation (2). The approximation $\varphi^{(2)} = \varphi_0 + \varepsilon\varphi_1 + \varepsilon^2\varphi_2$ satisfies (2) up to terms of order $O(\varepsilon)$. If φ is a solution of (2) with the initial values close to $\varphi^{(2)}|_{t=0}$, the estimates in [187] imply that φ and $\varphi^{(2)}$ are close for $t > 0$. Since $\varphi^{(2)}$ is close to φ_0, equation (10) can be considered as the averaging of the original equation (2).

Consider a stationary case when the original equation is

$$v_j \frac{\partial \varphi}{\partial x_j} + \frac{1}{\varepsilon} \sum_{q=0}^{\infty} \alpha_q \varepsilon^q \varphi - \frac{1}{\varepsilon} \sum_{q=0}^{\infty} \varepsilon^q P_q \varphi = \varepsilon \sum_{q=0}^{\infty} \varepsilon^q f_q.$$

Then we get the averaged zeroth-order equation

$$\frac{\partial}{\partial x_j}\left(A_{jk} \frac{\partial \varphi_0}{\partial x_k}\right) + A_k \frac{\partial \varphi_0}{\partial x_k} + A\varphi_0 = \{\langle f_0 \rangle\}.$$

If $\alpha_1 \; P_1 \equiv 0$, then the equation can be simplified:

$$A_k = 0, \; A = \{\langle \alpha_2 \rangle\}.$$

Let all α_q, $P_1 \equiv 0$ for $q > 0$, and let the α_0, K be independent of the fast variables ξ, i.e., the medium is homogeneous. Let, in addition, μ be equal to $\mu(|v|)$. Then (8) is satisfied for $N_k = -v_k$, $A_{jk} = -\{\langle v_j v_k \rangle\} = -\delta_{jk} A_0$, $A_0 = \int_{\mathbf{R}^s} |v_j|^2 \mu(|v_0|) dv$, and the averaged equation coincides with the diffusion approximation for a homogeneous medium: $\partial\varphi_0/\partial t - A_0 \Delta v = \{\langle f_0 \rangle\}$.

For the case where equation (1) maintains the original form after dimensionless scales have been introduced, the zeroth order averaged equation was obtained in [15].

§10 Eigenvalue Problems

We shall give an elementary example of averaging eigenvalue problems for equations with rapidly oscillating coefficients. Let $K(\xi)$ be a 1-periodic, infinitely differentiable function of the variable $\xi \in \mathbf{R}$, $0 < \kappa_1 \leqslant K(\xi) \leqslant \kappa_2 < \infty$.

Consider the Sturm–Liouville problem

$$Pu + \lambda u = 0, \quad P = \frac{d}{dx}\left(K\left(\frac{x}{\varepsilon}\right)\frac{d}{dx}\right), \qquad x \in (0, 1),$$
$$u(0) = u(1) = 0, \tag{1}$$

where $\varepsilon = 1/n \ll 1$, with n being a positive integer. By inverting P, we get

$$\frac{u(x)}{\lambda} = \int_0^1 G(x, y)\, u(y)\, dy;$$

here

$$G(x, y) = \begin{cases} \psi(x, 1)\,(\psi(0, 1))^{-1}\,\psi(0, y), & y \leqslant x, \\ \psi(0, x)\,(\psi(0, 1))^{-1}\,\psi(y, 1), & y \geqslant x, \end{cases} \tag{2}$$

$$\psi(a, b) = \int_a^b K^{-1}\left(\frac{t}{\varepsilon}\right) dt.$$

For the Sturm–Liouville problem

$$\begin{gathered} \widehat{P} v_0 + \widehat{\lambda} v_0 = 0, \quad \widehat{P} = \frac{d}{dx}\widehat{K}\frac{d}{dx}, \\ v(0) = v(1) = 0, \quad \widehat{K} = \langle K^{-1}\rangle^{-1}, \end{gathered} \tag{3}$$

we get the equivalent problem

$$\frac{v(x)}{\widehat{\lambda}} = \int_0^1 \widehat{G}(x, y)\, v(y)\, dy,$$

$$\widehat{G}(x, y) = \begin{cases} \widehat{\psi}(x, 1)\,(\widehat{\psi}(0, 1))^{-1}\,\widehat{\psi}(0, y), & y \leqslant x, \\ \widehat{\psi}(0, x)\,(\widehat{\psi}(0, 1))^{-1}\,\widehat{\psi}(y, 1), & y \geqslant x, \end{cases}$$

$$\widehat{\psi}(a, b) = \int_a^b \widehat{K}^{-1} dt.$$

Since $\psi(0, n\varepsilon) = \widehat{\psi}(0, n\varepsilon)$, $\psi(n\varepsilon, 1) = \widehat{\psi}(n\varepsilon, 1)$ for all integer n, then

$$\max_{x\in[0,1]} |\psi(0, x) - \widehat{\psi}(0, x)| = \max_{x\in[0,1]} |\psi(x, 1) - \widehat{\psi}(x, 1)| \leqslant c_1(\varkappa_1, \varkappa_2)\,\varepsilon,$$

where $c_1 < \frac{1}{2}\left(\frac{1}{\kappa_1} - \frac{1}{\kappa_2}\right)$.

This implies

$$\sqrt{\int_0^1\int_0^1 (G(x, y) - \widehat{G}(x, y))^2\, dx\, dy} \leqslant c_2(\varkappa_1, \varkappa_2)\,\varepsilon.$$

Note that $G(x, y) = G(y, x)$, $\widehat{G}(x, y) = \widehat{G}(y, x)$.

Let λ_i, $\hat{\lambda}_i$ $(i = 1, 2, \ldots)$ be eigenvalues of problems (1), (3) arranged in increasing order of magnitude. From the minimax property of eigenvalues, we see that

$$|1/\lambda_i - 1/\widehat{\lambda}_i| \leqslant c_2(\varkappa_1, \varkappa_2)\varepsilon.$$

This estimate extends without any changes to the case where u is a vector and $K(x/\varepsilon)$ is a symmetric matrix, while K may be piecewise continuous.

In the multi-dimensional case, the eigenvalue problem was considered in [168], and the one-dimensional problem with random coefficients was discussed in [17].

CHAPTER 5

General Formal Averaging Procedure

This chapter considers formal procedures of averaging general nonlinear differential equations and linear operator equations, which are of interest to the readers who concern themselves with possible generalizations and with applicability of the methods described herein to other models. In particular, the chapter deals with constructing averaged equations of infinite order of accuracy in ε. All the above-mentioned cases of averaging periodic problems (except for those treated in §§4.7–4.9) fit the framework of the general procedure. However, we preferred to consider first the great number of particular cases for a more graphic presentation. In addition, rigorous justification of the results in the case of boundary value problems calls for specific types of equations to be taken into account.

§1 Averaging Nonlinear Equations

Let Φ_r^m be the space of matrices of dimension $\underbrace{n \times \cdots \times n}_{r \text{ times}}$ whose elements are functions of the arguments ε, $x = (x_1, \dots, x_s)$, $\xi = (\xi_1, \dots, \xi_s)$, u, $w^{i_1 \dots i_l}$ $(1 \leqslant l \leqslant m)$, and are 1-periodic in

$$u = (u^1, \dots, u^n)^T, \quad w^{i_1 \dots i_l} = \varepsilon^{l-1} \frac{\partial^l u}{\partial x_{i_1} \dots \partial x_{i_l}}.$$

We denote by Ω_r^m the space of such matrices whose elements are functions of the arguments x, ξ, v, $v^{(i_1 \dots i_l)} = \frac{\partial^l v}{\partial x_{i_1} \dots \partial x_{i_l}}$, 1-periodic in ξ, $v = (v_1, \dots, v_n)^T$. By ψ_r^s, we shall mean the subspace of Ω_r^s independent of ξ. We set

$$\Omega_r = \bigcup_{m=1}^{\infty} \Omega_r^m, \quad \Omega^m = \bigcup_{r=1}^{\infty} \Omega_r^m, \quad \Psi_r = \bigcup_{m=1}^{\infty} \Psi_r^m, \quad \Psi^m = \bigcup_{r=1}^{\infty} \Psi_r^m,$$

$$\langle N \rangle = \int_{-1/2}^{1/2} \dots \int_{-1/2}^{1/2} N \, d\xi_1 \dots d\xi_s, \qquad N \in \Omega_r^m.$$

We consider the system of differential equations

$$L_\varepsilon(u) \equiv \frac{A}{\varepsilon} + \sum_{l=1}^{q} \varepsilon^{l-1} D^{(i_1, \dots, i_l)} A_{i_1 \dots i_l} = 0, \qquad A_{i_1 \dots i_l} \in \Phi_1^p; \tag{1}$$

If, as in (1), summation over repeated indices is not indicated by the summation sign, the summation from 1 to s is understood.

In what follows, we assume for simplicity that the vectors A are infinitely differentiable with respect to their arguments. In specific cases, the requirement for differentiability with respect to the variables ξ_q can be removed by determining a solution of (1) as a generalized solution in the sense of Sobolov.

For this procedure to be applicable to nonstationary problems, one of the variables x_i, for example x_1, should be interpreted as the time variable. The coefficients of the equation do not depend on the fast time variable $\xi_1 = x_1/\varepsilon$. However, this case fits into the scheme, since the function that is independent of any fast variable ξ_q can always be regarded as 1-periodic in ξ_q.

We seek an asymptotic expansion of the solution of (1) as

$$u^{(\infty)} \sim v + \sum_{k=1}^{\infty} \varepsilon^k N_k, \qquad N_k \in \Omega_1, \tag{2}$$

where $\xi_j = x_j/\varepsilon$, and $v = v(x)$ is a solution of the averaged equation.

Remark 1: Processes in periodic media with more elaborate equations of state than in the case of the problems of elasticity theory sometimes call for an asymptotic series to be sought in the form

$$u^{(\infty)} \sim \sum_{k=0}^{\infty} \varepsilon^k N_k, \qquad N_k \in \Omega_1,$$

where the zeroth-order term also depends on fast variables.

Let us expand $L_\varepsilon(u)$ into an asymptotic series in powers of ε. At this stage, v is regarded as a function of the arguments x_q and independent of ε. After combining similar terms, we get

$$L_\varepsilon (u^\infty) \sim \sum_{k=1}^{\infty} \varepsilon^{k-2} H_k, \qquad H_k \in \Omega_1.$$

We require the following relations to be satisfied

$$H_1 = 0, \quad H_k = h_k, \tag{3}$$

for $k > 1$, $h_k \in \Psi_1$.

The equation

$$\widehat{L}(v) \equiv \sum_{k=2}^{\infty} \varepsilon^{k-2} h_k \sim 0, \tag{4}$$

not containing the fast variables ξ_q, can be regarded as the average of (1); if v satisfies (4), then $u^{(\infty)}$, defined by (2), satisfies (1). A solution of (4) can be sought as a series in powers of ε:

$$v \sim \sum_{j=0}^{\infty} \varepsilon^j v_j(x).$$

For $z \in \Phi_r^m$, we denote by $\{z\}$ the element of Ω_r^m obtained by substituting into z the relations

$$\varepsilon = 0, \quad u = v, \quad w^j = \frac{\partial v}{\partial x_j} + \frac{\partial N_1}{\partial \xi_j},$$

$$w^{i_1 \dots i_l} = \frac{\partial^l N_1}{\partial \xi_{i_1} \dots \partial \xi_{i_l}} \quad \text{for} \quad l > 1.$$

In this notation

$$H_1 = \{A\} + \sum_{l=1}^{q} \frac{\partial^l \{A_{i_1 \dots i_l}\}}{\partial \xi_{i_1} \dots \partial \xi_{i_l}}.$$

We assume that the equation $H_1 = 0$ has a solution $N_1 \in \Omega_1$.

Note that x, v, $v^{(i_1 \dots i_l)}$ appear in this equation as parameters; we seek N_1 as a solution of the equation, which is 1-periodic in ξ. For $k > 1$, we have

$$H_k = LN_k + W_k,$$

$$LN = \sum_{l=0}^{q} \frac{\partial^l}{\partial \xi_{i_1} \dots \partial \xi_{i_l}} \left(\sum_{m=1}^{p} \left\{ \frac{\partial A_{i_1 \dots i_l}}{\partial w^{j_1 \dots j_m}} \right\} \frac{\partial^m N}{\partial \xi_{j_1} \dots \partial \xi_{j_m}} \right);$$

here $\partial A_{i_1 \dots i_l} / \partial w^{j_1 \dots j_m}$ is the Jacobian with the elements equal to derivatives of the components of the vector $A_{i_1 \dots i_l}$ with respect to the components of the vector $w^{j_1 \dots j_m}$, and W_k is defined by specifying $v, N_1, \dots, N_{k-1}$. After defining N_1 from the condition $H_1 = 0$, we successively define h_2, N_2, h_3, $N_3, \dots$ according to the following algorithm: We choose h_k each time so that the equation $H_k = h_k$ is solvable for $N_k \in \Omega_1$, and then find N_k from this equation.

Consider the case when the following condition is satisfied: The equation $LN = M(M \in \Omega_1)$ is solvable in the class $N \in \Omega_1$, if $\langle M \rangle = 0$. If we assume $h_k = \langle W_k \rangle$ for all k, the equation $LN_k + W_k = h_k = \langle W_k \rangle$ will satisfy the required solvability condition.

In §4.5 we have obtained an averaged equation of infinite order for a nonstationary problem, which contains differentiation with respect to t of arbitrarily high order. A similar situation arises in the present case if, for example, x_1 has the meaning of time. It is desirable, in physical interpretations of the averaged equation, that it does not contain differentiation with respect to this variable of order higher than the second. We now show how to obtain an averaged equation satisfying this condition.

The vectors N_k that are obtained successively as solutions of the equations (3) are defined ambiguously for each k: If N_k is a solution of the appropriate equation of (3), then $N_k + z_k(z_k \in \Psi_1)$ is also a solution of this equation. For definiteness, we choose z_k so that the condition $\langle N_k\rangle = 0$ is satisfied.

Let T be the space of matrices S of dimension $\underbrace{n \times \cdots \times n}_{r \text{ times}}$, for any possible r, representable as the direct product of the vectors $v^{(i_1 \ldots i_l)} (l > 1)$. We denote the largest of the orders of differentiation with respect to x_1 in these vectors by $g(S)$, and we denote the sum of all the orders of differentiation l corresponding to these vectors by $b(S)$. Let T_k be the subspace of matrices S satisfying the condition $b(S) \leqslant k$. By investigating the representation of W_k in terms of $N_1, \ldots, N_{k-1}$ and the form of the equations (3), we can see that N_k, h_k can be written as

$$N_k = \sum_j^{2k-2} N^j_{2k-2} S_j, \qquad N^j_{2k-2} \in \Omega^1,$$

$$h_k = \sum_j^{2k-2} h^j_{2k-2} S_j, \qquad h^j_{2k-2} \in \Psi^1;$$

where the symbol $\sum_j^l$ denotes summation over $S_j \in T_l$. Thus, (4) has the form

$$\sum_{k=2}^{\infty} \varepsilon^{k-2} \sum_j^{2k-2} h^j_{2k-2} S_j \sim 0.$$

Consider now the case where the matrix h^j_2, the factor of $S_j = \partial^2 v/\partial x_1^2$, is invertible. We denote it by R.

The algorithm for transforming (4) into the required form contains an induction in the set of parameters r and m: $2 < r$, $0 < m$; with respect to r, we go from small values to large ones, whereas for m, we go in the reverse order. Let (4) be transformed into a form not containing the nonzero terms $\varepsilon^{k-2} h^j_{2k-2} S_j$ for $k < r$, $1 < g(S_j)$ and also for $k = r$, $m < g(S_j)$ (except for the term $R\partial^2 v/\partial x^2$). For $r = 3$ and $m = 4$, the inductive hypothesis is satisfied.

Let us describe the step of the induction in r consisting of the steps of the induction in m. Among the matrices S_j with nonzero coefficients h^j_{2r-2}, we select one of the matrices with $g(S_j) = m$. By rearranging the indices in the appropriate convolution $h^j_{2r-2} S_j$, we represent it as

$$\overline{h}\overline{S} \circ \frac{\partial^{r+l} v}{\partial x_1^r \partial x_{i_2} \ldots \partial x_{i_{l+1}}}, \qquad i_2, \ldots, i_{l+1} \neq 1;$$

here $\circ$ denotes the direct product, and $\overline{h}, \overline{S}$ are matrices.

To equation (4) we apply the operator $\bigwedge$ defined by

$$\Lambda w \equiv w - \varepsilon^{r-2} \overline{h}\overline{S} \circ \frac{\partial^{m+l-2} (R^{-1} w)}{\partial x_1^{m-2} \partial x_{i_2} \ldots \partial x_{i_{l+1}}}.$$

We obtain an equation asymptotically equivalent to the original; moreover, the general number of factors $\frac{\partial^{m+l} v}{\partial x_1^m \partial x_{i_2} \ldots \partial x_{i_l}}$, $i_2, \ldots, i_l \neq 1$, in all the matrices $S_j \in T_{2r-2}$ appearing in the equation with the factor ε^{r-2}, is diminished by at least 1.

By continuing the process, we obtain an equation satisfying the requirement of the induction by m decreased by one. The step of the induction in m is terminated. By continuing these steps down to $m = 1$, we obtain an equation corresponding to r greater than one, i.e., we terminate the induction step in r. Passing to the limit as $r \to \infty$, we obtain the averaged equation with a single term involving differentiation with respect to x_1 of order higher than the first: $R\partial^2 v/\partial x_1^2$.

Sometimes, in constructing averaged equations which admit a physical interpretation, it may be advisable not to let the coefficients vanish, even for other matrices S containing the second-order differentiation with respect to x_1.

The averaging procedure we propose is formally applicable also in those cases where either $p = \infty$, $q = \infty$, or $p, q = \infty$, which correspond to pseudodifferential equations.

§2 Averaged Equations of Infinite Order for a Linear Periodic Medium and for the Equation of Moment Theory

Sections 4.5 and 5.1 present an averaged equation of infinite order for processes described by differential equations. A formal averaging of infinite order of accuracy will be constructed below for an arbitrary linear periodic medium. The formalism of the averaging, in particular, covers problems of elasticity theory and so-called systems with distributed parameters.

Let U_ε be a family of linear spaces which may depend on parameter ε; by U^ε (ε is fixed), we denote a space of elements u such that for almost all $x = (x_1, \dots, x_s)$ ($s \leqslant \infty$), the values of $u(x) \in U_\varepsilon$ are defined. The spaces F_ε and F^ε are defined in the same way.

REMARK 1: In describing a number of phenomena, it is advisable to introduce the space $U_\varepsilon(x)$ depending on x, and the space U^ε of elements u, such that the values of $u(x) \in U_\varepsilon(x)$ are defined for nearly all x.

We define the shift operator $S_m g(x) = g(x + m\varepsilon)$, $m = (m_1, \dots, m_s)$. The vector m is called an *integer one* if all m_i are integers.

An operator p mapping U^ε into F^ε *corresponds to a periodic structure* if

$$p(S_m u) = S_m(pu) \tag{1}$$

for any integer m. The set of operators satisfying (1) will be denoted by $\widetilde{P}_0^\varepsilon$. The subspace of linear operators which belong to $\widetilde{P}_0^\varepsilon$ will be denoted by P_0^ε.

For example, the operator of the elasticity theory,

$$p = L_\varepsilon = \frac{\partial}{\partial x_i}\left(A_{ij}(\xi)\frac{\partial}{\partial x_j}\right), \qquad \xi = (\xi_1, \dots, \xi_s), \quad \xi_q = x_q/\varepsilon, \tag{2}$$

considered in §4.4 satisfies (1) for the 1-periodic $A_{ij}(\xi)$; here U_ε is the space of vectors of dimension s. In terms of elasticity theory, condition (1) is treated in the following natural way: In a periodic structure, the field of stresses and forces shifts by $m\varepsilon$ when the translation pattern shifts by $m\varepsilon$.

The above-discussed dynamic problem for the system of equations of elastic theory

$$pu = f, \quad p \equiv \rho(\xi)\frac{\partial^2}{\partial t^2} - \frac{\partial}{\partial x_i}\left(A_{ij}(\xi)\frac{\partial}{\partial x_j}\right) \tag{3}$$

can be treated in various ways.

1. We may take U_ε to be the space of vectors $u(t)$ defined either for all the t or, for the Cauchy problem, for $t \geqslant t_0$.

2. The time t may be considered together with all spatial variables, assuming $t = x_{s+1}$. The coefficients of (3) do not depend on t, and therefore they are periodic in t with any period ε'. In the case of (3), for example, ε' can be taken to be equal to ε/c, where c is a typical perturbation velocity. We take U_ε to be the space of vectors of dimension s; the functions that belong to U^ε will be defined for all the collections $(x_1, \ldots, x_s, x_{s+1})$. For $m' = (m, m_{s+1}) = (m_1, \ldots, m_s, m_{s+1})$, we define the shift operator $S_{m'}(g(x,t)) = g(x + m\varepsilon, t + m_{s+1}\varepsilon)$. The operator of (3) satisfies (1) for any integer m'.

Consider the heat-propagation problem for a radiator-type system:

$$\frac{\partial}{\partial x_1}\left(K\frac{\partial w}{\partial x_1}\right) + \frac{\partial}{\partial x_2}\left(K\frac{\partial w}{\partial x_2}\right) = 0, \qquad K = K(\xi_1, x_2).$$

This problem was averaged in [11]. In terms of the present section, the problem can be treated as 1-periodic in ξ_1.

By U_ε we shall mean the space of functions $z(x_2)$ in one variable defined on $[d_1, d_2]$ and by U^ε the space of $u(x_1)$ ($x_1 \in [c_1, c_2]$) such that the values of $u(x_1) \equiv w(x_1, x_2)$ are defined for all x_1. Thus, $s = 1$.

The original equation is written in the operator form

$$pu = \frac{\partial}{\partial x_1}\left(K\frac{\partial u(x_1)}{\partial x_1}\right) + p^1 u(x_1) = 0,$$

$$p^1 u(x_1) \equiv \frac{\partial}{\partial x_2}\left(K(\xi_1, x_2)\frac{\partial w(x_1, x_2)}{\partial x_2}\right).$$

The operators p^1 and p satisfy condition (1).

REMARK 2: It is advisable to introduce more elaborate constructions for media with the periodicity domain slowly changing its properties (see §3.5). Let us define P^ε, with the set of operators $P(x, u)$ mapping $R^s \times U^\varepsilon$ into F^ε and having the following property: For a fixed y, the operator $p(y, u)$ belongs to P_0^ε. An example is the operator $\frac{\partial}{\partial x_i}(A_{ij}(x,\xi)\frac{\partial}{\partial x_j})$ with the coefficients A_{ij}, 1-periodic in the variables $\xi_q = x_q/\varepsilon$. The appropriate formalism of averaging is described in [10] (see §1).

The class of problems under study in this section includes systems of equations with distributed parameters, e.g., the system of equations

$$Lu \equiv \rho(\xi)\frac{\partial^2 u}{\partial t^2} - L_1 u, \quad L_1 u = \frac{1}{\varepsilon^{2+s}} \int_{-\infty}^{+\infty} \cdots \int_{-\infty}^{+\infty} K(x, y)\, u(y)\, dy, \tag{4}$$

where $y = (y_1, \ldots, y_s)$, $u = u(x)$ is a vector, the matrix $K(x,y)$ is of the form $K(x,y) = K_0(\xi, \lambda)$, $\xi = x/\varepsilon$, and $\lambda = y/\varepsilon$, and satisfies the condition

$$\int_{-\infty}^{+\infty} \cdots \int_{-\infty}^{+\infty} K_0(\xi, \lambda)\, d\lambda = 0. \tag{5}$$

That the requirement (5) is natural is suggested by the following considerations. If there is a displacement of the body as a whole in the direction of the axis e_l, no additional forces occur in the body; it means that

$$\int_{-\infty}^{+\infty} \cdots \int_{-\infty}^{+\infty} K_0(\xi, \lambda)\, e_l d\lambda = \left(\int_{-\infty}^{+\infty} \cdots \int_{-\infty}^{+\infty} K_0(\xi, \lambda)\, d\lambda \right) e_l = 0.$$

Since this relation holds for any l, we have (5).

Lemma 1. *If the condition*

$$K_0(\xi + m, \ \lambda + m) = K_0(\xi, \ \lambda) \tag{6}$$

holds for any integer vector m, then the operator L_1 satisfies (1).

Proof: From the definition of L_1 and S_m, we have

$$I_1 = L_1(S_m u(x)) = \varepsilon^{-s-2} \int_{-\infty}^{+\infty} \cdots \int_{-\infty}^{+\infty} K(x, y)\, u(y + m\varepsilon)\, dy =$$

$$= \varepsilon^{-2} \int_{-\infty}^{+\infty} \cdots \int_{-\infty}^{+\infty} K_0(\xi, \lambda)\, u(\lambda\varepsilon + m\varepsilon)\, d\lambda,$$

$$I_2 = S_m(L_1 u(x)) = \varepsilon^{-s-2} \int_{-\infty}^{+\infty} \cdots \int_{-\infty}^{+\infty} K(x + m\varepsilon, y)\, u(y)\, dy =$$

$$= \varepsilon^{-2} \int_{-\infty}^{+\infty} \cdots \int_{-\infty}^{+\infty} K_0(\xi + m, \lambda)\, u(\lambda\varepsilon)\, d\lambda = \varepsilon^{-2} \int_{-\infty}^{+\infty} \cdots \int_{-\infty}^{+\infty} K_0(\xi, \lambda - m)\, u(\lambda\varepsilon)\, d\lambda.$$

The latter equality stems form $K_0(\xi + m, \lambda) = K_0(\xi, \lambda - m)$ due to (6). By making the change of variables $\lambda - m = \lambda'$ in the latter integral, we get

General recommendation for averaging is as follows: Seek a solution of the equation as the asymptotic series

$$u^{(\infty)} \sim \sum_{l=0}^{\infty} \varepsilon^{a_l} u_l(x, \xi), \quad a_l < a_{l+1}; \tag{7}$$

where u_1 are 1-periodic functions and a_1 are quantities yet unknown. Expand $p(x, u)$ into a series in powers of ε:

$$p\left(x, u^{(\infty)}\right) \sim \sum_{l=0}^{\infty} \varepsilon^{b_l} H_l(x, \xi) \tag{8}$$

and by selecting u_l and a_l successively, obtain the relations $H_l(x, \xi) = \partial_{b_l, 0} f$.

In the examples considered above, the a_l and b_l have always been integers, but there may be a case when a_l and b_l are not integers. Let convection also be taken into account in the problem of heat exchange between a system of screens by radiation (see §3.5); namely, condition (3.6.4) is replaced by the condition

$$\left(K \frac{du}{dx}\right)\Big|_{n\varepsilon+0} = \left(K \frac{du}{dx}\right)\Big|_{n\varepsilon-0} = g(u)|_{n\varepsilon+0} - g(u)|_{n\varepsilon-0} +$$
$$+ a\left(u(n\varepsilon+0, t) - u(n\varepsilon-0, t)\right)^b, \quad a, b = \text{const}, \quad 1 < b < 2.$$

Then the expansion (7) and (8) will be carried out in powers of a_l of the form $i + jb$ ($i, j \geqslant 0$; *they* are integers). With large temperature drops, the coefficients a and b are considered as depending on the mean temperature:

$$a = a\left(\frac{u(n\varepsilon+0, t) + u(n\varepsilon-0, t)}{2}\right),$$
$$b = b\left(\frac{u(n\varepsilon+0, t) + u(n\varepsilon-0, t)}{2}\right), \quad 1 < b < 2.$$

Then an asymptotic expansion of the solution has to be sought as

$$u(x, t) \sim v(x, t) + \varepsilon v_1(x, t) + \int_{b_0}^{+\infty} \varepsilon^b v_2(b, x, t, \xi)\, db,$$
$$b_0 = \inf_u b(u).$$

The asymptotic equality is understood in the following sense: for any $d \geqslant b_0$,

$$u(x, t) - v(x, t) - \varepsilon v_1(x, t) - \int_{b_0}^{d} \varepsilon^b v_2(b, x, t, \xi)\, db = O\left(\varepsilon^d\right).$$

Going back to the case of the operator equation $pu = f$, we assume some regularity of the properties of this equation.

LEMMA 2. *If u is 1-periodic and $p \in p_0^\varepsilon$, then $S_m pu = pu$ for any integral vector m.*

PROOF: We have $pS_m pu = pu$. On the other hand, $pS_m u = S_m pu$ for $p \in P_0^\varepsilon$. Thus, $S_m pu = pu$, and pu also had periodicity property.

We denote by $P_{0,k}$ the subspace of operators $p \in P_0^\varepsilon$ which satisfy the following condition: If the element $u = u(\xi)$ is 1-periodic and does not depend explicitly on ε, then the element $\varepsilon^k pu$ (1-periodic in ξ under the condition of Lemma 2) does not depend explicitly on ε either (except for the dependence on ξ).

For example, the operator p given by (2) belongs to $P_{0,2}^\varepsilon$. The multiplication operator by a periodic matrix-valued function ($pu = A(\xi)u$) belongs to $P_{0,0}^\varepsilon$.

The operator p, $pu = \frac{\partial}{\partial x_i}(A_{ij}(\xi)\frac{\partial u}{\partial x_j}) + \frac{\partial}{\partial x_i}(A_i(\xi)u) + A(\xi)u$ does not belong to any $P_{0,k}^\varepsilon$ but is equal to the sum of three operators in $P_{0,2}^\varepsilon$, $P_{0,1}^\varepsilon$, and $P_{0,0}^\varepsilon$, respectively. Otherwise it can be represented as $pu = \varepsilon^{-2}p_2 + \varepsilon^{-1}p_1 + p_0$, where $p_i \in P_{0,0}^\varepsilon$. The formalism of averaging for the nonstationary problem corresponding to an operator of a somewhat more general form was considered in [9].

The averaging technique that we propose is formally applicable to linear operator equations of the form $pu = f$, with $f = f(x)$ being a function smooth in the variables x (with the characteristic size of variation much larger than ε), and $pu = \varepsilon^{-l}\sum_{j=-l}^{\infty}\varepsilon^j p_j$, where all the $p_j \in P_{0,0}^\varepsilon$.

We restrict ourselves to discussing this technique more specifically for the operator equation

$$pu = L^1\frac{\partial^2 u}{\partial t^2} - L^0 u = f, \quad L^1 \in P_{0,0}^\varepsilon, \quad L^0 \in P_{0,2}^\varepsilon, \tag{9}$$

which includes the equations (3) and (4) as particular cases.

Specifically, for $L^1 \equiv 0$, we obtain an equation, more general formally than (9), if we set $x = (x_1, \ldots, x_s, t)$.

An analysis of the general scheme (7) and (8) of the expansion of the solution shows that in this case (9), we can restrict ourselves to the same particular case (7) as in §4.5. Namely, it suffices to seek an asymptotic expansion of the solution in the form

$$u(x,t) \sim \sum_{r=0}^{\infty}\varepsilon^r\sum_{q=0}^{r}\sum_{|i|=r-q}N_{qi}(\xi)\,v^{(q,i)}, \qquad i = (i_1, \ldots, i_{r-q}), \tag{10}$$

where the N_{qi} are 1-periodic functions, $v = v(x,t)$ is a function with the characteristic size of variation much larger than ε, and

$$v^{(q,i)} \equiv v^{(q,i_1,\ldots,i_{r-q})} = \frac{\partial^r v}{\partial t^q \delta x_{i_1}\ldots\partial x_{i_{r-q}}}.$$

We regard the operators of differentiation in various x_j as being unpermutable; therefore, $v^{(0,1,2)} = \frac{\partial^2 v}{\partial x_1 \partial x_2}$ are not considered to be similar terms. This formalism offers simpler and more natural means of obtaining correct boundary conditions for the averaged equation in the case of boundary value problems (Chapter 9).

Summation on the right side of (10) is understood in the following sense: For each $r-q$, all the possible collections of integers $i_1, \dots, i_{r-q} \in \{1, \dots, s\}$ are chosen and the summation is made over these collections, i.e., (10) is understood as

$$u \sim \sum_{r=0}^{\infty} \varepsilon^r \sum_{q=0}^{r} \sum_{i_1,\dots,i_{r-q}=1}^{s} N_{qi_1\dots i_{r-q}}(\xi)\, v^{(q,i_1,\dots,i_{r-q})}.$$

We substitute this series into pu. Each of the summands in (10) is of the form $N(\xi)w(x,t)$. Therefore, to obtain an expansion of pu into a series in powers of ε, one must first expand in series the expressions $L^j(Nw)$; the dependence of w on t is discarded, since t acts here as a parameter.

Let x be a point at which a value of the operator pu is calculated; by y we shall mean the argument of the function to which the operator is applied. We expand the function $w(y)$ into the Taylor series with respect to the point $\overline{x}$:

$$w(y) \sim \sum_{n=0}^{\infty} \sum_{|i|=n} \frac{\left(y_{i_1} - \overline{x}_{i_1}\right) \cdots \left(y_{i_n} - \overline{x}_{i_n}\right)}{n!} w^{(i_1,\dots,i_n)}(\overline{x});$$

here the notation $w^{(i_1,\dots,i_n)}(\overline{x}) = \frac{\partial^n w}{\partial x_{i_1} \dots \partial x_{i_n}}\Big|_{\overline{x}}$ was introduced temporarily; the right side of the latter asymptotic equality is understood as in (10):

$$w(y) \sim \sum_{n=0}^{\infty} \sum_{i_1,\dots,i_n=1}^{s} \frac{\left(y_{i_1} - \overline{x}_{i_1}\right) \cdots \left(y_{i_n} - \overline{x}_{i_n}\right)}{n!} w^{(i_1,\dots,i_n)}(\overline{x}).$$

Then

$$L^j(Nw)\big|_x \sim$$

$$\sim \sum_{n=0}^{\infty} \sum_{|i|=n} \left(L^j\left(N\left(\frac{y}{\varepsilon}\right)\left(y_{i_1} - \overline{x}_{i_1}\right) \cdots \left(y_{i_n} - \overline{x}_{i_n}\right)\right)\right)\Big|_x \frac{w^{(i_1,\dots,i_n)}(\overline{x})}{n!}.$$

We set $\overline{x} = x$ and introduce the notation $\xi = x/\varepsilon$, $\overline{\xi} = \overline{x}/\varepsilon$, $\lambda = y/\varepsilon$; we have

$$L^j(Nw)\big|_x \sim \sum_{n=0}^{\infty} \sum_{|i|=n} \left(L^j\left(N(\lambda)\left(\lambda_{i_1} - \overline{\xi}_{i_1}\right) \cdots \right.\right.$$

$$\left.\left.\cdots \left(\lambda_{i_n} - \overline{\xi}_{i_n}\right)\right)\right)\Big|_{\overline{\xi}=\xi} \frac{w^{(i_1,\dots,i_n)}(x)}{n!}. \tag{11}$$

Here and elsewhere the operators L are applied to functions of the form $g(y,z)$, with z being a parameter; then $Lg = v(x,z)$ is a function of the arguments x, z, and by $(Lg(x,z))|_x$ or $Lg(x,z))|_\xi$, we mean $v(x,x)$. Let us demonstrate that the coefficients on the right side depend periodically on the variables ξ.

Lemma 3. *Let*

$$g(\lambda, \xi) = g(\lambda + m, \xi + m) \tag{12}$$

for any integer vector m and $L \in P_0^\varepsilon$. Then $h(\xi) = (Lg)|_x$ is a 1-periodic function of the variables ξ, that is, $h(\xi + m) = h(\xi)$ for any integer vector m.

Proof: By virtue of property (1) of the operators $L^j \in P_0$, we have

$$(LS_m g(\lambda, \pi))|_\xi = (Lg(\lambda, \pi))|_{\xi+m} \tag{13}$$

for any real-valued π, ξ and an integer vector m.

From (12) the equality $(Lg(\lambda + m, \pi + m))|_\xi = (Lg(\lambda, \pi))|_\xi$ holds, and assuming that $\pi = \xi$, we get

$$h(\xi) = (Lg(\lambda, \xi))|_\xi = (Lg(\lambda + m, \xi + m))|_\xi.$$

We set $\pi = \xi + m$ in (13); then we have

$$(Lg(\lambda + m, \xi + m))|_\xi = (Lg(\lambda, \xi + m))|_{\xi+m} = h(\xi + m)$$

(the latter equality follows from the definition of $h(\xi)$).

All the relations imply $h(\xi) = h(\xi + m)$. This proves the lemma.

We suppose that the $N(\xi)$ do not depend on ε explicitly; then the $N(\xi)(\lambda_{i_1} - \xi_{i_1})\dots(\lambda_{i_n} - \xi_{i_n})$ do not depend on ε either.

From the condition $L^1 \in P_{0,0}^\varepsilon$, $L^0 \in P_{0,2}^\varepsilon$ it follows that

$$L^1(N(\xi)(\lambda_{i_1} - \xi_{i_1})\cdots(\lambda_{i_n} - \xi_{i_n})),$$
$$\varepsilon^2 L^0(N(\lambda)(\lambda_{i_1} - \xi_{i_1})\cdots(\lambda_{i_n} - \xi_{i_n}))$$

do not depend on ε explicitly. Thus, in the case $L \in P_{0,0}^\varepsilon$, we see that

$$L(Nw) \sim \sum_{n=0}^{\infty} \varepsilon^n \sum_{|i|=n} \tau_i w^{(i)}(x),$$

where

$$\tau_i \equiv \tau_{i_1\dots i_n} = \frac{1}{n!}(L(N(\lambda)(\lambda_{i_1} - \xi_{i_1})\cdots(\lambda_{i_n} - \xi_{i_n})))|_\xi$$

are 1-periodic functions of the fast variables ξ, which do not depend explicitly on ε.

We substitute (10) into pu and use the conditions L^1, $\varepsilon^l L^0 \in P_{0,0}$, and relation (11). Eventually, we get

$$pu \sim \sum_{r=0}^{\infty} \varepsilon^{r-2} \sum_{q=0}^{r} \sum_{|i|=r-q} H_{qi}(\xi)\, v^{(q,i)}, \tag{14}$$

$$H_{qi} \equiv H_{qi_1\ldots i_{r-q}} = \sum_{k=0}^{r-q} \varepsilon^2 \left(L^0 \left(N_{qi_{k+1}\ldots i_{r-q}}(\lambda) \frac{(\lambda_{i_1} - \xi_{i_1}) \cdots (\lambda_{i_k} - \xi_{i_k})}{k!} \right)\right)\Big|_\xi -$$
$$- \sum_{k=0}^{r-q} \left(L^1 \left(N_{q-2\, i_{k+1}\ldots i_{r-q}}(\lambda) \frac{(\lambda_{i_1} - \xi_{i_1}) \cdots (\lambda_{i_k} - \xi_{i_k})}{k!} \right)\right)\Big|_\xi . \tag{15}$$

As above, it is understood that in (14), for each $r - q$, the summation is made with respect to $i_l, \ldots, i_{r-q}$ from 1 to s, $\lambda = y/\varepsilon$.

Let us see what form the right side of (15) assumes in the particular case of the system of equations of elasticity theory (3) considered in §4.5.

Then L^1 is the operator of multiplication by the scalar function $\rho(\xi)$. We have $(\rho(\lambda)N(\lambda)(\lambda_{i_1} - \xi_{i_1}) \cdots (\lambda_{i_k} - \xi_{i_k}))|_\xi = 0$, if $k > 0$, since all the factors $(\lambda_j - \xi_j)|_\xi$ are zero. Thus, what remains in the second sum is the summand corresponding to $k = 0$: $\rho(\xi) N_{q-2i_1\ldots i_{r-q}}(\xi)$. Since $L^0 u = \frac{\partial}{\partial x_l}\left(A_{lj}(\xi)\frac{\partial u}{\partial x_j}\right)$, we have

$$\varepsilon^2 \left(L^0 \left(N(\lambda)(\lambda_{i_1} - \xi_{i_1}) \cdots (\lambda_{i_k} - \xi_{i_k})\right)\right)\Big|_\xi =$$
$$= \frac{\partial}{\partial \lambda_l}\left(A_{lj}(\lambda) \frac{\partial}{\partial \lambda_j}\left(N(\lambda)(\lambda_{i_1} - \xi_{i_1}) \cdots (\lambda_{i_k} - \xi_{i_k})\right)\right)\Big|_\xi .$$

We simplify the right side for various values of k. For $k = 0$:

$$\varepsilon^2 (L^0 N(\lambda))\big|_\xi = \frac{\partial}{\partial \xi_l}\left(A_{lj}(\xi) \frac{\partial N(\xi)}{\partial \xi_j}\right);$$

for $k = 1$:

$$\varepsilon^2 \left(L^0 \left(N(\lambda)(\lambda_{i_1} - \xi_{i_1})\right)\right)\big|_\xi = \frac{\partial}{\partial \lambda_l}\left(A_{lj}(\lambda) \frac{\partial \left(N(\lambda)(\lambda_{i_1} - \xi_{i_1})\right)}{\partial \lambda_j}\right)\Big|_\xi =$$
$$= \frac{\partial}{\partial \xi_l}\left(A_{li_1}(\xi) N(\xi)\right) + A_{i_1 j} \frac{\partial N(\xi)}{\partial \xi_j};$$

for $k = 2$:

$$\varepsilon^2 \left(L^0 \left(N(\lambda)(\lambda_{i_1} - \xi_{i_1})(\lambda_{i_2} - \xi_{i_2})\right)\right)\big|_\xi = \left(A_{i_1 i_2}(\xi) + A_{i_2 i_1}(\xi)\right) N(\xi);$$

for $k > 2$:

$$\varepsilon^2 \left(L^0 \left(N(\lambda)(\lambda_{i_1} - \xi_{i_1}) \cdots (\lambda_{i_k} - \xi_{i_k})\right)\right)\big|_\xi = 0.$$

Thus, in the scalar case, as in §4.5,

$$H_{qi_1\dots i_{r-q}} = -\left(\frac{\partial}{\partial\xi_l}\left(A_{lj}\frac{\partial N_{qi_1\dots i_{r-q}}}{\partial\xi_j}\right) + \frac{\partial}{\partial\xi_j}\left(A_{ji_1}N_{qi_2\dots i_{r-q}}\right) + \right.$$
$$\left. + A_{i_1 j}\frac{\partial N_{qi_2\dots i_{r-q}}}{\partial\xi_j} + A_{i_1 i_2}N_{qi_3\dots i_{r-q}} - \rho N_{q-2i_1\dots i_{r-q}}\right)$$

holds.

The formalism of averaging consists of successively setting to zero the coefficients of negative powers of ε and making the coefficients H_{qi} of the positive powers of ε independent of the fast variables ξ_j; in other words, $H_{qi} = h_{qi}$, where $S_m h_{qi} = h_{qi}$ for any m, not necessarily integer. By way of illustration of important features observed in applying this formalism, we consider systems (3) and (4). The coefficient of ε^{-2} is equal to $H_0(\xi) = \varepsilon^2 L^0 N_0$, where N_0 is a solution of the equation $L^0 N_0 = 0$. In cases (3) and (4), we set $N_0 = E$, the unit operator (matrix); for (3), the equality $L^0 E = 0$ is satisfied automatically, and for (4) under condition (5).

Consider the terms of order ε^{-1}: $\varepsilon^{-1}(H_1 v^{(1)} + H_{0i_1}v^{(0,i_1)})$. We have

$$H_1 = \varepsilon^2 L^0 N_1,$$
$$H_{0i_1} = \varepsilon^2\left(L^0 N_{0i_1}(\lambda)\right)\big|_\xi + \varepsilon^2 L^0\left(N_0(\lambda)\left(\lambda_{i_1} - \xi_{i_1}\right)\right)\big|_\xi. \tag{16}$$

The condition $H_1 = 0$ holds if we take $N_1 \equiv 0$.

Suppose the equation $H_{0i_1} = 0$ is solvable for N_{0i_1}. For (3), the equation $H_{0i_1} = 0$ takes the form

$$\frac{\partial}{\partial\xi_l}\left(A_{lj}\frac{\partial N_{0i_1}}{\partial\xi_j}\right) + \frac{\partial}{\partial\xi_l}A_{li_1} = 0. \tag{17}$$

On condition that L^0 is elliptic, this problem is solvable, and the functions N_{0i_1} satisfying (17) do exist.

For (4) in (16), we have the equation

$$H_{0i_1} = \int_{-\infty}^{+\infty}\dots\int_{-\infty}^{+\infty} K_0(\xi,\lambda)\,N_{0i_1}(\lambda)\,d\lambda + \int_{-\infty}^{+\infty}\dots\int_{-\infty}^{+\infty} K_0(\xi,\lambda)\left(\xi_{i_1} - \lambda_{i_1}\right)d\lambda = 0.$$

Since $N_{0i_1}(\lambda)$ is periodic, the equation is written as

$$\int_0^1 \cdots \int_0^1 G(\xi, \lambda) N_{0i_1}(\lambda)\, d\lambda + H_{i_1}(\xi) = 0, \tag{18}$$

$$H_{i_1}(\xi) = \int_{-\infty}^{+\infty} \cdots \int_{-\infty}^{+\infty} K_0(\xi, \lambda)\left(\xi_{i_1} - \lambda_{i_1}\right) d\lambda,$$

$$G(\xi, \lambda) = \sum_{m \in \mathbf{Z}^s} K_0(\xi, \lambda + m).$$

Consider the case of the symmetric kernel $K_0(\xi, \lambda) = K_0(\lambda, \xi)$. Then the kernel $G(\xi, \lambda)$ is also symmetric:

$$G(\lambda, \xi) = \sum_{m \in \mathbf{Z}^s} K_0(\lambda, \xi + m) = \sum_{m \in \mathbf{Z}^s} K_0(\xi + m, \lambda).$$

From (6), $K_0(\xi + m, \lambda) = K_0(\xi, \lambda - m)$, and therefore,

$$G(\lambda, \xi) = \sum_{m \in \mathbf{Z}^s} K_0(\xi, \lambda - m) = \sum_{m \in \mathbf{Z}^s} K_0(\xi, \lambda + m) = G(\xi, \lambda).$$

This symmetry of $G(\xi, \lambda)$ implies

$$\langle G(\xi, \lambda)\rangle = \int_{-\infty}^{\infty} \cdots \int_{-\infty}^{\infty} K_0(\lambda, \xi)\, d\xi = 0.$$

It follows that

$$\left\langle \int_0^1 \cdots \int_0^1 G(\xi, \lambda) N(\lambda)\, d\lambda \right\rangle = \int_0^1 \cdots \int_0^1 \langle G(\xi, \lambda)\rangle N(\lambda)\, d\lambda = 0.$$

By applying the operator $\langle\cdot\rangle$ to (18), we get

$$\langle H_{i_1}\rangle = 0. \tag{19}$$

Thus, (19) is a necessary condition for (18) to be solvable in the case $K_0(\xi, \lambda) = K_0(\lambda, \xi)$. By (5), we have

$$\left\langle \int_{-\infty}^{+\infty} \cdots \int_{-\infty}^{+\infty} K_0(\xi, \lambda)\, \xi_{i_1} d\lambda \right\rangle = \left\langle \xi_{i_1} \left(\int_{-\infty}^{+\infty} \cdots \int_{-\infty}^{+\infty} K_0(\xi, \lambda)\, d\lambda \right) \right\rangle = 0.$$

We require that, in addition to (5), the condition

$$\left\langle \int_{-\infty}^{+\infty} \cdots \int_{-\infty}^{+\infty} K_0(\xi, \lambda)\, \lambda_{i_1} d\lambda \right\rangle = 0, \qquad i_1 = 1, \ldots, s.$$

hold. Then the necessary condition for (19) to be solvable will be satisfied.

We shall now consider the question of solvability of (18).

Let $\varphi_1, \dots, \varphi_n \dots$ be the eigenfunctions of the kernel G, which form, together with the eigenfunctions $\varphi = \text{const}$, a complete orthogonal system; let λ_j be their eigenvalues. Let $H_{i_1}(\xi) = -\sum_{j=1}^{\infty} A_{i_1 j}\varphi_j(\xi)$. Then

$$N_{0i_1}(\xi) = -\sum_{j=1}^{\infty} \frac{A_{i_1 j}}{\lambda_j} \varphi_j(\xi) \tag{20}$$

is a formal solution of (18). For this solution to make sense, it suffices that the series (20) converge in the L_2 norm, that is, that the condition $\sum_{j=1}^{\infty}(A_{i_1 j}/\lambda_j)^2 < \infty$ be satisfied. If this condition does not hold, the question of regularization of a solution of (18) arises.

This problem arose in [13] when we considered the problem of heat transition by means of radiation in a system of dust particles; a similar problem was discussed in [15] for the problem of averaging the transition equation in a periodic medium.

Let us turn to the general case (14), (15), under the additional assumption that $L^0 E = 0$, where E is the unit operator. Then for $N_0 = E$, we get

$$H_0 = 0, \quad \left(L^0 E \xi_{i_1}\right)\Big|_{\xi} = \xi_{i_1} L^0 E = 0.$$

Consequently, the equation $H_{0i_1} = 0$ is written as

$$H_{0i_1} = \varepsilon^2 \left(L^0 \left(N_{0i_1}(\lambda)\right) + L^0 \left(N_0(\lambda)\lambda_{i_1}\right)\right) = 0.$$

We shall assume that this equation has a solution for all $i_1 = 1, \dots, s$.

We have killed the coefficients of the negative powers of ε. Now we require that the coefficients of the positive powers of ε be independent of the fast variables:

$$H_{qi}(\xi) = h_{qi}.$$

This equation in N_{qi} is of the form

$$\varepsilon^2 L^0 N_{qi} = h_{qi} - T_{qi}(\xi), \tag{21}$$

where T_{qi} depends on N_{pj} with p, j such that $p \leqslant q$, $|j| \leqslant |i|$, $p + |j| < q + |i|$. If these N_{pj} are already known, we can try to find the N_{qi} from (21). Equation (21) is solvable but not with any right side $h_{qi} - T_{qi}(\xi)$. A solvability condition must also be satisfied. For the examples (3) and (4) considered above, it is $h_{qi} = \langle T_{qi}(\xi)\rangle$.

In other cases, the necessary solvability condition results in the relation of a more general form:

$$h_{qi} = \langle T_{qi}\Psi(\xi)\rangle; \tag{22}$$

$\Psi(\xi)$ does not depend on q, i. Furthermore, the formalism of averaging coincides with that for the differential equations discussed in §4.5. The averaged equation of infinite order

$$pu \sim \widehat{p}v \sim \sum_{r=2}^{\infty} \varepsilon^{r-2} \sum_{q=0}^{r} \sum_{|i|=r-q} h_{qi}v^{(q,i)} \sim f \tag{23}$$

is defined ambiguously. If $N_{qi}(\xi)$ is a solution of (21), then $N_{qi}(\xi)+cN_0(\xi)$ is also a solution of (21); therefore the process of defining N_{qi} is ambiguous. The choice of the summand $cN_0(\xi)$ affects the subsequent values of T_{qi} and consequently also affects h_{qi}.

Continuum mechanics considers a great number of models of moment theory [59, 98, 99, 152]. Equation (23) may be considered corresponding to some moment theory. And in contrast to other moment theories, we have proposed above an algorithm for determining the coefficients of equations of the moment theory on the basis of information about the microstructure of the medium specified by the operators L^0 and L^1.

In the same way as in §4.5, a solution of (23) can be sought as a regular series in ε:

$$v \sim \sum_{j=0}^{\infty} \varepsilon^j v_j(x, t).$$

We substitute this expansion into the equation $\hat{p}v - f \sim 0$ and simplify.

The operator $\hat{p}$ can be written as

$$\widehat{p} \sim \sum_{r=0}^{\infty} \varepsilon^r \widehat{p}_r, \quad \widehat{p}_r v = \sum_{q=0}^{r+2} \sum_{|i|=r+2-q} h_{qi}v^{(q,i)}.$$

Then $\hat{p}v - f$ is

$$\widehat{p}_0 v_0 - f + \sum_{r=1}^{\infty} \varepsilon^r \sum_{k=0}^{r} \widehat{p}_k v_{r-k}.$$

Thus, v_0 is defined from the equation

$$\widehat{p}_0 v_0 - f = 0, \tag{24}$$

and the rest of the v_r from

$$\widehat{p}_0 v_r + \sum_{k=1}^{r} p_k v_{r-k} = 0. \tag{25}$$

The set of equations (24), (25) can also be considered as a system of equations of the moment theory.

Only first approximations are usually considered, i.e., equations are constructed for $r \leqslant r_0$ when r_0 is small.

It may be inconvenient to interpret the general averaged equation in physical terms, since it contains derivatives with respect to t of order higher than the first. One can get rid of these derivatives in a number of ways, e.g., by selecting in some special manner the summands cN_0 in the functions N_{qi}.

We propose a specific algorithm for selecting constants such that all the $h_{qi} = 0$ for $q > 0$, except h_2, i.e., the averaged equation has the form

$$h_2 v^{(2)} + \sum_{l=2}^{\infty} \varepsilon^{l-2} \sum_{|i|=l} h_{0i} v^{(0,i)} \sim f.$$

We arrange the collections $(q, i_1, \ldots, i_l)$ in lexicographic order with respect to $(q+|i|, |i|)$, that is, N_{qi} always precede N_{rj} if $q+|i| < r+|j|$ or if $q+|i| = r+|j|$, $|i| < |j|$. We set $N_{qi}(\xi) = 0$ for odd q; then all the H_{qi} for the odd q vanish automatically according to (15). Consequently, all the h_{qi} for odd q also vanish. We reset the collections (q, i) successively and each time we determine N_{qi} up to the term cN_0, and then we select this term so that the h_{q+2i} (except for h_2) vanish.

The values of N_{0i} and h_{0i} are linked in a chain which is not affected by the choice of N_{qi}, h_{qi} for $q > 0$. Therefore, these values can be found independently of the others by specifying some criterium, e.g., the condition $\langle N_{0i} \rangle = 0$ for any i.

In constructing the averaged equation with predetermined properties, we need not bother with selecting the coefficients c in the cN_0 but may choose them arbitrarily; the results will be general equation (23). Then we proceed to transform it into the desired form by various manipulations, e.g., by multiplying from the left by the operators $\sum_{r=0}^{\infty} \varepsilon^r \sum_{q=0}^{r} \sum_{|i|=r-q} a_{qi} D^{q,i}$ or by making the change of variables of the form

$$v \sim \sum_{r=0}^{\infty} \varepsilon^r \sum_{q=0}^{r} \sum_{|i|=r-q} b_{qi} D^{q,i} w, \tag{26}$$

where a_0, $b_0 = E$, a_{qi}, b_{qi} $(|i| > 0)$ are constants.

Note that in the case $N_0 = E$, the change of variables (26) is equivalent to a new choice of the cN_0 in the functions N_{qi} for solving (21).

Consider the abstract problem corresponding to the heat equation in a periodic medium:

$$pu = \frac{\partial}{\partial t} L^1 u - L^0 u = f, \tag{27}$$

where L^1, L^0 are operators satisfying the same conditions as in (9).

We substitute the asymptotic expansion of the solution

$$u^{(\infty)} \sim \sum_{q=0}^{\infty} \sum_{|i|=0}^{\infty} \varepsilon^{2q+|i|} \overline{N}_{qi}(\xi) D^{q,i} v$$

into (27) and proceed as for (9). We get

$$pu \sim \sum_{q=0}^{\infty} \sum_{|i|=0}^{\infty} \varepsilon^{2q+|i|-2} \overline{H}_{qi}(\xi) D^{q,i} v,$$

$$\overline{H}_{qi} = \sum_{k=0}^{r-q} \varepsilon^2 \left(L^0 \left(\overline{N}_{qi_{k+1}\dots i_{r-q}}(\lambda) \frac{(\lambda_{i_1} - \xi_{i_1}) \cdots (\lambda_{i_k} - \xi_{i_k})}{k!} \right) \right) \Bigg|_{\xi} -$$
$$- \sum_{k=0}^{r-q} \left(L^1 \left(\overline{N}_{q-1 i_{k+1}\dots i_{r-q}}(\lambda) \frac{(\lambda_{i_1} - \xi_{i_1}) \cdots (\lambda_{i_k} - \xi_{i_k})}{k!} \right) \right) \Bigg|_{\xi}. \quad (28)$$

By comparing (15) and (28), we reach the following conclusions. If in expansion (10) we take $N_{qi}(\xi) = 0$ for all odd q and $N_{(2q)i}(\xi) = \overline{N}_{qi}(\xi)$, then $H_{qi}(\xi) = 0$ for all odd q and $H_{(2q)i}(\xi) = \overline{H}_{qi}(\xi)$.

Suppose that in averaging (9), we have taken $N_{qi}(\xi) = 0$ for all odd q, and from the constancy condition for H_{qi}, we have defined $N_{qi}(\xi)$ for even q; the result is the averaged equation

$$\widehat{p}v \sim \sum_{q=0}^{\infty} \sum_{|i|=0}^{\infty} \varepsilon^{2q+|i|-2} h_{(2q)i} D^{2q,i} v \sim f.$$

Then in the case of (27) for

$$u^{(\infty)} \sim \sum_{q=0}^{\infty} \sum_{|i|=0}^{\infty} \varepsilon^{2q+|i|} N_{(2q)i}(\xi) D^{qi} v$$

we automatically get

$$\widehat{p}v \sim \sum_{q=0}^{\infty} \sum_{|i|=0}^{\infty} \varepsilon^{2q+|i|-2} h_{(2q)i} D^{q,i} v \sim f$$

for the same values of the coefficients $h_{(2q)i}$ as in (23).

Notice that in terms of the asymptotic approach, different types of equations turn out to be equivalent. For example, suppose that for the averaged equation

$$\rho\left(\frac{x}{\varepsilon}\right) \frac{\partial u}{\partial t} - \frac{\partial}{\partial x} \left(K\left(\frac{x}{\varepsilon}\right) \frac{\partial u}{\partial x} \right) = f(x, t)$$

we get

$$h_{10}\frac{\partial v}{\partial t}+h_{02}\frac{\partial^2 v}{\partial x^2}+h_{20}\varepsilon^2\frac{\partial^2 v}{\partial t^2}+O(\varepsilon^3)=f.$$

For simplicity, we consider the case $f=0$ and discard the terms of order $O(\varepsilon^2)$. We get the hyperbolic equation

$$h_{10}\frac{\partial v}{\partial t}+h_{20}\varepsilon^2\frac{\partial^2 v}{\partial t^2}+h_{02}\frac{\partial^2 v}{\partial x^2}=0. \tag{29}$$

The question arises whether we should set the second-type initial value for this equation. Moreover, this equation involves a finite perturbation velocity, and another question may arise: Is the speed of the heat transition in a periodic medium also finite? These questions disappear if we approach the problem in a different way. By neglecting the terms of order $O(\varepsilon^2)$, we can write that $h_{10}\frac{\partial v}{\partial t}\approx -h_{02}\frac{\partial^2 v}{\partial x^2}$. By differentiating with respect to t, we get $h_{10}\frac{\partial^2 v}{\partial t^2}\approx -h_{02}\frac{\partial^3 v}{\partial t\partial x^2}$, whereas double differentiation with respect to x yields $h_{10}\frac{\partial^3 v}{\partial t\partial x^2}\approx -h_{02}\frac{\partial^4 v}{\partial x^4}$.

Thus, $\frac{\partial^2 v}{\partial t^2}\approx -\frac{h_{02}}{h_{10}}\frac{\partial^3 v}{\partial t\partial x^2}\approx \left(\frac{h_{02}}{h_{10}}\right)^2\frac{\partial^4 v}{\partial x^4}$. By substituting these representations of $\partial^2 v/\partial t^2$ into (29), we obtain two new equations up to terms of order $O(\varepsilon^3)$, which are equivalent to the original equation (29):

$$\begin{aligned} h_{10}\frac{\partial v}{\partial t}+h_{02}\frac{\partial^2 v}{\partial x^2}+h_{20}\left(\frac{h_{02}}{h_{10}}\right)^2\varepsilon^2\frac{\partial^4 v}{\partial x^4}&=0,\\ h_{10}\frac{\partial v}{\partial t}+h_{02}\frac{\partial^2 v}{\partial x^2}+h_{20}\frac{h_{02}}{h_{10}}\varepsilon^2\frac{\partial^3 v}{\partial t\,\partial x^2}&=0. \end{aligned} \tag{30}$$

The latter equation appears at first sight to be preferable, meaning that the correct formulation of the boundary value problem for this equation requires neither additional initial values nor additional boundary values.

At the same time, in terms of the asymptotic approach, the equations (30) are equivalent. Suppose the solution is $v(x,t)=v_0(x,t)+\varepsilon v_1(x,t)+\varepsilon^2 v_2(x,t)+O(\varepsilon^3)$ and the v_i are smooth functions. Then, in order to define $v_i(x,t)$ uniquely, it suffices to have one boundary value at each point of the boundary (see §9.5). Consider the case of the wave equation

$$\rho\left(\frac{x}{\varepsilon}\right)\frac{\partial^2 u}{\partial t^2}-\frac{\partial}{\partial x}\left(K\left(\frac{x}{\varepsilon}\right)\frac{\partial u}{\partial x}\right)=0.$$

The averaged equation of the order of accuracy $O(\varepsilon^3)$ is

$$h_{20}\frac{\partial^2 v}{\partial t^2}+h_{02}\frac{\partial^2 v}{\partial x^2}+\varepsilon\left(h_{21}\frac{\partial^3 v}{\partial t^2\partial x}+h_{03}\frac{\partial^3 v}{\partial x^3}\right)+$$
$$+\varepsilon^2\left(h_{40}\frac{\partial^4 v}{\partial t^4}+h_{22}\frac{\partial^4 v}{\partial t^2\partial x^2}+h_{04}\frac{\partial^4 v}{\partial x^4}\right)=0, \quad (31)$$
$$h_{20}=\langle\rho\rangle, \quad h_{02}=-\langle K^{-1}\rangle^{-1}=-\widehat{K}.$$

By selecting the average values in the functions N_{ij} and by multiplying the equation additionally by an operator of the form $E+\varepsilon(\alpha\partial/\partial t+\beta\partial/\partial x)$, we can make the coefficients h_{03}, h_{21} vanish. Since from (31) $\frac{\partial^2 v}{\partial t^2}=-\frac{h_{02}}{h_{20}}\frac{\partial^2 v}{\partial x^2}+O(\varepsilon^2)$,

$$\frac{\partial^4 v}{\partial t^2\partial x^2}\approx-\frac{h_{02}}{h_{20}}\frac{\partial^4 v}{\partial x^4}, \quad \frac{\partial^4 v}{\partial t^4}\approx\left(\frac{h_{02}}{h_{20}}\right)^2\frac{\partial^4 v}{\partial x^4}.$$

Therefore, (31) is equivalent to the equation

$$\widehat{\rho}\frac{\partial^2 v}{\partial t^2}-\widehat{K}\frac{\partial^2 v}{\partial x^2}+\varepsilon^2\widehat{q}\frac{\partial^4 v}{\partial x^4}=0, \quad \widehat{q}=h_{04}-h_{22}\frac{h_{02}}{h_{20}}+h_{40}\left(\frac{h_{02}}{h_{20}}\right)^2$$

up to terms of order $O(\varepsilon^3)$ (a more exact estimate yields accuracy $O(\varepsilon^4)$).

This kind of averaged equation was constructed in [34] for the case of piecewise constant ρ and K. If particular solutions of (31) are sought as $e^{i(\omega t-hx)}$, we get a variance equation which relates ω to h:

$$-\widehat{\rho}\omega^2+\widehat{K}h^2+\varepsilon^2\widehat{q}h^4=0.$$

The wave speed

$$\widehat{c}=h/\omega=1\Big/\sqrt{\widehat{K}/\widehat{\rho}+(\widehat{q}/\widehat{\rho})h^2}$$

depends on the wavelength $\lambda=2\pi/h$. This dependence is called *wave dispersion* [34, 35, 151].

If a boundary value problem is considered, it is more convenient to derive the averaged equation of the form

$$\widehat{\rho}\frac{\partial^2 v}{\partial t^2}-\widehat{K}\frac{\partial^2 v}{\partial x^2}-\widehat{r}\varepsilon^2\frac{\partial^4 v}{\partial t^2\partial x^2}=0,$$

similar to (30).

Remark 3: The formal constructions made in §§1 and 2 can be extended to a system of equations of theory of plates and shells [3–7, 50, 65, 192, 193].

§3 A Method of Describing Multi-Dimensional Periodic Media That Does Not Involve Separating Fast and Slow Variables

Section 3.12 proposed an approach to the investigation of one-dimensional periodic problems that did not involve separation into individual problems for fast and slow variables: Strain was sought as a function depending on both the fast and slow variables. This method is extended below to the case of multi-dimensional problems of elasticity theory.

We have the system of equations

$$\rho(\xi)\frac{\partial^2 u^i}{\partial t^2}-\frac{\partial}{\partial x_j}\sigma^i_j=0,\quad \sigma^i_j=a^{ik}_{jl}(\xi)\frac{\partial u^k}{\partial x_l},\tag{1}$$

where $\xi = x/\varepsilon$, $a^{ik}_{jl}(\xi)$ are 1-periodic functions satisfying the symmetry conditions (4.4.1); due to (4.4.1), we have $\sigma^i_j = \sigma^j_i$ for all i, j.

System (1) can be written as

$$\rho(\xi)\frac{\partial^2 u^i}{\partial t^2}-\frac{\partial}{\partial x_j}\sigma^i_j=0,\tag{2}$$

$$e^i_j=\frac{1}{2}\left(\frac{\partial u^i}{\partial x_j}+\frac{\partial u^j}{\partial x_i}\right)=b^{ik}_{jl}(\xi)\sigma^k_l,\quad i\leqslant j,\quad \sigma^k_l=\sigma^l_k.\tag{3}$$

We seek an asymptotic expansion of the solution in the form

$$\begin{aligned}u^i&=\overset{0}{u}{}^i(x,t,\xi)+\varepsilon\overset{1}{u}{}^i(x,t,\xi)+\dots,\\ \sigma^i&=\overset{0}{\sigma}{}^i(x,t,\xi)+\varepsilon\overset{1}{\sigma}{}^i(x,t,\xi)+\dots,\end{aligned}\tag{4}$$

where $\overset{k}{u}{}^i$, $\overset{k}{\sigma}{}^i$ are functions 1-periodic in ξ_q.

By substituting (4) into (3) and by equating the terms of orders ε^{-1} and ε^0 to zero, we get

$$\frac{1}{2}\left(\frac{\partial\overset{0}{u}{}^i}{\partial\xi_j}+\frac{\partial\overset{0}{u}{}^j}{\partial\xi_i}\right)=0,\quad i\leqslant j,\tag{5}$$

$$\overset{0}{e}{}^i_j+\overset{1}{e}{}^i_j-b^{ik}_{jl}\overset{0}{\sigma}{}^k_l=0,\quad i\leqslant j,\tag{6}$$

here

$$\overset{0}{e}{}^i_j=1/2\left(\partial\overset{0}{u}{}^i/\partial x_j+\partial\overset{0}{u}{}^j/\partial x_i\right),\quad \overset{1}{e}{}^i_j=1/2\left(\partial\overset{1}{u}{}^i/\partial\xi_j+\partial\overset{1}{u}{}^j/\partial\xi_i\right).$$

The solution of system (5) has the form $u^0 = a(x,t) + B(x,t)\xi$, where B is a skew-symmetric matrix. The 1-periodicity of u^0 in the variables ξ_q implies $B \equiv 0$; thus, $\overset{0}{u}{}^i = \overset{0}{u}{}^i(x,t)$.

We use the compatibility condition of [143] for the e^i_j:

$$\frac{\partial^2 e^i_j}{\partial x_m \partial x_r} + \frac{\partial^2 e^m_r}{\partial x_i \partial x_j} - \frac{\partial^2 e^i_m}{\partial x_j \partial x_r} - \frac{\partial^2 e^j_r}{\partial x_i \partial x_m} = 0.$$

Since $e^i_j \sim \overset{0}{e}{}^i_j + \overset{1}{e}{}^i_j = b^{ik}_{jl}\overset{0}{\sigma}{}^j_l$, it follows that

$$\frac{\partial^2 b^{ik}_{jl}(\xi)\,\overset{0}{\sigma}{}^k_l}{\partial \xi_m \partial \xi_r} + \frac{\partial^2 b^{mk}_{rl}(\xi)\,\overset{0}{\sigma}{}^k_l}{\partial \xi_i \partial \xi_j} - \frac{\partial^2 b^{ik}_{ml}(\xi)\,\overset{0}{\sigma}{}^k_l}{\partial \xi_j \partial \xi_r} - \frac{\partial^2 b^{jk}_{rl}(\xi)\,\overset{0}{\sigma}{}^k_l}{\partial \xi_i \partial \xi_m} = 0. \quad (7)$$

By applying $\langle\cdot\rangle$ to (6), we get

$$\overset{0}{e}{}^i_j = \left\langle b^{ik}_{jl}(\xi)\overset{0}{\sigma}{}^k_l \right\rangle. \quad (8)$$

By substituting (4) into (3) and by equating the coefficients of ε^{-1} and ε^0 to zero, we obtain the relations

$$\frac{\partial}{\partial \xi_j}\overset{0}{\sigma}{}^i_j = 0, \quad (9)$$

$$\rho(\xi)\frac{\partial^2 \overset{0}{u}{}^i}{\partial t^2} - \frac{\partial}{\partial x_j}\overset{0}{\sigma}{}^i_j - \frac{\partial}{\partial \xi_j}\overset{1}{\sigma}{}^i_j = 0. \quad (10)$$

We apply the operation $\langle\cdot\rangle$ of taking the average to (10); then we have

$$\langle\rho\rangle\frac{\partial^2 \overset{0}{u}{}^i}{\partial t^2} - \frac{\partial}{\partial x_i}\left\langle \overset{0}{\sigma}{}^i_j \right\rangle = 0. \quad (11)$$

The system of equations (7), (9), together with the conditions $\overset{0}{\sigma}{}^i_j = \overset{0}{\sigma}{}^j_i$, forms the well-known system of stress equations (in the variables ξ_q) [125]. A solution of the system in the class of periodic functions takes the form

$$\overset{0}{\sigma}{}^i_j = \overset{0}{n}{}^{ik}_{jl}(\xi)\,W^k_l(x,t); \quad (12)$$

here the $\overset{0}{n}{}^{ik}_{jl}$ with fixed k, l are some particular solutions of the system and the $W^k_l(x,t)$ are arbitrary functions independent of the fast variables. Condition (8) implies a relationship between $\overset{0}{e}{}^i_k$ and W^r_m, which is $\overset{0}{e}{}^i_j = \langle b^{ik}_{jl}\overset{0}{n}{}^{kr}_{lm}\rangle W^r_m$. Hence, we see that the functions W^r_m are expressed in terms of the strains $\overset{0}{e}{}^i_j$, and $W^r_m = C^{ri}_{mj}\overset{0}{e}{}^i_j$; by substituting this relationship into (12) and then into (11), we obtain the same averaged equations as in §4.5. In the above-described linear case, the fast and slow variables have separated and the above constructions do not differ much from similar constructions in §4.5.

Consider the case of the system of equations of elasticity theory, which corresponds to a geometrically linear and physically nonlinear model. There is a system of equations

$$\rho(\xi)\frac{\partial^2 u^i}{\partial t^2} - \frac{\partial \sigma^i_j}{\partial x_j} = 0, \quad \sigma^i_j = \sigma^j_i,$$
$$e^i_j = \frac{1}{2}\left(\frac{\partial u^i}{\partial x_j} + \frac{\partial u^j}{\partial x_i}\right) = b^i_j(\xi, \|\sigma^k_l\|). \tag{13}$$

By substituting the expansions (4) into (13) as above, we obtain equations (5) and the equations $\overset{1}{e}{}^i_j = b^i_j\left(\xi, \|\overset{0}{\sigma}{}^k_l\|\right) - \overset{0}{e}{}^i_j$. It follows from the conditions of (5) that $\overset{0}{u}{}^i = \overset{0}{u}{}^i(x,t)$, and therefore $\overset{0}{e}{}^i_j = \overset{0}{e}{}^i_j(x,t)$. From the compatibility conditions, similarly to (7), we get the system of equations

$$\frac{\partial^2 b^i_j\left(\xi, \left\|\overset{0}{\sigma}{}^k_l\right\|\right)}{\partial \xi_m \partial \xi_r} + \frac{\partial^2 b^m_r\left(\xi, \left\|\overset{0}{\sigma}{}^k_l\right\|\right)}{\partial \xi_i \partial \xi_j} - \frac{\partial^2 b^i_m\left(\xi, \left\|\overset{0}{\sigma}{}^k_l\right\|\right)}{\partial \xi_j \partial \xi_r} - \frac{\partial^2 b^j_r\left(\xi, \left\|\overset{0}{\sigma}{}^k_l\right\|\right)}{\partial \xi_i \partial \xi_m} = 0. \tag{14}$$

Instead of (8), we accordingly obtain the equations

$$\overset{0}{e}{}^i_j = \left\langle b^i_j\left(\xi, \left\|\overset{0}{\sigma}{}^i_j\right\|\right)\right\rangle, \tag{15}$$

with equations (9), (10) remaining the same.

We propose the following scheme of calculations for the solution of this problem (for definiteness, in the two-dimensional case). The approximate values of $\overset{0}{u}(x,t)$ are calculated at the points $(k_1h_1, k_2h_2, k\tau)$, where h_1, h_2, τ are steps with respect to x_1, x_2, t, and the values of $\overset{0}{\sigma}(x,t,\xi)$ are calculated at the points $((k_1+1/2)h_1,\ (k_2+1/2)h_2,\ k\tau)$.

The derivatives appearing in (11) are approximated up to the second order of accuracy, for example,

$$\left.\frac{\partial\langle\sigma\rangle}{\partial x_1}\right|_{k_1,k_2,k} \approx \frac{1}{2h_1}\Big(\big(\langle\sigma_{k_1+1/2,k_2+1/2,k}\rangle + \langle\sigma_{k_1+1/2,k_2+1/2,k}\rangle\big) - \big(\langle\sigma_{k_1-1/2,k_2+1/2,k}\rangle + \langle\sigma_{k_1-1/2,k_2-1/2,k}\rangle\big)\Big).$$

Let all the approximate values of $\overset{0}{u}$ and $\overset{0}{\sigma}$ be found for $k \leqslant n$; we determine $u^0_{k_1,k_2,n+1}$ from the approximation of the equations of motion (11). Next, by using the approximate formulas of the form

$$\left.\frac{\partial \overset{0}{u}}{\partial x_1}\right|_{(k_1+1/2,k_2+1/2,k)} \approx \frac{1}{2h_1}\left(\left(\overset{0}{u}_{k_1+1,k_2+1,k} + \overset{0}{u}_{k_1+1,k_2,k}\right) - \right.$$
$$\left. - \left(\overset{0}{u}_{k_1,k_2+1,k} + \overset{0}{u}_{k_1,k_2,k}\right)\right)$$

we find all the $\overset{0}{e}{}^i_{j(k_1+1/2,k_2+1/2,n+1)}$. From the equations (9), (14), and (15), we determine $\overset{0}{\sigma}{}^k_l|_{k_1+1/2,k_2+1/2,n+1)}$.

The above procedure of studying processes in periodic media is applicable to a greater number of problems than the basic procedure of the present book; for example, it can be applied to the types of problems involved in the theory of plasticity (see §3.12).

However, in the cases where the basic procedure is applicable, the one proposed in the present section appears to be less effective.

The application of the basic procedure calls for a solution of an auxiliary system of equations on a cell:

$$\frac{\partial}{\partial \xi_j}\sigma^i_j\left(\xi, \left\| \frac{\partial v^k}{\partial x_l} + \frac{\partial N^k}{\partial \xi_l} \right\|\right) = 0, \qquad i = 1, 2, \ldots, \tag{16}$$

for all $||\partial v^k/\partial x_1||$.

With the approach described here, similar cell problems (9), (14), (15) can only be solved for the values of $||\partial v^k/\partial x_l||$ which correspond to a specific solution of a problem, and this seems to suggest that such a procedure is more effective. But a more effective version can be proposed even for the original procedure. We determine the effective values $\hat{\sigma}^i_j(||\partial v^k/\partial x_l||)$ for $||\partial v^k/\partial x_l||$ corresponding to the initial conditions. We shall integrate numerically the averaged problem proceeding by steps in time. At each step, we find only those effective values of $\hat{\sigma}^i_j(||\partial v^k/\partial x_l||)$ that cannot be found by interpolation from the values found before. This procedure is more elaborate in organization but appears to be more effective as compared with those discussed above.

REMARK 1: In some cases, it may be interesting to investigate the response of a periodic medium to perturbations of a certain type, e.g., to a plane wave travelling in some fixed direction.

Consider the case when the wave propagates along the axis x_1. When the basic procedure is used, problem (16) has to be solved only for the matrices $||\partial v^k/\partial x_l||$ with the nonzero first column. In the case of the plane wave, for the averaged equation $v = v(\sum \alpha_i x_i - t/c)$, the auxiliary problem (16) on a cell should be solved only for the matrices $||\partial v^k/\partial x_l||$, where all the rows are proportional to the vector $\alpha = (\alpha_1, \ldots, \alpha_m)$. If a longitudinal plane wave travels along the axis x_1, problem (16) needs to be solved only for the matrices $||\partial v^k/\partial x_l||$ with a single nonzero element $\partial v^1/\partial x_1$. Accordingly, the number of the auxiliary cell problems (16) is reduced also in the linear case. For example, in the latter instance, it suffices to find only the first column of the matrix $N_1(\xi) = ||n_1^{ik}||$.

REMARK 2: It may be particularly convenient for the case of plane waves travelling along coordinate axes to use the idea of Yanenko about predicting the behaviour of a solution for small ε from the resultsof direct calculations for large ε. Due to the periodicity of the solution in x_2, x_3 with period ε, it suffices to restrict calculations to the parallelepiped $0 \leqslant x_2, x_3 \leqslant \varepsilon$. If the wave is longitudinal and the periodicity domain satisfies the symmetry condition (6.3.3) relative to the planes $\xi_2 = 0$, $\xi_3 = 0$, the calculations can be confined only to the domain $0 \leqslant \xi_2, \xi_3 \leqslant 1/2$, provided the solution is even with respect to the faces $\xi_2 = 0$, $\xi_2 = 1/2$, $\xi_3 = 0$, and $\xi_3 = 1/2$.

CHAPTER 6

Properties of Effective Coefficients. Relationship Among Local and Averaged Characteristics of a Solution

This chapter investigates properties of solutions of equations on a cell and of averaged equations and relates the average characteristics of the original problem to those of a solution of the averaged equation.

§1 Maintaining the Properties of Convexity and Symmetry of the Minimized Functional in Averaging

Consider the extremum problem for the functional

$$J(u) = \int_G \Phi\left(\frac{x}{\varepsilon}, x, \nabla_x u\right) dx, \quad x \in \mathbf{R}^s, \quad u\,|_{\partial G} = u_0(x);$$

here ∂G is the boundary of G, $u = (u^1, \dots, u^n)^T$, $\nabla_x u$ is the $n \times s$ matrix with the elements $\partial u^i/\partial x_j$, $\Phi(\xi, x, U)$ is a 1-periodic function of the arguments ξ, $u^i_j = \partial u^i/\partial x_j$.

We shall show that the formal averaging by the methods described in §§5.1 and 4.6 also leads to the extremum problem of a functional.

The system of Euler's equations in the function which realizes the extremum of J is of the form

$$\frac{\partial}{\partial x_m} a^k_m\left(\frac{x}{\varepsilon}, x, \nabla_x u\right) = 0,$$

$$k = 1, \dots, n, \quad a^k_m(\xi, x, U) = \frac{\partial \Phi(x, \xi, U)}{\partial u^k_m}.$$

The approximation to the solution is sought in §5.1 as

$$u^{(\infty)} \sim v + \varepsilon N_1 + \varepsilon^2 N_2 + \dots,$$

or component-wise

$$u^{i,\infty} \sim v^i + \varepsilon N^i_1 + \varepsilon^2 N^i_2 + \dots,$$

where, in particular, $N^i_1 = N^i_1(\xi, x, \nabla_x v_0)$. The system of equations from which the functions $N^i_1(\xi, x, y)$, 1-periodic in the variables ξ, are defined takes the form

$$\frac{\partial}{\partial \xi_m} A_m^k(\xi, x, y) = 0, \tag{1}$$

$A_m^k(\xi, x, y)$ is the abbreviated notation for $a_m^k(\xi, x, y + \nabla_\xi N_1(\xi, x, y))$. The averaged system in $\nabla_x v_0$ has the form

$$\frac{\partial}{\partial x_m} \widehat{a}_m^k(x, \nabla_x v_0) = 0; \tag{2}$$

here

$$\widehat{a}_m^k(x, y) = \langle A_m^k(\xi, x, y)\rangle. \tag{3}$$

Henceforth in this section, we assume that $\Phi(\xi, x, y)$ is twice continuously differentiable with respect to x, y_j^i; in addition, we suppose that a periodic solution of problem (1) exists and is differentiable for the required number of times with respect to x, y_j^i.

In the case of discontinuous coefficients, a solution of equation (1) is understood in the generalized sense:

$$\left\langle A_m^k(\xi, x, y) \frac{\partial \varphi^k}{\partial \xi_m} \right\rangle = 0 \tag{4}$$

for any 1-periodic function $\varphi(\xi) \in W_2^{1\,\mathrm{loc}}$.

Theorem 1. *The system of equations* (2) *is a system of Euler's equations for the functional*

$$\widehat{J}(v_0) = \int_G \widehat{\Phi}(x, \nabla_x v_0)\, dx,$$

where

$$\widehat{\Phi}(x, y) = \langle \Phi(\xi, x, y + \nabla_\xi N_1(\xi, x, y))\rangle.$$

We shall give the first proof in terms of differential relations, assuming in addition that Φ, N_1, and their derivatives with respect to the parameters y_j^i are differentiable with respect to the variables ξ_1.

The system of Euler's equations for the functional $\acute{J}$ is

$$\frac{\partial}{\partial x_m}\left(\frac{\partial \widehat{\Phi}(x, \nabla_x v_0)}{\partial v_m^k}\right) = 0, \quad v_m^k = \frac{\partial v_0^k}{\partial x_m};$$

and we have

$$\frac{\partial\widehat{\Phi}(x, y)}{\partial y_m^k} = \left\langle \frac{\partial}{\partial y_m^k} \Phi(\xi, x, y + \nabla_\xi N_1(\xi, x, y)) \right\rangle = \langle A_m^k(\xi, x, y)\rangle + \sigma_m^k,$$

where $\sigma_m^k = \left\langle A_q^p(\xi, x, y)\partial\frac{\partial N_1^p}{\partial \xi_q}/\partial y_m^k \right\rangle$. In this section, the summation over the repeated superscripts p, k, i, and r is made from 1 to n, and over the subscripts q, m, j, and l from 1 to s. The first summand is exactly equal to $\hat{a}_m^k$, and therefore, to prove the assertion that $\hat{a}_m^k = \partial\widehat{\Phi}/\partial y_m^k$, it suffices to show that $\sigma_m^k = 0$.

We transpose the differentiation with respect to ξ_q and y_m^k and integrate by parts with respect to the ξ_1; from (1), we get

$$\sigma_m^k = -\left\langle \left(\frac{\partial}{\partial \xi_q} A_q^p(\xi, x, y)\right) \frac{\partial N_1^p}{\partial y_m^k} \right\rangle = 0.$$

Thus,

$$\frac{\partial\widehat{\Phi}}{\partial y_m^k} = \widehat{a}_m^k,$$

where the $\hat{a}_m^k$ are given by (1), (3) according to the averaging procedure described in §§5.1 and 4.6. □

We can produce the same results by means of integral identities. In this case, the differentiability of Φ with respect to the ξ_q is not assumed. The averaged system of equations obtained by the method given in §§5.1 and 4.6 corresponds to the integral identity

$$\int_G \widehat{a}_m^k(x, \nabla_x v_0)\frac{\partial \varphi^k}{\partial x_m}\, dx = 0 \quad \forall \varphi \in \mathring{W}_2^1(G),$$

where the $\hat{a}_m^k$ are found from (3).

We consider the functional $\hat{J}$. The condition that

$$\frac{d}{dt}\hat{J}(v_0 + t\varphi)\big|_{t=0} = 0 \quad \forall \varphi \in \mathring{W}_2^1(G),$$

results in the equality

$$\int_G \frac{\partial\widehat{\Phi}}{\partial v_m^k}\frac{\partial \varphi^k}{\partial x_m}\, dx = 0,$$

that is,

$$\int_G \left\langle A_q^p(\xi, x, y)\, \partial\left(y_q^p + \frac{\partial N_1^p}{\partial \xi_q}(\xi, x, y)\right)\Big/\partial y_m^k \right\rangle\Bigg|_{y=\nabla_x v_0} \frac{\partial \varphi^k}{\partial x_m}\, dx = 0$$

$$\forall \varphi \in \overset{\circ}{W}{}_2^1(G).$$

These relations are rewritten as

$$\int_G \langle A_m^k(\xi, x, \nabla_x v_0)\rangle \frac{\partial \varphi^k}{\partial x_m}\, dx + \int_G \sigma_m^k \frac{\partial \varphi^k}{\partial x_m}\, dx = 0;$$

$$\sigma_m^k = \left\langle A_q^p(\xi, x, y)\, \partial \frac{\partial N_1^p(\xi, x, y)}{\partial y_m^k}\Big/\partial \xi_q \right\rangle\Bigg|_{y=\nabla_x v_0} = 0$$

due to (4). Thus, Euler's condition for the extremum of the functional $\hat{J}$ brings about the equations (2).

In some important cases, the functional J is convex, meaning that for $a_{mj}^{ki} = \partial^2\Phi(\xi, x, U)/\partial u_m^k \partial u_j^i$,

$$a_{mj}^{ki}\zeta_m^k\zeta_j^i \geqslant \varkappa\, \zeta_m^k\zeta_m^k, \quad \varkappa > 0, \tag{5}$$

for any x and a real tensor ζ_m^k. We shall show that this relation is also valid for the functional $\hat{J}$.

Theorem 2. *The tensor $\hat{a}_{mj}^{ki}(x, y) = \partial^2\hat{\Phi}(x, y)/\partial y_m^k \partial y_j^i$ has the following properties:*

1) *$\hat{a}_{mj}^{ki}(x, y) - \hat{a}_{jm}^{ik}(x, y)$ for any k, i, m, and j;*

2) *y satisfies the following inequality for any x:*

$$\hat{a}_{mj}^{ki}(x, y)\, \eta_m^k \eta_j^i \geqslant \varkappa \eta_m^k \eta_m^k$$

(with any matrix $\|\eta_m^k\|$).

Proof: 1) This property follows from

$$\hat{a}_{mj}^{ki} = \frac{\partial^2\hat{\Phi}}{\partial y_m^k \partial y_j^i} = \frac{\partial^2\hat{\Phi}}{\partial y_j^i \partial y_m^k} = \hat{a}_{jm}^{ik}.$$

2) We set $Y_m^k = y_m^k + \frac{\partial N_1^k(\xi, x, y)}{\partial \xi_m}$. For arbitrary $\|\eta_m^k\|$, $\|y_m^k\|$, we assume

$$\zeta_m^k = \frac{\partial Y_m^k}{\partial y_q^p}\eta_q^p$$

and substitute into (5). We get

$$a^{ki}_{mj}\left(\frac{\partial Y^k_m}{\partial y^p_q}\eta^p_q\right)\left(\frac{\partial Y^i_j}{\partial y^r_l}\eta^r_l\right)\geqslant \varkappa\sum_{k=1}^{n}\sum_{m=1}^{s}\left(\frac{\partial Y^k_m}{\partial y^p_q}\eta^p_q\right)^2.$$

By applying the operation $\langle\cdot\rangle$, we get

$$\beta^{pr}_{ql}\eta^p_q\eta^r_l\geqslant\varkappa\left\langle\sum_{k=1}^{n}\sum_{m=1}^{s}\left(\frac{\partial Y^k_m}{\partial y^p_q}\eta^p_q\right)^2\right\rangle;\tag{6}$$

here

$$\beta^{pr}_{ql}=\left\langle a^{ki}_{mj}\frac{\partial Y^k_m}{\partial y^p_q}\frac{\partial Y^i_j}{\partial y^r_l}\right\rangle.$$

By substituting $Y^k_m = y^k_m + \partial N^k_1/\partial\xi_m$, we see that

$$\beta^{pr}_{ql}=\left\langle a^{pi}_{qj}\frac{\partial Y^i_j}{\partial y^r_l}\right\rangle+\left\langle a^{ki}_{mj}\left(\partial\frac{\partial N^k_1}{\partial y^p_q}\Big/\partial\xi_m\right)\frac{\partial Y^i_j}{\partial y^r_l}\right\rangle.$$

We differentiate (4) with respect to y^r_l and set $\varphi^k=\partial N^k_1/\partial y^p_q$; we obtain

$$\left\langle a^{ki}_{mj}\frac{\partial Y^i_j}{\partial y^r_l}\left(\partial\frac{\partial N^k_1}{\partial y^p_q}\Big/\partial\xi_m\right)\right\rangle=0.$$

We substitute p for k and q for m in (3) and differentiate (3) with respect to $y^r_l=\partial v^r/\partial x_l$; the result is $\hat{a}^{pr}_{ql}=\langle a^{pi}_{qj}\partial Y^i_j/\partial y^t_l\rangle$. Thus,

$$\beta^{pr}_{ql}=\hat{a}^{pr}_{ql}.\tag{7}$$

We consider the term $\rho^k_l=\left\langle\left(\frac{\partial Y^k_l}{\partial y^p_q}\eta^p_q\right)^2\right\rangle$; by substituting $Y^k_l = y^k_l + \partial N^k_1/\partial\xi_l$, we have

$$\rho^k_l=\left\langle\left(\eta^k_l+\frac{\partial}{\partial\xi_l}\eta^p_q\frac{\partial N^k_1}{\partial y^p_q}\right)^2\right\rangle=\left\langle\left(\eta^k_l+\frac{\partial}{\partial\xi_l}\chi^k\right)^2\right\rangle;$$

here $\chi^k=\eta^p_q\partial N^k_1/\partial y^p_q$.

Since $\left\langle\eta^k_l\frac{\partial}{\partial\xi_l}\chi^k\right\rangle=\left\langle\frac{\partial}{\partial\xi_l}\eta^k_l\chi^k\right\rangle=0$ as an integral of a derivation of a periodic function, $\rho^k_l=(\eta^k_l)^2+\left\langle\left(\frac{\partial}{\partial\xi_l}\chi^k\right)^2\right\rangle\geqslant(\eta^k_l)^2$. This and (6), (7) imply that $\hat{a}^{pr}_{ql}\eta^p_q\eta^r_l\geqslant\kappa\eta^k_l\eta^k_l$.

In the case of small strains for $n = s$, the function $\Phi(\xi, x, U)$ can be represented as $\Phi(\xi, x, U) = W(\xi, x, e)$, where $e = (U + U^T)/2$ is the tensor of small strains ($e^k_m = (u^k_m + U^m_k)/2$, $W(\xi, x, e) = W(\xi, x, e^T)$. Since

$$a^{ki}_{mj}(\xi, x, U) = \\ = \frac{\partial^2 W(\xi, x, (U+U^T)/2)}{\partial u^k_m \partial u^i_j} = \frac{1}{4}\left(\frac{\partial}{\partial e^k_m} + \frac{\partial}{\partial e^m_k}\right)\left(\frac{\partial}{\partial e^j_i} + \frac{\partial}{\partial e^i_j}\right) W(\xi, x, e),$$

the tensor a^{ki}_{mj} has the following symmetry properties: $a^{ki}_{mj} = a^{mi}_{kj} = a^{kj}_{mi} = a^{ik}_{jm}$. Let

$$a^{ki}_{mj}(\xi, x, U)\zeta^k_m\zeta^i_j \geqslant \varkappa\zeta^k_m\zeta^k_m \tag{8}$$

for any ξ x, U and the symmetric tensor ζ^k_m. Here $\kappa > 0$ and is independent of ξ, x, U. In this case, the following theorem holds.

THEOREM 3. 1) *The function* $\widehat{\Phi}(x, y)$ *can be represented as*

$$\widehat{\Phi}(x, y) = \widehat{W}(x, z), \quad z = (y + y^T)/2,$$
$$\widehat{a}^{ki}_{mj}(x, y) = \frac{\partial^2 \widehat{W}(x, (y + y^T)/2)}{\partial y^k_m \partial y^i_j}.$$

2) $\widehat{a}^{ki}_{mj} = \widehat{a}^{mi}_{kj} = \widehat{a}^{kj}_{mi} = \widehat{a}^{ik}_{jm}$.
3) $\widehat{a}^{ki}_{mj}(x, y)\eta^k_m\eta^i_j \geqslant \varkappa\eta^k_m\eta^k_m$

for any symmetric tensor η^k_m.

PROOF: 1) we rewrite (1) as

$$\frac{\partial}{\partial \xi_m} a^k_m(\xi, x, y + \nabla_\xi N_1(\xi, x, y)) = 0, \tag{9}$$

or

$$\frac{\partial}{\partial \xi_m}\frac{\partial W}{\partial u^k_m}\left(\xi, x, \frac{1}{2}\left(y + y^T + \nabla_\xi N_1 + (\nabla_\xi N_1)^T\right)\right) = 0.$$

By making the change of parameter $z = (y + y^T)/2$, we get the family of equations with z as the parameter:

$$\frac{1}{2}\frac{\partial}{\partial \xi_m}\left(\frac{\partial}{\partial e^m_k} + \frac{\partial}{\partial e^k_m}\right) W(\xi, x, z + \mathrm{Def}_\xi \overline{N}_1(\xi, x, z)) = 0,$$

where

$$\mathrm{Def}_\xi \overline{N}_1 = \| d^k_m \| = \left\| \frac{1}{2}\left(\frac{\partial \overline{N}^k_1}{\partial \xi_m} + \frac{\partial \overline{N}^m_1}{\partial \xi_h}\right)\right\|.$$

The vector-function $N_1(\xi,x,y)$ (solution of (9)) can be found from the formula $N_1(\xi,x,y)=\overline{N}_1(\xi,x,(y+y^T)/2)$. (By assuming $\tilde{a}^{ki}_{mj}=\frac{\partial}{\partial y^i_j}(a^k_m(\xi,x,y+\nabla_\xi N_1))$, we have $\tilde{a}^{ki}_{mj}=1/4(\partial/\partial z^i_j+\partial/\partial z^j_i)(\partial/\partial e^k_m+\partial/\partial e^m_k)W(\xi,x,z+\mathrm{De}_\xi\,\overline{N}_1)$, therefore $\tilde{a}^{ki}_{mj}=\tilde{a}^{mi}_{kj}=\tilde{a}^{kj}_{mi}$). Thus,

$$\begin{aligned}\widehat{\Phi}(x,y)&=\langle\Phi(\xi,x,y+\nabla_\xi N_1(\xi,x,y))\rangle=\\&=\langle W(\xi,x,(y+y^T)/2+\mathrm{Def}_\xi\,\overline{N}_1(\xi,x,(y+y^T)/2))\rangle=\\&=\widehat{W}(x,z),\quad z=(y+y^T)/2.\end{aligned}$$

2) The chain of relations

$$\begin{aligned}\widehat{a}^{ki}_{mj}(x,y)&=\frac{\partial^2\widehat{W}(x,(y+y^T)/2)}{\partial y^k_m\partial y^i_j}=\\&=\frac{1}{4}\left(\frac{\partial}{\partial z^k_m}+\frac{\partial}{\partial z^m_k}\right)\left(\frac{\partial}{\partial z^i_j}+\frac{\partial}{\partial z^j_i}\right)\widehat{W}(x,z)\Big|_{z=(y+y^T)/2}=\\&=\widehat{a}^{mi}_{kj}(x,y)=\widehat{a}^{kj}_{mi}(x,y)=\widehat{a}^{ik}_{jm}(x,y)\end{aligned}$$

holds.

3) We set

$$Z^k_m=z^k_m+\frac{1}{2}\left(\frac{\partial\overline{N}^k_1}{\partial\xi_m}(\xi,x,z)+\frac{\partial\overline{N}^m_1}{\partial\xi_k}(\xi,x,z)\right);$$

and for any symmetric tensors z^k_m, η^k_m, we assume $\zeta^k_m=\frac{1}{2}\left(\frac{\partial}{\partial z^p_q}+\frac{\partial}{\partial z^q_p}\right)Z^k_m\eta^p_q$. The tensor ζ^k_m is symmetric. By substituting it into (8), we get

$$\begin{aligned}&a^{ki}_{mj}\left(\xi,x,\frac{1}{2}(y+y^T)+\mathrm{Def}_\xi\overline{N}_1\left(\xi,x,\frac{1}{2}(y+y^T)\right)\right)\zeta^k_m\zeta^i_j=\\&=\frac{1}{4}a^{ki}_{mj}\left(\left(\frac{\partial}{\partial z^p_q}+\frac{\partial}{\partial z^q_p}\right)Z^k_m\eta^p_q\right)\left(\left(\frac{\partial}{\partial z^r_l}+\frac{\partial}{\partial z^l_r}\right)Z^i_j\eta^r_l\right)\geqslant\\&\geqslant\frac{\varkappa}{4}\sum_{k,m=1}^{s}\left(\left(\frac{\partial}{\partial z^p_q}+\frac{\partial}{\partial z^q_p}\right)Z^k_m\eta^p_q\right)^2.\end{aligned}$$

By applying the operation $\langle\cdot\rangle$, we see that

$$\beta^{pr}_{ql}\eta^p_q\eta^r_l\geqslant\varkappa\left\langle\sum_{k,m=1}^{s}\left(\frac{1}{2}\left(\frac{\partial}{\partial z^p_q}+\frac{\partial}{\partial z^q_p}\right)Z^k_m\eta^p_q\right)^2\right\rangle.$$

As in proving Theorem 2, we get the equalities

$$\beta_{ql}^{pr} = \left\langle a_{mj}^{ki} \frac{\partial Z_j^i\left(\xi,\, x,\, \frac{1}{2}(y+y^T)\right)}{\partial y_l^r} \frac{\partial Z_m^k(\xi,\, x,\, (y+y^T)/2)}{\partial y_p^q} \right\rangle =$$

$$= \frac{1}{4}\left\langle a_{mj}^{ki} \frac{\partial\left(Y_i^j + Y_j^i\right)}{\partial y_l^r} \frac{\partial\left(Y_m^k + Y_k^m\right)}{\partial y_p^q} \right\rangle = \left\langle a_{mj}^{ki} \frac{\partial Y_i^j}{\partial y_l^r} \frac{\partial Y_m^k}{\partial y_p^q} \right\rangle = \widehat{a}_{ql}^{pr}.$$

We consider the summand

$$\rho_m^k = \left\langle \left(\frac{1}{2}\left(\frac{\partial}{\partial z_q^p} + \frac{\partial}{\partial z_p^q} \right) Z_m^k \eta_q^p \right)^2 \right\rangle$$

and transform it into

$$\rho_m^k = \frac{1}{4}\left\langle \left(\frac{\partial}{\partial y_q^p}\left(Y_m^k + Y_k^m\right)\eta_q^p \right)^2 \right\rangle =$$

$$= \frac{1}{4}\left\langle \left(\eta_m^k + \eta_k^m + \frac{\partial}{\partial \xi_m}\chi^k + \frac{\partial}{\partial \xi_k}\chi^m \right)^2 \right\rangle;$$

here $\chi^k = \eta_q^p \partial N_1^k / \partial y_q^p$.

Since

$$\eta_m^k = \eta_k^m, \quad \left\langle \eta_m^k \frac{\partial}{\partial \xi_m}\chi^k \right\rangle = 0,$$

we have $\rho_m^k \geqslant (\eta_m^k)^2$, and consequently, $\hat{a}_{ql}^{pr}\eta_q^p\eta_l^r \geqslant \kappa\eta_l^r\eta_l^r$. □

Remark 1: Euler's equation for the functional $J(u)$ considered here describes the strained state of a body free from external loads. For the volume forces distributed by the law $f(\xi, x)$, the functional $J(u)$ should be replaced by $J_1(u) = J(u) + f(\frac{x}{\varepsilon}, x)u$. The appropriate system of Euler's equations is

$$\frac{\partial}{\partial x_m} a_m^k\left(\frac{x}{\varepsilon},\, x,\, \nabla_x u\right) - f\left(\frac{x}{\varepsilon},\, x\right) = 0.$$

The averaged functional is

$$\widehat{J}_1(v_0) = \widehat{J}(v_0) + \langle f\rangle v_0;$$

and the averaged system is

$$\frac{\partial}{\partial x_m} \widehat{a}_m^k(x,\, \nabla_x v_0) - \langle f(\xi,\, x)\rangle = 0.$$

Here $\hat{J}$ and $\hat{a}_m^k$ are the same as for $f \equiv 0$.

Remark 2: Consider the case of the quadratic functional

$$J(u) = \int\limits_G \left(\frac{1}{2} \left(A_{ij}\left(\frac{x}{\varepsilon}, x\right) \frac{\partial u}{\partial x_j}, \frac{\partial u}{\partial x_i} \right) + \left(f\left(\frac{x}{\varepsilon}, x\right), u \right) \right) dx,$$

where the $A_{ij}(\xi, x)$ are $n \times n$ matrices, $f(\xi, x)$ is an n-dimensional vector, the elements of A_{ij}, and f are 1-periodic functions smooth in x and piecewise smooth in ξ (infinitely differentiable everywhere in $\mathbf{R}^{2s}$ of some infinitely smooth surfaces $\sum_1$), and $A_{ij} = A_{ji}^T$ for any i, j.

Then we get

$$\widehat{J}(v) = \int\limits_G \left(\frac{1}{2} \left(\widehat{A}_{ij} \frac{\partial v}{\partial x_j}, \frac{\partial v}{\partial x_i} \right) + (\langle f \rangle, v) \right) dx$$

and the averaged system of equations is

$$\frac{\partial}{\partial x_i} \left(\widehat{A}_{ij} \frac{\partial v}{\partial x_j} \right) - \langle f \rangle = 0.$$

We denote the elements of the matrices A_{ij} by a_{ij}^{kl}. From Theorem 2 follows another theorem.

THEOREM 2′. 1) *The equality*

$$\widehat{A}_{ij} = \widehat{A}_{ji}^T \quad \forall i, j$$

holds.

2) *Let*

$$a_{ij}^{kl} \zeta_i^k \zeta_j^l \geqslant \varkappa \zeta_i^k \zeta_i^k$$

for any x and a real-valued tensor ζ_i^k. Then

$$\widehat{a}_{ij}^{kl} \eta_i^k \eta_j^l \geqslant \varkappa \eta_i^k \eta_i^k$$

for any x and a real-valued tensor η_i^k.

Theorem 3 can be reformulated in the following way.

THEOREM 3′. *Let $n = s$, $a_{ij}^{kl} = a_{kj}^{il} = a_{il}^{kj} = a_{ji}^{lk}$ for all the i, j, k, l, and $a_{ij}^{kl} \zeta_i^k \zeta_j^l \geqslant \kappa \zeta_i^k \zeta_i^k$ for any x and a real-valued symmetric tensor ζ_i^k. Then*

$$\widehat{a}_{ij}^{kl} = \widehat{a}_{kj}^{il} = \widehat{a}_{il}^{kj} = \widehat{a}_{ji}^{lk}$$

for any i, j, k, l and,

$$\widehat{a}_{ij}^{kl} \eta_i^k \eta_j^l \geqslant \varkappa \eta_i^k \eta_i^k$$

for any x and a real-valued symmetric tensor η_i^k.

Note that, in the case of the quadratic functional, the assumptions that Φ is smooth in y are automatically valid.

In conclusion, we give another relation which can be used to estimate the $\hat{a}_{kk}^{ii}$. In particular, these estimates are used in [14] to investigate the speed of sound in mixtures.

By $W_2^{1(j,k)}$ we denote the set of column-vectors $m(\xi)$ that satisfy the following conditions: for any domain $G \subset R^3$

$$m_l(\xi) \in W_2^1(G), \qquad l = 1, 2, 3,$$

the $m_l(\xi)$ are periodic in all the ξ_q except for ξ_j (with period 1), $m_l|_{\xi_j=1} - m_l|_{\xi_j=0} = \delta_{kl}$ (for fixed values of the other ξ_q).

By analogy with Remark 4.1.3, we can show that

$$\widehat{a}_{kk}^{ii} = \inf_{m \in W_2^{1(k,i)}} \left\langle \frac{\partial m'}{\partial \xi_r} A_{rq} \frac{\partial m}{\partial \xi_q} \right\rangle.$$

§2 On the Principle of Equivalent Homogeneity

This section ascertains the relationship between local characteristics of state in a periodic medium and those of the homogeneous medium described by the averaged equation.

1. In the case of the boundary value problem for the elliptic system of equations in §§4.1 and 4.4, we obtained the relation $\|u - v_0\|_{L_2(G)} = 0(\sqrt{\varepsilon})$, and for a single equation a stronger estimate is true: $\|u - v_0\|_{C(G)} = O(\varepsilon)$. The estimate

$$\|u - v_0\|_{L_2(G)} = O(\sqrt{\varepsilon}),$$

was obtained for the hyperbolic boundary value problem in §4.5. The problems considered above include stationary and nonstationary boundary value problems for the system of equations of elasticity theory as a particular case.

Thus, the asymptotic equality

$$u \approx v_0. \tag{1}$$

holds.

2. For the hyperbolic boundary value problem, the estimate (4.5.10) implies

$$\left\| \frac{\partial u}{\partial t} - \frac{\partial v_0}{\partial t} \right\|_{L_2(G \times (0,t_0))} = O(\sqrt{\varepsilon}).$$

Thus,

$$\frac{\partial u}{\partial t} \approx \frac{\partial v_0}{\partial t}.$$

3. In order to estimate the closeness of stresses and energy, it is advisable to introduce the operator of local averaging

$$\{f\}|_{x,t} = \varepsilon^{-s} \int_{\Pi_x^\varepsilon} f(y, t)\, dy;$$

here Π_x^ε is the set of points y such that $|y_q - x_q| \leqslant /2$ for all q. In the case of stationary problems, the dependence on the parameter t is omitted.

Sections 4.1 and 4.4 also yielded the estimate $\|u - u^{(1)}\|_{W_2^1(G)} = O(\sqrt{\varepsilon})$, and consequently

$$\left\| \frac{\partial u}{\partial x_i} - \frac{\partial u^{(1)}}{\partial x_i} \right\|_{L_2(G)} = O(\sqrt{\varepsilon}). \tag{2}$$

In what follows, we will need two auxiliary assertions.

LEMMA 1. *For any vector $f \in L_2(G)$,*

$$\| \{f\} \|_{L_2(G_\varepsilon)} \leqslant \| f \|_{L_2(G)};$$

here G_ε is the set of points y such that

$$\Pi_y^\varepsilon = \{x \in \mathbf{R}^s \,|\, |x_i - y_i| \leqslant \varepsilon/2 \ \forall i\} \subset G.$$

PROOF: By setting $f^2 = (f, f)$, we have

$$\{f\}^2|_x = \left(\frac{1}{\varepsilon^s} \int_{\Pi_x^\varepsilon} f(y)\, dy \right)^2 \leqslant \frac{1}{\varepsilon^s} \int_{\Pi_x^\varepsilon} f^2(y)\, dy$$

according to the Cauchy–Buniakowski–Schwarz inequality. Therefore,

$$\| \{f\} \|_{L_2(G_\varepsilon)}^2 \leqslant \frac{1}{\varepsilon^s} \int_{G_\varepsilon} \left(\int_{\Pi_x^\varepsilon} f^2(y)\, dy \right) dx = \frac{1}{\varepsilon^s} \int_{G_\varepsilon} \int_{\Pi_0^\varepsilon} f^2(y + x)\, dy\, dx.$$

By changing the order of integration, we present the last integral as

$$I = \frac{1}{\varepsilon^s} \int_{\Pi_0^\varepsilon} \int_{\{z | z - y \in G_\varepsilon\}} f^2(z)\, dz\, dy.$$

By considering the embedding $\{z | z - y \in G_\varepsilon\} \subset G$, where $y \in \Pi_0^\varepsilon$, we see that

$$I \leqslant \frac{1}{\varepsilon^s} \int_{\Pi_0^\varepsilon} \int_G f^2(z)\, dz\, dy = \int_G f^2(z)\, dz.$$

□

From (2), and due to Lemma 1, we get

$$\left\| \left\{ \frac{\partial u}{\partial x_j} \right\} - \left\{ \frac{\partial u^{(1)}}{\partial x_j} \right\} \right\|_{L_2(G_\varepsilon)} = O(\sqrt{\varepsilon}). \tag{3}$$

We recall that $u^{(1)} = v_0 + \varepsilon N_{i_1} \frac{\partial v_0}{\partial x_{i_1}}$, and therefore,

$$\frac{\partial u^{(1)}}{\partial x_j} = \frac{\partial v_0}{\partial x_j} + \frac{\partial N_{i_1}}{\partial \xi_j} \frac{\partial v_0}{\partial x_{i_1}} + \varepsilon N_{i_1} \frac{\partial^2 v_0}{\partial x_j \partial x_{i_1}}. \tag{4}$$

Lemma 2. *Let $N(\xi)$ be a 1-periodic function of the variables ξ_q, let $N(\xi) \in L_2(Q)$, with Q being the unit cube, and let $g(x) \in C^1(\Pi_x)$. Then the inequality*

$$\left| \left\{ N\left(\frac{x}{\varepsilon}\right) g(x) \right\} - \langle N(\xi) \rangle g(x) \right| \leqslant \frac{s}{2} \max_{1 \leqslant i \leqslant s} \max_{\Pi_x^\varepsilon} \left| \frac{\partial g}{\partial x_i} \right| \langle N^2(\xi) \rangle^{1/2} \varepsilon$$

holds.

Proof: We have $\{Ng\}|_x = a_1 + a_2$, where

$$\begin{aligned} a_1 &= \varepsilon^{-s} \int_{\Pi_x^\varepsilon} N\left(\frac{y}{\varepsilon}\right) g(x)\, dy = \langle N(\xi) \rangle g(x), \\ a_2 &= \varepsilon^{-s} \int_{\Pi_x^\varepsilon} N\left(\frac{y}{\varepsilon}\right) (g(y) - g(x))\, dy, \end{aligned} \tag{5}$$

$$| g(y) - g(x) | \leqslant \sum_{j=1}^{s} \max_{\Pi_x^\varepsilon} \left| \frac{\partial g}{\partial x_j} \right| \frac{\varepsilon}{2} \leqslant \frac{s}{2} \max_{1 \leqslant j \leqslant s} \max_{\Pi_x^\varepsilon} \left| \frac{\partial g}{\partial x_j} \right| \varepsilon.$$

Then by the Cauchy–Buniakowski–Schwarz inequality,

$$| a_2 | \leqslant \frac{s}{2} \max_{1 \leqslant j \leqslant s} \max_{\Pi_x^\varepsilon} \left| \frac{\partial g}{\partial x_j} \right| \langle N^2 \rangle^{1/2} \varepsilon. \tag{6}$$

The assertion of Lemma 2 follows from (5), (6). □

The relation $N_{i_1} \in W_2^1(Q)$ implies that $\langle \nu^2 \rangle < \infty$ for any element ν of the matrices N_{i_1} and for any $j \langle (\partial \nu / \xi_j)^2 < \infty$.

Since $v_0 \in C^3(\overline{G})$, then from Lemma 2, we have

$$\left\| \left\{ \frac{\partial N_{i_1}}{\partial \xi_j} \frac{\partial v_0}{\partial x_{i_1}} \right\} \right\|_{C(\bar{G}_\varepsilon)} = O(\varepsilon); \tag{7}$$

and in the same way,

$$\left\| \left\{ N_q \frac{\partial^2 v_0}{\partial x_q \partial x_{i_1}} \right\} - \langle N_q \rangle \frac{\partial^2 v_0}{\partial x_q \partial x_{i_1}} \right\|_{C(\bar{G}_\varepsilon)} = O(\varepsilon). \tag{8}$$

By using Lemma 2 for $N \equiv 1$, $g = \partial v_0 / \partial x_j$, we get

$$\left\| \left\{ \frac{\partial v_0}{\partial x_j} \right\} - \frac{\partial v_0}{\partial x_j} \right\|_{C(\bar{G}_\varepsilon)} = O(\varepsilon). \tag{9}$$

Relation (4) implies

$$\left\{ \frac{\partial u^{(1)}}{\partial x_j} \right\} = \left\{ \frac{\partial v_0}{\partial x_j} \right\} + \left\{ \frac{\partial N_{i_1}}{\partial \xi_j} \frac{\partial v_0}{\partial x_{i_1}} \right\} + \varepsilon \left\{ N_q \frac{\partial^2 v_0}{\partial x_q \partial x_j} \right\}.$$

We derive from (7)–(9) that

$$\left\| \left\{ \frac{\partial u^{(1)}}{\partial x_j} \right\} - \frac{\partial v_0}{\partial x_j} \right\|_{L_2(G_\varepsilon)} = O(\sqrt{\varepsilon}).$$

It directly follows from this and (3) that

$$\left\| \left\{ \frac{\partial u}{\partial x_j} \right\} - \frac{\partial v_0}{\partial x_j} \right\|_{L_2(G_\varepsilon)} = O(\sqrt{\varepsilon}).$$

Thus, we have

$$\left\{ \frac{\partial u}{\partial x_j} \right\} \approx \frac{\partial v_0}{\partial x_j}; \tag{10}$$

and accordingly, for the strains

$$e_{ij} = \frac{1}{2} \left(\frac{\partial u_i}{\partial x_j} + \frac{\partial u_j}{\partial x_i} \right), \quad e_{ij}^0 = \frac{1}{2} \left(\frac{\partial v_i^0}{\partial x_j} + \frac{\partial v_j^0}{\partial x_i} \right),$$

where $v_0 = (v_1^0, \dots, v_n^0)^T$, the approximate equality

$$\{e_{ij}\} \approx e_{ij}^0.$$

holds.

For the dynamic problem of elasticity theory, these relations are obtained in the same fashion from estimate (4.5.10).

4. It follows from (2):

$$\left\| A_{ij}\frac{\partial u}{\partial x_j} - A_{ij}\frac{\partial u^{(1)}}{\partial x_j} \right\|_{L_2(G)} = O(\sqrt{\varepsilon}).$$

By substituting $\partial u^{(1)}/\partial x_j$ from (4), as in §4.4, we get

$$A_{ij}\frac{\partial u^{(1)}}{\partial x_j} = \tilde{A}_{ij}\frac{\partial v_0}{\partial x_j} + \varepsilon A_{ij}N_{i_1}\frac{\partial^2 v_0}{\partial x_j \partial x_{i_1}}; \tag{11}$$

here $\tilde{A}_{ij} = A_{ij} + A_{ij_1}\partial N_j/\partial \xi_{j_1}$. Following the above procedure for (10), we derive

$$\left\| \left\{ A_{ij}\frac{\partial u}{\partial x_j} \right\} - \langle \tilde{A}_{ij} \rangle \frac{\partial v_0}{\partial x_j} \right\|_{L_2(G_\varepsilon)} = O(\sqrt{\varepsilon}).$$

The vectors $\sigma_i = A_{ij}\partial u/\partial x_j$ comprise the stress tensor for the original problem.

Since $\langle \tilde{A}_{ij} \rangle = \hat{A}_{ij}$, the vectors $\sigma_i^0 = \langle \hat{A}_{ij}\partial v_0/\partial x_j$ comprise the stress tensor for the averaged equation. So the latter relation is written as

$$\{\sigma_i\} \approx \sigma_i^0. \tag{12}$$

We obtain a stronger estimate to derive averaged boundary conditions at the level of plausible reasoning in the case of the second boundary value problem (see the accurate derivation in Chapter 9).

We denote by $\Pi_{y,m}^{\varepsilon}$ the section of the cube Π_y^{ε} by plane $x_m = y_m$ and define the operator $\{\cdot\}_m$ by

$$\{f\}_m|_x = \varepsilon^{1-s} \int\limits_{\Pi_{x,m}^{\varepsilon}} f(y)\, d^m y, \quad d^m y = \prod_{j \neq m} dy_j.$$

Similarly, for the 1-periodic $f(\xi)$, we set

$$\langle f \rangle_m |_{\xi_m} = \int_0^1 \dots \int_0^1 f(\xi)\, d^m \xi, \quad d^m \xi = \prod_{j \neq m} d\xi_j.$$

Further constructions are rigorous under the assumption that the A_{ij} are differentiable with respect to the variables ξ_q. The justification of final derivations for the piecewise smooth A_{ij} is analogous to the reasoning in §4.1.

Relation (4.4.4), $\frac{\partial}{\partial \xi_i}\left(A_{ij}\frac{\partial(N_{i_1}|\xi_{i_1}E)}{\partial \xi_j}\right) = 0$, can be represented by the definition of A_{ij} as

$$\frac{\partial}{\partial \xi_i}\tilde{A}_{ii_1} = 0. \tag{13}$$

We redenote i_1 by j and apply the operator $\langle\cdot\rangle_m$ to both sides. The integral $\langle \partial\tilde{A}_{ij}/\partial\xi_i\rangle_m$ (no summation over i) is zero for $i \neq m$ as an integral of a derivation of a periodic function; therefore $\frac{\partial}{\partial \xi_m}\langle\tilde{A}_{mj}\rangle_m = 0$, that is, $\langle\tilde{A}_{mj}\rangle_m = \text{const}$. Since $\hat{A}_{mj} \equiv \int_0^1 \langle\tilde{A}_{mj}\rangle_m d\xi_m$, we have $\langle\tilde{A}_{mj}\rangle_m = A_{mj}$.

Now we apply the operator $\{\cdot\}_m$ rather than $\{\cdot\}$ to (11). Then, by the same reasoning as in Subsections 3 and 4, we get $\|\{\sigma_i\}_m - \sigma_i^0\|_{L_2(G_\varepsilon)} = 0(\sqrt{\varepsilon})$, that is,

$$\{\sigma_i\}_m \approx \sigma_i^0. \tag{14}$$

5. Let us make the same constructions for the functionals of kinetic energy and potential energy. By assuming $v_0 \in C^3$ from (4.5.10), we see due to Lemmas 1 and 2 that

$$\left\|\left\{\rho\left(\frac{\partial u}{\partial t}\right)^2\right\} - \hat{\rho}\left(\frac{\partial v_0}{\partial t}\right)^2\right\|_{L_1(G_\varepsilon\times(0,t_0))} = O(\sqrt{\varepsilon}),$$

that is,

$$\left\{\rho\left(\frac{\partial u}{\partial t}\right)^2\right\} \approx \hat{\rho}\left(\frac{\partial v_0}{\partial t}\right)^2. \tag{15}$$

Consider the functionals of potential energy

$$w(u) = (A_{ij}\partial u/\partial x_j,\ \partial u/\partial x_i),$$
$$w_0(v_0) = (\hat{A}_{ij}\,\partial v_0/\partial x_j,\ \partial v_0/\partial x_i).$$

For brevity, we shall make formal transformations without estimating the discarded terms and shall only indicate the final estimate of accuracy of the transformations. We have

$$w(u) \approx w\left(u^{(1)}\right) = \left(\tilde{A}_{ij}\frac{\partial u^{(1)}}{\partial x_j}, \frac{\partial u^{(1)}}{\partial x_i}\right) \approx \left(\tilde{A}_{ij}\frac{\partial v_0}{\partial x_j}, \frac{\partial v_0}{\partial x_i} + \frac{\partial N_{i_1}}{\partial \xi_i}\frac{\partial v_0}{\partial x_{i_1}}\right),$$

and therefore,

$$\{w(u)\} = \left(\langle \widetilde{A}_{ij}\rangle \frac{\partial v_0}{\partial x_j}, \frac{\partial v_0}{\partial x_i}\right) + \left\langle\left(\widetilde{A}_{ij} \frac{\partial v_0}{\partial x_j}, \frac{\partial N_{i_1}}{\partial \xi_i} \frac{\partial v_0}{\partial x_{i_1}}\right)\right\rangle.$$

By integrating by parts with respect to the variables ξ_q in the second summand, we get

$$\left\langle\left(\frac{\partial}{\partial \xi_i} \widetilde{A}_{ij} \frac{\partial v_0}{\partial x_j}, \frac{\partial v_0}{\partial x_{i_1}}\right)\right\rangle = 0$$

(due to (13)). Since $\langle \tilde{A}_{ij}\rangle = \hat{A}_{ij}$, it follows that

$$\{w(u)\} \approx \left(\widehat{A}_{ij} \frac{\partial v_0}{\partial x_j}, \frac{\partial v_0}{\partial x_i}\right) = w_0(v_0);$$

the estimate of accuracy of this relation is

$$\|\{w(u)\} - w_0(v_0)\|_{L_1(G_\varepsilon)} = O(\sqrt{\varepsilon}).$$

Thus,

$$\{w(u)\} \approx w_0(v_0). \tag{16}$$

Relations (1), (10), (12), (14), and (16) can be justified also for parabolic problems.

In the nonlinear case (see §1), similar relations can be obtained under the assumption that the expansion $u \sim v_0(x) + \varepsilon N_1(\xi, x, \nabla_x v) + \dots$ is true in a sufficiently strong norm, where v_0, N_1 satisfy (4.6.17) and (4.6.14)–(4.6.16), respectively. As in the linear case, we have (10) and

$$\left\{\frac{\partial u_i}{\partial x_j} + \frac{\partial u_j}{\partial x_i}\right\} \approx \frac{\partial v_i^0}{\partial x_j} + \frac{\partial v_j^0}{\partial x_i}.$$

We consider the components a_m^k of the stress tensor:

$$a_m^k(\xi, x, \nabla_x u) \approx a_m^k(\xi, x, \nabla_x v_0 + \nabla_\xi N_1(\xi, x, \nabla_x v_0)); \tag{17}$$

since

$$\{a_m^k(\xi, x, \nabla_x v_0 + \nabla_\xi N_1(\xi, x, \nabla_x v_0))\} \approx \widehat{a}_m^k(x, \nabla_x v_0),$$

we have

$$\{a_m^k(\xi, x, \nabla_x u)\} \approx \hat{a}_m^k(x, \nabla_x v_0).$$

By applying $\langle\cdot\rangle_m$ to (1.1) and substituting

$$y = \nabla_x v_0 + \nabla_\xi N_1(\xi, x, \nabla_x v_0),$$

we get

$$\frac{\partial}{\partial \xi_m} \langle a_m^k(\xi, x, \nabla_x v_0 + \nabla_\xi N_1(\xi, x, \nabla_x v_0))\rangle_m = 0.$$

Hence, we derive the approximate equality

$$\{a_m^k(\xi, x, \nabla_x u)\}_m \approx \hat{a}_m^k(x, \nabla_x v_0).$$

Following from Theorem 1.1, the relation

$$\{\Phi(\xi, x, \nabla_x u)\} \approx \hat{\Phi}(x, \nabla_x v_0),$$

is an analogue of (16).

The approximate equalities (15) also hold in the nonstationary case. The above relations can be interpreted as a justification of the hypothesis of equivalent homogeneity.

In contrast to this hypothesis, the procedure which we propose makes it possible not only to determine the average characteristics e_{ij}^0, σ_i^0, w_0 but also to obtain local fluctuations e_{ij}, σ_i w (see, for example, (17)).

§3 The Symmetry Properties of Effective Coefficients and Reduction of Periodic Problems to Boundary Value Problems

The periodicity domain often has a certain symmetry that can make the search for effective coefficients much easier due to the following features:

1. Some of them vanish automatically.
2. Solving auxiliary problems (4.4.4) on a period reduces to solving problems in a smaller domain or to solving standard boundary value problems.

For convenience, we consider the cube $\{-1/2 < \xi_1, \dots, \xi_s \leqslant 1/2\}$ as a periodicity domain and denote the integral over the cube by $\langle\cdot\rangle$.

We assume the following conditions to be satisfied: If the $F_i (i = 0, 1, \dots, s)$ are 1-periodic matrices from $L_2(Q)$ and $\langle F_0\rangle = 0$, then the equation

$$L_{\xi\xi} N = F_0(\xi) + \frac{\partial F_i}{\partial \xi_i}, \quad F_i \in L_2(Q),$$

is uniquely solvable in the class of the 1-periodic matrices $N \in W_2^1(Q)$ satisfying $\langle N \rangle = 0$.

We denote by S_h the operator of reflection in the plane $\xi_h = 0$:

$$S_h\xi = ((-1)^{\delta_{h1}}\xi_1, \dots, (-1)^{\delta_{hs}}\xi_s);$$

here

$$\delta_{hj} = \begin{cases} 1 & \text{for} \quad h = j, \\ 0 & \text{for} \quad h \neq j. \end{cases}$$

For an arbitrary 1-periodic function $\varphi(\xi)$ in W_2^1, the equality

$$\frac{\partial}{\partial \xi_j} \varphi(S_h\xi) = (-1)^{\delta_{hj}} \frac{\partial \varphi(\eta)}{\partial \eta_j}\bigg|_{\eta = S_h\xi}. \tag{1}$$

holds.

The two types of symmetry of the operator $L_{\xi\xi}$ in $\xi_h = 0$ are of interest here.

A. The equality

$$A_{ij}(S_h\xi) = (-1)^{\delta_{hi}+\delta_{hj}} A_{ij}(\xi) \tag{2}$$

holds for any i, j.

B. The elements a_{ij}^{kl} of the matrices A_{ij} satisfy the relations

$$a_{ij}^{kl}(S_h\xi) = (-1)^{\delta_{hi}+\delta_{hj}+\delta_{hk}+\delta_{hl}} a_{ij}^{kl}(\xi) \tag{3}$$

for any i, j, k, l (no summation over these indices!).

The first type of symmetry in the scalar case corresponds to the diffusion (heat transition) problem in a medium symmetric in the plane ξ_h and to the second type to problems of elasticity theory.

THEOREM 1. *Let the N_q be 1-periodic solutions of the system*

$$L_{\xi\xi}N_q + \partial A_{iq}/\partial \xi_i = 0, \tag{4}$$

which satisfy the condition $\langle N_q \rangle = 0$ and let condition (2) be valid. Then N_h is odd with respect to $\xi + h = 0$, and the remaining N_q are even, that is,

$$N_q(S_h\xi) = (-1)^{\delta_{hq}} N_q(\xi), \tag{5}$$

all the $\hat{A}_{qh} = 0$ and all the $A_{hq} = 0$ for $q \neq h$.

PROOF: For the arbitrary 1-periodic function φ from $W_2^1(Q)$ and from (1), (2), we have

$$A_{ij}\frac{\partial}{\partial \xi_j}\varphi(S_h\xi) = (-1)^{\delta_{hi}+\delta_{hj}}A_{ij}(S_h\xi)(-1)^{\delta_{hj}}\frac{\partial\varphi(\eta)}{\partial\eta_j}\bigg|_{\eta=S_h\xi} =$$

$$= (-1)^{\delta_{hi}}\left(A_{ij}(\eta)\frac{\partial\varphi(\eta)}{\partial\eta_j}\right)\bigg|_{\eta=S_h\xi}.$$

By using (1) again, we get

$$L_{\xi\xi}\varphi(S_h\xi) = (-1)^{\delta_{hi}}\frac{\partial}{\partial\xi_i}\left(\left(A_{ij}(\eta)\frac{\partial\varphi(\eta)}{\partial\eta_j}\right)\bigg|_{\eta=S_h\xi}\right) = L_{\eta\eta}\varphi(\eta)\Big|_{\eta=S_h\xi}. \quad (6)$$

The equalitites

$$\frac{\partial}{\partial\eta_i}A_{iq}(\eta)\bigg|_{\eta=S_h\xi} = (-1)^{\delta_{hi}}\frac{\partial}{\partial\xi_i}A_{iq}(S_h\xi) = (-1)^{\delta_{hq}}\frac{\partial}{\partial\xi_i}A_{iq}(\xi); \quad (7)$$

hold: one due to (1) and the other due to (2).

We write (4) at the point $S_h\xi$:

$$L_{\eta\eta}N_q(\eta)\,|_{\eta=S_h\xi} + \frac{\partial A_{iq}(\eta)}{\partial\eta_i}\bigg|_{\eta=S_h\xi} = 0.$$

From (6), (7), we get

$$L_{\xi\xi}N_q(S_h\xi) + (-1)^{\delta_{hq}}\frac{\partial}{\partial\xi_i}A_{iq}(\xi) = 0.$$

Multiply this by $(-1)^{\delta_{hq}}$ and subtract it from (4):

$$L_{\xi\xi}\left(N_q - (-1)^{\delta_{hq}}N_q(S_h\xi)\right) = 0.$$

For an arbitrary 1-periodic function $\varphi(\xi)$ from $L_2(Q)$, the following is true:

$$\langle\varphi(\xi)\rangle = \langle\varphi(S_h\xi)\rangle. \quad (8)$$

Since $\langle N_q\rangle = 0$, it follows from (8) that

$$\left\langle N_q - (-1)^{\delta_{hq}}N_q(S_h\xi)\right\rangle = 0.$$

The uniqueness of solution (4) satisfying this condition implies the validity of the equality

$$N_q(\xi) - (-1)^{\delta_{hq}}N_q(S_h\xi) \equiv 0.$$

Assertion (5) is proved.

Let us investigate the symmetry properties of the coefficients $\tilde{A}_{hq}$ and $\tilde{A}_{qh}$. Since $\tilde{A}_{hq} = A_{hq} + A_{hj}\partial N_q/\partial\xi_j$, it follows from (5) that

$$\left.\frac{\partial N_q(\eta)}{\partial\eta_j}\right|_{\eta=S_h\xi} = (-1)^{\delta_{hj}+\delta_{hq}}\frac{\partial N_q(\xi)}{\partial\xi_j}.$$

This and (2) imply

$$\tilde{A}_{hq}(S_h\xi) = (-1)^{\delta_{hh}+\delta_{hq}}A_{hq}(\xi) + (-1)^{\delta_{hh}+\delta_{hj}}A_{hj}(\xi)(-1)^{\delta_{hj}+\delta_{hq}}\frac{\partial N_q}{\partial\xi_j} =$$
$$= (-1)^{1+\delta_{hq}}\tilde{A}_{hq}(\xi). \quad (9)$$

So the $\tilde{A}_{hq}$ are odd in ξ_h for $q \neq h$ (and even in ξ_h for $q = h$); consequently, $\langle\tilde{A}_{hq}\rangle = \hat{A}_{hq} = 0$ for $q \neq h$.

From (5) we have $\partial N_h(\eta)/\partial\eta_j|_{\eta=S_h\xi} = (-1)^{\delta_{hj}+1}\partial N_h(\xi)/\partial\xi$, and therefore we obtain in just the same way $\tilde{A}_{qh}(S_h\xi) = (-1)^{\delta_{hq}+1}\tilde{A}_{hq}(\xi)$; consequently, $\hat{A}_{qh} = 0$ for $q \neq h$. □

REMARK 1: Since N_h is odd in ξ_h, it follows that $N_h|_{\xi_h} = 0$. Condition (2) and the 1-periodicity of the coefficients imply the similar symmetry property with respect to the plane $\xi_h = 1/2$. Therefore, $N_h|_{\xi_h=1/2} = 0$.

Thus, instead of seeking a 1-periodic solution of (4), we can look for a solution of (4) periodic in all the $\xi_q (q \neq h)$, which would satisfy the boundary values $N_h|_{\xi_h=d} = 0$, $d = 1/2$. With the discontinuous A_{ij}, the term $\partial A_{ih}/\partial\xi$ in (4) contains δ-functions. Therefore, it is more convenient numerically to seek the matrix-valued function $M_h = N_h + \xi_h E$ periodic in all $\xi_q (q \neq h)$, which would satisfy the homogeneous equation $L_{\xi\xi}M_h = 0$ and the conditions $M_h|_{\xi_h=d} = dE$, $d = 0, 1/2$. The coefficient A_{hh} is given by

$$\hat{A}_{hh} = 2\int_0^1 \ldots \int_0^1 \left(\int_0^{1/2} A_{hj}\frac{\partial M_h}{\partial\xi_j}\,d\xi_h\right)\prod_{q\neq h} d\xi_q.$$

THEOREM 2. *Let condition (2) be satisfied for $s-1$ values of h. Then the operator $\hat{L}$ is of the form $\hat{L} = \frac{\partial}{\partial x_i}\hat{A}_{ii}\frac{\partial}{\partial x_i}$.*

PROOF: Let $i \neq j$. Under the conditions of the theorem, (2) holds either for $h = i$ or for $h = j$, and therefore $\hat{A}_{ij} = 0$ from Theorem 1. □

Consider the case when condition (2) is satisfied for all h. Then from Theorem 1, for all q, h we have

$$N_q(S_h\xi) = (-1)^{\delta_{hq}}N_q(\xi)$$

and problems on the period (4) reduce to boundary value problems in the domain $0 < \xi_1, \ldots, \xi_s < 1/2$ for the matrices $M_q = N_q + \xi_q E$:

$$L_{\xi\xi}M_q = 0, \quad M_q|_{\xi_q=d} = dE, \quad d = 0, 1/2,$$
$$A_{hj}\frac{\partial M_q}{\partial \xi_j}\bigg|_{\xi_h=d} = 0, \quad d = 0, 1/2, \quad h \neq q. \tag{10}$$

Let us verify (10) for $\xi_h = 0$; we have $A_{hj}(S_h\xi) = (-1)^{\delta_{hj}+1}A_{hj}(\xi)$ from (2). If $\xi_h = 0$, then $S_h\xi = \xi$, and therefore $A_{hj}(\xi) = (-1)^{\delta_{hj}+1}A_{hj}(\xi)$, that is, $(1+(-1)^{\delta_{hj}})A_{hj}(\xi) = 0$. Consequently, $A_{hj}(\xi)|_{\xi_h=0} = 0$ for $h \neq j$. It follows that $A_{hj}\partial M_q/\partial\xi_j|_{\xi_h=0} = A_{hh}\partial M_q/\partial\xi_h|_{\xi_h=0}$; for $h \neq q$ we see that $\partial M_q/\partial \xi_h = \partial N_q/\partial\xi_h$, $\partial N_q/\partial\xi_h|_{\xi_h=0} = 0$, since N_q is even in ξ_h. Consequently, all the $A_{hj}\partial M_q/\partial\xi_j|_{\xi_h=0}$ vanish.

The coefficients $\hat{A}_{qq}$ are given by

$$\hat{A}_{qq} = 2^s \int_0^{1/2} \cdots \int_0^{1/2} A_{qj}\frac{\partial M_q}{\partial \xi_j} d\xi_1 \ldots d\xi_s.$$

REMARK 2: In the scalar case, given an invariant medium for all possible transformations $\xi_i \leftrightarrow \xi_j$ and condition A satisfied for all the h, the averaged operator is $(\hat{a}\partial/\partial x_i)\partial/\partial x_i$.

We now consider symmetry of the second type (3). Let the n_q^{lr} be the elements of the matrices N_q. Relations (4) are assigned component-wise in the form

$$\frac{\partial}{\partial \xi_i}\left(a_{ij}^{kl}\frac{\partial n_q^{lr}}{\partial \xi_j}\right) + \frac{\partial}{\partial \xi_i}a_{iq}^{kr} = 0 \tag{11}$$

and split into systems of equations in the components n_q^{lr} of the vectors $n_q^r = (n_q^{1r}, \ldots, n_q^{nr})^T$, with r and q being fixed.

THEOREM 3. *Under condition* (3), *the symmetry property*

$$n_q^{lr}(S_h\xi) = (-1)^{\delta_{hq}+\delta_{hr}+\delta_{hl}} n_q^{lr}(\xi) \tag{12}$$

is valid, and $\hat{a}_{iq}^{kr} = 0$ *if* $\delta_{hi}+\delta_{hq}+\delta_{hk}+\delta_{hr}$ *is odd.*

To prove the first assertion, consider the system of equations (11) for fixed r and q. In just the same way as in Theorem 1, we can derive the relation (no summation over k, q, r!)

$$\frac{\partial}{\partial \xi_i}\left(a_{ij}^{kl}\frac{\partial n_q^{lr}(S_h\xi)}{\partial \xi_j}\right) + (-1)^{\delta_{hr}+\delta_{hq}+\delta_{hk}}\frac{\partial}{\partial \xi_i}a_{iq}^{kr}(\xi) = 0.$$

It follows that

$$\frac{\partial}{\partial \xi_i}\left(a_{ij}^{kl}\frac{\partial\left(n_q^{lr}(S_h\xi)-(-1)^{\delta_{hr}+\delta_{hq}+\delta_{hl}}n_q^{lr}(\xi)\right)}{\partial \xi_j}\right)=0,$$

and by assuming that solution (4) is unique, we obtain (12).

By assumption, $\hat{a}_{iq}^{kr}=\langle\tilde{a}_{iq}^{kr}\rangle$, where $\tilde{a}_{iq}^{kr}=a_{iq}^{kr}+a_{ij}^{kl}\partial n_q^{lr}/\partial\xi_j$. Since

$$\frac{\partial n_q^{lr}(S_h\xi)}{\partial \xi_j}=(-1)^{\delta_{hq}+\delta_{hl}+\delta_{hr}+\delta_{hj}}\frac{\partial n_q^{lr}(\xi)}{\partial \xi_j}$$

due to (12), we use (3) to get

$$\tilde{a}_{iq}^{kr}(S_h\xi)=(-1)^{\delta_{hi}+\delta_{hq}+\delta_{hk}+\delta_{hr}}\tilde{a}_{ij}^{kr}(\xi).$$

Thus, if $\delta_{hi}+\delta_{hq}+\delta_{hk}+\delta_{hr}$ is odd, then $\tilde{a}_{iq}^{kr}$ is odd in ξ_h and therefore $\hat{a}_{iq}^{kr}=0$.

As in A, problems on a period (4) reduce to problems on a half period.

We consider the case of $n=s$ (n is the dimension of the matrices A_{ij}).

THEOREM 4. *Let condition (3) be satisfied for $s-1$ values of h. Then $\hat{a}_{ij}^{kl}=0$ if $\delta_{ri}+\delta_{rj}+\delta_{rk}+\delta_{rl}$ is odd for some r.*

PROOF: If this expression is odd, then one or three of i, j, k, l coincide with r. In this case, one or three of them coincide with some other number r_2. From the condition of the theorem, either for $h=r$ or for $h=r_2$, condition (3) is satisfied. Therefore $\hat{a}_{ij}^{kl}=0$. □

REMARK 3: If condition (3) is valid for all the h, problems (4) reduce to boundary value problems in the domain $0<\xi_1,\dots,\xi_s<1/2$.

The truth of the first assertion in Theorem 3 implies

$$n_q^{lr}(S_h\xi)=(-1)^{\delta_{hl}+\delta_{hr}+\delta_{hq}}n_q^{lr}(\xi),$$

$$\left(a_{hj}^{il}(\eta)\frac{\partial n_q^{lr}(\eta)}{\partial \eta_j}\right)\Bigg|_{\eta=S_h\xi}=(-1)^{\delta_{hi}+\delta_{hr}+\delta_{hq}+1}a_{hj}^{il}(\xi)\frac{\partial n_q^{lr}(\xi)}{\partial \xi_j}.$$

As in A, it follows that

$$n_q^{lr}|_{\xi_h=0}=0,\qquad \text{for}\qquad \delta_{hl}+\delta_{hr}+\delta_{hq}\qquad \text{odd},\quad (13)$$

$$\mathcal{L}_h^i n_q^r|_{\xi_h=0}=a_{hj}^{il}\frac{\partial n_q^{lr}}{\partial \xi_j}\Bigg|_{\xi_h=0}=0,\quad \text{for}\quad \delta_{hi}+\delta_{hr}+\delta_{hq}+1\qquad \text{odd}.\quad (14)$$

We consider the three-dimensional case $n=s=3$ and write down boundary conditions for the vectors n_1^1 and n_2^1. From (13), (14), we have

$$n_1^{11},\ \mathscr{L}_1^2 n_1^1,\ \mathscr{L}_1^3 n_1^1\big|_{\xi_1=0} = 0,$$
$$\mathscr{L}_2^1 n_1^1,\ n_1^{21},\ \mathscr{L}_2^3 n_1^1\big|_{\xi_2=0} = 0, \qquad (15)$$
$$\mathscr{L}_3^1 n_1^1,\ \mathscr{L}_3^2 n_1^1,\ n_1^{31}\big|_{\xi_3=0} = 0;$$

the same evenness condition for the planes $\xi_h = 1/2$ yields the same conditions on all the faces $\xi_h = 1/2$. Conditions (15) are called *third-type boundary conditions,* wherein normal translation and tangent components of the stress tensor on each face are equal to zero.

The following boundary conditions are valid for the vector n_1^2:

$$\mathscr{L}_1^1 n_1^2,\ n_1^{22},\ n_1^{32}\big|_{\xi_1=0} = 0,$$
$$n_1^{12},\ \mathscr{L}_2^2 n_1^2,\ n_1^{32}\big|_{\xi_2=0} = 0,$$
$$\mathscr{L}_3^1 n_1^2,\ \mathscr{L}_3^2 n_1^2,\ n_1^{32}\big|_{\xi_3=0} = 0;$$

the same relations occur for $\xi_h = 1/2$.

A third-type boundary condition is set on the faces parallel to the plane $\xi_3 = 0$. The boundary condition on the other faces is not standard: The vector of tangent translations is equal to zero as is the normal component of the stress tensor.

Passing now to equations for the vectors m_1^r with the components $m_q^{lr} = n_q^{lr} + \xi_q \delta_{rl}$ satisfying the homogeneous system $L_{\xi\xi} m_q^r = 0$, we have

$$a_{hj}^{il} \frac{\partial \xi_q \delta_{rl}}{\partial \xi_j} = a_{hj}^{il} \delta_{qj} \delta_{rl} = a_{hq}^{ir},$$
$$a_{hq}^{ir}(S_h \xi) = (-1)^{\delta_{hq}+\delta_{hi}+\delta_{hr}+1} a_{hq}^{ir}(\xi),$$

and therefore, when $\mathcal{L}_h^i n_q^r|_{\xi_h=0} = 0$, also $a_{hq}^{ir}|_{\xi_h=0} = 0$.

Thus we have obtained the boundary value problems

$$\mathscr{L}_{\xi\xi} m_1^1 = 0,$$
$$m_1^{11} = d,\ \ \mathscr{L}_1^2 m_1^1,\ \ \mathscr{L}_1^3 m_1^1\big|_{\xi_1=d} = 0,$$
$$\mathscr{L}_2^1 m_1^1,\ \ m_1^{21},\ \ \mathscr{L}_2^3 m_1^1\big|_{\xi_2=d} = 0,$$
$$\mathscr{L}_3^1 m_1^1,\ \mathscr{L}_3^2 m_1^1,\ m_1^{31}\big|_{\xi_3=d} = 0,\quad d = 0,\ 1/2,$$
$$\mathscr{L}_{\xi\xi} m_1^2 = 0,$$
$$m_1^{22} = d,\ \mathscr{L}_1^1 m_1^2,\ m_1^{32}\big|_{\xi_1=d} = 0,\quad m_1^{12},\ \mathscr{L}_2^2 m_1^2,\ m_1^{32}\big|_{\xi_2=d} = 0,$$
$$\mathscr{L}_3^1 m_1^2,\ \mathscr{L}_3^2 m_1^2,\ m_1^{32}\big|_{\xi_3=d} = 0,\quad d = 0,\ 1/2.$$

The boundary value problems for the rest of the m_q^r are similar and are obtained from those written above by replacing r and q in the indices.

REMARK 4: Let the medium be invariant with respect to all possible transformations $\xi_j \leftrightarrow \xi_i$ and let conditions B be satisfied for all h. Then the averaged equations have the same symmetry, and therefore the nonzero coefficients $\hat{a}_{kl}^{ij}$ include only the coefficients $\hat{a}_{ii}^{ii}$, all equal; $\hat{a}_{ik}^{ik}$ $(i \neq k)$, all equal; and $\hat{a}_{ki}^{ik} = \hat{a}_{kk}^{ii}$ $(i \neq k)$, also equal. To define them, it suffices to solve two boundary value problems for the vectors m_1^1 and m_2^1.

REMARK 5: Consider a medium reinforced with fibers of some material with the generatrix of the fibers along the axis ξ_3, the materials of the matrix and the fibers being isotropic. Then the a_{ij}^{kl} do not depend on ξ_3 and have the form

$$a_{ij}^{hl}(\xi) = \lambda(\xi)\,\delta_{ih}\delta_{jl} + \mu(\xi)\,(\delta_{il}\delta_{jh} + \delta_{ij}\delta_{hl}), \tag{16}$$

where

$$\lambda(\xi) = \begin{cases} \lambda_B, & \xi \in \mathscr{B}, \\ \lambda_M, & \xi \in Q \setminus \mathscr{B}, \end{cases} \qquad \mu(\xi) = \begin{cases} \mu_B, & \xi \in \mathscr{B}, \\ \mu_M, & \xi \in Q \setminus \mathscr{B}, \end{cases}$$

where λ_B, μ_B are the Lamé coefficients of the fiber, and λ_M, μ_M are matrices. In other words, only the following elements are nonzero:

$$a_{ii}^{ii} = \lambda + 2\mu, \quad a_{ik}^{ik} = \lambda, \quad a_{ii}^{kk} = a_{ik}^{ki} = \mu, \qquad i \neq k.$$

Further, it is only the symmetry of the periodicity cell with respect to the transformations $\xi_1 \leftrightarrow -\xi_1$, $\xi_2 \leftrightarrow -\xi_2$, $\xi_1 \rightarrow \xi_2$ that is important, whereas it is unimportant that the cross-section of the fiber is a circle.

The system of equations (4) splits into nine systems in the vectors n_q^r $(q, r = 1, 2, 3)$. However, by Theorem 4, only the elements $\hat{a}_{ii}^{ii}$, $\hat{a}_{ik}^{ik}$, and $\hat{a}_{ik}^{ki}$ are zero. Moreover, since the medium is invariant with respect to the substitution $\xi_1 \leftrightarrow \xi_2$, all the elements produced by replacing the indices that take on values 1 and 2 (that is $\hat{a}_{22}^{11} = \hat{a}_{22}^{22}$, $\hat{a}_{13}^{13} = \hat{a}_{23}^{23}$, and $\hat{a}_{13}^{31} = \hat{a}_{23}^{32}$) are equal. So it is necessary to determine $\hat{a}_{11}^{11}$, $\hat{a}_{21}^{21}$, $\hat{a}_{21}^{12}$; $\hat{a}_{31}^{13}$, $\hat{a}_{13}^{13}$, and $\hat{a}_{33}^{33}$. Since $\hat{a}_{iq}^{kr} = \langle a_{iq}^{kr} + a_{ij}^{kl}\partial n_q^{lr}/\partial\xi_j\rangle$, it suffices to find the vectors n_1^1, n_1^2, n_1^3, and n_3^3.

Since all the coefficients are independent of ξ_3 (the embedding has the form of fibers extended along the axis ξ_3), then if φ is independent of ξ_3, we have

$$L_{\xi\xi}\varphi = (L_1\widehat{\varphi},\, L_2\varphi_3)^T, \quad \varphi = (\widehat{\varphi},\, \varphi_3)^T.$$

Here L_1 is a two-dimensional operator of elasticity theory:

$$(L_1\hat{\varphi})_k = \sum_{i,j,l=1}^{2} \frac{\partial}{\partial \xi_i}\left(a_{ij}^{kl}\frac{\partial \varphi_l}{\partial \xi_j}\right), \qquad k=1,2,$$

L_2 is an elliptic operator:

$$L_2\varphi_3 = \sum_{i=1}^{2}\frac{\partial}{\partial \xi_i}\left(a_{ii}^{33}\frac{\partial \varphi_3}{\partial \xi_i}\right) = \sum_{i=1}^{2}\frac{\partial}{\partial \xi_i}\left(\mu\frac{\partial \varphi_3}{\partial \xi_i}\right).$$

In fact, we have

$$(L_{\xi\xi}\varphi)_k = \sum_{i,j=1}^{2}\frac{\partial}{\partial \xi_i}\left(a_{ij}^{kl}\frac{\partial \varphi_l}{\partial \xi_j}\right) = \sum_{i,j,l=1}^{2}\frac{\partial}{\partial \xi_i}\left(a_{ij}^{kl}\frac{\partial \varphi_l}{\partial \xi_j}\right) + \sum_{i,j=1}^{2}\frac{\partial}{\partial \xi_i}\left(a_{ij}^{k3}\frac{\partial \varphi_3}{\partial \xi_j}\right).$$

For $k=1,2$ the second summand vanishes by (16) and it is the first one that vanishes if $k=3$; besides, $a_{ij}^{33}=0$ for $i\neq j$.

An analysis of solutions of the system (4), $n_1^r = (\hat{n}_1^r, n_1^{3r})$ $(r=1,2,3)$ and $n_3^3 = (\hat{n}_3^3, n_3^{33})$, suggests that $n_1^{31} = n_1^{32} = n_3^{33} = 0$, $\hat{n}_1^3 = (n_1^{13}, n_1^{23})^T = 0$. Thus, in order to determine all the a_{iq}^{kr}, it is necessary to solve three two-dimensional problems of elasticity theory and one elliptic equation in the domain $0<\xi_1$, $\xi_2<1/2$. The boundary conditions on the boundary of the domain are set on the basis of the even–odd conditions as above.

The boundary value problems for the vectors n_1^r $(r=1,2,3)$ can be rewritten in another form if we pass to the vectors $m_1^r = n_1^r + \xi_1 e_r$, where e_r is the unit vector of the rth coordinate axis:

$$L_1\hat{m}_1^r = 0, \qquad r=1,2,$$

$$m_1^{11}\big|_{\xi_1=d} = d,\quad \tau_{12}^{(1)}\big|_{\xi_1=d} = 0,\quad \tau_{21}^{(1)},\ m_1^{21}\big|_{\xi_2=d} = 0, \qquad d=0,1/2,$$

$$m_1^{22}\big|_{\xi_1=d} = d,\quad \tau_{11}^{(2)}\big|_{\xi_1=d} = 0,\ m_1^{12},\ \tau_{22}^{(2)}\big|_{\xi_2=d} = 0, \qquad d=0,1/2.$$

Here

$$\tau_{ik}^{(r)} = \sum_{j,l=1}^{2} a_{ij}^{kl}\frac{\partial m_1^{lr}}{\partial \xi_j}.$$

From the $m_1^r = (\hat{m}_1^r, 0)^T$, we define the elements

$$\tilde{a}_{i1}^{hr} = a_{ij}^{kl}\frac{\partial m_1^{lr}}{\partial \xi_j},\quad \hat{a}_{i1}^{kr} = 4\int_0^{1/2}\!\!\int_0^{1/2}\tilde{a}_{i1}^{kr}\,d\xi_1 d\xi_2, \qquad r=1,2.$$

In order to find $m_1^3 = (0,0,m_1^{33})^T$, we have the problem

$$L_2 m_1^{33} = 0, \quad m_1^{33}\big|_{\xi_1 = d} = d, \quad \frac{\partial m_1^{33}}{\partial \xi_2}\bigg|_{\xi_2 = d} = 0, \qquad d = 0, 1/2.$$

As a result of the calculations, we determine all the coefficients $\hat{a}_{iq}^{kr}$ except for $\hat{a}_{33}^{33}$.

The search for $\hat{a}_{33}^{33}$ involves solving the system of equations

$$L_1 \widehat{n}_3^3 + F = 0, \quad F = (\partial\lambda/\partial\xi_1, \ \partial\lambda/\partial\xi_2)^T. \tag{17}$$

When the λ are discontinuous, the right side has δ-functions; therefore, it is desirable to pass to another vector which would be consistent with a homogeneous system of equations.

Let the vector $v = (v_1, v_2)^T$ satisfy the system of equations (17) but with different constant term

$$g = 2(\partial(\lambda + \mu)/\partial\xi_1, \ \partial(\lambda + \mu)/\partial\xi_2)^T.$$

Since λ, μ are piecewise constant, $\hat{n}_3^3$ and v are related by $\hat{n}_3^3 = ([\lambda]/2[\lambda + \mu])v$, where $[\varphi] = \varphi_B - \varphi_M$ is a jump of the piecewise constant function.

We can prove this statement as follows. Since the constant term enters into the system of equations for n_3^3 and v in a divergent fashion, we can add or subtract an arbitrary constant under the sign of the derivative and pass to the piecewise constant functions $\overline{\lambda}(\xi) = \lambda(\xi) - \lambda_M$, $\overline{\mu}(\xi) = \mu(\xi) - \mu_M$. Since

$$\overline{\lambda}(\xi) = \begin{cases} [\lambda], & \xi \in \mathscr{B}, \\ 0, & \xi \in Q \setminus \mathscr{B}, \end{cases} \qquad \overline{\mu}(\xi) = \begin{cases} [\mu], & \xi \in \mathscr{B}, \\ 0, & \xi \in Q \setminus \mathscr{B}, \end{cases}$$

the resultant right sides $(\partial\overline{\lambda}/\partial\xi_1, \partial\overline{\lambda}/\partial\xi_2)^T$ and $2(\partial(\overline{\lambda} + \overline{\mu})/\partial\xi_1, \partial(\overline{\lambda} + \overline{\mu})/\partial\xi_2)^T$ differ just by the factor $[\lambda]/2[\lambda + \mu]$. This implies the required relation.

The vector

$$w = \begin{pmatrix} w_1 \\ w_2 \end{pmatrix} = \begin{pmatrix} v_1 + \xi_1 \\ v_2 + \xi_2 \end{pmatrix}$$

is consistent with the homogeneous system of equations (17). So

$$n_3^{i3} = \frac{(w_i - \xi_i)[\lambda]}{2[\lambda + \mu]}, \qquad i = 1, 2,$$

where w now satisfies the homogeneous system of equations. The boundary conditions are set for w from those for n_3^3:

$$n_3^{13}, \ \tau_{12}^{(3)}\big|_{\xi_1 = d} = 0, \quad \tau_{21}^{(3)}, \ n_3^{23}\big|_{\xi_2 = d} = 0, \qquad d = 0, 1/2.$$

Here

$$\tau_{ik}^{(3)} = \sum_{j,l=1}^{2} a_{ij}^{kl} \frac{\partial n_3^{l3}}{\partial \xi_j}.$$

The coefficient $\tilde{a}_{33}^{33}$ is given by

$$\tilde{a}_{33}^{33} = \frac{[\lambda]}{2\,[\lambda+\mu]} \sum_{l,j=1}^{2} a_{3j}^{3l} \frac{\partial w_l}{\partial \xi_j} + \left(a_{33}^{33} - \frac{[\lambda]}{2\,[\lambda+\mu]} \sum_{j=1}^{2} a_{3j}^{3j} \right) =$$

$$= \frac{\lambda\,[\lambda]}{2\,[\lambda+\mu]} \left(\frac{\partial w_1}{\partial \xi_1} + \frac{\partial w_2}{\partial \xi_2} \right) + (\lambda + 2\mu) - \frac{\lambda\,[\lambda]}{[\lambda+\mu]}.$$

Remark 6: Not all the vectors $n_q^r = (n_q^{1r}, n_q^{2r}, n_q^{3r})$ were found above; the remaining functions n_q^{lr} are found from

$$n_q^{lr}(\xi) = n_r^{lq}(\xi), \qquad l = 1, 2, 3, \quad r \neq q,$$

$$n_q^{lr}(\xi_1, \xi_2, \xi_3) = n_{q_1}^{l_1 r_1}(\xi_2, \xi_1, \xi_3),$$

where the set of indices (l_1, r_1, q_1) is obtained from the set (l, r, q) by substituting 2 for all the values of l and vice versa (it is assumed that in the set (l, r, q), at least one index is 1 or 2).

To sum up, we conclude that taking into consideration specific properties of the periodicity domain, particularly the symmetry properties, substantially reduces the amount of work required to determine the local and effective characteristics of a composite material (see also [26]).

§4 Agreement Between Theoretically Predicted Values of Effective Coefficients and Those Determined by an Ideal Experiment

Experiments to determine effective coefficients and the elements of the matrices $\hat{A}_{ij}$ use a rectangular specimen which occupies at rest the volume $0 \leqslant x_q \leqslant d_q$ $(q = 1, 2, 3)$. The specimen is subjected to loads of certain type and the values of $\hat{A}_{ij}$ are determined from the relation between stresses imposed on the faces and strains. In what follows, we assume that all the $\nu_q = d_q/\varepsilon$ are integers.

For definiteness, we consider the definition of the elements of the matrices $\hat{A}_{ij}$. From (4.4.4), $L_{\xi\xi}(N_1(\xi) + \xi_1 E) = 0$, and consequently the matrix

$$M_1^0(x) = \left(N_1\left(\frac{x}{\varepsilon}\right) + \frac{x_1}{\varepsilon} E\right)\varepsilon = x_1 E + \varepsilon N_1\left(\frac{x}{\varepsilon}\right)$$

is a solution of the system

$$\frac{\partial}{\partial x_i}\left(A_{ij}\left(\frac{x}{\varepsilon}\right)\frac{\partial M_1^0}{\partial x_j}\right) = 0; \tag{1}$$

its column-vectors $m_1^{0,q}$ also satisfy (1).

The vector $\hat{a}_{11}^1$ is found in the experiment as the average over the faces $x_1 = 0$, $x_1 = d_1$ of the vector of normal stresses σ_1, under the strain $u = (u^1, u^2, u^3)^T$, satisfying the conditions

$$\begin{gathered} u^1(0, x_2, x_3) = 0, \quad u^1(d_1, x_2, x_3) = d_1, \\ u^2(x_1, 0, x_3) = u^2(x_1, d_2, x_3) = u^3(x_1, x_2, 0) = u^3(x_1, x_2, d_3) = 0. \end{gathered} \tag{2}$$

on the specimen's boundary.

We consider the strain of the specimen, which corresponds to the displacement vector $u = m_1^1$. If all the elements of the column n_1^1 are uniformly bounded, then

$$m_1^1 = x_1 + O(\varepsilon)$$

and therefore,

$$(m_1^1 - u_1^1)|_\Gamma = O(\varepsilon).$$

The first column of the stress tensor under this strain is $\sigma_1 = A_{1j}\partial m_1^1/\partial\xi_j$. From §2, we have $\langle A_{1j}\partial m_1^1/\partial\xi_j\rangle_1 = \hat{a}_{11}^1$, and therefore

$$\int_0^{d_2}\int_0^{d_3} \sigma_1\Big|_{x_1 = x_1^0} dx_2 dx_3 = d_2 d_3 \hat{a}_{11}^1 \quad \forall x_1^0.$$

Thus, the average value of σ_1 over any cross-section and, in particular, over the cross-sections $x_1 = 0$ and $x_1 = d$, is $\hat{a}_{11}^1$. In the same way, we see that the average values of the vectors of normal stresses to the faces x_2, x_3 are equal to $\hat{a}_{12}^1$, $\hat{a}_{13}^2$, respectively.

The second column of the matrix M_1^0 satisfies the relation $m_1^{02} \approx (0, x_2, 0)^T$ and corresponds to shears along the axis x_2; in this case, the columns $\hat{a}_{11}^2$, $\hat{a}_{12}^2$, $\hat{a}_{13}^2$ coincide with the average values of the vectors of normal stresses to the corresponding faces x_1, x_2, x_3. Similar correspondence is established for the other vectors $\hat{a}_{ij}^q$.

The above-described theoretical experiment may appear inconclusive, since under the strains, the faces of the specimen remain flat only with accuracy up to $O(\varepsilon)$.

Consider the case when the periodic structure satisfies the symmetry condition (3.3) for all h. Using the property of the vector n_1^1 (see §3), we see that m_1^1 now satisfies conditions (2). Thus, the specimen remains a parallelepiped under this strain.

CHAPTER 7

Composite Materials Containing High-Modulus Reinforcement

This chapter presents mathematical models of composite materials where elastic, thermal, or other physical characteristics of the reinforcement differ greatly from those of the matrix. High strength and rigidity and good thermal insulation properties of these materials have made them suitable for many applications in engineering today (see [52, 66, 91, 146, 149, 150]). The mathematical modelling of composites with high-modulus reinforcement is distinguished above all by the presence of the large dimensionless parameter ω, which is the order of the ratio of the reinforcement's particular characteristic to that of the matrix. Thus, for plastics reinforced with fibers of carbon, boron, or silicon carbide, the ratio of Young's modulus of the reinforcement E_B to that of the matrix E_M is of order 100 (see [52]). There is a great distinction between the characteristics of air and the matrix in porous media, such as rubber and metals. This chapter is concerned mainly with obtaining explicit formulas for the effective characteristics of these materials in the multi-dimensional case (see the principle of splitting in §5).

Consider the equation

$$\sum_{j=1}^{2} \frac{\partial}{\partial x_j}\left(K\left(\frac{x_1}{\varepsilon}, \frac{x_2}{\varepsilon}\right)\frac{\partial u}{\partial x_j}\right) = 0, \tag{1}$$

where $K(\xi_1, \xi_2)$ is a 1-periodic function equal to K_B for $(\xi_1, \xi_2) \in \mathcal{B}$, and K_M for $(\xi_1, \xi_2) \in q \backslash \mathcal{B}$ (see Fig. 19).

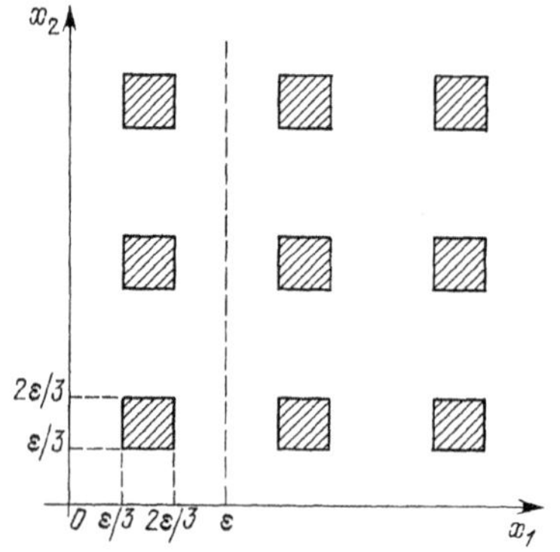

Figure 19

Let $K_B \gg K_M$, and let K_M be of order 1; then $K_B = \omega \overline{K}_B$, where $\overline{K}_B$ is the normed coefficient of order 1 and ω is the large dimensionless parameter

($\omega \gg 1$).

Hence

$$K(\xi) = \begin{cases} \omega \overline{K}_B & \text{for } (\xi_1, \xi_2) \in \mathscr{B}, \\ K_M & \text{for } (\xi_1, \xi_2) \in \mathscr{M}. \end{cases} \tag{2}$$

Equation (1) with coefficient (2) is characterized by the presence of both the small parameter ε and the large parameter ω.

The large parameter makes numerical solution of problems on a cell (4.5.21) more difficult. At the same time as the ratio $\lambda - K_B/K_M$ increases, the quality of the known explicit estimates for the effective characteristics, e.g., Hill's "forks," deteriorates (see Fig. 20 [190]).

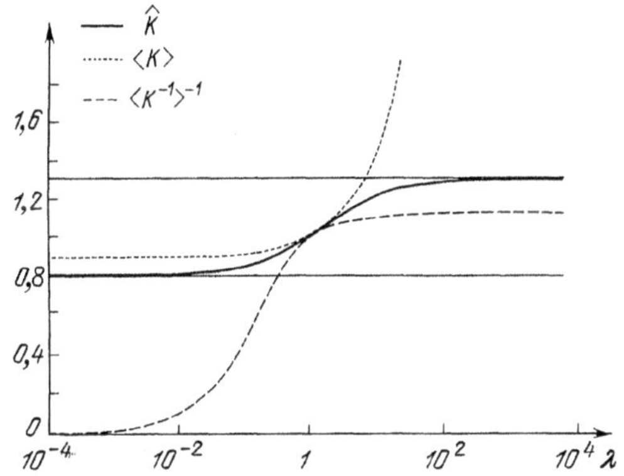

Figure 20

On the other hand, it will be shown in what follows that asymptotic expansions of the cell problems can be constructed in this case. The constructions result in explicit formulas for the principal part (in ω) of the coefficients of the averaged equations (see §5). The explicit formulas make it possible to handle the optimization of the properties of these materials more efficiently.

§1 The Stationary Field in a Layered Material

We shall start the presentation of the asymptotic method for calculating the effective characteristics of composites having high-modulus reinforcement with a simple example. Consider problems (2.4.1)–(2.4.4) with the periodic coefficient $K(\xi)$ of the form

$$K(\xi) = \begin{cases} \omega K_1 & \text{for } \xi \in [0, 1] \setminus (\xi^1, \xi^2), \\ K_2 & \text{for } \xi \in (\xi^1, \xi^2), \end{cases}$$

where $0 < \xi^1 < \xi^2 < 1$, $\omega \gg 1$, and K_1, K_2 are positive integers independent of ω and ε.

An asymptotic solution as $\varepsilon \downarrow 0$ and $\omega \to +\infty$ is constructed in two stages: First we formally ignore the dependence on ω and construct the asymptotics $u^{(2)}$

from §§2.3 and 2.4 in ε; then we construct the asymptotics in ω of the solutions $N_1(\xi)$ and $N_2(\xi)$ of the cell problems ((2.4.10)–(2.4.13), $i = 1,2$). Since the procedure of constructing $u^{(2)}$ was described in §§2.3 and 2.4, we proceed directly to the second stage. The problem on a cell for the periodic function N_1 has the form

$$\frac{d}{d\xi}\left(K(\xi)\left(\frac{dN_1(\xi)}{d\xi}+1\right)\right)=0, \quad \xi\in[0,1], \quad \xi\neq\xi^j, \quad j=1,2, \tag{1}$$

$$[N_1(\xi)]=\left[K(\xi)\left(\frac{dN_1(\xi)}{d\xi}+1\right)\right]=0, \quad \xi=\xi^j, \quad j=1,2, \tag{2}$$

$$N_1(0)=0. \tag{3}$$

We seek an asymptotic solution of problem (1)–(3) as

$$\widetilde{N}_1(\xi)=\begin{cases} N^1_{1,0}(\xi)+\omega^{-1}N^1_{1,1}(\xi) & \text{for } \xi\in[0,1]\setminus(\xi^1,\xi^2), \\ N^2_{1,1}(\xi)+\omega^{-1}N^2_{1,2}(\xi) & \text{for } \xi\in(\xi^1,\xi^2). \end{cases} \tag{4}$$

By substituting (4) into equation (1) and conditions (2), (3) and by simplifying, we get

$$\omega\frac{d}{d\xi}\left(K_1\left(\frac{dN^1_{1,0}}{d\xi}+1\right)\right)+\frac{d}{d\xi}\left(K_1\frac{dN^1_{1,1}}{d\xi}\right)=0, \quad \xi\in[0,1]\setminus(\xi^1,\xi^2), \tag{5}$$

$$\frac{d}{d\xi}\left(K_2\left(\frac{dN^2_{1,1}}{d\xi}+1\right)\right)+\omega^{-1}\frac{d}{d\xi}\left(K_2\frac{dN^2_{1,2}}{d\xi}\right)\sim 0, \quad \xi\in(\xi^1,\xi^2), \tag{6}$$

$$\omega K_1\left(\frac{dN^1_{1,0}}{d\xi}+1\right)+\left(K_1\frac{dN^1_{1,1}}{d\xi}-K_2\left(\frac{dN^2_{1,1}}{d\xi}+1\right)\right)-$$

$$-\omega^{-1}K_2\frac{dN^2_{1,2}}{d\xi}=0, \quad \xi=\xi^1,\ \xi^2, \tag{7}$$

$$(N^1_{1,0}-N^2_{1,1})+\omega^{-1}(N^1_{1,1}-N^2_{1,2})=0, \quad \xi=\xi^1,\xi^2, \tag{8}$$

$$N^1_{1,0}+\omega^{-1}N^1_{1,1}=0, \quad \xi=0. \tag{9}$$

We select the functions $N^1_{1,0}$, $N^1_{1,1}$, $N^2_{1,1}$, and $N^2_{1,2}$ in order to satisfy (5), (7)–(9) exactly, and (6) up to the term of order ω^{-1}. Then $\overline{N}_1(\xi)$ satisfies (2) and (3) exactly, and (1) up to $O(\omega^{-1})$.

We require that the factor of each power of ω (except for ω^{-1} in (6)) vanish to obtain the following algorithm for recurrent definition of $N_{1,k}$.

1. Find $N^1_{1,0}$ from the equation

$$\frac{d}{d\xi}\left(K_1\left(\frac{dN^1_{1,0}}{d\xi}+1\right)\right)=0, \quad \xi\in[0,1]\setminus(\xi^1,\xi^2), \tag{10}$$

and from the conditions

$$K_1\left(\frac{dN^1_{1,0}}{d\xi}+1\right)=0 \quad \text{for} \quad \xi=\xi^1,\,\xi^2, \tag{11}$$

$$N^1_{1,0}(0)=0, \tag{12}$$

and also find conditions for the 1-periodicity of $N^1_{1,0}$.

The function $-\xi$ for $\xi \in [0,\xi^1]$, $1-\xi$ for $\xi \in [\xi^2,1]$ is a solution of problems (10)–(12).

2. Find $N^2_{1,1}$ from the equation

$$\frac{d}{d\xi}\left(K_2\left(\frac{dN^2_{1,1}}{d\xi}+1\right)\right)=0, \quad \xi \in (\xi^1,\,\xi^2), \tag{13}$$

and the conditions

$$N^2_{1,1}=N^1_{1,0} \quad \text{for} \quad \xi=\xi^1,\,\xi^2. \tag{14}$$

The function $N^2_{1,1}=(\xi-\xi^1)/(\xi^2-\xi^1)-\xi$ is a solution of problem (13), (14).

3. Find $N^1_{1,1}$ from the equations

$$\frac{d}{d\xi}\left(K_1\,\frac{dN^1_{1,1}}{d\xi}\right)=0, \quad \xi \in [0,1]\setminus(\xi^1,\,\xi^2), \tag{15}$$

from the conditions

$$\begin{gathered} K_1\,\frac{dN^1_{1,1}}{d\xi}=K_2\left(\frac{dN^2_{1,1}}{d\xi}+1\right) \quad \text{for } \xi=\xi^1,\,\xi^2, \\ N^1_{1,1}(0)=0 \end{gathered} \tag{15}$$

and from the periodicity condition.

The functions

$$N^1_{1,1}(\xi)=\begin{cases} \dfrac{K_2}{K_1}\,\dfrac{\xi}{\xi^2-\xi^1} & \text{for} \quad \xi \in [0,\,\xi^1], \\ \dfrac{K_2}{K_1}\,\dfrac{\xi-1}{\xi^2-\xi^1} & \text{for} \quad \xi \in [\xi^2,\,1]. \end{cases}$$

are a solution of problem (15), (16).

4. Find $N^2_{1,2}(\xi) \in C^{(\infty)}(\xi_1,\xi_2)$ from the conditions

$$N^2_{1,2}=N^1_{1,1}, \quad K_2\,\frac{dN^2_{1,2}}{d\xi}=0$$

for $\xi = \xi^1, \xi^2$, that is, for example,

$$N_{1,2}^2 = N_{1,1}^1(\xi^1) + \frac{1}{2}\left(1 - \cos\left(\frac{\xi - \xi^1}{\xi^2 - \xi^1}\pi\right)\right)\left(N_{1,1}^1(\xi^2) - N_{1,1}^1(\xi^1)\right).$$

Note that the elementary problems of the above algorithm do not depend on the parameter ω. In this case, an analytical solution is possible.

By substituting $\widetilde{N}_1(\xi)$ for $N_1(\xi)$ into (2.4.10)–(2.4.13) with $i = 2$, we find the asymptotic solution $\widetilde{N}_2(\xi)$ of this problem:

$$\widetilde{N}_2(\xi) = -\int_0^{\xi} \widetilde{N}_1(t)\,dt.$$

The averaged problem (leading term) is

$$\widetilde{K}\frac{d^2\tilde{v}_0}{dx^2} = f(x), \quad \widetilde{K} = \frac{K_2}{\xi_2 - \xi_1},$$
$$\tilde{v}_0(0) = g_1, \quad \tilde{v}_0(l) = g_2.$$

By substituting the function

$$\tilde{u}^{(2)}(x) = \tilde{v}_0(x) + \varepsilon\widetilde{N}_1\left(\frac{x}{\varepsilon}\right)\frac{d\tilde{v}_0}{dx} + \varepsilon^2\widetilde{N}_2\left(\frac{x}{\varepsilon}\right)\frac{d^2\tilde{v}_0}{dx^2}$$

into (2.4.1)–(2.4.4), we verify that $\tilde{u}^{(2)}(x)$ satisfies conditions (2.4.2), (2.4.3) exactly, equation (2.4.1) up to terms of order $O(\omega^{-1}/\varepsilon)$, and condition (2.4.4) to $O(\varepsilon)$.

Consider problems (2.4.2)–(2.4.4) for $f \equiv 0$, $g_1 = 0$, $g_2 = 1$, $l = 1$, $\xi^1 = 1/4$, and $\xi^2 = 3/4$.

We expand the exact solution $u(x)$ from §5 of the Introduction into a series in ω^{-1}. By comparing the expansion of $u(x)$ with $\tilde{u}^{(1)}(x) = \tilde{v}_0(x) + \varepsilon\widetilde{N}_1(x/\varepsilon)d\tilde{v}_0/dx$, we obtain the estimates

$$\max_{[0,1]} |u(x) - \tilde{v}(x)| \leqslant \varepsilon/2, \tag{17}$$

$$\max_{[0,1]}\left|K\left(\frac{x}{\varepsilon}\right)\frac{du}{dx} - K\left(\frac{x}{\varepsilon}\right)\frac{d\tilde{u}^{(1)}}{dx}\right| < \frac{2K_2^2}{\omega K_1}(\pi + 1), \tag{18}$$

$$\max_{[\varepsilon/2,1-\varepsilon/2]}\left|\varepsilon^{-1}\int_{x-\varepsilon/2}^{x+\varepsilon/2} K\left(\frac{t}{\varepsilon}\right)\frac{du(t)}{dt}\,dt - \widetilde{K}\frac{d\tilde{v}_0(x)}{dx}\right| < \frac{2K_2^2}{\omega K_1}. \tag{19}$$

Estimates (17)–(19) give an idea of the magnitude of error of the averaging method for the composites with high-modulus reinforcements. For example, in (17) the accuracy of 10^{-2} is attained for a composite consisting of 101 layers.

Inequalities (18), (19) estimate the absolute difference between the exact and average values of the flows, and the difference between the average flow of the original problem and that of the averaged problem. These inequalities ascertain the inverse relationship between the above errors and the ratio of the heat conductivity coefficients of the layers. Thus, assuming $K_1 = K_2 = 1$, we see that to achieve the accuracy 7×10^{-2} in (18) and (19), it suffices to have the difference in the values of the layer characteristics by a factor of 120 and 35, respectively. It is this jump of the values of the coefficients that above was said to be typical of most high-modulus composite materials. Of course, in more complex problems, e.g., multi-dimensional ones, the proportionality factors between the error and the values of ε and ω^{-1} depend on many parameters and often cannot be obtained explicitly as it was done in the estimates (17)–(19). In these cases, one has to be content with the order of the error as $\varepsilon \to 0$, $\omega \to +\infty$, and the estimates such as (17)–(19) can be used as models.

Similarly to the stationary problem of thermal conductivity, investigation can be made also into the stationary problem of elasticity theory for a laminated composite (see §2.5).

Problems (1)–(3) discussed in the present Section belong to the class of so-called "rigid problems" which describe processes in strongly inhomogeneous media [42, 204]. Some results on the rigid problems were produced by [114, 115, 119, 141, 145, 214].

An approximate solution of (1)–(3) of the form (4) has been constructed in this section. A complete asymptotic expansion of the solution of the cell problem can be constructed. This will be done in the next section for the multidimensional case.

§2 Composite Materials with Grains for Reinforcement

In this section, the foregoing procedure of constructing an asymptotic solution is applied to a nonstationary system of partial differential equations.

We consider the case of highly heat-conducting grains where the methodology of constructing an asymptotic expansion of a solution to the problem on a cell is the most transparent.

Problems of elasticity of multidimensionally reinforced materials have some complicating aspects which will be discussed in §4. In particular, an additional condition for applicability of the method, $\omega\varepsilon^2 \ll 1$, arises in this case.

1. Suppose we have a system of partial differential equations of the form

$$L_{\omega,\varepsilon}u \equiv R\left(\frac{x}{\varepsilon}\right)\frac{\partial^2 u}{\partial t^2} + S\left(\frac{x}{\varepsilon}\right)\frac{\partial u}{\partial t} - \frac{\partial}{\partial x_r}\left(A_{rm}^{(\omega)}\left(\frac{x}{\varepsilon}\right)\frac{\partial u}{\partial x_m}\right) = f(x, t), \quad x \in \mathbf{R}^s. \qquad (1)$$

Here R, S, $A_{rm}^{(\omega)}$ are $n \times n$ matrices whose elements are 1-periodic functions $\xi_q = x_q/\varepsilon$, with the $A_{rm}^{(\omega)}$ depending on the large parameter ω:

$$A_{rm}^{(\omega)}(\xi) = \begin{cases} \omega A_{rm}^{1}(\xi) & \text{for } \xi \in G_1, \\ A_{rm}^{2}(\xi) & \text{for } \xi \in G_2 = Q \setminus \overline{G}_1, \end{cases}$$

where Q is the unit cube in $\mathbf{R}^s$: $Q = \{\xi\mathbf{R}^s | 0 < \xi_l < 1, \ l = 1, \ldots, s\}$, G_1 is a bounded subdomain Q, $\partial G_1 \cap \partial Q = \emptyset$ and u, f are n-dimensional column-vectors.

It is assumed that the surface ∂G_1 and the coefficients of R, S, $A_{rm}^{(\omega)}$ are infinitely smooth everywhere on $\overline{Q} \setminus \partial G_1$ up to ∂G_1, and $f \in C^\infty(\mathbf{R}^s)$.

On the coefficient's discontinuity surfaces, the standard relations

$$[u] = [\partial u/\partial \nu] = 0, \tag{2}$$

hold. Here $\partial u/\partial \nu$ is a derivative with respect to the conormal $A_{rm}^{(\omega)} \partial u/\partial x_m n_r$, and $n = (n_1, \ldots, n_s)$ is the normal to ∂G_1, which is exterior relative to G_1. In the scalar case when $R \equiv 0$, equation (1) with conditions (2) can describe thermal processes in composites reinforced with highly heat-conducting grains or with unidirectional fibers. A two-dimensional formulation of the problems ($s = 2$) is considered in the latter case (see Figs. 1 and 2).

2. The asymptotics of the solution of equation (1) is sought as the series (4.5.12):

$$u^{(\infty)} \sim \sum_{p,q,l=0}^{\infty} \varepsilon^{p+q+l} \sum_{|i|=l} N_{qi}^{p} D^{qi} v.$$

Here the matrices N_{qt}^{p} and the vector v depend, in general, on the parameter ω.

As in §1, the $u^{(\infty)}$ is constructed in two stages. At the first stage, we substitute $u^{(\infty)}$ into equation (1) and the contact conditions (2) formally assuming the N_{qi}^{p} and v to be independent of ω. By the same reasoning as in Subsection 2, §4.5 we obtain cell problems (4.5.21), (4.5.18) for the periodic matrices $N_{qi}^{p}(\xi)$. The problems contain large parameter ω.

3. At the second stage, the asymptotic solutions of the cell problems (4.5.21) and (4.5.18) are constructed as $\omega \to +\infty$. We seek the N_{qi}^{p} as the series

$$N_{qi}^{p}(\xi) \sim \begin{cases} \sum\limits_{k=0}^{\infty} \omega^{-k} N_{qi,k}^{p1}(\xi) & \text{for } \xi \in G_1, \\ \sum\limits_{k=0}^{\infty} \omega^{-k+1} N_{qi,k}^{p2}(\xi) & \text{for } \xi \in G_2, \end{cases} \tag{3}$$

where the $N_{qi,k}^{pj}(\xi)$ are 1-periodic functions.

It can be shown that in this case the T_{qi}^{p} and h_{qi}^{p} can be represented as

$$T_{qi}^{p}(\xi) \sim \sum_{k=0}^{\infty} \omega^{-k+1} T_{qi,k}^{p}(\xi), \quad h_{qi}^{p} \sim \sum_{k=0}^{\infty} \omega^{-k+1} h_{qi,k}^{p},$$

where

$$h^p_{qi_1\dots i_l,k} = \Big\langle -\overline{A}_{i_1j}\frac{\partial N^p_{qi_2\dots i_l,k}}{\partial \xi_j} - \overline{A}_{i_1i_2}N^p_{qi_3\dots i_l,k} + \\ + RN^p_{q-2i_1\dots i_l,\overline{k}} + SN^{p-1}_{q-1i_1\dots i_l,\overline{k}}\Big\rangle,$$

$$T^p_{qi_1\dots i_l,k}(\xi) = -\overline{A}_{i_1j}\frac{\partial N^p_{qi_2\dots i_l,k}}{\partial \xi_j} - \frac{\partial}{\partial \xi_j}\left(\overline{A}_{ji_1}N^p_{ji_2\dots i_l,k}\right) - \\ - \overline{A}_{i_1i_2}N^p_{qi_3,\dots i_l,k} + RN^p_{q-2i_1\dots i_l,\overline{k}} + SN^{p-1}_{q-1i_1\dots i_l,\overline{k}}, \tag{4}$$

with $N^{p1}_{qi_1\dots i_l,-1} = 0$; here

$$N^p_{qi,k} = \begin{cases} N^{p1}_{qi,k} & \text{for } \xi \in G_1, \\ N^{p2}_{qi,k} & \text{for } \xi \in G_2, \end{cases} \qquad \overline{k} = \begin{cases} k-1 & \text{for } \xi \in G_1, \\ k & \text{for } \xi \in G_2, \end{cases}$$

$$\overline{A}_{rm}(\xi) = \begin{cases} A^1_{rm} & \text{for } \xi \in G_1, \\ A^2_{rm} & \text{for } \xi \in G_2. \end{cases}$$

Consider the fixed collection $(p, q, i) = (p, q, i_1, \dots, i_l)$ for $p + q + |i| > 0$. We shall define the $N^{p,j}_{qi,k}$ successively for $k = 0, 1, \dots$. We substitute (3) into (4.5.21) and (4.5.18), simplify, and equate to zero the factors of all power of ω. We obtain an algorithm for recurrent definition of the $N^{pj}_{qi,k}$.

Let all the $N^{pj}_{qi,k}$ be defined for $k \leqslant K_0 - 1$. We define $N^{p2}_{qi,k_0}(\xi)$ as a 1-periodic solution of the problem

$$-L_2N^{p2}_{qi,K_0} + T^p_{qi,K_0} = h^p_{qi,K_0}, \quad \xi \in \mathbf{R}^s \setminus \widetilde{G}_1, \tag{5}$$

$$N^{p2}_{qi,K_0} = N^{p1}_{qi,K_0-1}, \quad \xi \in \partial\widetilde{G}_1, \tag{6}$$

(here $\widetilde{G}_1$ is a periodic continuation of G_1 to $\mathbf{R}^s$) and then find N^{pi}_{qi,K_0} from the equation

$$-L_1N^{p1}_{qi,K_0} + T^p_{qi,K_0} = h^p_{qi,K_0}, \quad \xi \in G_1, \tag{7}$$

with the boundary condition

$$\frac{\partial N^{p1}_{qi,K_0}}{\partial \nu^1} = -\left(A^1_{ji_1}N^{p1}_{qi_2\dots i_l,K_0}\big|_{\partial G_1} - A^2_{ji_1}N^{p2}_{qi_2\dots i_l,K_0}\big|_{\partial G_1}\right)n_j + \\ + \frac{\partial N^{p2}_{qi,K_0}}{\partial \nu^2}, \quad \xi \in \partial G_1. \tag{8}$$

Here $i = (i_1, \dots i_l)$,

$$L_j = \frac{\partial}{\partial \xi_r}\left(A^j_{rm}\frac{\partial}{\partial \xi_m}\right), \quad \frac{\partial}{\partial \nu^j} = A^j_{rm}(\xi)\, n_r(\xi)\, \frac{\partial}{\partial \xi_m},$$

and the $n_r(\xi)$ are cosines of the angles between the exterior (relative to G_1) normal n to ∂G_1 and the axis ξ_r. Note that it follows from (7), (8) for $K_0 = 0$, $p = q = 0$, and $l = 1$ that $N^{01}_{oi_1,0} = -\xi_{i_1}E + C_{i_1}$, where E is the unit matrix and C_{i_1} is an arbitrary constant matrix which is selected in what follows to be equal to zero.

When an asymptotic solution of problem (4.5.21), (4.5.18) is constructed, it is assumed that a periodic solution from W^1_2 of problem (5), (6) exists and is unique, that this solution has a derivative with respect to the conormal from $L_2(\partial G_1)$, and that there exists a unique solution from $W^1_2(G_1)$ of problem (7), (8) up to an arbitrary constant term under condition (4).

The necessity of condition (4) for the Neumann problem (7), (8) to be solvable stems from the equality

$$-\int\limits_{\partial G_1} \frac{\partial N^{p2}_{qi,K_0}}{\partial \nu^2}\, ds = \int\limits_{G_0}\left(T^p_{qi,K_0} - h^p_{qi,K_0}\right) d\xi,$$

which follows from (5). The above assumptions are valid, in particular, for the scalar elliptic operator $L_{\xi\xi}$ with bounded coefficients smooth in a neighbourhood of ∂G from every side of ∂G_1 and up to ∂G_1.

The above constructions have to be modified for a system of equations of elasticity theory due to the following: Condition (4) alone is insufficient from problem (7), (8) to be solvable—the orthogonality of the right side to rigid-body displacement is also needed. On the other hand, the degree of indefiniteness of the solution is higher, with the solution being defined up to an arbitrary rigid-body displacement.

4. It can be shown that under the condition $\langle N^p_{qi,k}\rangle = 0$ for any k,

$$\sum_{j\in \mathrm{inv}\, i} N^{p2}_{qj,0} = 0, \quad \sum_{j\in \mathrm{inv}\, i} h^p_{qj,0} = 0;$$

here the sum consists of all possible permutations j of the multi-index i. (The proof is by induction in $p+q+l$; we sum the relations (4)–(8) with respect to all possible permutations of a fixed multi-index i, obtain boundary value problems for $N^{p,d}_{q(i)} = \sum_{j\in \mathrm{inv}\, i} N^{p,d}_{qj,0}$, and ascertain that

$$N^{p,1}_{q\{i\}} = \delta_{p+q,0}\frac{(-1)^l}{\alpha_1!\ldots\alpha_s!}\xi_1^{\alpha_1}\ldots\xi_s^{\alpha_s}E, \quad N^{p,2}_{q\{i\}} = 0, \quad \sum_{j\in \mathrm{inv}\, i} h^p_{qj,0} = 0).$$

By substituting the expansion of h^p_{qi} in powers of ω^{-1} into (4.5.19), we get the asymptotic equation

$$\sum_{\substack{p,q,l\geqslant 0\\ p+q+l\geqslant 2}} \sum_{|i|=l} \sum_{k=1}^{\infty} \varepsilon^{p+q+l-2}\omega^{-k+1} h^{p}_{qi,k} D^{qi} v - f(x,t) \sim 0. \tag{9}$$

We seek a solution of (9) in the form

$$v \sim \sum_{j,k=0}^{\infty} \varepsilon^{j}\omega^{-k} v_{j,k}(x,t),$$

where the $v_{j,k}$ are independent of ω and ε. To define the $v_{j,k}$, we use a recurrent chain of equations of the form

$$\hat{L}_0 v_{j,k} = f_{j,k}(x,t),$$

where $\hat{L}_0$ is the principal part of the operator in (9) acting on v, $f_{j,k}(x,t)$ is a function depending on $v_{q,r}(q+r<j+k)$, and $f_{00}=f$.

The coefficients of $\hat{L}_0$ are constant:

$$\widehat{L}_0 = \widehat{R}^0 \frac{\partial^2}{\partial t^2} + \widehat{S}^0 \frac{\partial}{\partial t} - \frac{\partial}{\partial x_r} \widehat{A}^0_{rm} \frac{\partial}{\partial x_m}.$$

We give the algorithm for calculating the coefficients of $\hat{L}_0$. We have $\widehat{R}^0 = \langle R\rangle$, $\widehat{S}^0 = \langle S\rangle$. To calculate $\hat{A}^0_{rm}$, we first solve the two boundary value problems:

a) For $\widetilde{N}^{(2)}_{i_1}(\xi)$ $(\widetilde{N}^{(2)}_{i_1} = N^{02}_{0i_1,1})$:

$$-L_2\left(\widetilde{N}^{(2)}_{i_1} + \xi_{i_1}E\right) = 0, \quad \xi \in \mathbf{R}^s \setminus \widetilde{G}_1,$$
$$\widetilde{N}^{(2)}_{i_1} + \xi_{i_1}E = 0, \quad \xi \in \partial\widetilde{G}_1,$$

the $\widetilde{N}^{(2)}_{i_1}$ are 1-periodic matrix-valued functions.

b) For $\widetilde{N}^{(2)}_{i_1}(\xi)$ $(\widetilde{N}^{(1)}_{i_1} = N^{01}_{0i_1,1})$:

$$-L_1\widetilde{N}^{(1)}_{i_1} = 0, \quad \xi \in G_1,$$
$$\frac{\partial \widetilde{N}^{(1)}_{i_1}}{\partial \nu^1} = \frac{\partial}{\partial \nu^2}\left(\widetilde{N}^{(2)}_{i_1} + \xi_{i_1}E\right), \quad \xi \in \partial G_1,$$
$$\widehat{A}^0_{rm} = \sum_{j=1}^{2} \int_{G_j} A^j_{rk} \frac{\partial \widetilde{N}^{(j)}_m}{\partial \xi_k}\, d\xi + \int_{G_2} A^2_{rm}\, d\xi.$$

Passing to the limit as $\omega \to +\infty$, we can show from Theorem 4.1.1. that if the coefficients A^j_{rm} satisfy the properties (4.1.3) with the constant κ_1 independent of ω, then the averaged coefficients $\hat{A}^0_{rm}$ satisfy the same relations. In addition, it follows from the property $R>0$ for $S>0$ that $R>0$ for $\widehat{S}>0$.

5. Suppose

$$u^{(K,J)} = \sum_{\substack{p,q,l\geqslant 0 \\ p+q+l\leqslant K+1}} \varepsilon^{p+q+l} \sum_{|i|=l} N_{qi}^{p(J)} D^{qi} v^{(K,J)},$$

$$N_{qi}^{p(J)} = \begin{cases} \sum_{k=0}^{J} \omega^{-k} N_{qi,k}^{p1}(\xi), & \xi \in G_1, \\ \sum_{k=0}^{J} \omega^{-k+1} N_{qi,k}^{p2}(\xi), & \xi \in G_2, \end{cases}$$

$$v^{(K,J)} = \sum_{j=0}^{K-1} \sum_{r=0}^{J-1} \varepsilon^{j} \omega^{-r} v_{j,r}(x,t).$$

Then $u^{(K,J)}$ satisfies equation (1) and conditions (2) up to terms of order $O(\omega^{-j}/\varepsilon + \omega\varepsilon^{K} + \omega^{-J+1})$, that is,

$$L_{\omega,\varepsilon} u^{(K,J)} - f = O(\omega^{-J}/\varepsilon + \omega\varepsilon^{K} + \omega^{-J+1}), \quad [u^{(K,J)}] = 0,$$
$$[\partial u^{(K,J)}/\partial\nu] = O(\omega^{-J+1} + \omega\varepsilon^{K}).$$

Thus, if $\omega^{-1} = O(\varepsilon^{\alpha})$, $\varepsilon = O(\omega^{-\beta})$ $(\alpha, \beta > 0)$, then by selecting sufficiently large J and K, we can obtain the asymptotic solution $u^{(K,J)}$ that satisfies (1) with accuracy $O(\varepsilon^{N})$.

6. Given $n = 1$, the hypersurface $\partial G_1 \in C^{\infty}$, $A_{rm}^{j}(\xi) \in C^{\infty}(\overline{G}_j(x) \in C^{\infty}(\overline{G})$, where G is a bounded domain in $\mathbf{R}^s$ with smooth boundary, we consider the Dirichlet problem

$$L_{\omega,\varepsilon} u \equiv \frac{\partial}{\partial x_r}\left(A_{rm}^{(\omega)} \frac{\partial u}{\partial x_m}\right) = f(x),$$
$$u|_{\partial G} = \varphi(x), \quad A_{rm}^{j} = A_{mr}^{j}, \quad \varkappa_1 |\eta|^2 \leqslant A_{rm}^{j} \eta_r \eta_m,$$
$$\eta \in \mathbf{R}^s, \quad 0 < \varkappa_1, \quad \varphi(x) \in C^{\infty}(\partial G).$$

Then the construction procedure described in paragraphs 1–5 can be carried out, and relying on estimates like the one based on the generalized maximum principle (see [81, 82]), we can obtain

$$\max_{x \in \overline{G}} |u - v_{0,0}| = O\left(\varepsilon + \omega^{-1}\right)$$

provided $\omega^{-1} = O(\varepsilon^{\alpha})$, $\varepsilon = O(\omega^{-\beta})$ $(\alpha, \beta > 0)$; here $v_{0,0}$ is a solution of the averaged zeroth-order equation

$$\hat{L}_0 v_{0,0} = f(x), \quad x \in G.$$

7. This paragraph deals with the case of the bounded domain G_1 whose boundary has an empty intersection with the boundary of Q. Then the tensor $\hat{A}_{rm}^{0}$, which characterizes the effective heat-conducting properties of the mate-

rial, has the order 1, whereas the appropriate coefficients of the filler have the order $\omega \gg 1$. This means that reinforcement with high-modulus grains or unidirectional fibers does not bring about any substantial increase in the effective characteristics of the composite in the directions that are orthogonal to those of the reinforcement (the effective properties of fibrous materials in the direction of reinforcement have not been considered up to now, since a two-dimensional problem was investigated with $s = 2$).

In the case of $\partial G_1 \cap \partial Q \neq \emptyset$, when the periodic continuation $\partial \widetilde{G}_1$ belongs to C^∞, problem (7), (8) must be supplemented by the 1-periodicity condition for $N^{p1}_{qi,k0}$; if, in this case, each component of ∂G_1 is contained in a finite number of periodic cells independent of ε, all the $h^p_{qi,0}$ are still equal to zero; otherwise, some of the $h^p_{qi,0}$ may be nonzero, and the averaged equation begins with terms of order ω:

$$\widetilde{L}v = \sum_{\substack{p,q,l\geqslant 0\\ p+q+l\geqslant 2}} \sum_{|i|=l} \sum_{k=0}^{\infty} \varepsilon^{p+q+l-2}\omega^{-k+1} h^p_{qi,k} D^{qi} v \sim f(x,t). \qquad (9')$$

The periodic matrices $N^{01}_{0i_1,0}$ are found from the equations

$$-L_1\left(N^{01}_{0i_1,0} + \xi_{i_1}E\right) = 0, \quad \xi \in G_1, \qquad (10)$$

with the condition

$$\frac{\partial}{\partial \nu^1}\left(N^{01}_{0i_1,0} + \xi_{i_1}E\right) = 0, \quad \xi \in \partial G_1 \setminus \partial Q. \qquad (11)$$

The leading term of the operator L from (9′) is of the form

$$\widehat{L}_0 = -\frac{\partial}{\partial x_r}\left(\widehat{A}^0_{rm}\frac{\partial}{\partial x_m}\right),$$

where

$$\widehat{A}^0_{rm} = \omega \int_{G_1} A^1_{rj}(\xi)\frac{\partial}{\partial \xi_j}\left(N^{01}_{0m,0} + \xi_m E\right) d\xi. \qquad (12)$$

Now the $N^0_{0i_1,0}$ may not be equal to $-\xi_{i_1}E$, and some of the $\hat{A}^0_{rm}$ (nonzero) have the order of the appropriate characteristics of the reinforcement $\omega \gg 1$.

As an example, consider a composite reinforced by unidirectional cylindrical fibers parallel to the x_3-axis and forming a periodic structure along the x_1- and x_2- axes (with period ε). We assign period ε also to the variable x_3. Then we pass to the variables $\xi_q = x_q/\varepsilon$. Let the cube Q intersect only one fiber occupying the cylinder G_1, with its generatrix parallel to ξ_3. In the scalar case for $n = 1$, $A^1_{rm} = \delta_{rm}K (K = \text{const})$, problem (10), (11) is solved analytically:

$N^{01}_{0i_1,0} = -\xi_{i_1}$ for $i_1 \neq 3$, and $N^{01}_{03,0} = 0$ for $i_1 = 3$. For the principal part of A_{rm} of order ω and considering (12), we obtain $\hat{A}^0_{rm} = \omega\delta_{3r}\delta_{m3}K \cdot \operatorname{mes} G_1$; here $\operatorname{mes} G_1$ is the volume fraction of the fiber.

Thus, of all the components of the effective heat-conduction tensor only $\hat{A}^0_{33}$ has the order ω and is calculated explicitly. The other components (of order 1) can be determined by the same algorithm as in the two-dimensional case, but in order to realize the algorithm, problems (5), (6), (7) and (8) should be solved numerically. Note that these problems do not contain parameters ω and ε. A large spread in values of $\hat{A}^0_{rm}$ is the reason for the strong anisotropy of the material reinforced with unidirectional fibers.

The results of this section in the case $\partial Q \cap \partial G_1 \neq \emptyset$ are applicable to the description of processes in high-modulus matrices reinforced with low-modulus fibers.

§3 Dissipation of Waves in Layered Media

Let a composite material have a periodic laminated structure, and let it occupy the half-space $x_1 > 0$, with the axis Ox_1 being perpendicular to the planes that divide the layers. The subscript 1 of x is omitted.

By §3.3, the equation

$$L_{\omega,\varepsilon}u \equiv R^{(\omega)}\left(\frac{x}{\varepsilon}\right)\frac{\partial^2 u}{\partial t^2} + S^{(\omega)}\left(\frac{x}{\varepsilon}\right)\frac{\partial u}{\partial t} - \\ - \frac{\partial}{\partial x}\left(A^{(\omega)}\left(\frac{x}{\varepsilon}\right)\frac{\partial u}{\partial x}\right) = 0, \quad x > 0, \quad t > 0, \quad (1)$$

describes propagation of electromagnetic waves in a laminated medium, u denotes E_2 or E_3, $R = \epsilon$, $S = \sigma$, and $A = \mu^{-1}$. Here $R^{(\omega)}(\xi)$, $S^{(\omega)}(\xi)$, and $A^{(\omega)}(\xi)$ are 1-periodic functions of the form

$$R^{(\omega)}(\xi) = \begin{cases} R^1, & \xi \in [0,1]\setminus(a_1, a_2), \\ \omega^{-1}R^2, & \xi \in (a_1, a_2), \end{cases}$$

$$S^{(\omega)}(\xi) = \begin{cases} S^1, & \xi \in [0,1]\setminus(a_1, a_2), \\ \omega^{-1}S^2, & \xi \in (a_1, a_2), \end{cases}$$

$$A^{(\omega)}(\xi) = \begin{cases} A^1, & \xi \in [0,1]\setminus(a_1, a_2), \\ \omega^{-1}A^2, & \xi \in (a_1, a_2), \end{cases}$$

where a_i, R^i, S^i, A^i $(i = 1, 2)$ are positive numbers, $a_1 < a_2 < 1$ and $\omega \gg 1$.

The characteristics of the layers differ (in order) by a factor of ω as before. To reduce equation (1) to the form similar to (2.1), it suffices to multiply (1) by ω.

Suppose that for $x = 0$ the boundary value

$$u|_{x=0} = \varphi(t), \quad (2)$$

is set, where $\varphi(t)$ is an infinitely differentiable function with compact support, all the derivatives of which $d^l\varphi/dt^l$ at $t = 0$ for $l \geqslant 0$, vanish, and suppose the initial conditions are homogeneous:

$$u|_{t=0} = \partial u/\partial t|_{t=0} = 0. \tag{3}$$

The solution of problem (1)–(3) is understood in the generalized sense. We construct an asymptotic expansion of the solution of (1)–(3) as $\varepsilon \to +0$, $\omega \to +\infty$, and for $\varepsilon = O(\omega^{-1/2-\delta})$, $\delta > 0$.

As in §2, we construct the asymptotics $u^{(\infty)}$ in two stages. At the first stage, the procedure from Subsection 2 of §4.5 is realized, and the second state involves the construction of the asymptotic expansion (2.3) of solutions of cell problems (4.5.21) and (4.5.18). The averaged equation (2.9) is

$$\sum_{\substack{p,q,l\geqslant 0\\ p+q+l\geqslant 2}} \sum_{k=0}^{\infty} \varepsilon^{p+q+l-2}\omega^{-k}h^p_{ql,k}\frac{\partial^{q+l}v}{\partial t^q\partial x^l} \sim 0. \tag{4}$$

Substitution of the expansions (4.5.12) and (2.3) into the boundary condition (2) yields

$$v + \sum_{\substack{p,q,l\geqslant 0\\ p+q+l\geqslant 1}} \varepsilon^{p+q+l}\sum_{k=0}^{\infty}\omega^{-k}N^{p1}_{ql,k}(0)\frac{\partial^{q+l}v}{\partial t^q\partial x^l} \sim \varphi(t), \quad x = 0. \tag{5}$$

We supplement equation (4) and the boundary value (5) with the initial value

$$v|_{t=0} = \partial v/\partial t|_{t=0} = 0.$$

The ambiguity of solutions of problem (2.7), (2.8) implies an ambiguity of definition of $N^{p1}_{ql,k}$: $N^{p1}_{ql,k} = \overline{N}^p_{ql,k} + C^{p1}_{ql,k}$, where the $C^p_{ql,k}$ are some constants. We select them under the condition

$$N^{p1}_{ql,k}(0) = 0, \qquad p + q + l > 0;$$

then condition (5) changes to

$$v|_{x=0} = \varphi(t).$$

After the change of coordinates $\eta = \sqrt{\omega}x$, (4) takes the form

$$\sum_{\substack{p,q,l\geqslant 0\\ p+q+l\geqslant 2}} \sum_{k=0}^{\infty} \varepsilon^{p+q+l-2}\omega^{-k+l/2}h^p_{ql,k}\frac{\partial^{q+l}v}{\partial t^q\partial \eta^l} \sim 0,$$

that is,

$$\sum_{p,q=0}^{\infty}\ \sum_{l=\max\{0,2-p-q\}}^{\infty}\ \sum_{k=0}^{\infty} \varepsilon^{p+q-2}(\varepsilon\sqrt{\omega})^{l}\,\omega^{-k}h^{p}_{ql,k}\frac{\partial^{q+l}v}{\partial t^{q}\partial\eta^{l}} \sim 0. \qquad (6)$$

Upon calculating some of the $h^{p}_{ql,k}$, we obtain the principal part of equation (6):

$$\widehat{L}_0 v \equiv (a_2-a_1)\left(R^1\frac{\partial^2 v}{\partial t^2}+S^1\frac{\partial v}{\partial t}-\frac{A^2}{(a_2-a_1)(1-a_2+a_1)}\frac{\partial^2 v}{\partial\eta^2}\right)\sim-\Delta; \quad (7)$$

here

$$\Delta=\sum_{\substack{p,q,l\geqslant 0\\ p+q+l>2}} \varepsilon^{p+q+l-2}\sum_{k=0}^{\infty}\omega^{-k+l/2}h^{p}_{ql,k}\frac{\partial^{q+l}v}{\partial t^{q}\partial\eta^{l}}+$$
$$+\left(\sum_{k=2}^{\infty}\omega^{-k+1}h^{0}_{02,k}\right)\frac{\partial^2 v}{\partial\eta^2}+\omega^{-1}(1-a_2+a_1)\left(R^2\frac{\partial^2 v}{\partial t^2}+S^2\frac{\partial v}{\partial t}\right)=$$
$$=O\left(\varepsilon^{\gamma/2}+\omega^{-1/2}\right)$$

under the conditions $v|_{\eta=0}=\varphi(t)$, $v|_{t=0}=\frac{\partial v}{\partial t}|_{t=0}=0$.

An asymptotic solution of (7) is sought as the series

$$v\sim\sum_{s,l,k=0}^{\infty}\varepsilon^{s}\omega^{-k}(\varepsilon\sqrt{\omega})^{l}\,v_{slk}(\eta,t),$$

where the $v_{slk}(\eta,t)$ are independent of ω and ε. To define the v_{slk}, we use a recurrent chain of problems of the form

$$\widehat{L}_0 v_{slk}=f_{slk}(\eta,t),\quad v_{slk}|_{\eta=0}=\varphi(t)\,\delta_{s0}\delta_{l0}\delta_{k0},$$
$$v_{slk}|_{t=0}=\left.\frac{\partial v_{slk}}{\partial t}\right|_{t=0}=0,$$

where $f_{slk}(\eta,t)$ is a function defined by specifying $v_{\overline{slk}}$ $(\overline{s}+\overline{l}+\overline{k}<s+l+k)$, $f_{000}=0$. Thus, v_{000} satisfies (7) for $\Delta=0$.

By using the standard technique of energy estimates (see, for example, [76]), we get

$$\max_{t\in[0,T]}\left\|u(x,t)-v_{000}(\sqrt{\omega}\,x,t)\right\|^2_{L_2((0,+\infty))}=O(\varepsilon\sqrt{\omega}+\omega^{-1}).$$

Thus, $u(x,t)\approx v_{000}(\sqrt{\omega x},t)$. If $|v(0,t)|\leqslant Q\leqslant\infty$, then v_{000} decreases exponentially in η: $|v(\eta,t)|<c_1Qe^{-c_2\eta}$. This follows from Green's function associated with the given boundary value problem. Hence, for $\eta\geqslant\eta_0$, the solution $v_{000}(\eta,t)$ is negligible; therefore $u(x,t)$ is also negligible for $x\geqslant\eta_0/\sqrt{\omega}$ $(\eta_0>0)$.

By comparison, when the wave travels through any of the homogeneous media with the characteristics (R^1, S^1, A^1) or $(\omega^{-1}R^2, \omega^{-1}S^2, \omega^{-1}A^2)$ its amplitude decreases essentially only at the distance of order 1. Thus, the periodic layered medium under consideration is highly absorbing.

§4 High-Modulus 3D Composite Materials

The composites reinforced with unidirectional fibers, which were discussed in §2, had a low rigidity in the directions orthogonal to the fibers. Multi-dimensional arrangements of reinforcement have been used to improve rigidity in these directions (see [52]).

By isolating the unity periodicity cube and by identifying its opposite faces, we obtain the torus Q (see the Supplement); in this case, differently oriented fibers occupy the subdomains $G_{1,1}, \ldots, g_{1,M} \subset Q$ (M is the number of the reinforcement directions). Thus, for 3D-composites, the formulation of the problem differs from that in §2 by the absence of connectedness of the set G_1 : $G_1 = \bigcup_{l=1}^{M} G_{1,l}$, $\overline{G}_{1,l} \cap \overline{G}_{1,j} = \emptyset$ for $l \neq j$. In particular, these conditions are satisfied by the domain shown in Fig. 21. Note that the geometry of the domains $G_{1,l}$ can be quite complicated (see [52]).

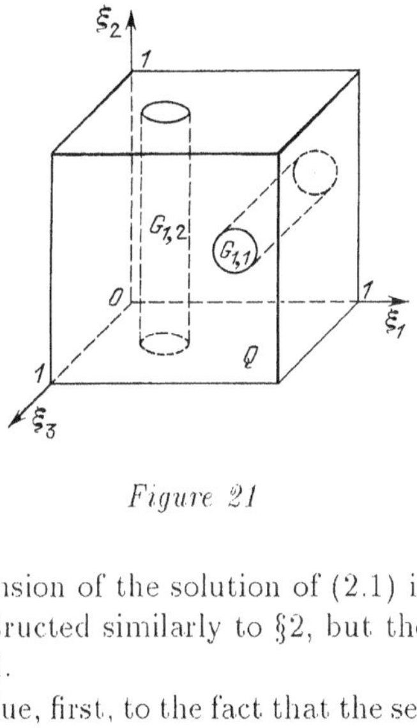

Figure 21

The asymptotic expansion of the solution of (2.1) in the case of the many-component set G_1 is constructed similarly to §2, but the algorithm for the construction has to be altered.

The alternations are due, first, to the fact that the set G_1 is many-connected, and second, to the fact that there is no requirement (imposed in §2) that only constants can be solutions of the homogeneous problem $L_1 N = 0$, $\xi \in G_1$, $\partial N/\partial \nu^1 = 0$, and $\xi \in \partial G_1$. The absence of this requirement means that we now can consider the system of equations of elasticity theory. Thus, problem (2.7), (2.8) is solved independently in each of the components of G_1, and the solution is defined up to an arbitrary solution of the homogeneous problem; the degree of ambiguity increasing due to the greater number of arbitrary constants. On the other hand, the solvability conditions (orthogonality of the right side of (2.7), (2.8) to nonzero solutions of the homogeneous problem) are written for each of the components of $G_{1,j}$, and this will produce the required number of equations.

The need to satisfy the solvability conditions for the problems (2.7), (2.8) in each of the components of $G_{1,j}$ causes the order in ω of the solution N^p_{qi} and, consequently, of the constants h^p_{qi} to increase (with the growth of $p+q+|i|$).

Therefore, explicit calculations of the zeroth approximation of the averaged operator become possible only under the additional restriction $\varepsilon\sqrt{\omega} \ll 1$. This condition is often met by actual materials.

Consider the system of equations

$$L_{\omega,\varepsilon}u \equiv R^{(\omega)}\left(\frac{x}{\varepsilon}\right)\frac{\partial^2 u}{\partial t^2} + S^{(\omega)}\left(\frac{x}{\varepsilon}\right)\frac{\partial u}{\partial t} - \\ - \frac{\partial}{\partial x_r}\left(A^{(\omega)}_{rm}\left(\frac{x}{\varepsilon}\right)\frac{\partial u}{\partial x_m}\right) = \omega f(x,t),\ x \in \mathbf{R}^s,\ t > 0, \quad (1)$$

with the contact conditions (2.2). The introduction of the factor ω on the right side is due to normalization of the solution. Here $R^{(\omega)}(\xi)$, $S^{(\omega)}(\xi)$, and $A^{(\omega)}_{rm}(\xi)$ are 1-periodic $n\times n$ matrices:

$$\{R^{(\omega)},\ S^{(\omega)},\ A^{(\omega)}_{rm}\} = \begin{cases} \{\omega R^1(\xi),\ \omega S^1(\xi),\ \omega A^1_{rm}(\xi)\}, & \xi \in G_1, \\ \{R^2(\xi),\ S^2(\xi),\ A^2_{rm}(\xi)\}, & \xi \in G_2 = Q\setminus \overline{G}_1. \end{cases}$$

The manifold $\partial\widetilde{G}_1, C^\infty$, and the coefficients $R^j(\xi)$, $S^j(\xi)$, and $A^j_{rm}(\xi)$ are assumed to be smooth everywhere on $\mathbf{R}^s\backslash\partial\widetilde{G}_1$ up to $\partial\widetilde{G}_1$, $f\in C^\infty$, where $\widetilde{G}_1$ is a periodic continuation of C_1 onto $\mathbf{R}^s$, and besides, $A^j_{rm} = (A^j_{mr})^T$.

The construction of the f.a.s. $u^{(\infty)}$, as in §1–3, is done in two stages. We seek the matrices N^p_{qi} in the form

$$N^p_{qi}(\xi) \sim \begin{cases} \sum\limits_{k=-\mathrm{ent}((p+q+l)/2)}^{\infty} \omega^{-k} N^{p1}_{qi,k}(\xi), & \xi \in G_1, \\ \sum\limits_{k=-\mathrm{ent}((p+q+l)/2)+1}^{\infty} \omega^{-k+1} N^{p2}_{qi,k}(\xi), & \xi\in G_2, \end{cases} \quad (2)$$

where the $N^{pj}_{qi,k}$ are 1-periodic and $\mathrm{ent}(a)$ is the integer (entire) part of a. Suppose asymptotic solutions of the form (2) were constructed for the matrices N^p_{qi}, with $p+q+|i| \leqslant J_0 - 1$. Then for $p+q+i \leqslant J_0$, we have

$$T^p_{qi}(\xi) \sim \begin{cases} \sum\limits_{k=-\mathrm{ent}((p+q+l-1)/2)}^{\infty} \omega^{-k+1} T^{p1}_{qi,k}(\xi), & \xi\in G_1, \\ \sum\limits_{k=-\mathrm{ent}((p+q+l-1)/2)+1}^{\infty} \omega^{-k+1} T^{p2}_{qi,k}(\xi), & \xi \in G_2, \end{cases}$$

$$h^p_{qi} = \sum_{k=-\mathrm{ent}((p+q+l-1)/2)}^{\infty} \omega^{-k+1} h^p_{qi,k}; \quad (3)$$

here

$$T^{p\sigma}_{qi_1\dots i_l}(\xi)=$$

$$=-\left(A^{\sigma}_{i_1 j}\frac{\partial N^{p\sigma}_{qi_2\dots i_l,k}}{\partial\xi_j}+\frac{\partial}{\partial\xi_j}\left(A^{\sigma}_{ji_1}N^{p\sigma}_{qi_2\dots i_l,k}\right)+A^{\sigma}_{i_1 i_2}N^{p\sigma}_{qi_3\dots i_l,k}\right)+$$

$$+R^{\sigma}N^{p\sigma}_{q-2,i_1\dots i_l,k}+S^{\sigma}N^{p-1,\sigma}_{q-1,i_1\dots i_l,k}$$

(no summation over σ),

$$h^{p}_{qi_1\dots i_l,k}=\sum_{\sigma=1}^{2}\int_{G_\sigma}\left(-A^{\sigma}_{i_1 j}\frac{\partial N^{p\sigma}_{qi_2\dots i_l,k}}{\partial\xi_j}-A^{\sigma}_{i_1 i_2}N^{p\sigma}_{qi_3\dots i_l,k}+\right.$$

$$\left.+R^{\sigma}N^{p\sigma}_{q-2,i_1\dots i_l,k}+S^{\sigma}N^{p-1,\sigma}_{q-1,i_1\dots i_l,k}\right)d\xi,\qquad(4)$$

the matrices N with the index k outside the limits of the summation in (2) are assumed to be zero. The matrices $N^{p2}_{qi,k}$ and $N^{p1}_{qi,k}$ are solutions of problems (2.5), (2.6) and (2.7), (2.8), respectively.

With $k=-\operatorname{ent}((p+q+l-1/2)$, $p+q+l$ being odd, we get from (4)

$$h^{p}_{qi_1\dots i_l,k}=-\int_{G_1}A^{1}_{i_1 j}\,\partial N^{p1}_{qi_2\dots i_l,k}/\partial\xi_j d\xi.$$

Let us prove that the last integral is equal to zero. The function $N^{p1}_{qi_2\dots i_l,k}$ satisfies equation (2.7) and condition (2.8). Since $p+q+l$ are odd, we have

$$T^{p1}_{qi_2\dots i_l,k}\equiv 0,\quad h^{p}_{qi_2\dots i_l,k}=0,$$

$$N^{p1}_{qi_3\dots i_l,k}=N^{p2}_{qi_3\dots i_l,k}=N^{p2}_{qi_2\dots i_l,k}\equiv 0,$$

$$k=-\operatorname{ent}((p+q+l-1)/2).$$

Consequently, $N=N^{p1}_{qi_2\dots i_l,k}$ is a solution of the problem

$$L_1N=0,\quad \xi\in G_1,\quad \partial N/\partial\nu^1=0,\quad \xi\in\partial G_1.\qquad(5)$$

We assume that any solution of the homogenous problem (5) satisfies the condition $A^1_{i_1 j}\partial N/\partial\xi_j\equiv 0$ ($\xi\in G_1$). The validity of this condition can be verified directly for the case where L_1 is a scalar elliptic operator or the operator of elasticity theory (then $N=\text{const}$, or N is a rigid-body displacement).

Then, for the odd $p+q+l$, the constant $h^{p}_{qi,-\operatorname{ent}((p+q+l-1)/2)}$ is zero, and so after the passage from $p+q+l$ to $p+q+l$, the order of h^p_{qi} in ω does not increase. But when the passage is from $p+q+l$ to $p+q+l+1$, the order of h^p_{qi} in ω increases by 1. This enables us to rewrite (3) as

$$h^p_{qi} \sim \sum_{k=-\mathrm{ent}((p+q+l)/2)+1}^{\infty} \omega^{-k+1} h^p_{qi,k}. \tag{6}$$

Now we give an algorithm for constructing $N^{pj}_{qi,k}$ with $p+q+l = J_0$. Let each of the problems

$$L_1\varphi = 0, \quad \xi \in G_{1,j},$$
$$\partial\varphi/\partial\nu^1 = 0, \quad \xi \in \partial G_{1,j},$$

have its own space of nonzero vector solutions with the basis $\overline{\Phi}^{-j} = (\varphi^j_1, \dots, \varphi^j_{\overline{r}_j})$ (all the vector-constants are included into this basis).

We introduce the auxiliary problems (no summation over j, r)

$$\begin{aligned} L_2\psi^j_r &= 0, \quad \xi \in G_2, \\ \psi^j_r &= \delta_{\tau j}\varphi^j_r, \quad \xi \in \partial G_{1,\tau}, \qquad \tau = 1, \dots, M. \end{aligned} \tag{7}$$

with Ψ^j_r being a 1-periodic vector-function, and let $\overline{\Psi}^j = (\Psi^j_1, \dots, \Psi^j_{r_j})$ be an $n \times \overline{r}_j$ matrix composed of the solutions of this problem.

We describe the step of the recursive construction of the matrices $N^{pj}_{qi,k}$. Suppose all the matrices $N^p_{qi,k}$ were found for $k \leqslant K_0 - 1$, with the matrix N^{p1}_{qi,K_0-1} having been defined on each of the domains $G_{1,j}$ up to an arbitrary nonzero solution of problem (5), that is,

$$N^{p1}_{qi,K_0-1} = \overline{N}^{p1j}_{qi,K_0-1} + \overline{\Phi}^j\overline{C}_j \tag{8}$$

for $\xi \in G_{1,j}$, where $n\overline{N}^{p1j}_{qi,K_0-1}$ is orthogonal on $G_{1,j}$ to all the solutions of (5), and $\overline{C}_j$ is an $\overline{r}_j \times n$ matrix of indeterminate coefficients of φ^j_r.

Let $\overline{N}^{p2}_{qi,K_0}$ be a solution of equation (2.5) with the condition $\overline{N}^{p2}_{qi,K_0} = \overline{N}^{p1j}_{qi,K_0-1}$, $\xi \in G_{1,j}$. Then

$$N^{p1}_{qi,K_0} = \overline{N}^{p2}_{qi,K_0} + \sum_{j=1}^{M} \overline{\Psi}^j\overline{C}_j. \tag{9}$$

Let problems (2.7), (2.8) be solvable according to Fredholm, i.e., the equalities

$$\begin{aligned} \sum_{\tau=1}^{M} \int_{\partial G_{1,j}} \overline{\Phi}^{jT} \frac{\partial\overline{\Psi}^\tau}{\partial\nu^2}\overline{C}_\tau ds &= \int_{G_{1,j}} \overline{\Phi}^{jT}\left(T^{p1}_{qi_1\dots i_l,k} - h^p_{qi_1\dots i_l,k}\right) d\xi - \\ &- \int_{\partial G_{1,j}} \overline{\Phi}^{jT}\left(\frac{\partial N^{p2}_{qi_1\dots i_l,k}}{\partial\nu^2} - \sum_{k=1}^{s}\left(A^1_{ki_1}N^{p1}_{qi_2\dots i_l,k}\big|_{\partial G_{1,j}} - \right.\right. \\ &\left.\left. - A^2_{ki_1}N^{p2}_{qi_2\dots i_l,k}\big|_{\partial G_{1,j}}\right) n_k\right) ds \end{aligned} \tag{10}$$

are necessary and sufficient conditions for their solutions to exist.

We find the matrices $\overline{C}_\tau$ from the system of algebraic equations (10). We define the matrices N^{p1}_{qi,K_0-1} and N^{p2}_{qi,K_0} completely from (8), (9) and N^{p1}_{qi,K_0} up to an arbitrary matrix-valued eigenfunction by solving problem (2.7), (2.8). This problem splits into M independent problems for the domains $G_{1,j}$ ($j = 1,\dots, M$). The recursion step is completed. The basis of the recursion: $K_0 = -\operatorname{ent}((p+q+l)/2$. When $p+q+l$ is even, we have

$$T^{p1}_{qi,K_0} \equiv 0, \quad h^{p}_{qi,K_0} = 0,$$
$$N^{p1}_{qi_2\dots i_l,K_0} = N^{p2}_{qi_2\dots i_l,K_0} = N^{p2}_{qi,K_0} \equiv 0.$$

Thus, for $p+q+l$ even, N^{p1}_{qi,K_0} is some solution of (5). When $p+q+l$ is odd,

$$T^{p1}_{qi,K_0} = -A^1_{i_1 j}\frac{\partial N^{p1}_{qi_2\dots i_l,K_0}}{\partial \xi_j} - \frac{\partial}{\partial \xi_j}\left(A^1_{ji_1} N^{p1}_{qi_2\dots i_l,K_0}\right). \tag{11}$$

The first term in the right side vanishes, because

$$N^{p1}_{qi_2\dots i_l,-\operatorname{ent}((p+q+l)/2)} = N^{p1}_{qi_2\dots i_l,-\operatorname{ent}((p+q+l-1)/2)}$$

is a solution of (5); hence N^{p1}_{qi,K_0} can be found from the conditions

$$\begin{aligned} L_1 N^{p1}_{qi,K_0} &= -\frac{\partial}{\partial \xi_j}\left(A^1_{ji_1} N^{p1}_{qi_2\dots i_l,-\operatorname{ent}((p+q+l-1)/2)}\right), & \xi \in G_1,\\ \frac{\partial}{\partial \nu_1} N^{p1}_{qi,K_0} &= -A^1_{ji_1} N^{p1}_{qi_2\dots i_l,-\operatorname{ent}((p+q+l-1)/2)} n_j, & \xi \in \partial G_1. \end{aligned} \tag{12}$$

Let us verify the solvability conditions for (12):

$$-\int_{G_{1,\rho}} (\overline{\Phi}^\rho)^T \frac{\partial}{\partial \xi_j}\left(A^1_{ji_1} N^{p1}_{qi_2\dots i_l,k}\right) d\xi + \int_{\partial G_{1,\rho}} (\overline{\Phi}^\rho)^T \left(A^1_{ji_1} N^{p1}_{qi_2\dots i_l,k} n_j\right) ds =$$
$$= \int_{G_{1,\rho}} \frac{\partial}{\partial \xi_j} (\overline{\Phi}^\rho)^T A^1_{ji_1} N^{p1}_{qi_2\dots i_l,k} d\xi = \int_{G_{1,\rho}} \left(\left(N^{p1}_{qi_2\dots i_l,k}\right)^T A^1_{i_1 j}\frac{\partial \overline{\Phi}^\rho}{\partial \xi_j}\right)^T d\xi = 0,$$
$$k = -\operatorname{ent}((p+q+l-1)/2).$$

The following assumptions were made when the f.a.s. of the cell problem (4.5.21), (4.5.18) was constructed: The existence of a unique 1-periodic solution of problem (2.5), (2.6), the solvability of problem (2.7), (2.8) according to Fredholm, the existence of a derivative with respect to the conormal from $L_2(\partial G_1)$ for the solutions of these problems, and the solvability of the algebraic system

(10). These conditions hold for elliptic equations and for the system of equations of elasticity theory with piecewise smooth coefficients (see §5).

By substituting expansion (6) for the coefficients of the averaged system in (4.5.19), we get

$$\sum_{\substack{p,q,l\geqslant 0\\ p+q+l\geqslant 2}} \sum_{|i|=l} \sum_{m=0}^{\infty} (\varepsilon^2\omega)^{(p+q+l)/2-1}\omega^{1-m-\delta((p+q+l)/2)} \times$$
$$\times h^{p}_{qi,m-\mathrm{ent}((p+q+l)/2)+1} D^{qi} v - \omega f(x,\,t) \sim 0, \quad \delta(a) = a - \mathrm{ent}\,(a). \quad (13)$$

An asymptotic solution of (13) is sought in the form

$$v \sim \sum_{j=0}^{\infty} \sum_{k=0}^{\infty} (\varepsilon^2\omega)^{j/2}\omega^{-k/2} v_{j,k}(x,\,t),$$

with $v_{j,k}$ being independent of ω and ε. To define $V_{j,k}$, we use a recurrent chain of equations of the form

$$\frac{1}{\omega}\widehat{L}_0 v_{j,k} = f_{j,k}(x,\,t). \quad (14)$$

Here $\hat{L}_0$ is the principal part of the operator in (13) acting on v, $f_{j,k}(x,t)$ is a function that depends on $v_{q,r}(q+r<j+k)$, $f_{0,0}=f$,

$$\widehat{L}_0 = \widehat{R}^0 \frac{\partial^2}{\partial t^2} + \widehat{S}^0 \frac{\partial}{\partial t} - \frac{\partial}{\partial x_r}\left(\widehat{A}^0_{rm}\frac{\partial}{\partial x_m}\right),$$
$$\widehat{R}^0 = \omega \sum_{j=1}^{M} \int_{G_{1,j}} R^1(\xi)\,d\xi, \quad \widehat{S}^0 = \omega\sum_{j=1}^{M}\int_{G_{1,j}} S^1(\xi)\,d\xi, \quad (15)$$
$$\widehat{A}^0_{rm} = \omega\sum_{j=1}^{M}\int_{G_{1,j}} A^1_{r\tau}(\xi)\frac{\partial}{\partial\xi_\tau}\left(N^{01}_{0m,0} + \xi_m E\right)d\xi,$$

$N^{01}_{0m,0}$ is a solution of problem (2.10), (2.11). If the chain of problems (14) is successively solvable, then the f.a.s. of (1), (2.2) is constructed completely.

Remark 1: The matrices $N^{p1}_{qi,k}$ are defined up to an arbitrary constant summand $C^p_{qi,k}$. We shall select this term each time (for $p+q+|i|\leqslant 1$) so that

$$\int_{G_1} N^{p1}_{qi,k}d\xi = -\int_{G_2} N^{p2}_{qi,k+1}d\xi.$$

Then $\langle N^p_{qi}\rangle = O(\omega^{-d})$, with an arbitrary d; as in §4.2, it can be proved that under the condition $\langle v_{j,k}\rangle_T = 0$ for any $k>0$, the truncated sum of the series $u^{(\infty)}$ (see §5) has the mean over the cube $(0,T)^s$ of order $O(\varepsilon^h)$.

Remark 2: Consider a composite reinforced with unidirectional cylindrical fibers parallel to the x_3-axis and forming an ε-periodic structure in x_1 and x_2 (see §2) such that the torus Q intersects only one "fiber."

With the A^1_{rm} constant, problem (2.10), (2.11) can be solved analytically: for $i_1 \neq 3$ we have $N^{01}_{0i_1,0} = -\xi_{i_1} E$; for $i_1 = 3$ the matrix-valued function $N^{01}_{03,0}$ is sought in the form $C_1\xi_1 + C_2\xi_2$, and C_1 and C_2 are selected so that condition (2.11) is satisfied. Thus, they are found from the system of equations

$$\sum_{j=1}^{2} A^1_{1j}C_j = -A^1_{13}, \quad \sum_{j=1}^{2} A^1_{2j}C_j = -A^1_{23}. \tag{16}$$

It can be proved that in the cases when L_1 is a scalar elliptic operator ($n = 1$) or the operator of linear elasticity theory ($n = 2,3$) (see [155]), the algebraic system (16) has a solution C_1, C_2. By substituting these $N^{01}_{0i_1,0}$ into (15) for the principal part $\hat{A}_{rm}$ of order ω, we get the following relation

$$\widehat{A}^0_{rm} = \omega\delta_{3r}\delta_{m3} \int_{G_1} \left(\sum_{j=1}^{2} A^1_{3j}C_j + A^1_{33} \right) d\xi =$$

$$= \omega\delta_{3r}\delta_{m3} \operatorname{mes} G_1 \left(A^1_{31}C_1 + A^1_{32}C_2 + A^1_{33} \right), \tag{17}$$

where $\operatorname{mes} G_1$ is the volume fraction of the fiber. If the fiber is made of an isotropic material and A^1_{rm} is the elasticity tensor of the fiber, it can be verified that

$$A^1_{31}C_1 + A^1_{32}C_2 + A^1_{33} = \begin{bmatrix} 0 & 0 & 0 \\ 0 & 0 & 0 \\ 0 & 0 & E_B \end{bmatrix}, \tag{18}$$

where E_B is Young's modulus of the fiber. As in §2, it turned out that only one component of the effective elasticity tensor a^{33}_{33} is of order ω and can be calculated explicitly. But in order to define the principal part of the other components, problems (2.5)–(2.8) must be solved numerically.

§5 The Splitting Principle for the Averaged Operator for 3D High-Modulus Composites

This section investigates the properties of the principal part of the averaged operator $\hat{L}_0$, which was found in §4. It can be proved that the assumptions about the properties of the original operators L_j made in constructing the f.a.s. of the cell problem hold with respect to the scalar elliptic operators and the systems of equations of elasticity theory. An explicit formula is given for isotropic fibers, and the question of the minimum number of the reinforcement directions required to get effective rigidity of order ω with respect to arbitrary deformations.

1. Consider problem (4.1), (2.2) with the coefficients (and the domains $G_{1,j}$) satisfying the conditions imposed in the preceding section. By L_j we denote the operators $\partial(A^j_{rm}\partial/\partial\xi_m)/\partial\xi_r$, with L_j being independent of ω and ε ($j = 1,2$). We shall assume that the L_j satisfy all the conditions of §4 required to realize the

procedure of constructing the asymptotic solution $u^{(\infty)}$ of problem (4.1). The averaged equation (4.13) was obtained, and the leading term of order ω of the operator $\hat{L}$ (operator $\hat{L}_0$ from (4.15)) was found in §4. We set

$$\widehat{R}^j = \omega \int\limits_{G_{1,j}} R^1(\xi)\,d\xi, \quad \widehat{S}^j = \omega \int\limits_{G_{1,j}} S^1(\xi)\,d\xi,$$

$$\widehat{A}^j_{rm} = \omega \int\limits_{G_{1,j}} A^1_{r\tau}(\xi)\frac{\partial}{\partial \xi_\tau}(N^{01}_{0m,0} + \xi_m E)\,d\xi,$$

$$\widehat{L}^j_0 = \widehat{R}^j \frac{\partial^2}{\partial t^2} + \widehat{S}^j \frac{\partial}{\partial t} - \widehat{A}^j_{rm}\frac{\partial^2}{\partial x_r \partial x_m}.$$

Each of the operators $\hat{L}^j_0$ is the leading term of order ω of an appropriate averaged operator for problem (4.1), (2.2) if $G_1 = G_{1,j}$.

The above can be formulated as the following statement, which naturally can be called the *splitting principle for the averaged operator*: The leading term of order ω of the averaged operator for problem (4.1), (2.2) with $G_1 = \bigcup_{j=1}^{M} G_{1,j}$ is a sum of the leading terms of the averaged operators of this problem for $G_1 = G_{1,j}$, that is, $\hat{L}_0 = \sum_{j=1}^{M} \hat{L}^j_0$. In the case where $G_{1,j}$ is a domain occupied by the fiber of the jth direction, the splitting principle means that to calculate the averaged operator of a spatially reinforced medium requires adding up M averaged operators of the media with unidirectional arrangements of reinforcement, with the fibers being oriented in the jth direction ($j = 1, \dots, M$).

2. Justification of the splitting principle. Let $G_{1,j}$ for any $j = 1, \dots, M$ be a cylinder whose cross-section is the two-dimensional domain β_j with smooth boundary, and $\overline{G}_{1,l} \cap \overline{G}_{1,j} = \emptyset$ when $l \neq j$.

We consider three cases: a) hyperbolic ($R^j > 0$, $S^j = 0$, $j = 1, \dots, M$); b) parabolic $R^j > 0$, $S^j > 0$, $j = 1, \dots, M$); and c) elliptic $R^j = S^j = 0$, $j = 1, \dots, M$). The initial values for case a) are

$$u\,|_{t=0} = \frac{\partial u}{\partial t}\bigg|_{t=0} = 0; \tag{1a}$$

for case b), the value is

$$u\,|_{t=0} = 0 \tag{1b}$$

and it is assumed that $f(x,t)$ with $t = 0$ satisfies the condition $\partial^l f/\partial t^l|_{t=0} = 0$ for any $l = 0, 1, \dots$; and in case c), f does not depend on t.

We set

$$u^{(K)} = \sum_{\substack{p,q,l \geqslant 0 \\ p+q+l \leqslant 2K+2}} \varepsilon^{p+q+l} \sum_{|i|=l} N_{qi}^{p(K)} D^{qi} v^{(K)},$$

where

$$N_{qi}^{p(K)} = \begin{cases} \displaystyle\sum_{k=-\mathrm{ent}((p+q+l)/2)}^{K+1} \omega^{-k} N_{qi,k}^{p1}(\xi), & \xi \in G_1, \\ \displaystyle\sum_{k=-\mathrm{ent}((p+q+l)/2)+1}^{K+1} \omega^{-k+1} N_{qi,k}^{p2}(\xi), & \xi \in G_2, \end{cases}$$

$$v^{(K)} = \sum_{j,r=0}^{2K} (\varepsilon^2\omega)^{j/2}\omega^{-r/2} v_{jr}, \; j-r \quad \text{even}.$$

Theorem 1. *Let the scalar case ($n = 1$) be considered, and let the conditions*

$$\begin{aligned} \varkappa_1 \sum_{r=1}^{3} \eta_r^2 &\leqslant A_{rm}^j(\xi)\,\eta_r\eta_m, \\ \omega\varkappa_2 \sum_{r=1}^{3} \eta_r^2 &\leqslant \widehat{A}_{rm}^0 \eta_r\eta_m \end{aligned} \tag{2}$$

hold for any $\xi \in \mathbf{R}^3$, η_1, η_2, $\eta_3 \in \mathbf{R}$, $A_{rm}^j = A_{mr}^j$, and κ_1, κ_2 being positive constants independent of ξ, η_q, ε, and ω. Let $\omega^{-1} = O(\varepsilon^\alpha)$, $\varepsilon = O(\omega^{-1/2-\beta})$ $(\alpha, \beta > 0)$, and let a solution of problem (4.1), (1) be sought in the class of functions T-periodic in x_q ($T \sim 1$, T is a multiple of ε), and let f have the same periodicity in space variables, while in case c) $\langle f(x)\rangle_T = 0$, $\langle u(x)\rangle_T = 0$. Then the procedure of constructing an asymptotic solution can be carried out for any integer $K > 0$. The following estimates are valid,

$$\|u^{(K)} - u\|_W = O(\omega(\varepsilon^2\omega)^K + \omega^{-K}), \tag{3}$$

where in case a) $W = W_2^1(Q_T \times (0, t_1))$, in case b) $W = W_2^{1,0}(Q_T \times (0, t_1))$, and in case c) $W = W_2^1(Q_T)$, $Q_T = \{x \in \mathbf{R}^3 | 0 < x_l < T,\ l = 1, 2, 3\}$.

The splitting principle is valid for the principal part of the averaged operator $\hat{L}_0$, and for the solution $v_{0,0}$ of the principal part of the averaged equation

$$\widehat{L}_0 v_{0,0} = \omega f(x, t) \tag{4}$$

with the initial values

$$v_{0,0}|_{t=0} = \left.\frac{\partial v_{0,0}}{\partial t}\right|_{t=0} = 0 \;(\textit{ in case a})),$$

$$v_{0,0}|_{t=0} \;(\textit{ in case b}))$$

the estimate

$$\| v_{0,0} - u \|_{L_2(Q_T \times (0,t_1))} = O(\varepsilon^2\omega + \omega^{-1} + \varepsilon). \tag{5}$$

holds.

THEOREM 2. *Let $n = 1$, let G be a bounded domain in $\mathbf{R}^3$, $\partial G \in C^\infty$, $R^j = S^j = 0$ (case c)), $f(x) \in C^\infty(\overline{G})$, $\omega^{-1} = O(\varepsilon^\alpha)$, $\varepsilon = O(\omega^{-1/2-\beta})$ $(\beta > 0)$, let $u(x)$ be a solution of equation (1) with the boundary value*

$$u|_{\partial G} = \varphi(x), \quad \varphi(x) = C^\infty(\partial G),$$

and let $v_{0,0}$ be a solution of equation (4) with the boundary value $v_{0,0}|_{\partial G} = \varphi(x)$. Then under the assumptions of Theorem 1 on coefficients A^j_{rm}, $\hat{A}^0_{rm}$:

$$\max_{\overline{G}} |u - v_{0,0}| = O(\varepsilon^2\omega + \omega^{-1} + \varepsilon). \tag{6}$$

Next, consider either the case a) ($R^j > 0$, $S^j = 0$) or the case c) (stationary).

THEOREM 3. *Let 3×3 matrices $A^j_{rm} = \|a^{j,kl}_{rm}\|$ and $\hat{A}^0_{rm} = \|\hat{a}^{0kl}_{rm}\|$ satisfy the conditions*

$$\begin{aligned} \varkappa_1 \sum_{r,k=1}^{3} \eta^2_{rk} &\leqslant a^{j,kl}_{rm}(\xi)\, \eta_{rk}\eta_{ml}, \\ \omega\varkappa_2 \sum_{r,k=1}^{3} \eta^2_{rk} &\leqslant \hat{a}^{0kl}_{rm}\, \eta_{rk}\eta_{ml} \end{aligned} \tag{2'}$$

for any $\eta_{rl} = \mathbf{R}(\eta_{rk} = \eta_{kr})$ and for any $\xi \in \mathbf{R}^3$ (κ_1 and κ_2 are independent of ξ, η_{rk}, ω, and ε). Assume also that the symmetry conditions

$$a^{j,kl}_{rm}(\xi) = a^{j,rl}_{km}(\xi) = a^{j,rm}_{hl}(\xi) = a^{j,lh}_{mr}(\xi).$$

are satisfied. Let $\omega^{-1} = O(\varepsilon^\alpha)$, $\varepsilon = O(\omega^{-1/2-\beta})$ $(\alpha, \beta > 0)$, and let a solution of problem (4.1), (1) be sought in the class of vector-functions T-periodic in $x_q(T \sim 1)$, with f having the same periodicity in space variables, while in case c) $\langle f \rangle = 0$ and $\langle u \rangle = 0$. Then the procedure of constructing the asymptotic solution $u^{(K)}$ is realizable, and for any integer $K > 0$, the estimate

$$\left\| u^{(K)} - u \right\|_W = O\left(\omega\,(\varepsilon^2\omega)^K + \omega^{-K}\right)$$

holds.

The splitting principle holds for the principal part of the averaged operator $\hat{L}_0$, whereas for the solution $v_{0,0}$ of the principal part of the averaged equation (4) with the initial values

$$v_{0,0}\,|_{t=0} = \left.\frac{\partial v_{0,0}}{\partial t}\right|_{t=0} = 0 \quad (\ \textit{in case a)})$$

estimate (5) is valid.

REMARK 1: The approximation $\tilde{u}^{(1)} = v_{00} + H^0_{0i_1,0}\partial v_{00}/\partial x_{i_1}$ is of interest for the applications, where $N^0_{0i_1,0}(\xi)$ is a function which coincides with $N^{0j}_{0i_1,0}(\xi)$ for $\xi \in G_j$ $(j = 1, 2)$. From Theorems 1 and 3

$$\left\| \frac{1}{\omega} A^{(\omega)}_{rm}\left(\frac{x}{\varepsilon}\right)\frac{\partial}{\partial x_m}\left(\widetilde{u}^{(1)} - u\right)\right\|_{L_2(\widetilde{Q}_T)} = O\left(\varepsilon^2\omega + \omega^{-1} + \varepsilon\right),$$

where $\widetilde{Q}_T = Q_T$ in case c), and $Q_T = Q_T \times (0, t_1)$ in cases a) and b). The normalization is natural here, since stresses in the case under consideration are of order ω.

The proof of Theorems 1–3 is constructed according to the following pattern. The existence of solutions of problem (2.5), (2.6) and problem (2.7), (2.8) under condition (4.10) is proved by using the standard technique (see [76, 155]). Relations (3) and (5) are derived by means of a priori estimates for hyperbolic, parabolic, and elliptic equations (see [51, 76, 105]). Estimate (6) is proved on the basis of the generalized maximum principle. The proof of the solvability of the system of equations (4.10) was constructed by Kosarev (see also [68]). It can be assumed without loss of generality that the first n columns (n is the dimension of u) of each matrix $\overline{\Phi}^j$ are constants forming the unit matrix E, and for the remaining $\overline{r}_j - n$ columns (if $\overline{\Phi}^j \neq E$), there exists a point $\xi^j \in G_{1,j}$ such that they vanish there (the latter condition will be satisfied if the vector equal to the value of each remaining column at the point ξ^j is subtracted from this column). Then the first n scalar equations of system (4.10) (produced by substituting E for $\overline{\Phi}^1$, $j = 1$ in (4.10)) are the consequence of the other equations and of relation (4.4). Thus, assuming the first n rows of the matrix $\overline{C}_1$ to be zero, we shall reduce the system (4.10) to the system

$$\sum_{\tau=1}^{M}\int\limits_{\partial G_{1,j}} \Phi^{jT}\frac{\partial\Psi^{\tau}}{\partial v^2}C_{\tau}ds = \int\limits_{G_{1,j}} \Phi^{jT}\left(T^{p1}_{qi_1\ldots i_l,h} - h^p_{qi_1\ldots i_l,h}\right)d\xi -$$

$$-\int\limits_{\partial G_{1,j}} \Phi^{jT}\left(\frac{\partial \overline{N}^{p2}_{qi_1\ldots i_l,h}}{\partial v^2} - \left(A^1_{ri_1}N^{p1}_{qi_2\ldots i_l,h}\,|_{\partial G_{1,j}} - A^2_{ri_1}N^{p2}_{qi_2\ldots i_l,h}\,|_{\partial G_{1,j}}\right)n_r\right)ds_{\nu}$$

$$j = 1, \ldots, M, \quad (7)$$

where the matrices Φ^1, Ψ^1, and C_1^T are obtained from $\overline{\Phi}^1$, $\overline{\Psi}^1$, and $\overline{C}_1^T$ by discarding the first n columns, $\Phi^j = \overline{\Phi}^j$, $\Psi^j = \overline{\Psi}^j$, and $C_j = \overline{C}_j$ with $j > 1$. When $\overline{\Phi}^1 = E$, the indices j and τ in system (7) run from 2 to M.

We denote the right side of system (7) by B and set

$$C=(C_1,\ldots,C_M)^T,\quad \Phi=(\Phi^1,\ldots,\Phi^M),$$

$$\Psi=(\Psi^1,\ldots,\Psi^M),\quad A^j_\tau=-\int\limits_{\partial G_{1,j}}\Phi^{jT}\frac{\partial\Psi^\tau}{\partial\nu^2}\,ds,$$

$$A=\begin{bmatrix}A^1_1 & \ldots & A^1_M\\ \cdot & \cdot\,\cdot\,\cdot & \cdot\\ A^M_1 & \ldots & A^M_M\end{bmatrix}=-\sum_{j=1}^{M}\int\limits_{\partial G_{1,j}}\Psi^T\frac{\partial\Psi}{\partial\nu^2}\,ds,$$

since $\Psi^j|_{\partial G_{1,r}}=\delta_{jr}\Phi^j|_{\partial G_{1,r}}$ (see (4.7)).

Then system (7) can be represented as $AC=B$, with the sizes of A, B, and C being

$$-AC=B,$$

We shall prove that the matrix A is symmetric and positive-definite. Indeed, for any $\bar{r}$-dimensional vectors α and β,

$$(A\alpha,\beta)=\left(-\sum_{j=1}^{M}\int\limits_{\partial G_{1,j}}\Psi^T\frac{\partial\Psi}{\partial\nu^2}\alpha\,ds,\ \beta\right)=$$

$$=-\sum_{j=1}^{M}\int\limits_{\partial G_{1,j}}(\Psi\beta)^T\frac{\partial(\Psi\alpha)}{\partial\nu^2}\,ds=-\sum_{j=1}^{M}\int\limits_{\partial G_{1,j}}\left(\frac{\partial(\Psi\alpha)}{\partial\nu^2},\ \Psi\beta\right)ds=$$

$$=\int\limits_{G_2}\left(A^2_{rm}\frac{\partial(\Psi\alpha)}{\partial\xi_m},\ \frac{\partial(\Psi\beta)}{\partial\xi_r}\right)d\xi=$$

$$=\int\limits_{G_2}\left(\frac{\partial(\Psi\alpha)}{\partial\xi_m},\ A^2_{mr}\frac{\partial(\Psi\beta)}{\partial\xi_r}\right)d\xi=(A\beta,\alpha)=(\alpha,A\beta),$$

$$(A\alpha,\alpha)=\int\limits_{G_2}\left(A^2_{rm}\frac{\partial(\Psi\alpha)}{\partial\xi_m},\ \frac{\partial(\Psi\alpha)}{\partial\xi_r}\right)d\xi=\int\limits_{Q}\left(A_{rm}\frac{\partial(\tilde{\Psi}\alpha)}{\partial\xi_m},\ \frac{\partial(\tilde{\Psi}\alpha)}{\partial\xi_r}\right)d\xi\geqslant 0;$$

here

$$A_{rm}=\begin{cases}A^1_{rm} & \text{for}\ \ \xi\in G_1,\\ A^2_{rm} & \text{for}\ \ \xi\in G_2,\end{cases}$$

$$\tilde{\Psi}^\tau=\begin{cases}\delta_{\tau j}\Phi^\tau & \text{for}\ \ \xi\in G_{1,j},\\ \Psi^\tau & \text{for}\ \ \xi\in G_2.\end{cases}$$

From the inequality $\kappa_1 \sum_{r=1}^{3} \eta_r^2 \leqslant A_{rm}^j(\xi)\eta_r\eta_m$ for a single equation and from Corn's inequality for the system of equations of elasticity theory, we get $(A\alpha, \alpha) \geqslant c\|\nabla_\xi(\widetilde{\Psi}\alpha)\|_{L_2(Q)}^2$.

Consequently, if $(A\alpha, \alpha) = 0$, then $\widetilde{\Psi}\alpha \approx$ const, and since $\widetilde{\Psi}(\xi^1) = 0$, then $\widetilde{\Psi}\alpha \equiv 0$. Because the column vectors of the matrix $\widetilde{\Psi}$ are linearly independent, $\alpha = 0$. The positive-definiteness of A and, consequently, the solvability of system (4.10) are thus proved.

3. The case of isotropic fibers. The splitting principle can help to answer the question on the number of reinforcement directions necessary and sufficient for the leading term of the averaged operator $\hat{L}_0$ to maintain the type of the original operator. As far as the system of equations of elasticity theory is concerned, this question corresponds to that of the number of directions, which would ensure high rigidity with small strains of any type.

Let $L_{\omega,\varepsilon}$ be the operator of the isotropic theory of elasticity, $n = s = 3$, let R^1 and A_{rm}^1 be constant in any domain $G_{1,j}$,

$$R^1 > 0, \quad S = 0, \quad A_{rm}^1 = \| a_{rm}^{lq} \|,$$

$$a_{rm}^{lq} = \frac{E_j}{2(1+\nu_j)}\left(\frac{2\nu_j}{1-2\nu_j}\delta_{rl}\delta_{mq} + \delta_{rm}\delta_{lq} + \delta_{rq}\delta_{ml}\right),$$

and let $\omega E_k > 0$ and $\nu_j > 0$ be Young's modulus and the Poisson ratio, respectively, in the domain $G_{1,j}$. Suppose, as before, that $G_{1,j}$ for any j is a cylinder whose cross-section is the two-dimensional domain β_j with a smooth boundary, and the generatrix has the direction cosines γ_1^j, γ_2^j, γ_3^j. Then, by using (4.17), (4.18), we can demonstrate that

$$\hat{L}_0^j = \omega \int\limits_{G_{1,j}} R^1 d\xi \frac{\partial^2}{\partial t^2} - \omega\theta_j \frac{\partial}{\partial x_r} \| \hat{b}_{rm}^{j,lq} \| \frac{\partial}{\partial x_m} \tag{8}$$

(no summation over j); here

$$\theta_j = E_j \operatorname{mes} G_{1,j}, \quad \hat{b}_{rm}^{j,lq} = \gamma_r^j\gamma_l^j\gamma_m^j\gamma_q^j. \tag{9}$$

THEOREM 4. *With $M \leqslant 5$, the operator $\hat{L}_0$ does not satisfy (2′) for any type of reinforcement. When $M = 6$, the directions of reinforcement can be chosen so that $\hat{L}_0$ satisfies (2′).*

PROOF: By the splitting principle,

$$\hat{A}_{rm}^0 = \omega \sum_{j=1}^{M} E_j \operatorname{mes} G_{1,j} \| \hat{b}_{rm}^{j,lq} \|.$$

(If all the fibers are parallel to one of the coordinate planes, this formula coincides with the one proposed in [223]; experimental studies of effective moduli of composites for some arrangements of high-modulus fiber reinforcement are described in [150].)

Consider the quadratic form

$$I = \left(\omega \sum_{j=1}^{M} E_j^{j} \operatorname{mes} G_{1,j} \hat{b}_{rm}^{j,lq}\right) \eta_{rl}\eta_{mq};$$

for $\eta_{rl} = \eta_{lr}$:

$$I = \sum_{j=1}^{M} \omega E_j \operatorname{mes} G_{1,j} \gamma_r^j \gamma_l^j \gamma_m^j \gamma_q^j \eta_{rl}\eta_{mq} = \sum_{j=1}^{M} \omega E_j \operatorname{mes} G_{1,j} \left(\sum_{r,l=1}^{3} \gamma_r^j \gamma_l^j \mu_{rl}\right)^2.$$

Consequently, $I = 0$ if and only if for $j = 1, 2, \ldots, M$ the equations

$$\sum_{r,l=1}^{3} \gamma_r^j \gamma_l^j \eta_{rl} = 0. \tag{10}$$

hold.

Relations (10) form a system of linear equations in the η_{rl}. Since $\eta_{rl} = \eta_{lr}$, then (10) can be written as

$$\Gamma\eta = 0,$$

where Γ is an $M \times 6$ matrix, $\eta = (\eta_{11}, \eta_{22}, \eta_{33}, \eta_{12}, \eta_{13}, \eta_{23})^T$, and the jth row of Γ is of the form $((\gamma_1^j)^2, (\gamma_2^j)^2, (\gamma_3^j)^2, 2\gamma_1^j\gamma_2^j, 2\gamma_1^j\gamma_3^j, 2\gamma_2^j\gamma_3^j)$.

If $\operatorname{rank}\Gamma < 6$, the system of relations (10) has a nonzero solution, and then the operator $\hat{\Lambda}_0 = \partial(\hat{A}_{rm}^0 \partial/\partial x_m)/\partial x_r$ does not satisfy condition (2′). This means that the composite material does not have rigidity of order ω under strains corresponding to a nonzero solution of (10). In particular, $\operatorname{rank}\Gamma < 6$, i.e., the number of equations is less than the number of the unknowns.

In the case of $M = 6$, arrangements of reinforcements can be shown with $\operatorname{rank}\Gamma < 6$ and with $\operatorname{rank}\Gamma = 6$.

Lemma 1. *Let $M = 6$, and suppose no four vectors γ^j belong to the same plane, the three vectors γ^1, γ^2, γ^4 belong to the same plane, and the vectors γ^3, γ^5, γ^6 do not belong to the same plane. Then* $\operatorname{rank}\Gamma = 6$.

Proof: Make the nondegenerate transformation $x = Ay$ of the coordinates of the axes, whereby γ^1 becomes the direction vector of the $y_{\overline{1}}$-axis, γ^2 becomes the direction vector of the $y_{\overline{2}}$-axis, and γ^3 of the $y_{\overline{3}}$-axis.

Maintain the old notation for the coordinates of the γ^j and note that $\gamma_3^4 = 0$, because γ^1, γ^2, γ^4 belong to the same plane. Then

$$\det\Gamma = 8\gamma_1^4\gamma_2^4\gamma_3^5\gamma_3^6\left(\gamma_1^5\gamma_2^6 - \gamma_1^6\gamma_2^5\right);$$

$\gamma_1^4, \gamma_2^4 \neq 0$, since otherwise γ^4 is proportional to one of the vectors γ^1, γ^2, γ^3; $\gamma_3^5 \neq 0$, since otherwise the vectors γ^1, γ^2, γ^4, γ^5 belong to the same plane; by the same token, $\gamma_3^6 \neq 0$; and $\gamma_1^5\gamma_2^6 - \gamma_1^6\gamma_5^2 \neq 0$, because otherwise the vectors γ^3,

γ^5, γ^6 belong to the same plane. So $\det\Gamma \neq 0$ and $\operatorname{rank}\Gamma = 6$. Lemma 1 and consequently Theorem 1 are proved.

On the other hand, we can give an example of the collection of the vectors γ^j that seem to be in the same position (no three vectors belong to the same plane) when $\operatorname{rank}\Gamma = 5$:

$$\|\gamma_r^j\| = \begin{bmatrix} 1 & 0 & 0 & 2/3 & 6/7 & 15/19 \\ 0 & 1 & 0 & 2/3 & 3/7 & 10/19 \\ 0 & 0 & 1 & 1/3 & 2/7 & 6/19 \end{bmatrix}.$$

By analogy with Theorem 4, it can be proved that in the scalar case, the minimum of the reinforcement directions for which the operator $\hat{\Lambda}_0$ is elliptic (provided the original operator is uniformly elliptic) equals 3.

The conditions of Lemma 1 are satisfied, for example, by the following reinforcement arrangement: Consider an arbitrary nondegenerate tetrahedron taking the direction vectors of its edges as the γ^j. For example, let γ^1, γ^2, γ^3 be basis vectors,

$$\gamma^4 = \frac{\gamma^1 \pm \gamma^2}{\sqrt{2}}, \quad \gamma^5 = \frac{\gamma^1 \pm \gamma^3}{\sqrt{2}}, \quad \gamma^6 = \frac{\gamma^2 \pm \gamma^3}{\sqrt{2}}.$$

When the new reinforcement directions are added, new rows are "attached" to the matrix, and so its rank does not diminish.

Therefore, the arrangement of reinforcement with nine vectors, including γ^1, γ^2, γ^3 as basis ones, and

$$\gamma^{4,5} = \frac{\gamma^1 \pm \gamma^2}{\sqrt{2}}, \quad \gamma^{6,7} = \frac{\gamma^1 \pm \gamma^3}{\sqrt{2}}, \quad \gamma^{8,9} = \frac{\gamma^2 \pm \gamma^3}{\sqrt{2}}$$

also determines the averaged system satisfying condition (2′). This reinforcement system contains all the directions parallel to the edges and diagonals of the faces of the unit cube.

Let us investigate the properties of this reinforcement system under the additional assumption that its properties are symmetric:

$$\begin{gathered} \theta_1 = \theta_2 = \theta_3 = \Theta^1, \\ \theta_4 = \theta_5 = \theta_6 = \theta_7 = \theta_8 = \theta_9 = \Theta^2; \end{gathered} \tag{11}$$

it follows directly from (9) that $a_{kl}^{0iq} = 0$ when there is one number among i, q, k, and l, which is different from all the others; in addition, all the $\hat{a}_{ii}^{0ii}$ are equal to one another, and all the a_{qq}^{0ii} with $i \neq q$ are equal to one another.

Lemma 2 (on Cauchy relations). *If (l_2, q_2, r_2, m_2) is a permutation of (l_1, q_1, r_1, m_1), then for any reinforcement system*

$$\widehat{a}^{0l_1q_1}_{r_1m_1} = \widehat{a}^{0l_2q_2}_{r_2m_2}.$$

PROOF: It follows from (9) that under the conditions of Lemma 2,

$$\widehat{b}^{jl_1q_1}_{r_1m_1} = \widehat{b}^{jl_2q_2}_{r_2m_2};$$

and after multiplying by θ_j and summation over j, we get the assertion of Lemma 2. □

From Lemma 2, we now see that for the arrangement of the reinforcement in question and under condition (11), the following relations hold: all the $\hat{a}^{0ii}_{ii}$ are equal to $\alpha\omega = (\theta^1 + \theta^2)\omega$, all the $\hat{a}^{0ij}_{ji}$ and $\hat{a}^{0ij}_{ij}$ with $i \neq j$ are equal to $\beta\omega = \theta^2\omega/2$, and the remaining $\hat{a}^{0ij}_{hl}$ are 0.

It can be verified for $\theta^1 = \theta^2/2$ that $\alpha = 3\beta$. Then the operator $\hat{\Lambda}_0$ is the operator of the isotropic elasticity theory with $\hat{\lambda} = \hat{\mu} = \beta\omega$. Therefore, the effective Poisson ratio is

$$\widehat{\nu} = \frac{\widehat{\lambda}}{2(\widehat{\lambda} + \widehat{\mu})} = \frac{1}{4}.$$

A more general result is valid.

THEOREM 5. *Let* Λ_0 *be the operator of the isotropic elasticity theory. Then the appropriate Poisson ratio is* 1/4.

PROOF: For the operator of the isotropic elasticity theory $\hat{\lambda} = \hat{a}^{012}_{12}$, $\hat{\mu} = \hat{a}^{012}_{21}$. From Lemma 2, we see that $\lambda = \mu$, and therefore $\nu = 1/4$. □

When $\operatorname{rank}\Gamma < 6$, the above-described splitting principle makes it possible to calculate explicitly the effective characteristics of the material that corresponds to strains of certain type. The lack of explicit formulas for the other characteristics cannot be considered an important drawback of the constructions. The point is that a composite material is usually under loading of a certain type, and the material is engineered with a high (of order ω) rigidity precisely for these loads.

Consider an elementary model problem of heat conduction where a homogeneous medium is reinforced with a unidirectional system of highly conducting fibers having a large heat capacity ratio, and the direction of reinforcement is the x_1-axis. Then the averaged system of equations is of the form

$$\widehat{S}\,\frac{\partial v}{\partial t} - \widehat{A}_{11}\,\frac{\partial^2 v}{\partial x_1^2} - \sum_{(i,j)\neq(1,1)} \widehat{A}_{ij}\,\frac{\partial^2 v}{\partial x_i \partial x_j} = 0, \tag{12}$$

where $\hat{S}$, $\hat{A}_{11}$ are calculated by explicit formulas (see §2).

We now consider a heat wave travelling in the direction $\gamma = (\gamma_1, \gamma_2, \gamma_3)$, $\|\gamma\| = 1$. Substituting $v = \overline{v}(y,t)$, $y = \gamma_j x_j$ into (12), we get the equation

$$\widehat{S}\,\frac{\partial \overline{v}}{\partial t} - \left(\widehat{A}_{11}\gamma_1^2 + \sum_{(i,j)\neq(1,1)} \widehat{A}_{ij}\gamma_i\gamma_j\right)\frac{\partial^2 \overline{v}}{\partial y^2} = 0.$$

For the principal part of the equation, this implies

$$\widehat{S}^0\,\frac{\partial v_0}{\partial t} - \widehat{A}^0_{11}\gamma_1^2\,\frac{\partial^2 v_0}{\partial y^2} = 0.$$

Thus, we have obtained in an explicit form the value of the effective heat-conductivity coefficient for any direction of wave propagation not orthogonal to the x_1-axis ($\gamma_1 \neq 0$).

The main advantage of the splitting principle is that it simplifies the procedure of calculating the leading term of the averaged operator (see (8)), since in the general case ($\omega \sim 1$), defining the averaged operator requires numerical solution of a series of three-dimensional problems of elasticity theory on a cell. Formula (8) can be used to calculate the parameters M, γ_r^j, E_j, and $\operatorname{mes} G_{1,j}$ optimizing some efficiency function $\Phi(\hat{R}, \hat{A}_{rm})$.

REMARK 2: The condition $\overline{G}_{1,l} \cap \overline{G}_{1,j} \neq \emptyset$ for $l \neq j$ means the absence of contact between the fibers, but the final formulation of the splitting principle remains valid even if there is a point contact between them.

REMARK 3: Actual composite materials often have matrices with more complicated laws of loading than in the linear case considered above. Our constructions do not cover this case. However, the independence of the principal part of the averaged equation makes the hypothesis that the splitting principle is valid also for such matrices quite feasible. For instance, the procedure of constructing the f.a.s. set forth in §4 can also be extended to the nonlinear case.

REMARK 4: The splitting principle allows explicit calculation not only of the effective characteristics but local stresses in fibers as well. It is for the fibers of the jth direction that Remark 1 implies the approximate formula

$$\omega^{-1}\sigma_i^k = E_j\gamma_i^j\gamma_k^j \sum_{q,l=1}^{3} \gamma_q^j\gamma_l^j \hat{e}_q^l + O\left(\varepsilon + \varepsilon^2\omega + \omega^{-1}\right),$$

$$\hat{e}_q^l = 0{,}5\left(\frac{\partial v_{00}^l}{\partial x_q} + \frac{\partial v_{00}^q}{\partial x_l}\right)$$

(which is true at a sufficiently great distance from the boundary of a sample), i.e., the fibers are only under tension or compression. This feature makes it possible to approach the determination of the criterium of strength of high-modulus composite materials when breaking starts from fibers.

REMARK 5: The above-described theory differs from the well-known studies of the Riga and Kiev schools (see [52, 70, 91, 149]) in that it does without mechanical assumptions concerning the structure of the stress field in the composite.

At the same time, it is important that the reinforcement is assumed to be of high modulus ($\omega \gg 1$). All the results are derived directly from the equations of mathematical physics written for the composite. (The necessity of the condition $\varepsilon^2\omega \ll 1$ was shown by Kulikovsky and Eglit.)

REMARK 6: When a fibrous composite is loaded, the effect of instability of the fibers may arise [28, 29], which requires further discussion.

REMARK 7: In some cases, the second of the inequalities (2″) is not required to carry out the procedure from §4 and to prove Theorem 3. Thus, in case c), $f = f_0(x_1)$, it suffices to require that $\det \|\Delta_{kl}\| =\neq 0$, where

$$\Delta_{kl} = \sum_{j=1}^{M} E_j \theta_j (\gamma_1^j)^2 \gamma_k^j \gamma_l^j.$$

CHAPTER 8

AVERAGING PROCESSES IN SKELETAL STRUCTURES

Frameworks have been widely adopted in engineering today, primarily for building structures (trusses of bridges and industrial installations, frameworks of housing, supports of electric power lines, etc.). The methods of solving the system of equations of the strength of materials theory presently used for calculating the frameworks have the drawback that the number of equations increases with the number of nodes of the skeletal structure. This greatly impedes the calculation of skeletal structures with a large number of nodes, especially when investigating nonstationary and nonlinear processes.

To study the motion of ground water and migration of substances in the soil, a capillary system is often used as a geometric model (see [122, 128]), which can be regarded as a "framework."

In this chapter a method is proposed for calculating processes in skeletal structures which is based on averaging the partial differential equations describing these processes.

The splitting principle for the averaged operator, similar to that considered in §7.5, is formulated and justified.

The chapter is arranged as follows. Section 1 considers the simplest two-dimensional rectangular framework consisting of two periodic systems of thin strips stretched in the coordinate directions. An elliptic equation with constant coefficients defined on this framework is averaged. As small parameters, we first take the side ε of a periodic cell of the framework, as before, and second, the ratio μ of the width of the strip to the period ε.

A theorem on the closeness of the original problem to the averaged one is stated and proved. Section 2 gives a definition of an s-dimensional framework consisting of thin bars in M directions, assuming that the strips (beams) go through the entire framework (see Fig. 22).

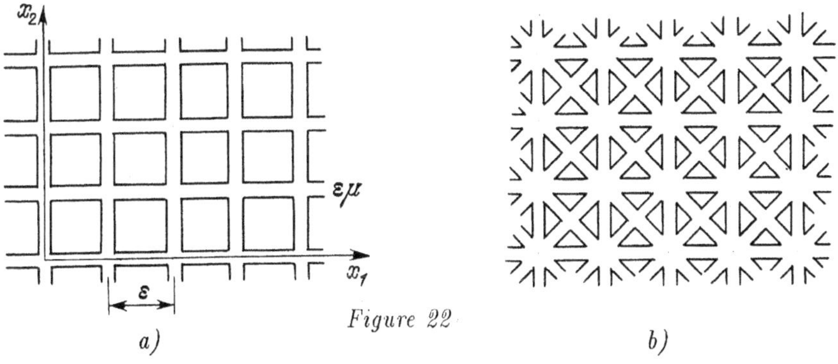

Figure 22

Section 3 is central to this chapter. It considers a nonlinear, nonstationary system of partial differential equations of the second order defined on the framework from §2. The splitting principle for the averaged operator, similar to the one considered in §7.5, is formulated for this system. It allows calculation of the coefficients of the averaged equations by explicit formulas; for the system of equations of elasticity theory with the number of the directions < 6, some coefficients cannot be calculated explicitly.

As in Chapter 7, the framework in this case has a low (of order $O(\mu)$) rigidity under some types of strains.

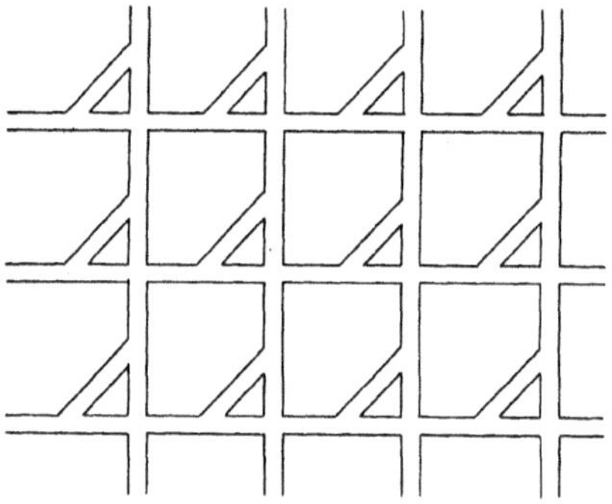

Figure 23

Next, in Subsection 8, a definition of framework, more general than in §2, is formulated, which does not require the bars to go through the framework (see Fig. 23). The splitting principle is justified for such frameworks. In the last subsection of §3, asymptotics of order $O(\varepsilon^K/\mu + \sqrt{\mu})$ are constructed. This yields an averaged equation of infinite order in ε. All its coefficients can be calculated by explicit formulas with accuracy $O(\mu)$. The coefficients of higher orders contain, in particular, transverse elastic characteristics of skeletal girders (see Fig. 24) (such terms as "laced beams" or "lattice beams" are also used in mechanics, see [2]). A rigorous definition of this type of geometric domains is given in §4. Problems of constructing continual models for grid shells and plates were investigated earlier in the book [131] (see also references in [131]).

As an example, let us consider linear elastic vibrations of the truss, shown in Fig. 25, under the effect of the field of forces distributed by the law $f = (f_1(x_1, t), 0, 0)^T$. The restraints in all the nodes of the truss are rigid, and all the translations for $x_1 = 0$ and $x_1 = 1 > 0$ are equal to the given vector-functions $g_0(t) = (g_0^1(t), 0, 0)^T$ and $g_l(t) = (g_l^1(t), 0, 0)^T$, respectively. The truss consists of a recurrent element (see Fig. 26), where $AB = AC = \varepsilon \ll 1$.

It is assumed in this example that the horizontal and inclined bars have cross-sectional areas F, whereas the cross-sectional areas of the vertical bars are $F/2$, and the truss is made of isotropic material.

We denote the part of the space occupied by the truss by Φ and the boundary of Φ by $\partial\Phi$. Then the problem of elasticity theory describing elastic vibrations takes the form

$$\rho \frac{\partial^2 u}{\partial t^2} - \sum_{r,m=1}^{3} \frac{\partial}{\partial x_r}\left(A_{rm} \frac{\partial u}{\partial x_m}\right) = f(x_1, t),$$
$$x = (x_1, x_2, x_3) \in \Phi, \quad t > 0,$$
$$\sum_{r,m=1}^{3} n_r A_{rm} \frac{\partial u}{\partial x_m} = 0, \quad x \in \partial\Phi \setminus (\{x_1 = 0\} \cup \{x_1 = l\}),$$
$$u|_{x_1=0} = g_0(t), \quad u|_{x_1=l} = g_l(t),$$

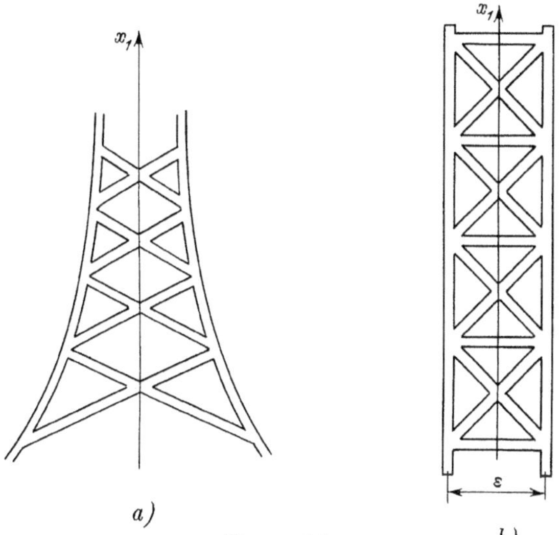

Figure 24

where ρ is density, A_{rm} is the rigidity tensor of the material, $A_{rm} = ||a_{rm}^{pq}|| = ||\lambda\delta_{rp}\delta_{mq} + \zeta(\delta_{rm}\delta_{pq} + \delta_{rq}\delta_{mp})||$, and n_r is the cosine of the angle between the normal to $\partial\Phi$ and the axis x_r. If the truss was at rest at the initial instant of time, the initial conditions were

$$u|_{t=0} = \partial u/\partial t|_{t=0} = 0, \quad x \in \Phi.$$

The averaged problem for the principal part of the average longitudinal displacement v_1 is (see §4)

$$2F(1+\sqrt{2})\rho \frac{\partial^2 v_1}{\partial t^2} - F\overline{E}\sqrt{2} \frac{\partial^2 v_1}{\partial x_1^2} =$$
$$= 2F(1+\sqrt{2}) f_1, \quad x_1 \in (0, l), \quad t > 0, \tag{1}$$
$$v_1|_{x_1=0} = g_0(t), \quad v_1|_{x_1=l} = g_l(t), \tag{2}$$
$$v_1 = \partial v_1/\partial t = 0, \quad t = 0, \quad x_1 \in (0, l); \tag{3}$$

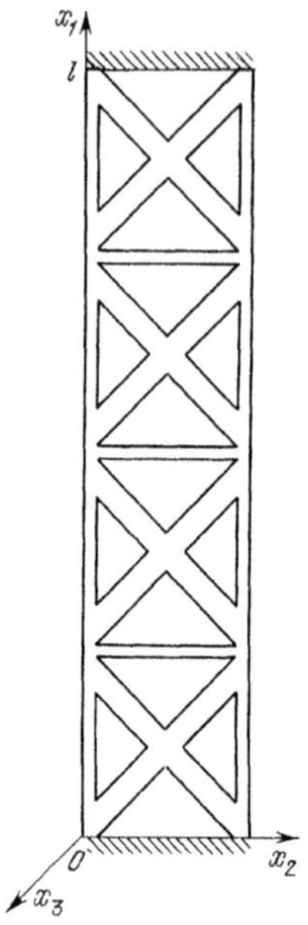

Figure 25

here $\overline{E} = (3\lambda + 2\zeta)\zeta(\lambda + \zeta)^{-1}$ is Young's modulus, and λ, ζ are the Lamé coefficients of the material.

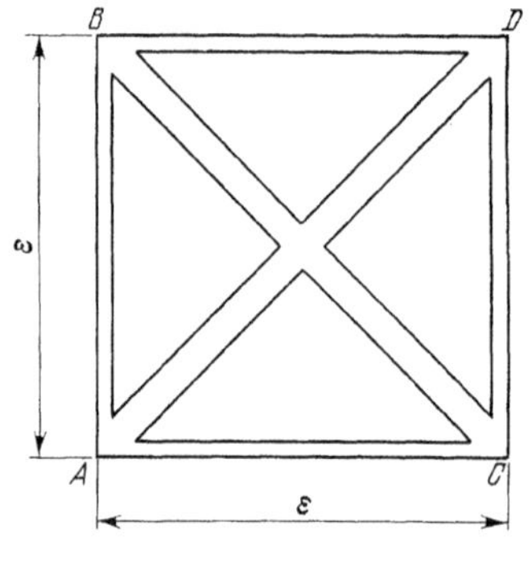

Figure 26

Problems (1)–(3) describe longitudinal vibrations of a bar made of a hypothetical material such that the bar's longitudinal elastic properties are equivalent to the effective properties of the whole truss (as $\varepsilon \to 0$, $F \to 0$).

Problems (1)–(3) are solved analytically (see [151]). The average longitudinal stress is found from the formula

$$\widehat{\sigma}_{11} = F\overline{E}\sqrt{2}\widehat{e}_{11}, \quad \widehat{e}_{11} = \partial v_1/\partial x_1.$$

The transverse strain of the truss is

$$\widehat{e}_{22} = (1 - \sqrt{2})\frac{\partial v_1}{\partial x_1} = (1 - \sqrt{2})\,\widehat{e}_{11}.$$

The method proposed in §4 also allows to approximately determine the microfields of the stresses σ_{ij}. Thus,

$$\sigma_{11}^{(1)} \approx \overline{E}\widehat{e}_{11}, \quad \sigma_{rm}^{(1)} \approx 0 \quad \text{for} \quad (r, m) \neq (1, 1); \tag{4_1}$$

$$\sigma_{11}^{(2)} \approx \frac{1}{4}\overline{E}\,(\widehat{e}_{11} + \widehat{e}_{22}), \quad \sigma_{21}^{(2)} = \sigma_{12}^{(2)} \approx \frac{1}{4}\overline{E}\,(\widehat{e}_{11} + \widehat{e}_{22}),$$
$$\sigma_{22}^{(2)} \approx \frac{1}{4}\overline{E}\,(\widehat{e}_{11} + \widehat{e}_{22}); \tag{4_2}$$

$$\sigma_{22}^{(3)} \approx \overline{E}\widehat{e}_{22}, \quad \sigma_{rm}^{(3)} \approx 0 \quad \text{for} \quad (r, m) \neq (2, 2); \tag{4_3}$$

$$\sigma_{11}^{(4)} \approx \frac{1}{4}\overline{E}\,(\widehat{e}_{11} + \widehat{e}_{22}), \quad \sigma_{21}^{(4)} = \sigma_{12}^{(4)} \approx \frac{1}{4}\overline{E}\,(\widehat{e}_{11} + \widehat{e}_{22}),$$
$$\sigma_{22}^{(4)} \approx \frac{1}{4}\overline{E}\,(\widehat{e}_{11} + \widehat{e}_{22}). \tag{4_4}$$

The superscript corresponds to the numbers indicating the direction number in Fig. 27.

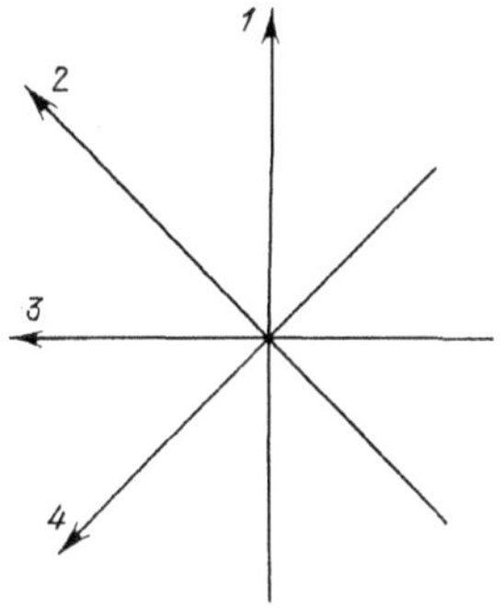

Figure 27

Note that in §§3 and 4 the closeness of the exact and the averaged solutions is justified in a weak sense (see Theorem 4.1). Rigorous justification of (1)–(4) requires that the Cauchy inequality should be obtained with the constant depending on μ and ε in a particular fashion.

The averaged equations for transverse (along the x_2-axis) vibrations in the presence of transverse loads are written out explicitly by means of a fairly cumbersome procedure of asymptotic expansions, whose formalism is outlined in Subsection 9, §3. We do not consider vibrations along the x_3-axis.

Note that in considering a nonperiodic framework (see, for example, Fig. 24a), we can sometimes make the change of variables to transform it into a periodic framework. In this case, the coefficients of the equation will depend smoothly on space variables, and so the problem can be averaged by using the splitting principle.

Section 5 discusses the ways of refining the splitting principle.

In §6 a full asymptotic expansion in ε and μ is carried out for solutions of hyperbolic, parabolic, and elliptic equations defined on the intersection of a rectangular framework with a bounded domain (parallelepiped). The main mathematical difficulties here are involved in constructing and matching functions of the boundary-layer type in neighbourhoods of the framework nodes as well as in a neighbourhood of the boundary of the parallelepiped. Note that with finite μ, the full asymptotic solution in powers of ε can be constructed only in the case of a layer (see [109]).

Section 7 presents equations with random coefficients and the frameworks where any "bar" may be missing with a certain probability, regardless of the other bars. These problems arise in simulating processes in capillary and fissured media with random properties.

§1 An Example of Averaging a Problem on the Simplest Framework

Let the mesh s_ε be defined on a plane with the coordinates of the nodes $(k_1\varepsilon, k_2\varepsilon) = k\varepsilon$, where $k = (k_1, k_2)$ is a real vector and $\varepsilon \ll 1$ is a small parameter consisting of the straight lines $\Pi^1_{k_2}$ and $\Pi^1_{k_2}$ running through the nodes $(k_1\varepsilon, k_2\varepsilon)$ and parallel to the coordinate axes x_1 and x_2, respectively. By $B^1_{k_2}$ and $B^2_{k_1}$, we denote strips consisting of points less than $\mu\varepsilon$ away from $\Pi^1_{k_2}$ and $\Pi^2_{k_1}$, where $\mu \ll 1$ is the second small parameter. The union of all the strips $B^1_{k_2}$ and $B^2_{k_1}$ is called the rectangular periodic framework B:

$$B = \bigcup_{k=-\infty}^{+\infty} (B^1_k \cup B^2_k).$$

Thus, B is a system of thin strips stretched in each of the coordinate directions (see Fig. 22a).

Let G be a domain with the boundary $\partial G \in c^\infty$, which is independent of ε and μ. In the domain $G \cap B$, we consider the equation

$$Pu \equiv \frac{\partial}{\partial x_r}\left(A_{rm}\frac{\partial u}{\partial x_m}\right) + f(x) = 0 \tag{1}$$

with the boundary conditions

$$\partial u/\partial \mathrm{v} = 0 \quad \text{for} \quad x \in \partial B \cap G, \tag{2}$$

$$u = 0 \quad \text{for} \quad x \in \partial G \cap \overline{B}; \tag{3}$$

here $x = (x_1, x_2)$, the A_{rm} are constant numbers, and the quadratic form $A_{rm}\eta_r\eta_m$ is symmetric and positive-definite. The symbol $\partial/\partial\nu$ means the derivative along the conormal: $\partial/\partial\nu - n_r A_{rm}\partial/\partial x_m$, where n_r is the cosine of the angle between the exterior normal and the axis x_r. Since the exterior normal ∂B is always collinear to one of the axes, $\partial/\partial\nu = \pm A_{2m}\partial/\partial x_m$ when part of ∂B coincides with ∂B^1_k, and $\partial/\partial\nu = \pm A_{1m}\partial/\partial x_m$ when ∂B coincides with ∂B^2_k; the sign is selected to match the direction of the exterior normal to ∂B. In the isotropic case, $A_{rm} = \delta_{rm} K$, $\partial/\partial\nu = K\partial/\partial n$. It is assumed that $f(x) \in C^1(\overline{G})$.

Problems (1)–(3) simulate a stationary thermal field in the skeletal structure $B \cap G$ under the conditions of thermal insulation on the boundary of B and constant temperature on the boundary of G, with A_{rm} being the heat-conductivity tensor for the material of which the framework is made ("cut out"). For $A_{rm} = \delta_{rm} D$, (1)–(3) may be interpreted also as a problem of diffusion of a substance in fissured rock (see [140, 167]) filled with water, with D being the diffusion coefficient of the substance in the water. In this case, the framework B models a rectangular system of cracks $2\mu\varepsilon$ wide, filled with water. This model is very approximate, of course, since the system of cracks in rock has a much less regular structure (see [140]). The model of a random framework to be discussed in §7 is more suitable for its description.

Numerical solution of problems (1)–(3), with $\varepsilon \ll 1$, $\mu \ll 1$, is very difficult, since the step size of the mesh must have an order much less than ε (see Introduction). The realization of the procedure from Chapter 4 is also impeded, since the problem on a cell depends on the small parameter μ, and in order to solve it numerically, we must select the step size of the mesh to be much less than μ. Hence, an asymptotic investigation of the problem on a cell is needed (see §4.1). It turns out that a solution of this problem can be found analytically with accuracy $O(\mu)$ in the norm W_2^1. It becomes possible to find explicit formulas for the coefficients of the principal part of the averaged operator.

THEOREM 1. *Let $u(x)$ be a solution of problems (1)–(3) and let $v_0(x)$ be a solution of the problem*

$$\frac{\partial}{\partial x_1}\left(\widehat{A}_{11}\frac{\partial \tilde{v}_0}{\partial x_1}\right) + \frac{\partial}{\partial x_2}\left(\widehat{A}_{22}\frac{\partial \tilde{v}_0}{\partial x_2}\right) + f(x) = 0, \quad x \in G, \tag{4}$$

$$\tilde{v}_0|_{\partial G} = 0; \tag{5}$$

here

$$\widehat{A}_{11} = (A_{11} - A_{12}A_{22}^{-1}A_{21})/2, \quad \widehat{A}_{22} = (A_{22} - A_{21}A_{11}^{-1}A_{12})/2. \tag{6}$$

Then

$$\frac{1}{\sqrt{M}}\|u - \tilde{v}_0\|_{L_2(B\cap G)} = O(\sqrt{\varepsilon + \mu}), \tag{7}$$

where $M = \operatorname{mes}(G \cap B) = O(\mu)$ is the area of the domain $G \cap B$.

The idea of Theorem 1 is that if we assign to problems (1)–(3) in the inhomogeneous domain $B \cap G$ the problem (4), (5) in the uniform domain G and calculate $\hat{A}_{rr}$ by (6), the solution of problem (4), (5) is close to that of the original problem in the norm in $L_2(B \cap G)$. The normalization factor $1/\sqrt{M}$ on the left side of (7) is necessary because $\|u\|_{L_2(B\cap G)}$ is a value of order $O(\sqrt{M})$.

Notice that equation (1) being elliptic implies $\hat{A}_{rr} > 0$.

We start the proof of Theorem 1 with a heuristic construction of an approximate solution of problems (1)–(3).

1. Let $u^{(2)}(x)$ be the second approximation of an asymptotic solution of (1), (2), with μ finite. We recall the formalism of its construction (see §§4.1 and 4.3), with $u^{(2)}(x)$ being sought in the form

$$u^{(2)}(x) = v_0(x) + \varepsilon \sum_{i_1=1}^{2} N_{i_1}\left(\frac{x}{\varepsilon}\right)\frac{\partial v_0}{\partial x_{i_1}} + \varepsilon^2 \sum_{i_1,i_2=1}^{2} N_{i_1 i_2}\left(\frac{x}{\varepsilon}\right)\frac{\partial^2 v_0}{\partial x_{i_1}\partial x_{i_2}}, \tag{8}$$

where $v_0(x)$ is a solution of the averaged equation, which depends only on x and μ, and N_{i_1}, $N_{i_1 i_2}$ are 1-periodic functions dependent only on $\xi = x/\varepsilon$ and μ.

We substitute (8) into (1), (2), considering formally $u^{(2)}$ independent of μ. We get

$$\begin{aligned} Pu^{(2)} = {} & \varepsilon^{-1} \sum_{i_1=1}^{2} L_{\xi\xi} N_{i_1} \frac{\partial v_0}{\partial x_{i_1}} + \\ & + \left(\sum_{i_1,i_2=1}^{2} \left(L_{\xi\xi} N_{i_1 i_2} + \sum_{j=1}^{2} 2A_{i_1 j} \frac{\partial N_{i_2}}{\partial \xi_j} + A_{i_1 i_2} \right) \frac{\partial^2 v_0}{\partial x_{i_1}\partial x_{i_2}} + f(x) \right) + \\ & + \varepsilon \sum_{i_1,i_2,i_3=1}^{2} \left(\sum_{j=1}^{2} 2A_{i_1 j} \frac{\partial N_{i_2 i_3}}{\partial \xi_j} + A_{i_1 i_2} N_{i_3} \right) \frac{\partial^3 v_0}{\partial x_{i_1}\partial x_{i_2}\partial x_{i_3}} + \\ & + \varepsilon^2 \sum_{i_1,i_2,i_3,i_4=1}^{2} A_{i_1 i_2} N_{i_3 i_4} \frac{\partial^4 v_0}{\partial x_{i_1}\partial x_{i_2}\partial x_{i_3}\partial x_{i_4}}, \quad x \in B \cap G; \end{aligned} \tag{9}$$

$$\begin{aligned} \frac{\partial u^{(2)}}{\partial \nu} = {} & \sum_{i_1=1}^{2} \frac{\partial}{\partial \nu_\xi} (N_{i_1} + \xi_{i_1}) \frac{\partial v_0}{\partial x_{i_1}} + \\ & + \varepsilon \sum_{i_1,i_2=1}^{2} \left(\frac{\partial}{\partial \nu_\xi} N_{i_1 i_2} + \sum_{j=1}^{2} n_j A_{j i_1} N_{i_2} \right) \frac{\partial^2 v_0}{\partial x_{i_1}\partial x_{i_2}} + \\ & + \varepsilon^2 \sum_{i_1,i_2,i_3=1}^{2} \sum_{j=1}^{2} n_j(\xi) A_{j i_1} N_{i_2 i_3} \frac{\partial^3 v_0}{\partial x_{i_1}\partial x_{i_2}\partial x_{i_3}}, \quad x \in \partial B \cap G, \end{aligned} \tag{10}$$

where $\partial/\partial\nu_\xi = n_j(\xi) A_{jm} \partial/\partial\xi_m$, and $n_j(\xi)$ is the cosine of the angle between the exterior normal and the axis ξ_j (in the variables ξ).

We require that the first and second summands in (9) and (10) vanish. We obtain the following algorithm for defining N_{i_1} and $N_{i_1 i_2}$:

1. Find N_{i_1} as a 1-periodic solution of the problem

$$L_{\xi\xi} N_{i_1} = 0, \quad \xi \in B_\xi, \tag{11}$$

$$\frac{\partial}{\partial \nu_\xi}(N_{i_1} + \xi_{i_1}) = 0, \quad \xi \in \partial B_\xi; \tag{12}$$

here B_ξ is the image of the domain B in the variables $\xi = x/\varepsilon$. A periodic solution of (11), (12) from $W_2^1(B_\xi \cap Q)$, where Q is the square $\{\xi \in R^2 | -1/2 < \xi_l < 1/2\}$, exists and can be defined up to an arbitrary constant summand (see Supplement).

2. Find $N_{i_1 i_2}$ as a 1-periodic solution of the problem

$$L_{\xi\xi} N_{i_1 i_2} = -2A_{i_1 j} \frac{\partial N_{i_2}}{\partial \xi_j} - A_{i_1 i_2} + h_{i_1 i_2}, \quad \xi \in B_\xi, \tag{13}$$

$$\frac{\partial N_{i_1 i_2}}{\partial \nu_\xi} = -n_j A_{j i_1} N_{i_2}, \quad \xi \in \partial B_\xi, \tag{14}$$

where $h_{i_1 i_2}$ is a constant determined from the condition for the existence of the solution of (13), (14) (see Supplement):

$$\operatorname{mes}(B_\xi \cap Q)\, h_{i_1 i_2} = \int_{B_\xi \cap Q} \left(2A_{i_1 j} \frac{\partial N_{i_2}}{\partial \xi_j} + A_{i_1 i_2}\right) d\xi - \int_{\partial B_\xi \cap Q} n_m A_{m i_1} N_{i_2} ds;$$

here $\operatorname{mes}(B_\xi \cap Q)$ is the area of $B_\xi \cap Q$; hence,

$$h_{i_1 i_2} = \frac{1}{\operatorname{mes}(B_\xi \cap Q)} \int_{B_\xi \cap Q} A_{i_1 j} \frac{\partial N_{i_2}}{\partial \xi_j} d\xi + A_{i_1 i_2} = \left\langle A_{i_1 j} \frac{\partial N_{i_2}}{\partial \xi_j} + A_{i_1 i_2} \right\rangle^{(B_\xi \cap Q)}. \tag{15}$$

And for v_0, this yields the averaged equation

$$h_{i_1 i_2} \frac{\partial^2 v_0}{\partial x_{i_1} \partial x_{i_2}} + f(x) = 0, \quad x \in G. \tag{16}$$

If equation (16) is supplemented with the condition

$$v_0 |_{\partial G} = 0, \tag{17}$$

then $u^{(2)}(x)$ formally satisfies condition (1) and condition (3) with accuracy $O(\varepsilon)$, and condition (2) with accuracy $O(\varepsilon^2)$.

Further proof of Theorem 1 consists of the following stages: We construct the asymptotics $\widetilde{N}_{i_1}$ of the solution N_{i_1} to problem (11), (12) (as $\mu \downarrow 0$), prove the estimate of the closeness of $\widetilde{N}_{i_1}$ and N_{i_1}, derive an estimate of the functional

$$\Phi(\varphi) = \int_{B\cap G} A_{rm} \frac{\partial \widetilde{u}^{(2)}}{\partial x_m} \frac{\partial \varphi}{\partial x_r}\, dx - \int_{B\cap G} f\varphi\, dx,$$

introduce a correction in the vicinity of ∂G to correct $\tilde{u}^{(2)}$ and obtain an estimate of the closeness of u and $\tilde{u}^{(2)}$ (here $\tilde{u}^{(2)}$ is a function obtained by replacing v_0 in (8) by its principal part as $\mu \downarrow 0 \tilde{v}_0$).

2. We use the dependence of B_ξ on the small parameter μ to construct the approximate solution $\widetilde{N}_{i_1}(\xi)$ of problem (11), (12).

To problem (11), (12) we assign independent problems on each of the strips $B_\xi^1 = \{\xi \in \mathbf{R}^2 || \xi_2| < \mu\}$, $B_\xi^2 = \{\in \mathbf{R}^2 || \xi_1| < \mu\}$:

$$L_{\xi\xi} \widetilde{N}_{i_1}^j = 0 \quad \text{for} \quad \xi \in B_\xi^j, \tag{18}$$

$$\frac{\partial}{\partial \nu_\xi} (\widetilde{N}_{i_1}^j + \xi_{i_1}) = 0 \quad \text{for} \quad \xi \in \partial B_\xi^j, \tag{19}$$

where the $\widetilde{N}_{i_1}^j$ are 1-periodic in ξ_j $(j = 1, 2)$; problems (18), (19) have solutions $\widetilde{N}_{i_1}^j = d_{i_1}^j \xi_{\bar{j}}$. Here $\bar{j} = 3 - j$, that is $\bar{j} = 1$ for $j = 2$, $\bar{j} = 2$ for $j = 1$, and the $d_{i_1}^j$ are found from (19) by the formulas

$$d_{i_1}^1 = -A_{22}^{-1}(A_{21}\delta_{i_1 1} + A_{22}\delta_{i_1 2}), \quad d_{i_1}^2 = -A_{11}^{-1}(A_{12}\delta_{i_1 2} + A_{11}\delta_{i_1 1}). \tag{20}$$

After matching $\widetilde{N}_{i_1}^1$ and $\widetilde{N}_{i_1}^2$ on $B_\xi \cap Q$:

$$\widetilde{N}_{i_1}(\xi) = \sum_{j=1}^{2} \widetilde{N}_{i_1}^j \tilde{\rho}\left(\frac{\xi_j}{\mu}\right), \quad \xi \in B_\xi \cap Q,$$

where $\tilde{\rho}(t)$ is an infinitely differentiable function of the real variable t, which satisfies the conditions

$$\tilde{\rho}(t) = \begin{cases} 0 & \text{for} \quad |t| \leqslant 1, \\ 1 & \text{for} \quad |t| \geqslant 2, \end{cases} \quad |\tilde{\rho}| \leqslant 1, \quad \tilde{\rho}(t) = \tilde{\rho}(-t)$$

(Fig. 28). It follows from (20) that

$$\widetilde{N}^j_{i_1} = \begin{cases} -\xi_{i_1} & \text{for} \quad i_1 \neq j, \\ -A^{-1}_{\bar{i}_1 \bar{i}_1} A_{\bar{i}_1 i_1} \xi_{\bar{i}_1} & \text{for} \quad i_1 = j. \end{cases} \tag{21}$$

We continue $\widetilde{N}_{i_1}(\xi)$ from $B_\xi \cap Q$ periodically onto all the B_ξ. The proof will be given below for the estimate

$$\|\nabla_\xi (\widetilde{N}_{i_1} - N_{i_1})\|^2_{L_2(B_\xi \cap Q)} = O(\mu) \operatorname{mes}(B_\xi \cap Q), \tag{22}$$

which shows that $\widetilde{N}_{i_1}$ is really the approximate solution of (11), (12). Estimate (12) and the Cauchy–Buniakowski–Schwarz inequality (see [76]) imply that if $\widetilde{N}_{i_2}$ is substituted for N_{i_2} in (15), then $h_{i_1 i_2}$ changes to $O(\sqrt{\mu})$, whence

$$h_{i_1 i_2} = \delta_{i_1 i_2} \widehat{A}_{i_1 i_2} + O(\sqrt{\mu}). \tag{23}$$

Thus the coefficients of (4) approximate those of (16).

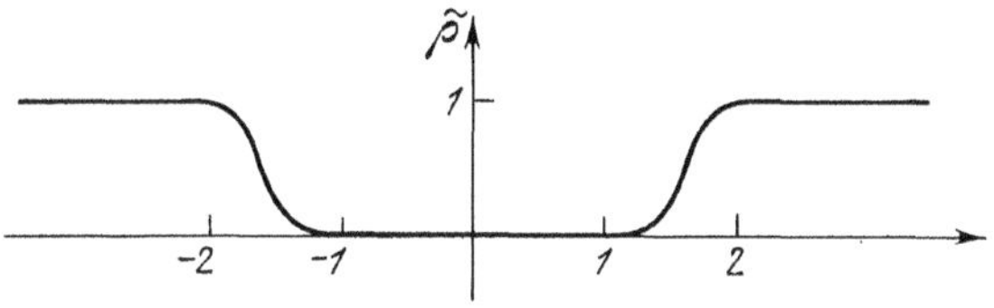

Figure 28

3. Let us prove estimate (22). Suppose $\varphi(\xi)$ is an arbitrary 1-periodic function from $W_2^{1\,\mathrm{loc}}$. Consider the integral

$$I_1 = \int\limits_{B_\xi \cap Q} A_{rj} \frac{\partial}{\partial \xi_j} (\widetilde{N}_{i_1} - N_{i_1}) \frac{\partial \varphi}{\partial \xi_r} \, d\xi;$$

since N_{i_1} is the exact solution of problem (11), (12),

$$I_1 = \int\limits_{B_\xi \cap Q} A_{rj} \frac{\partial}{\partial \xi_j} (\widetilde{N}_{i_1} + \xi_{i_1}) \frac{\partial \varphi}{\partial \xi_r} \, d\xi =$$

$$= \sum_{q = i_1, \bar{i}_1} \int\limits_{B^q_\xi \cap Q} A_{rj} \frac{\partial}{\partial \xi_j} (\widetilde{N}_{i_1} + \xi_{i_1}) \frac{\partial \varphi}{\partial \xi_r} \, d\xi -$$

$$- \int\limits_{B^1_\xi \cap B^2_\xi} A_{ij} \frac{\partial (\widetilde{N}_{i_1} + \xi_{i_1})}{\partial \xi_j} \frac{\partial \varphi}{\partial \xi_r} \, d\xi.$$

Since for $\xi \in B^{I_1}_\xi$

$$\widetilde{N}_{i_1}(\xi) = \widetilde{N}_{i_1}^{I_1}\tilde{\rho}(\xi_{I_1}/\mu) = -\xi_{i_1}\tilde{\rho}(\xi_{I_1}/\mu)$$

then $A_{rj}\partial(\widetilde{N}_{i_1} + \xi_{i_1}/\partial\xi_j$ for $\xi \in B_\xi$ can be represented as $A_{rj}\widetilde{N}_{i_1}^{\bar{i}_1}\partial\tilde{\rho}/\partial\xi_j + A_{rj}\delta_{ji_1}(1-\tilde{\rho})$. In addition, $\widetilde{N}_{i_1}(\xi) \equiv 0$ for $\xi \in B_\xi^1 \cap B_\xi^2$, and therefore $A_{rj}\partial(\widetilde{N}_{i_1} + \xi_{i_1})/\partial\xi_j = A_{rj}\delta_{ji_1}$. So we have

$$I_1 = \int\limits_{B_\xi^{i_1}\cap Q} \sum_{r,j=1}^{2} A_{rj}\frac{\partial}{\partial\xi_j}(\widetilde{N}_{i_1} + \xi_{i_1})\frac{\partial\varphi}{\partial\xi_r}\,d\xi - \int\limits_{B_\xi^1\cap B_\xi^2} \sum_{r,j=1}^{2} A_{rj}\delta_{ji_1}\frac{\partial\varphi}{\partial\xi_r}\,d\xi +$$
$$+ \int\limits_{B_\xi^{\bar{i}_1}\cap Q} \sum_{r,j=1}^{2}\left(A_{rj}\widetilde{N}_{i_1}^{\bar{i}_1}\frac{\partial\tilde{\rho}(\xi_{\bar{i}_1}/\mu)}{\partial\xi_j} + A_{rj}\delta_{ji_1}\left(1-\tilde{\rho}\left(\frac{\xi_{\bar{i}_1}}{\mu}\right)\right)\right)\frac{\partial\varphi}{\partial\xi_r}\,d\xi.$$

The last two integrals can be represented as

$$\int\limits_{B_\xi\cap Q} \psi_r\frac{\partial\varphi}{\partial\xi_r}\,d\xi, \quad \|\psi_r\|^2_{L_2(B_\xi\cap Q)} = O(\mu)\,\mathrm{mes}\,(B_\xi \cap Q). \tag{24}$$

We write the integral

$$I_2 = \int\limits_{B_\xi^{i_1}\cap Q} \sum_{r,j=1}^{2} A_{rj}\frac{\partial}{\partial\xi_j}(\widetilde{N}_{i_1} + \xi_{i_1})\frac{\partial\varphi}{\partial\xi_r}\,d\xi$$

by analogy with I_1 (no summation over i_1): $I_2 = I_3 + I_4$, where

$$I_3 = \sum_{r,j=1}^{2}\int\limits_{B_\xi^{i_1}\cap Q} A_{rj}\tilde{\rho}\left(\frac{\xi_{i_1}}{\mu}\right)\frac{\partial}{\partial\xi_j}\left(\widetilde{N}_{i_1}^{i_1} + \xi_{i_1}\right)\frac{\partial\varphi}{\partial\xi_r}\,d\xi,$$
$$I_4 = \sum_{r,j=1}^{2}\int\limits_{B_\xi^{i_1}\cap Q}\left(A_{rj}\frac{\partial}{\partial\xi_j}\tilde{\rho}\left(\frac{\xi_{i_1}}{\mu}\right)\widetilde{N}_{i_1}^{i_1} + \delta_{ji_1}A_{rj}\left(1-\tilde{\rho}\left(\frac{\xi_{i_1}}{\mu}\right)\right)\right)\frac{\partial\varphi}{\partial\xi_r}\,d\xi.$$

The integral I_4 is of the form (24), whereas

$$I_3 \stackrel{\text{def}}{=} \int\limits_{B_\xi^{i_1}\cap Q} \sum_{r,j=1}^{2} A_{rj}\tilde{\rho}\left(\frac{\xi_{i_1}}{\mu}\right)\frac{\partial}{\partial\xi_j}\left(\widetilde{N}_{i_1}^{i_1} + \xi_{i_1}\right)\frac{\partial\varphi}{\partial\xi_r}\,d\xi =$$
$$= \int\limits_{B_\xi^{i_1}\cap Q} \sum_{j=1}^{2} A_{i_1j}\frac{\partial}{\partial\xi_j}\left(\widetilde{N}_{i_1}^{i_1} + \xi_{i_1}\right)\frac{\partial\varphi}{\partial\xi_{i_1}}\,d\xi +$$
$$+ \int\limits_{B_\xi^{i_1}\cap Q} \sum_{j=1}^{2} A_{i_1j}\frac{\partial}{\partial\xi_j}\left(\widetilde{N}_{i_1}^{i_1} + \xi_{i_1}\right)\left(\tilde{\rho}\left(\frac{\xi_{i_1}}{\mu}\right) - 1\right)\frac{\partial\varphi}{\partial\xi_{i_1}}\,d\xi.$$

Here we have made use of the fact that by construction the $\widetilde{N}_{i_1}^{i_1}$ satisfy the relation (the summation is over j):

$$A_{rj}\frac{\partial}{\partial\xi_j}\left(\widetilde{N}_{i_1}^{i_1}+\xi_{i_1}\right)=0 \quad \text{for} \quad r\neq i_1.$$

Since $A_{i_1 j}\partial(\widetilde{N}_{i_1}^{i_1}+\xi_{i_1})/\partial\xi_j$ = const and the function φ is 1-periodic, the integration by parts yields

$$\int\limits_{B_\xi^{i_1}\cap Q}\sum_{j=1}^{2}A_{i_1 j}\frac{\partial}{\partial\xi_j}\left(\widetilde{N}_{i_1}^{i_1}+\xi_{i_1}\right)\frac{\partial\varphi}{\partial\xi_{i_1}}\,d\xi=0.$$

Consequently, I_3 can also be represented as (24).

Thus, I_1 is of the form (24). By substituting $\varphi=\widetilde{N}_{i_1}-N_i$ into I_1 and by taking into account the equivalence of the norms

$$\|\nabla_\xi\varphi\|_{L_2(B_\xi\cap Q)}=\left\|\sqrt{\frac{\partial\varphi}{\partial\xi_j}\frac{\partial\varphi}{\partial\xi_j}}\right\|_{L_2(B_\xi\cap Q)},\quad \sqrt{\int\limits_{B_\xi\cap Q}A_{rm}\frac{\partial\varphi}{\partial\xi_m}\frac{\partial\varphi}{\partial\xi_r}\,d\xi},$$

we get (22) and hence (23).

4. Let the N_{i_1} satisfy the conditions $\int\limits_{B_\xi\cap Q}N_{i_1}\,d\xi=0$. By using (21), the Poincaré–Friedrichs inequality, and estimate (22), we get

$$\|N_{i_1}\|^2_{L_2(B_\xi\cap Q)}=O(\mu)\operatorname{mes}(B_\xi\cap Q). \tag{25}$$

Now we derive the estimate

$$\|N_{i_1 i_2}\|^2_{W_2^1(B_\xi\cap Q)}=O(1)\operatorname{mes}(B_\xi\cap Q). \tag{26}$$

from the integral identity for problem (13), (14).

We shall prove that if $N_{i_1}(\xi)$, $N_{i_1 i_2}(\xi)$ are the solutions of problems (11), (12) and (13), (14), and $\tilde{v}_0(x)$ is the solution of problem (4), (5), then $\tilde{u}^{(2)}(x)$ produced by substituting $\tilde{v}_0(x)$ for $v_0(x)$ in (8) satisfies the integral identity

$$\int\limits_{B\cap G}A_{rm}\frac{\partial\widetilde{u}^{(2)}}{\partial x_m}\frac{\partial\varphi}{\partial x_r}\,dx=\int\limits_{B\cap G}f\varphi\,dx+\int\limits_{B\cap G}\psi_j\frac{\partial\varphi}{\partial x_j}\,dx+\int\limits_{B\cap G}\psi_0\varphi\,dx, \tag{27}$$

with

$$\| \psi_j \|^2_{L_2(B\cap G)} = O(\mu + \varepsilon^2)\, \mathrm{mes}\,(B \cap G), \quad j = 0, 1, 2, \tag{28}$$

and $\varphi(x) \in C^\infty(\overline{G})$ has a compact support.

We represent the left side of (27) as

$$\left\{ \sum_{i_2=1}^{2} \int_{B\cap G} \sum_{i_1,j=1}^{2} A_{i_1 j} \frac{\partial}{\partial \xi_j}(N_{i_2} + \xi_{i_2}) \frac{\partial}{\partial x_{i_1}} \left(\frac{\partial \tilde{v}_0}{\partial x_{i_2}} \varphi \right) dx \right\} +$$

$$+ \left\{ - \sum_{i_2=1}^{2} \int_{B\cap G} \sum_{i_1,j=1}^{2} A_{i_1 j} \frac{\partial}{\partial \xi_j}(N_{i_2} + \xi_{i_2}) \frac{\partial^2 \tilde{v}_0}{\partial x_{i_1} \partial x_{i_2}} \varphi\, dx + \right.$$

$$\left. + \varepsilon \sum_{i_1,i_2=1}^{2} \int_{B\cap G} \sum_{m=1}^{2} \left(\sum_{j=1}^{2} A_{mj} \frac{\partial}{\partial \xi_j} N_{i_1 i_2} + A_{m i_1} N_{i_2} \right) \frac{\partial}{\partial x_m} \left(\frac{\partial^2 \tilde{v}_0}{\partial x_{i_1} \partial x_{i_2}} \varphi \right) dx \right\} -$$

$$- \varepsilon \sum_{i_1,i_2=1}^{2} \int_{B\cap G} \sum_{m=1}^{2} \left(\sum_{j=1}^{2} A_{mj} \frac{\partial}{\partial \xi_j} N_{i_1 i_2} + A_{m i_1} N_{i_2} \right) \frac{\partial^3 \tilde{v}_0}{\partial x_m \partial x_{i_1} \partial x_{i_2}} \varphi\, dx +$$

$$+ \varepsilon^2 \sum_{i_1,i_2,i_3=1}^{2} \int_{B\cap G} \sum_{m=1}^{2} A_{m i_1} N_{i_2 i_3} \frac{\partial^3 \tilde{v}_0}{\partial x_{i_1} \partial x_{i_2} \partial x_{i_3}} \frac{\partial \varphi}{\partial x_m} dx. \tag{29}$$

By virtue of the integral identities for problems (11), (12) and (13), (14), the first expression in braces vanishes, and the second expression is

$$- \int_{B\cap G} \sum_{i_1,i_2=1}^{2} h_{i_1 i_2} \frac{\partial^2 \tilde{v}_0}{\partial x_{i_1} \partial x_{i_2}} \varphi\, dx =$$

$$= - \int_{B\cap G} \left(\hat{A}_{11} \frac{\partial^2 \tilde{v}_0}{\partial x_1^2} + \hat{A}_{22} \frac{\partial^2 \tilde{v}_0}{\partial x_2^2} \right) \varphi\, dx + \int_{B\cap G} \theta \varphi\, dx =$$

$$= - \int_{B\cap G} f \varphi\, dx + \int_{B\cap G} \theta \varphi\, dx,$$

$$\| \theta \|^2_{L_2(B\cap G)} = O(\mu)\, \mathrm{mes}\,(B \cap G).$$

Considering the estimates (25), (26), we can now get (27), (28).

5. The approximation $\tilde{u}^{(2)}$ does not satisfy the boundary condition (3) exactly. Therefore it should be corrected near ∂G. Let $\chi(x)$ be an infinitely differentiable function whose support belongs to the ε-neighbourhood of the boundary of the domain G, $\chi|_{\partial G} = 1$, $|\chi| \leqslant 1$, $\|\varepsilon \frac{\partial \chi}{\partial x_j}\|_{C(\overline{G})} = O(1)$.

The estimates (25), (26) imply

$$\left\| \sum_{m=1}^{2} A_{rm} \left(\varepsilon \sum_{i_1=1}^{2} N_{i_1} \frac{\partial \widetilde{v}_0}{\partial x_{i_1}} + \varepsilon^2 \sum_{i_1, i_2=1}^{2} N_{i_1 i_2} \frac{\partial^2 \widetilde{v}_0}{\partial x_{i_1} \partial x_{i_2}} \right) \frac{\partial \chi}{\partial x_m} \right\|^2_{L_2(B \cap G)} = \\ = O(\varepsilon(\mu + \varepsilon^2)) \operatorname{mes}(B \cap G),$$

$$\left\| \sum_{r,m=1}^{2} A_{rm} \frac{\partial \left(\widetilde{u}^{(2)} - \widetilde{v}_0 \right)}{\partial x_m} \chi \right\|^2_{L_2(B \cap G)} = O(\varepsilon) \operatorname{mes}(B \cap G),$$

where $\tilde{v}_0$ is a solution of problem (4), (5). The integral

$$I = \int_{B \cap G} A_{rm} \frac{\partial \left(\left(\widetilde{u}^{(2)} - \widetilde{v}_0 \right) \chi \right)}{\partial x_m} \frac{\partial \varphi}{\partial x_r} \, dx$$

can be represented as

$$I = \int_{B \cap G} \psi_j \frac{\partial \varphi}{\partial x_j} \, dx, \tag{30}$$

where

$$\| \psi_j \|^2_{L_2(B \cap G)} = O(\varepsilon) \operatorname{mes}(B \cap G). \tag{31}$$

The relations (27)–(31) result in the estimate

$$\left\| u - \widetilde{v}_0 - (1 - \chi) \left(\widetilde{u}^{(2)} - \widetilde{v}_0 \right) \right\|^2_{W_2^1(B \cap G)} = O(\varepsilon + \mu) \operatorname{mes}(B \cap G), \tag{32}$$

which implies the assertion of Theorem 1.

The estimate (32) suggests the validity of the relation

$$\left\| u - \widetilde{v}_0 - \varepsilon \widetilde{N}_{i_1} \left(\frac{x}{\varepsilon} \right) \frac{\partial \widetilde{v}_0}{\partial x_{i_1}} \right\|^2_{W_2^1(B \cap G)} = O(\varepsilon + \mu) \operatorname{mes}(B \cap G), \tag{33}$$

which guarantees the proximity of the flows

$$\left\| A_{rj} \frac{\partial u}{\partial x_j} - A_{rj} \frac{\partial}{\partial x_j} \left(\widetilde{v}_0 + \varepsilon \widetilde{N}_{i_1} \left(\frac{x}{\varepsilon} \right) \frac{\partial \widetilde{v}_0}{\partial x_{i_1}} \right) \right\|^2_{L_2(B \cap G)} = \\ = O(\varepsilon + \mu) \operatorname{mes}(B \cap G). \tag{34}$$

Section 6 presents the asymptotic expansion of the above problems in powers of ε and μ.

REMARK 1: The term $\varepsilon N_{i_1 i_2} \partial^2 v_0 / \partial x_{i_2} \partial x_{i_2}$ in the approximation $u^{(2)}$ was introduced for uniformity with Chapter 4; actually, the approximation $u^{(1)} = v_0 +$

$\varepsilon N_{i_1} \partial v_0 / \partial x_{i_1}$ might suffice, which might shorten the proof of Theorem 1 (see §3).

REMARK 2: The averaged equation (4) can be written more naturally as

$$\frac{\partial}{\partial x_1}\left(4\mu \widehat{A}_{11} \frac{\partial \widetilde{v}_0}{\partial x_1}\right) + \frac{\partial}{\partial x_2}\left(4\mu \widehat{A}_{22} \frac{\partial \widetilde{v}_0}{\partial x_2}\right) = 4\mu f, \tag{35}$$

since the heat conductivity of the framework is proportional to the cross-section of the latter, and the amount of heat emitted on any periodicity cell is equal to $\varepsilon^2(\text{mes}(B \cap Q) f(x) + O(\varepsilon)) \approx \varepsilon^2 4\mu f(x)$.

The estimates (7), (33), (34) show that (35) describes the field in a hypothetical uniform medium which is equivalent asymptotically to the framework. The coefficients of (35) can be used to calculate the effective characteristics of the framework.

§2 A Geometric Model of a Framework

The concept of a rectangular framework can be generalized to the three-dimensional case. Let the mesh S_ε be given in $\mathbf{R}^3$, with the node coordinates $(k_1\varepsilon, k_2\varepsilon, k_3\varepsilon) = k\varepsilon$, k_j being integers, consisting of the straight lines $\Pi^1_{k_2k_3}$, $\Pi^2_{k_1k_3}$, and $\Pi^3_{k_1k_2}$ running through the nodes $(k_1\varepsilon, k_2\varepsilon, k_3\varepsilon)$ and parallel to the coordinate axes x_1, x_2, and x_3, respectively. By $B^1_{k_2k_3}$, $B^2_{k_1k_3}$, and $B^3_{k_1k_2}$, we denote cylinders made up of points spaced less than $\mu\varepsilon$ away from $\Pi^1_{k_2k_3}$, $\Pi^2_{k_1k_3}$, and $\Pi^3_{k_1k_2}$, where $\mu \ll 1$. The union of all the cylinders

$$B = \bigcup_{k,l=-\infty}^{\infty} (B^1_{kl} \cup B^2_{kl} \cup B^3_{kl})$$

is called *rectangular periodic framework* B.

To generalize the model of a framework, we may consider the cylinder B^j_{kl} with cross-sections in the form of some domains $\beta_j \subset \mathbf{R}^2$; we may add cylinders (strips in the two-dimensional case) oriented in noncoordinate directions (see Fig. 22b), and finally, we may consider an s-dimensional framework.

Let us formulate the definition of a periodic s-dimensional framework. The concept of a random framework will be introduced in §7.

DEFINITION 1: Let $\beta_1, \dots, \beta_J$ be bounded domains in $\mathbf{R}^{s-1}$ with a piecewise smooth boundary, let B_j $(j = 1, \dots, J)$ be cylinders, suppose $B_j = \{x \in \mathbf{R}^s | (x_2/\varepsilon\mu, \dots, x_3/\varepsilon\mu) \in \beta_j, x_1 \in \mathbf{R}\}$, and B^α_{hj} is the cylinder obtained from B_j by orthogonal transformation with the matrix α^T, $\alpha = \|\alpha_{il}\|$, and by translation by $\varepsilon h = (\varepsilon h_1, \dots, \varepsilon h_s)$. We shall call the domain

$$B = \bigcup_{j=1}^{J} \bigcup_{\alpha \in \Delta_j} \bigcup_{h \in H_{j\alpha}} B^\alpha_{hj},$$

an *s-dimensional framework*, where Δ_j is a set of orthogonal $s \times s$ matrices, $H_{j\alpha} \subset \mathbf{R}^s$, and ε, μ are small parameters, with Δ_j and $H_{j\alpha}$ being independent of ε and μ and not containing two collections $(\alpha^{(1)}, h^{(1)})$ and $(\alpha^{(2)}, h^{(2)})$ with the same first columns $\alpha^{(1)}$ and $\alpha^{(2)}$ for $h^{(1)} = h^{(2)}$ (this requirement forbids intersection of two cylinders B^{α}_{hj} with parallel generatrices for sufficiently small μ). The domain B is assumed to be connected and periodic with period ε in each of the variables x_l, whereas the intersection of B with the periodicity cube $Q_0 = \{x \in \mathbf{R}^s | 0 < x_l < \varepsilon,\ l = 1, \ldots, s\}$ is connected and can be represented as the finite union

$$B \cap Q_0 = \bigcup_{p=1}^{I} B^{\alpha_p}_{h_p j_p} \cap Q_0. \tag{1}$$

In conclusion of this section, we define another geometric object to be used below. Let $B^{\alpha+}_{hj}$ be the maximal subset of B^{α}_{hj}, any cross-section of which by a plane perpendicular to the generatrix of B^{α}_{hj} is free of points of the other B^{α}_{hj}. The cylinders which are connected components of the set $B^{\alpha+}_{hj}$ are called *sections*. Figure 29 shows an example of the framework sections (the domain occupied by the sections is shaded).

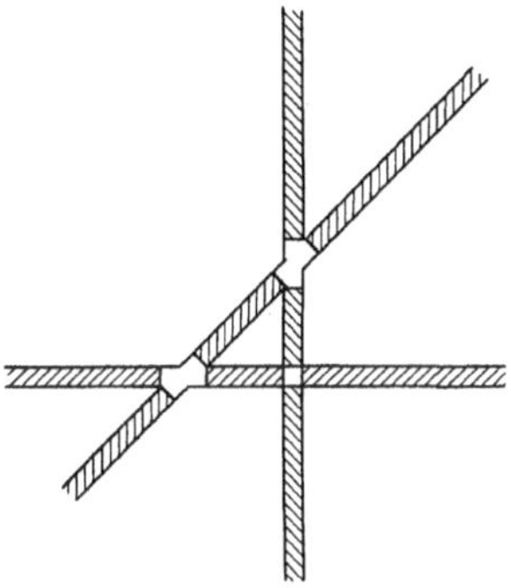

Figure 29

§3 The Splitting Principle for the Averaged Operator for a Periodic Framework

1. We begin the presentation of the splitting principle for the averaged operator with the example from §1. We multiply the averaged equation (1.16) by $\text{mes}(B_\xi \cap Q)$ (see Remark 1.1) and get

$$\Lambda v_0 \equiv - \sum_{i_1, i_2 = 1}^{2} \hat{b}_{i_1 i_2} \frac{\partial^2 v_0}{\partial x_{i_1} \partial x_{i_2}} = \text{mes}\,(B_\xi \cap Q)\, f(x),$$

where

$$\hat{b}_{i_1 i_2} = \int\limits_{B_\xi \cap Q} \left(A_{i_1 j} \frac{\partial N_{i_2}}{\partial \xi_j} + A_{i_1 i_2} \right) d\xi,$$

where N_{i_2} is a solution of the "problem on a cell" (1.11), (1.12). By $\hat{L}_0$ we denote the principal part of the averaged operator Λ as $\mu \to 0$ (see (1.35)).

The averaged operator along the strip B^i_ξ is the operator

$$\Lambda^i = - \sum_{i_1, i_2 = 1}^{2} \hat{b}^i_{i_1 i_2} \frac{\partial^2}{\partial x_{i_1} \partial x_{i_2}}, \qquad i = 1, 2, \tag{2}$$

where

$$\hat{b}^i_{i_1 i_2} = \int\limits_{B^i_\xi \cap Q} \left(A_{i_1 j} \frac{\partial \widetilde{N}^i_{i_2}}{\partial \xi_j} + A_{i_1 i_2} \right) d\xi, \tag{3}$$

$\widetilde{N}^i_{i_2}$ is a solution of the "problem on a strip" (1.18), (1.19). We denote the principal part of the averaged operator along the strip B^i_ξ – Λ^i as $\mu \to 0$ by $\hat{L}^i_0$.

Note that $\hat{L}^i_0$ coincides with the leading terms of the averaged operators for "soft" media reinforced with "rigid" unidirectional strips oriented in the ith coordinate direction (see Chapter 7).

The splitting principle for the averaged operator states that the leading term of the averaged operator for problem (1.1)–(1.3) is the sum of the leading terms of the averaged operators along B^i_ξ, that is,

$$\hat{L}_0 = \hat{L}^1_0 + \hat{L}^2_0.$$

Relations (2), (3), and (1.35) imply that

$$\hat{b}^i_{i_1 i_2} = \delta_{i_1 i} \delta_{i_2 i} \hat{\hat{A}}_{ii} = 2\delta_{i_1 i} \delta_{i_2 i} \left(A_{ii} - A_{i\bar{i}} A^{-1}_{\bar{i}\bar{i}} A_{\bar{i} i} \right) \mu,$$

whence $\hat{L}^i_0 = \hat{\hat{A}}_{ii} \frac{\partial^2}{\partial x_i^2}$ (no summation over i). Thus, for this example, the splitting principle is a corollary of Theorem 1.1.

2. We consider the two-dimensional framework B of a more general form, which consists of strips of I directions (see (2.1)). In this case, the averaged operator along the strip is naturally the operator

$$\hat{L}^{\alpha_p}_{h_p j_p} = - \sum_{i_1, i_2 = 1}^{2} \hat{b}^{\alpha_p}_{i_1 i_2, h_p j_p} \frac{\partial^2}{\partial x_{i_1} \partial x_{i_2}},$$

where

$$\hat{b}^{\alpha_p}_{i_1 i_2, h_p j_p} = \int\limits_{B^{\alpha_p}_{\xi h_p j_p} \cap Q} \left(A_{i_1 k} \frac{\partial \tilde{N}^{\alpha_p}_{i_2, h_p j_p}}{\partial \xi_h} + A_{i_1 i_2} \right) d\xi, \tag{4}$$

$\tilde{N}^{\alpha_p}_{i_2, h_p j_p}(\xi)$ is a bounded linear solution of the "problem on a strip,"

$$L \tilde{N}^{\alpha_p}_{i_2, h_p j_p} = 0 \quad \text{for} \quad \xi \in B^{\alpha_p}_{\xi h_p j_p}, \tag{5}$$

$$\frac{\partial}{\partial \nu_\xi} \left(\tilde{N}^{\alpha_p}_{i_2, h_p j_p} + \xi_{i_2} \right) = 0 \quad \text{for} \quad \xi \in \partial B^{\alpha_p}_{\xi h_p j_p}, \tag{6}$$

$B^{\alpha_p}_{\xi h_p j_p}$ is the image of the strip $B^{\alpha_p}_{h_p j_p}$ in the transformation of $\xi = x/\varepsilon$. We maintain the notation $\hat{L}^{\alpha_p}_{h_p j_p}$ for the principal part (as $\mu \to 0$) of the operator along $B^{\alpha_p}_{h_p j_p}$.

By analogy with Theorem 1.1, it can be proved that the solution of problem (1.1)–(1.3) in the norm in $L_2(B \cup G)$ is close to that of the averaged problem

$$\hat{L}_0 \tilde{v}_0 = \operatorname{mes}(B_\xi \cap Q) f, \quad x \in G,$$

$$\tilde{v}_0 \big|_{\partial G} = 0, \quad \hat{L}_0 = \sum_{p=1}^{I} \hat{L}^{\alpha_p}_{h_p j_p}.$$

To calculate the coefficients of $\hat{L}^{\alpha_p}_{h_p j_p}$ from (4), we should find a linear solution of problem (5), (6). This problem can be reduced to problem (1.18), (1.19) by substituting $\tilde{\xi} = \alpha_p(\xi - h_p)$. The solution of problem (1.18), (1.19) is given by (1.21).

3. In the case of the nonlinear system of equations of elasticity theory

$$\sum_{r=1}^{s} \frac{\partial}{\partial x_r} A_r(\nabla_x u) = f(x), \quad x \in B \cap G, \qquad \sum_{r=1}^{s} A_r(\nabla_x u) n_r = 0, \quad x \in \partial B \cap G,$$

where s is the dimension of space, and the A_r are n-dimensional vectors, the problem on a cell is of the form (see §4.6)

$$\frac{\partial}{\partial \xi_r} A_r(\nabla_\xi N_1 + \nabla_x v) = 0, \quad \xi \in B_\xi,$$

$$n_r A_r(\nabla_\xi N_1 + \nabla_x v) = 0, \quad \xi \in \partial B_\xi;$$

and the problem on a strip $B^{\alpha_p}_{h_p j_p}$ is of the form

$$\frac{\partial}{\partial \xi_r} A_r \left(\nabla_\xi \widetilde{N}^{\alpha_p}_{1,h_p j_p} + \nabla_x v \right) = 0, \quad \xi \in B^{\alpha_p}_{\xi h_p j_p}, \tag{7}$$

$$n_r A_r \left(\nabla_\xi \widetilde{N}^{\alpha_p}_{1,h_p j_p} + \nabla_x v \right) = 0, \quad \xi \in \partial B^{\alpha_p}_{\xi h_p j_p}, \tag{8}$$

where $\nabla_x v$ is the $n \times s$ matrix parameter and $\widetilde{N}^{\alpha_p}_{1,h_p j_p} = \overline{X}_i(\nabla_x v)(\xi_i - h^i_p)$ is a bounded function linear in ξ_i. The operator

$$\widehat{L}^{\alpha_p}_{h_p j_p} = \frac{\partial}{\partial x_r} \int\limits_{B^{\alpha_p}_{\xi h_p j_p} \cap Q} A_r \left(\nabla_\xi \widetilde{N}^{\alpha_p}_{1,h_p j_p} + \nabla_x v \right) d\xi.$$

is called the *averaged operator* along $B^{\alpha_p}_{h_p j_p}$. Problem (7), (8) can be also reduced to a similar problem in $B^E_{\xi_0 j_p}$ by substituting $\widetilde{\xi} = \alpha_p(\xi - h_p)$.

4. We give the formulation of the splitting principle for the averaged operator for a nonstationary, nonlinear system of partial differential equations and prove the theorem to justify the principle.

Let G be a domain in $\mathbf{R}^s$ such that ∂G is piecewise smooth and intersects with no more than K_0/ε^{s-1} sections. Consider the following problem:

$$Lxu \equiv R(x,t)\frac{\partial^2 u}{\partial t^2} + S(x,t)\frac{\partial u}{\partial t} - \frac{\partial}{\partial x_i} A_i(\nabla_x u, x, t) = f(x,t), \quad x \in B \cap G, \tag{9}$$

$$\frac{\partial u}{\partial \nu} \equiv A_i(\nabla_x u, x, t)\, n_i(x) = 0, \quad x \in \partial B \cap G, \tag{10}$$

$$u = 0, \quad x \in \partial G \cap \overline{B}, \tag{11}$$

$$u|_{t=0} = \left.\frac{\partial u}{\partial t}\right|_{t=0} = 0, \quad R > 0, \qquad u|_{t=0} = 0, \quad R = 0, \quad S > 0, \tag{12}$$

where $R(x,t)$ and $S(x,t)$ are $n \times n$ matrix-valued functions, $u(x,t)$, $f(x,t)$, $A_i(y,x,t)$ $(i = 1, \dots, s)$ are n-dimensional vector-functions, $\nabla_x u = \|\partial u_k/\partial x_l\|$ is an $n \times s$ matrix, $y = \|y_{kl}\|$ is an $n \times s$ matrix-argument, f, R, S are assumed to be continuously differentiable, and A_i to be twice continuously differentiable, and $n_i(x)$ is the cosine of the angle between the axis $0x_i$ and the exterior normal vector $n(x)$. By formally taking $R = S = 0$, we can consider the stationary problem (9)–(11). For linear elasticity,

$$A_i(\nabla_x u, x, t) = \sum_{j=1}^{s} A_{ij}(x,t)\frac{\partial u}{\partial x_j},$$

where the $A_{ij}(x,t)$ are $n \times n$ matrices satisfying the conditions for strong ellipticity and symmetry (see [76, 155]), and $R = \rho E$, $S = 0$. For $n = s$, the strong ellipticity and symmetry conditions have the form

$$
\begin{gathered}
a_{ij}^{kl}=a_{kj}^{il}=a_{il}^{kj}=a_{ji}^{lk}, \\
a_{ij}^{kl}\eta_{ik}\eta_{jl}\geqslant \varkappa_1\eta_{ik}\eta_{ik} \quad \text{for} \quad \eta_{ik}=\eta_{ki}.
\end{gathered} \tag{13}
$$

In the isotropic homogeneous case, $a_{ij}^{kl}=\lambda\delta_{ki}\delta_{jl}+\zeta(\delta_{kl}\delta_{ij}+\delta_{kj}\delta_{il})$, where λ, ζ are Lamé coefficients and ρ, λ, ζ are constant.

Let B_{hj}^{α} be any of the cylinders contained in the union (2.1). We make the change of variables $\tilde{x}=\alpha x$. In the new variables, the expression for $L_x u$ assumes the form

$$
L_{\tilde{x}}u=\widetilde{R}(\tilde{x},t)\frac{\partial^2 u}{\partial t^2}+\widetilde{S}(\tilde{x},t)\frac{\partial u}{\partial t}-\frac{\partial}{\partial\tilde{x}_i}\widetilde{A}_i\left(\nabla_{\tilde{x}}u,\tilde{x},t\right),
$$

where $\widetilde{R}(\tilde{x},t)=R(\alpha^T\tilde{x},t)$, $\widetilde{S}(\tilde{x},t)=S(\alpha^T\tilde{x},t)$, $\tilde{A}_i(\tilde{y},\tilde{x},t)=\alpha_{ij}A_j(\alpha^T y^T,\alpha^T\tilde{x},t)$. We introduce $s-1$ unknown n-dimensional vectors (altogether $(s-1)n$ unknown numbers) $X_2,\dots,X_2$, and the n-dimensional vector-valued parameter Y_1, and define $X_2,\dots,X_s$ from the system of $(s-1)n$ algebraic equations

$$
\begin{gathered}
\widetilde{A}_2(Y_1,X_2,\dots,X_s,\tilde{x},t)=0, \\
\cdots\cdots\cdots\cdots \\
\widetilde{A}_s(Y_1,X_2,\dots,X_s,\tilde{x},t)=0.
\end{gathered} \tag{14}
$$

Here $\tilde{x}$, t are also parameters, and so the solution $X_2,\dots,X_s$ of system (14) depends on the parameters Y_1, $\tilde{x}$, t.

Since we speak about the averaged equation of zeroth order (or, more precisely, about its principal part as $\mu\downarrow 0$) throughout the chapter, except for §6 and Subsection 9 herein, we shall denote the solution $\tilde{v}_0$ of the principal part of the averaged zero-order equation by v.

The system of algebraic equations (14) is a consequence of the problem on a strip

$$
\frac{\partial}{\partial\xi_r}A_r\left(\nabla_\xi\widetilde{N}_{1,hj}^{\alpha}+\nabla_x v,x,t\right)=0,\quad \xi\in B_{\xi hj}^{\alpha}, \tag{15}
$$

$$
n_r A_r\left(\nabla_\xi\widetilde{N}_{1,hj}^{\alpha}+\nabla_x v,x,t\right)=0,\quad \xi\in\partial B_{\xi hj}^{\alpha}, \tag{16}
$$

if a solution of problem (15), (16) is sought in the form

$$
\widetilde{N}_{1,h_j}^{\alpha}(\tilde{\xi})=\sum_{j=2}^{s}\left(X_j\left(\frac{\partial v}{\partial\tilde{x}_1},\tilde{x},t\right)-\frac{\partial v}{\partial\tilde{x}_j}\right)\tilde{\xi}_j,
$$

where $\tilde{x}=\alpha x$, $\tilde{\xi}=\alpha(\xi-h_p)$. The parameter Y_1 corresponds to $\partial v/\partial\tilde{x}_1$.

Let X_i be a twice continuously differentiable function of the arguments Y_1, $\tilde{x}$, t. The operator $L_{\tilde{x}}^{\alpha,h,j}$ defined by

$$\widehat{L}_{\widetilde{x}}^{\alpha,h,j} v = M_{hj}^{\alpha} \left(\widetilde{R}(\widetilde{x}, t) \frac{\partial^2 v}{\partial t^2} + \widetilde{S}(\widetilde{x}, t) \frac{\partial v}{\partial t} - \right.$$
$$\left. - \frac{\partial}{\partial \widetilde{x}_1} \widetilde{A}_1 \left(\frac{\partial v}{\partial \widetilde{x}_1}, X_2 \left(\frac{\partial v}{\partial \widetilde{x}_1}, \widetilde{x}, t \right), \ldots, X_s \left(\frac{\partial v}{\partial \widetilde{x}_1}, \widetilde{x}, t \right), \widetilde{x}, t \right) \right),$$

where M_{hj}^{α} is the principal part of $\mathrm{mes}(B_{hj}^{\alpha} \cap Q_0)/\varepsilon^s$ (that is, $\mu^{s-1} \lim_{\mu \to 0} \mathrm{mes}(B_{hj}^{\alpha} \cap Q_0)/\varepsilon^s \mu^{s-1}$), is called the *averaged operator along* B_{hj}^{α}.

By returning to the variables x, we obtain the averaged operator along B_{hj}^{α} in the original variable

$$\widehat{L}_{x}^{\alpha,h,j} v = M_{hj}^{\alpha} \left(R(x, t) \frac{\partial^2 v}{\partial t^2} + S(x, t) \frac{\partial v}{\partial t} - \sum_{k=1}^{s} \alpha_{1k} \frac{\partial}{\partial x_k} \times \right.$$
$$\times \left(\widetilde{A}_1 \left(\sum_{k=1}^{s} \alpha_{1k} \frac{\partial v}{\partial x_k}, X_2 \left(\sum_{k=1}^{s} \alpha_{1k} \frac{\partial v}{\partial x_k}, \alpha x, t \right), \ldots \right. \right.$$
$$\left. \left. \left. \ldots, X_s \left(\sum_{k=1}^{s} \alpha_{1k} \frac{\partial v}{\partial x_k}, \alpha x, t \right), \alpha x, t \right) \right) \right) =$$
$$= M_{hj}^{\alpha} \left(R(x, t) \frac{\partial^2 v}{\partial t^2} + S(x, t) \frac{\partial v}{\partial t} - \sum_{k=1}^{s} \frac{\partial}{\partial x_k} \times \right.$$
$$\times \left(\alpha_{1k} \sum_{j=1}^{s} \alpha_{1j} A_j \left(\alpha^T \left(\sum_{k=1}^{s} \alpha_{1k} \frac{\partial v}{\partial x_k}, X_2 \left(\sum_{k=1}^{s} \alpha_{1k} \frac{\partial v}{\partial x_k}, \alpha x, t \right), \ldots \right. \right. \right.$$
$$\left. \left. \left. \left. \ldots, X_s \left(\sum_{k=1}^{s} \alpha_{1k} \frac{\partial v}{\partial x_k}, \alpha x, t \right) \right) \right)^T, x, t \right) \right) \right). \quad (17)$$

The operator $\hat{L}_x^{\alpha,h,j}$ is the principal part of the operator

$$\int\limits_{B_{\xi hj}^{\alpha} \cap Q} \left(R(x, t) \frac{\partial^2}{\partial t^2} + S(x, t) \frac{\partial}{\partial t} - \frac{\partial}{\partial x_r} A_r \left(\nabla_{\xi} \widetilde{N}_{1,h_j}^{\alpha} (\nabla_{x}\cdot, x, t) + \nabla_{x}\cdot, x, t \right) \right) d\xi,$$

where $\widetilde{N}_{1,h_j}(\nabla_x v, x, t)$ is a solution of the problem on a strip (15), (16).

In the case of the linear system of equations of elasticity theory, system (14) reduces to a linear system of $s-1$ equations in $s-1$ unknown $n \times n$ matrices $C_2(\tilde{x}, t), \ldots, C_s(\tilde{x}, t)$:

$$-\sum_{l=2}^{s} \widetilde{A}_{2l}(\widetilde{x}, t)\, C_l = \widetilde{A}_{21}(\widetilde{x}, t),$$
$$\cdots\cdots\cdots\cdots\cdots \tag{14'}$$
$$-\sum_{l=2}^{s} \widetilde{A}_{sl}(\widetilde{x}, t)\, C_l = \widetilde{A}_{s1}(\widetilde{x}, t),$$

where $\tilde{A}_{jl}(\tilde{x},t) = \sum_{i,k=1}^{s} \alpha_{ji}a_{lk}A_{ik}(\alpha^T\tilde{x},t)$, $\tilde{x}$, t are parameters, in the notation (13) $X_l = C_lY_1$.

The operator $\hat{L}_{\tilde{x}}^{\alpha,h,j}$ has the form

$$M_{hj}^{\alpha}\left(R(\alpha^T\widetilde{x}, t)\frac{\partial^2}{\partial t^2} + S(\alpha^T\widetilde{x}, t)\frac{\partial}{\partial t} - \frac{\partial}{\partial \widetilde{x}_1}\left(\widehat{A}(\widetilde{x}, t)\frac{\partial}{\partial \widetilde{x}_1}\right)\right),$$

where $\hat{A} = \tilde{A}_{11} + \sum_{l=2}^{s} \tilde{A}_{1l}C_l$, and in particular, $\hat{A} = \tilde{A}_{11} = \tilde{A}_{12}\tilde{A}_{22}^{-1}\tilde{A}_2$ for $s = 2$, while for $n = s = 2$ and

$$\widetilde{A}_{11} = \begin{pmatrix} a & g \\ g & b \end{pmatrix},\quad \widetilde{A}_{22} = \begin{pmatrix} b & e \\ e & c \end{pmatrix},\quad \widetilde{A}_{12} = \begin{pmatrix} g & d \\ b & e \end{pmatrix},\quad \widetilde{A}_{21} = \begin{pmatrix} g & b \\ d & e \end{pmatrix}$$

we have

$$\widehat{A} = \begin{pmatrix} \widehat{E} & 0 \\ 0 & 0 \end{pmatrix},$$

where

$$\widehat{E} = a - \frac{g^2c - 2gde + bd^2}{bc - e^2}.$$

Returning to the variables x, we get

$$\widehat{L}_x^{\alpha,h,j} = M_{hj}^{\alpha}\left(R(x, t)\frac{\partial^2}{\partial t^2} + S(x, t)\frac{\partial}{\partial t} - \sum_{r=1}^{s}\frac{\partial}{\partial x_r}\left(\alpha_{1r}\sum_{l=1}^{s}\alpha_{1l}\widehat{A}(\alpha x, t)\frac{\partial}{\partial x_l}\right)\right).$$

Under conditions (13), we can prove the solvability of (14′).

The splitting principle for the averaged operator states that the principal part of the averaged operator $\hat{L}_x$ of problem (9)–(12) can be represented as a sum of averaged operators along $B_{h_pj_p}^{\alpha_p}$, that is,

$$\widehat{L}_x = \sum_{p=1}^{I} \widehat{L}_x^{\alpha_p,h_p,j_p}. \tag{18}$$

5. To problem (9)–(12), we assign the averaged problem

$$\widehat{L}_x v = Mf(x, t), \quad x \in G, \tag{19}$$

$$v|_{\partial G} = 0, \tag{20}$$

$$v|_{t=0} = \frac{\partial v}{\partial t}\Big|_{t=0} = 0, \quad R > 0, \qquad v|_{t=0} = 0, \quad R = 0, \quad S > 0, \tag{21}$$

where M is the principal part of $\mathrm{mes}(B \cap Q_0/\varepsilon^s)$.

For each cylinder B_{hj}^{α}, we define the function

$$\chi_0\left(\frac{\widetilde{x}_1}{\varepsilon\mu}\right) = \chi_0\left(\frac{1}{\varepsilon\mu}\sum_{j=1}^{s} \alpha_{1j}x_j\right) = \chi\left(\frac{x}{\varepsilon\mu}\right)$$

equal to zero everywhere outside the sections and also on the sections that have common points with ∂G; in all remaining sections outside the $\varepsilon\mu$-neighbourhood of the bases of the sections, we set $\chi_0(\tilde{x}_1/\varepsilon\mu) = 1$, and we complete the definition of $\chi_0(\tilde{x}_1/\varepsilon\mu)$ in $\varepsilon\mu$-neighbourhoods of the bases so that $\chi_0(t) \in C^\infty(R)$, $0 \leqslant \chi_0(t) \leqslant 1$. We set $u^{(1)} = v + \varepsilon N_0(\tilde{x}/\varepsilon, \nabla_{\tilde{x}} v, \tilde{x}, t)\chi_0(\tilde{x}_1/\varepsilon\mu)$, where

$$N_0\left(\frac{\widetilde{x}}{\varepsilon}, \nabla_{\widetilde{x}} v, \widetilde{x}, t\right) = \sum_{j=2}^{s}\left(X_j\left(\frac{\partial v}{\partial \widetilde{x}_1}, \widetilde{x}, t\right) - \frac{\partial v}{\partial \widetilde{x}_j}\right)\left(\frac{\widetilde{x}_j}{\varepsilon} - \alpha h\right), \tag{22}$$

i.e., in the variables x

$$u^{(1)} = v + \varepsilon N\left(\frac{x}{\varepsilon}, \nabla_x v, x, t\right)\chi\left(\frac{x}{\varepsilon\mu}\right), \tag{23}$$

where $N(x/\varepsilon, \nabla_x v, x, t)$ is the product of the matrix with the columns

$$0,\ X_2\left(\sum_{k=1}^{s}\alpha_{1k}\frac{\partial v}{\partial x_k}, \alpha x, t\right) - \sum_{k=1}^{s}\alpha_{2k}\frac{\partial v}{\partial x_k}, \ldots$$

$$\ldots, X_s\left(\sum_{k=1}^{s}\alpha_{1k}\frac{\partial v}{\partial x_k}, \alpha x, t\right) - \sum_{k=1}^{s}\alpha_{sk}\frac{\partial v}{\partial x_k}$$

by the vector $\alpha(x/\varepsilon - h)$.

For any cylinder B_{hj}^{α}, the function $N = \widetilde{N}_{1,hj}^{\alpha}$ is a solution of the problem on a strip (15), (16). The solution of problem (9)–(12) satisfies the following identity: for any vector-valued function $\varphi(x,t)$ ($\varphi|_{\partial G} = 0$) that belongs to the space $W_2^1(B \cap G)$ for each fixed t,

$$\int_{B\cap G}\left(R\frac{\partial^2 u}{\partial t^2}\varphi + S\frac{\partial u}{\partial t}\varphi + A_i(\nabla_x u, x, t)\frac{\partial \varphi}{\partial x_i} - f\varphi\right)dx = 0.$$

THEOREM 1. (Splitting Principle). *Let $\hat{L}_x$ be defined by (17), (18), let v be a solution of problems (19)–(21), and suppose that at all points of G, the function v has all partial derivatives up to the third order which are bounded by a constant c. Then for any function $\varphi(x,t)$, such that $\varphi(x,t) \in W_2^1(B \cap G)$ at each t (and $\varphi|_{\partial G} = 0$) and for any t, the following inequality holds:*

$$\Delta \equiv \frac{1}{\sqrt{\operatorname{mes}(B\cap G)}}\left|\int_{B\cap G}\left(R\frac{\partial^2(u^{(1)}-u)}{\partial t^2}\varphi + S\frac{\partial(u^{(1)}-u)}{\partial t}\varphi + \right.\right.$$

$$\left.\left. + \sum_{i=1}^{s}\left(A_i(\nabla_x u^{(1)}, x, t) - A_i(\nabla_x u, x, t)\right)\frac{\partial\varphi}{\partial x_i}\right)dx\right| \leqslant$$

$$\leqslant c_1(t)\left(\sqrt{\varepsilon} + \sqrt{\mu}\right)\|\varphi\|_{W_2^1(B\cap G)}, \qquad (24)$$

where $u^{(1)}$ is the vector-function defined on any beam B^{α}_{hj} by (23), ε, $\mu \to 0$, and $c_1(t) < \infty$ is independent of ε and μ.

REMARK 1: The quantity Δ in (24) is a characteristic of the error of problems (9)–(12). In the linear case, (24) may yield the closeness of $u^{(1)}$ and u in the norms of Chapter 1.

Namely, we have

COROLLARY 1: Let $n = 1$, $A_i(\nabla_x u, x, t) = \sum_{j=1}^{s} A_{ij}\frac{\partial u}{\partial x_k}$, and suppose that R, S, A_{ij} are constants, $A_{ij} = A_{ji}$, $\sum_{i,j=1}^{s} A_{ij}\eta_i\eta_j \geqslant \kappa\sum_{i=1}^{s}\eta_i^2$ for any $\eta_1, \dots, \eta_s$, and $f(x,t) \in C^\infty(\overline{G} \times [0,\infty))$. Let either a) $R > 0$, $S \geqslant 0$ or b) $R = 0$, $S > 0$ or c) $R = S = 0$, whereas in the nonstationary case, the conditions $f(x,0) = \frac{\partial f}{\partial t}(x,0) = 0$ are satisfied, but in the stationary case f is independent of t. Then problem (14) is solvable, there exists a unique solution v of problem (19)–(21), and the estimates

$$\frac{1}{\sqrt{\operatorname{mes}(B\cap G)}}\|u - v\|_{L_2(B\cap G)} \leqslant c_2\left(\sqrt{\varepsilon} + \sqrt{\mu}\right),$$

$$\frac{1}{\sqrt{\operatorname{mes}(B\cap G)}}\|u - u^{(1)}\|_W \leqslant c_3\left(\sqrt{\varepsilon} + \sqrt{\mu}\right)$$

hold for sufficiently small ε, μ; here

$$W = \begin{cases} W_2^1(B\cap G\times(0,\tau)) & \text{in case a)},\\ W_2^{1,0}(B\cap G\times(0,\tau)) & \text{in case b)},\\ W_2^1(B\cap G) & \text{in case c)}.\end{cases}$$

To prove Corollary 1, it suffices to substitute into (24) the function $\varphi = \partial(u^{(1)} - u)/\partial t$ in case a) and the function $\varphi = u^{(1)} - u$, in cases b) and c) and then to integrate (24) over t between 0 and T [cases a) and b)].

Considering the initial conditions, we get:
in case a)

$$\sup_{T\in[0,\tau]} \frac{1}{\sqrt{\operatorname{mes}(B\cap G)}}\left(\int_{B\cap G} \frac{1}{2} R\left(\frac{\partial(u^{(1)}-u)}{\partial t}\right)^2\Big|_{t=T} dx + \int_0^T\int_{B\cap G} S\left(\frac{\partial(u^{(1)}-u)}{\partial t}\right)^2 dx\,dt + \frac{1}{2}\sum_{i,j=1}^{s}\int_{B\cap G} A_{ij}\frac{\partial(u^{(1)}-u)}{\partial x_j}\frac{\partial(u^{(1)}-u)}{\partial x_i}\Big|_{t=T} dx\right) \leqslant$$

$$\leqslant c_1(\sqrt{\varepsilon}+\sqrt{\mu})\,\|u^{(1)}-u\|_{W_2^{1,0}(B\cap G\times(0,\tau))};$$

in case b)

$$\sup_{T\in[0,\tau]} \frac{1}{\sqrt{\operatorname{mes}(B\cap G))}}\left(\int_{B\cap G} \frac{1}{2} S\,(u^{(1)}-u)^2\Big|_{t=T} dx + \sum_{i,j=1}^{s}\int_0^T\int_{B\cap G} A_{ij}\frac{\partial(u^{(1)}-u)}{\partial x_j}\frac{\partial(u^{(1)}-u)}{\partial x_i} dx\,dt\right) \leqslant$$

$$\leqslant c_1(\sqrt{\varepsilon}+\sqrt{\mu})\,\|u^{(1)}-u\|_{W_2^{1,0}(B\cap G\times(0,\tau))};$$

in case c)

$$\frac{1}{\sqrt{\operatorname{mes}(B\cap G)}}\int_{B\cap G}\sum_{i,j=1}^{s} A_{ij}\frac{\partial(u^{(1)}-u)}{\partial x_j}\frac{\partial(u^{(1)}-u)}{\partial x_i} dx \leqslant$$

$$\leqslant c_1(\sqrt{\varepsilon}+\sqrt{\mu})\,\|u^{(1)}-u\|_{W_2^{1}(B\cap G)}.$$

Taking into account the inequalities

$$\int_{B\cap G} A_{ij}\frac{\partial(u^{(1)}-u)}{\partial x_j}\frac{\partial(u^{(1)}-u)}{\partial x_i} dx \geqslant c_4\,\|u^{(1)}-u\|^2_{W_2^1(B\cap G)},$$

$$\int_0^T\int_{B\cap G} S\left(\frac{\partial(u^{(1)}-u)}{\partial t}\right)^2 dx\,dt \geqslant 0,$$

where c_4 is independent of ε and μ, we can obtain the assertion of Corollary 1.

A more general definition of a framework as compared with §2 will be introduced in Subsection 8, and the proof of an analogue of Theorem 1 (Theorem 4) will be given for such structures. Theorem 1 follows from Theorem 4.

For the system of equations of elasticity theory, stresses are calculated by the approximate formulas $\sigma_i \approx A_i(\nabla_x u^{(1)}, x, t)$.

REMARK 2: The splitting principle for the averaged operator can be generalized to the case where the A_i may depend on u: $A_i(\nabla_x u, u, x, t)$.

6. For the framework shown in Figure 22b, in the linear isotropic case, system (19) has the form

$$-\mu\bar{E}\frac{\sqrt{2}}{2}\left((1+\sqrt{2})\frac{\partial^2 v_1}{\partial x_1^2}+2\frac{\partial^2 v_2}{\partial x_1\partial x_2}+\frac{\partial^2 v_1}{\partial x_2^2}\right)=2\mu(1+\sqrt{2})f_1,$$

$$-\mu\bar{E}\frac{\sqrt{2}}{2}\left(\frac{\partial^2 v_2}{\partial x_1^2}+2\frac{\partial^2 v_1}{\partial x_1\partial x_2}+(1+\sqrt{2})\frac{\partial^2 v_2}{\partial x_2^2}\right)=2\mu(1+\sqrt{2})f_2,$$

where $\overline{E} = (3\lambda + 2\zeta)\zeta/(\lambda + \zeta)$.

7. The splitting principle for the averaged operator allows us to obtain in the case of the system of equations of elasticity theory an explicit expression for the averaged elasticity tensor (7.5.8), (7.5.9). By analogy with §7.5, we can derive from this formula the following theorems.

THEOREM 2. *Let the framework B consist of cylinders in I directions, and let L_x be the operator of the isotropic elasticity theory (see §7.5). Then for $I \leqslant 5$, the operator L_x does not satisfy (13). For $I = 6$, it is possible to choose the directions of the cylinders making up B so that $\hat{L}_x$ satisfies (13).*

THEOREM 3. *If $\hat{L}_x$ is the operator of the isotropic elasticity theory, then the effective Poisson ratio is 1/4.*

8. The definition of a periodic framework introduced in §2 covers the structures which are "run through" by each cylinder B^{α}_{hj}. Yet the skeletal structure in Figure 23 will not come under Definition 2.1. We introduce a more general definition of a framework.

Let e^{α}_h be an s-dimensional vector obtained from a vector $i(h, \alpha)$ of the axis $0x_1$ by orthogonal transformation of $\mathbf{R}^s$ with matrix α^T and by translation by $h = (h_1, \dots, h_s)$. Let

$$\mathscr{B} = \bigcup_{\alpha\in\Delta}\ \bigcup_{h\in H_\alpha} e^{\alpha}_h$$

be such that any two segments e^{α}_h can have only one common point which is the end point for both segments. The end points e^{α}_h are called *nodes* $\Delta \in \mathbf{R}^s$, $H_\alpha \subset \mathbf{R}^s$, with Δ and H_α being independent of μ, ε.

We assume that the system of the segments, $\mathcal{B}$, is connected and periodic in all the x_i with period 1, with the periodicity cube $Q = \{x \in \mathbf{R}^s | -1/2 \leqslant x_i \leqslant 1/2\}$ intersecting a finite number of segments from $\mathcal{B}$: $e^{\alpha_1}_{h_1}, \dots, e^{\alpha_I}_{h_I}$. By e_h we will mean the segment $\{x \in \mathbb{R}^s | x/\varepsilon \in e^{\alpha}_h\}$ produced from e^{α}_h by homothetic contraction of space in $1/\varepsilon$ times. To each of the segments $e^{\alpha}_{h_i}$ $(i = 1, \dots, I)$, we assign the cylinder $B^{\alpha_i}_{h_i j_i}$ and denote by $\widetilde{B}^{\alpha_i}_{h_i j_i}$ the part of $B^{\alpha_i}_{h_i j_i}$ contained between the two planes passing through the ends of segment $e^{\alpha_i}_{h_i}\varepsilon$ and perpendicular to it (we assume that the bases belong to $\widetilde{B}^{\alpha_i}_{h_i j_i}$). By *skeletal structure* B, we shall mean the set of interior points of the periodic continuation (with the periodicity cube $(0,\varepsilon)^s$) to $\mathbf{R}^s$ of the union $\bigcup\limits_{i=1}^{I}\widetilde{B}^{\alpha_i}_{h_i j_i}$ while assuming that $B\cap Q_0$ is connected.

The maximal subsets of the cylinders $\widetilde{B}^{\alpha}_{hj}$ for which any cross-section by the plane perpendicular to the generatrix of $\widetilde{B}^{\alpha}_{hj}$ is free of points of other cylinders we call *sections*. By S_0, we denote the set of sections and let $c_0\varepsilon\mu > d$, where d is the maximum diameter of the connected subsets of the set $B\backslash S_0$, with c_0 being independent of ε and μ.

Let $B^{\alpha}_{hj}(B^{\alpha++}_{hj})$ be a part of $\widetilde{B}^{\alpha}_{hj}$ contained between the planes spaced by $c_0\varepsilon\mu$ (by $(c_0+1)\varepsilon\mu$) from the bases of $\widetilde{B}^{\alpha}_{hj}$. We introduce the notation

$$B^{+} = \bigcup_{\alpha\in\Delta}\bigcup_{h\in H_\alpha} B^{\alpha+}_{hj_{\alpha h}},\quad B^{++} = \bigcup_{\alpha\in\Delta}\bigcup_{h\in H_\alpha} B^{\alpha++}_{hj_{\alpha h}}.$$

The domain $B\backslash\overline{B}^{++}$ is called the *nodal domain*.

Let G be a bounded domain in $\mathbf{R}^s$ with the piecewise smooth boundary ∂G. For simplicity, we also assume that ∂G has an empty intersection with the nodal domain. We consider problem (9)–(12) under the same assumptions with regard to the coefficients as above.

We shall describe the procedure of constructing the averaged operator. To each segment $e\subset\mathcal{B}$, we assign a collection of n-dimensional vectors $X^e_0(y,x,t)$, $\ldots, X^e_s(y,x,t)\in C^2$, which satisfies the following four conditions:

1)

$$A_i\left(X^e_1(y,x,t),\ldots,X^e_s(y,x,t),x,t\right)\nu^j_i = 0, \tag{25a}$$

where $j = 1,\ldots,s-1$, and the vectors $\nu^1 = (\nu^1_1,\ldots,\nu^1_s),\ldots\nu^{s-1} = (\nu^{s-1}_1,\ldots,\nu^{s-1}_s)$ form the basis in $(s-1)$-dimensional space orthogonal to e, $y=(y_1,\ldots,y_s)$.

2) Let the segments $e_1,\ldots,e_q$ have a common end point; then

$$\sum_{j=1}^{q}\sum_{i=1}^{s} A_i\left(X^{e_j}_1(y,x,t),\ldots,X^{e_j}_s(y,x,t),x,t\right)\eta^j_i \operatorname{mes}\widetilde{\beta}_j = 0, \tag{25b}$$

where $(\eta^j_1,\ldots,\eta^j_s)$ are the direction cosines e_j, and $\tilde{\beta}_j$ is the cross-section of a cylinder with the axis e_j by the hyperplane orthogonal to this axis.

3) The vector-function defined on each segment $e\subset\mathcal{B}$ by

$$\sum_{i=1}^{s}\left(X^e_i(y,x,t)-y_i\right)\xi_i + X^e_0(y,x,t)$$

(for $\xi\in e$) is a continuous 1-periodic function $(\xi_1,\ldots,\xi_s)\in\mathcal{B}$.

4) The functions equal to $X^e_i(y,x,t)$ for $\xi\in e\subset\mathcal{B}$, which are piecewise

constant in the variables ξ, are 1-periodic.

By the averaged operator along the e_j, we will mean the operator $\hat{L}_x^j$:

$$\hat{L}_x^j v = M_j \Bigg(R(x,t)\frac{\partial^2 v}{\partial t^2} + S(x,t)\frac{\partial v}{\partial t} - \\ - \frac{\partial}{\partial x_i}\Big(A_i\big(X_1^{e_j}(\nabla_x v, x, t), \ldots, X_s^{e_j}(\nabla_x v, x, t), x, t\big)\Big)\Bigg),$$

where $M_j = \text{mes}\,\beta_j \varepsilon^{-s} \times$ length $(e_j \cap Q)$. We set $\hat{L}_x = \sum_{j=1}^{I} \hat{L}_x^j$.

Let $\chi(x/\varepsilon\mu)$ be a function equal to 0 in $b \backslash b^+$ and to 1 in the domain B^{++}, $0 \leqslant \chi \leqslant 1$, $\chi \in C^\infty$, and $|\partial\chi(z)/\partial z_i| < c_1$, where c_1 is independent of ε and μ.

Let $\Psi(x/\varepsilon)$ be a function equal to 0 in the ε-neighbourhood of ∂G and to 1 outside the 2ε-neighbourhood of ∂G, $|\partial\Psi(\xi)/\partial\xi_i| < c_1$, $0 \leqslant \Psi \leqslant 1$, and $\Psi \in C^\infty$.

We define the function

$$u^{(1)} = v + \varepsilon\Bigg(N^1\left(\frac{x}{\varepsilon}, \nabla_x v, x, t\right)\chi\left(\frac{x}{\varepsilon\mu}\right) + \\ + N^2\left(\frac{x}{\varepsilon}, \nabla_x v, x, t\right)\left(1 - \chi\left(\frac{x}{\varepsilon\mu}\right)\right)\Bigg)\Psi\left(\frac{x}{\varepsilon}\right); \quad (26)$$

here

$$N^1(\xi, y, x, t) = \sum_{i=1}^{s}\big(X_i^e(y,x,t) - y_i\big)\xi_i + X_0^e(y,x,t),$$

if $\varepsilon\xi$ belongs to the section with the axis segment e_ε, $N^1 = 0$ outside the sections, and $N^2(\xi, y, x, t)$ is a function piecewise constant in ξ, equal to

$$\sum_{i=1}^{s}\big(X_i^e(y,x,t) - y_i\big)\xi_i^0 + X_0^e(y,x,t)$$

in the $3c_0\mu\varepsilon$-neighbourhood of the nodal point $x^0 = \xi^0\varepsilon$, which is the end point for e, and equal to 0 in the rest of B ($1 - \chi = 0$ on the discontinuity surfaces of N^2).

THEOREM 4. *The assertion of Theorem 1 is valid for the frameworks defined in this subsection if $u^{(1)}$ is constructed according to the above-described procedure.*

PROOF:

$$\left|\frac{\partial}{\partial\xi_i}\big(N^1\chi + N^2(1-\chi)\big)\right| = \left|\frac{\partial N^1}{\partial\xi_i}\chi + N^1\frac{\partial\chi}{\partial\xi_i} - N^2\frac{\partial\chi}{\partial\xi_i}\right| = \\ = \left|\big(X_i^e(y,x,t) - y_i\big)\chi + (N^1 - N^2)\frac{\partial\chi}{\partial\xi_i}\right| \leqslant c_5,$$

with c_5 being independent of ε and μ.

By I_1 we denote the integral

$$\int\limits_{B\cap G}\left(R\frac{\partial^2 u^{(1)}}{\partial t^2}\varphi+S\frac{\partial u^{(1)}}{\partial t}\varphi+A_i\left(\nabla_x u^{(1)},\,x,\,t\right)\frac{\partial\varphi}{\partial x_i}\right)dx.$$

We have $I_1 = I_3 + \Delta_1$,

$$I_2=\int\limits_{B^{++}\cap G}\left(R\frac{\partial^2 u^{(1)}}{\partial t^2}\varphi+S\frac{\partial u^{(1)}}{\partial t}\varphi+\sum_{i=1}^{s}A_i\left(\nabla_x u^{(1)},\,x,\,t\right)\frac{\partial\varphi}{\partial x_i}\right)dx,$$

$$|\Delta_1|\leqslant c_6\sqrt{\operatorname{mes}((B\setminus B^{++})\cap G)}\,\|\varphi\|_{W_2^1(B\cap G)}\leqslant$$

$$\leqslant c_7\sqrt{\mu}\,\|\varphi\|_{W_2^1(B\cap G)}\sqrt{\operatorname{mes}(B\cap G)},$$

that is $\Delta_1 = O(\sqrt{\mu})\|\varphi\|_{W_2^1(B\cap G)}\sqrt{\operatorname{mes}(B\cap G)}$.

The boundedness of the derivatives $\partial^l N_i/\partial t^l$ $(l, i = 1, 2)$ implies

$$I_2=\int\limits_{B^{++}\cap G}\left(R\frac{\partial^2 v}{\partial t^2}\varphi+S\frac{\partial v}{\partial t}\varphi+\sum_{i=1}^{s}A_i\left(\nabla_x u^{(1)},\,x,\,t\right)\frac{\partial\varphi}{\partial xi}\right)dx+$$

$$+O(\varepsilon)\,\|\varphi\|_{W_2^1(B\cap G)}\sqrt{\operatorname{mes}(B\cap G)}.$$

Considering the smooth dependence of the functions A_i on its arguments, we have

$$I_3=\int\limits_{B^{++}\cap G}A_i\left(\nabla_x u^{(1)},\,x,\,t\right)\frac{\partial\varphi}{\partial x_i}\,dx=$$

$$=\int\limits_{B^{++}\cap G}A_i\left(\nabla_x v+\nabla_\xi\,(N^1\Psi),\,x,\,t\right)\frac{\partial\varphi}{\partial x_i}\,dx+O(\varepsilon)\sqrt{\operatorname{mes}(B\cap G)}\,\|\varphi\|_{W_2^1(B\cap G)}.$$

From

$$\left\|\frac{\partial}{\partial\xi_i}\left((1-\Psi)\,N^1\right)\right\|_{L_2(B^{++}\cap G)}=O\left(\sqrt{\varepsilon}\right)\sqrt{\operatorname{mes}(B\cap G)}$$

we get

$$I_3=\int\limits_{B^{++}\cap G}A_i\left(\nabla_x v+\nabla_\xi N^1,\,x,\,t\right)\frac{\partial\varphi}{\partial x_i}dx+O\left(\sqrt{\varepsilon}\right)\sqrt{\operatorname{mes}(B\cap G)}\,\|\varphi\|_{W_2^1(B\cap G)}.$$

Consequently,

$$I_3 = -\int\limits_{B^{++}\cap G} \frac{\bar\partial}{\partial x_i} A_i\left(\nabla_x v + \nabla_\xi N^1,\ x,\ t\right)\varphi\, dx +$$

$$+ \int\limits_{\partial B^{++}\cap G} A_i\left(\nabla_x v + \nabla_\xi N^1,\ x,\ t\right) n_i \varphi\, ds + O\left(\sqrt{\varepsilon}\right)\sqrt{\operatorname{mes}(B\cap G)}\,\|\varphi\|_{W_2^1(B\cap G)}.$$

Due to property 1) of the X_i^e (see (25a)), the last integral is

$$I_4 = \int\limits_{\partial B^{++}\cap B\cap G} A_i\left(\nabla_x v + \nabla_\xi N^1,\ x,\ t\right) n_i \varphi\, ds =$$

$$= \sum_{x^0:\Pi_{x^0}\cap G\neq\varnothing}\ \int\limits_{\partial\Pi_{x^0}\cap\partial B^{++}\cap G} A_i\left(\nabla_x v + \nabla_\xi N^1,\ x,\ t\right) n_i\varphi\, ds,$$

where the summation is made with respect to all the nodal points x^0 such that the connected component Π_{x^0} of the set $\overline{B\backslash B^{++}}$, which contains x^0, belongs to $\overline{G}$.

For each node x^0, we set

$$\langle\varphi\rangle_{\Pi_{x^0}} = \frac{1}{\operatorname{mes}\left(\Pi_{x^0}\right)}\int\limits_{\Pi_{x^0}} \varphi\, dx.$$

We have $\varphi(x) = \langle\varphi\rangle_{\Pi_{x^0}} + (\varphi - \langle\varphi\rangle_{\Pi_{x^0}})$.

LEMMA 1.

$$\left\|\varphi - \langle\varphi\rangle_{\Pi_{x^0}}\right\|_{L_2\left(\partial\Pi_{x^0}\cap\partial B^{++}\right)} \leqslant c_8\sqrt{\varepsilon\mu}\left\|\sqrt{\frac{\partial\varphi}{\partial x_i}\frac{\partial\varphi}{\partial x_i}}\right\|_{L_2\left(\Pi_{x^0}\right)},$$

where c_8 does not depend on φ, ε, and μ.

PROOF: In the integral

$$i = \int\limits_{\partial\Pi_{x^0}\cap\partial B^{++}} \left|\varphi - \langle\varphi\rangle_{\Pi_{x^0}}\right|^2 dx$$

make the change $\eta = x/\varepsilon\mu$; we get

$$i = (\varepsilon\mu)^{s-1}\int\limits_{\Sigma_\eta}\left|\varphi(\eta) - \langle\varphi\rangle_{\Pi_{x^0}}\right|^2 d\eta,$$

where $\sum_\eta$ is the image of the surface $\partial\Pi_{x^0}\cap\partial B^{++}$ and Π_η is the image of Π_{x^0} under the transformation $\eta = x/\varepsilon\mu$. Note that $\sum_\eta$ is independent of ε and μ. The estimate is

$$\int_{\Sigma_\eta} \left| \varphi(\eta) - \langle\varphi\rangle_{\Pi_{x^0}} \right|^2 d\eta \leqslant c \int_{\Pi_\eta} \frac{\partial\varphi}{\partial\eta_j} \frac{\partial\varphi}{\partial\eta_j} d\eta,$$

where c is independent of ε and μ.

By making the inverse change, we get

$$\int_{\Pi_\eta} \frac{\partial\varphi}{\partial\eta_j} \frac{\partial\varphi}{\partial\eta_j} d\eta = (\varepsilon\mu)^{-s} \int_{\Pi_{x^0}} (\varepsilon\mu)^2 \frac{\partial\varphi}{\partial x_j} \frac{\partial\varphi}{\partial x_j} dx = (\varepsilon\mu)^{-s+2} \int_{\Pi_{x^0}} \frac{\partial\varphi}{\partial x_j} \frac{\partial\varphi}{\partial x_j} dx.$$

Thus,

$$i \leqslant c\varepsilon\mu \int_{\Pi_{x^0}} \frac{\partial\varphi}{\partial x_j} \frac{\partial\varphi}{\partial x_j} dx.$$

By taking the square root of the two sides of the inequality, we obtain the assertion of Lemma 1.

LEMMA 2.

$$\left| \langle\varphi\rangle_{\Pi_{x^0}} \right| \leqslant c_9 \| \varphi \|_{L_2(\Pi_{x^0})} \Big/ \sqrt{\varepsilon\mu \operatorname{mes}\left(\partial\Pi_{x^0} \cap \partial B^{++}\right)}.$$

PROOF: We have

$$\left|\langle\varphi\rangle_{\Pi_{x^0}}\right| = \frac{1}{\operatorname{mes}\Pi_{x^0}} \left| \int_{\Pi_{x^0}} \varphi\, dx \right| \leqslant \frac{1}{\sqrt{\operatorname{mes}\Pi_{x^0}}} \| \varphi \|_{L_2(\Pi_{x^0})} \leqslant$$

$$\leqslant c_9 \frac{\| \varphi \|_{L_2(\Pi_{x^0})}}{\sqrt{\varepsilon\mu \operatorname{mes}\left(\partial\Pi_{x^0} \cap \partial B^{++}\right)}}.$$

The estimate

$$\operatorname{mes}\Pi_{x^0} \geqslant \frac{1}{c_9^2}\, \varepsilon\mu \operatorname{mes}\left(\partial\Pi_{x^0} \cap \partial B^{++}\right),$$

have been used here with $c_9 > 0$ independent of ε, μ, and φ.

The smoothness of A_i implies that

$$\int\limits_{\partial\Pi_{x^0}\cap\partial B^{++}} A_i\left(\nabla_x v + \nabla_\xi N^1,\, x,\, t\right) n_i \varphi\, ds =$$

$$= \int\limits_{\partial\Pi_{x^0}\cap\partial B^{++}} A_i\left(\nabla_x v\big|_{x=x^0} + \nabla_\xi N^1\big|_{x=x^0},\, x^0,\, t\right) n_i \varphi\, ds +$$

$$+\, O(\varepsilon\mu) \int\limits_{\partial\Pi_{x^0}\cap\partial B^{++}} |\varphi|\, ds =$$

$$= \int\limits_{\partial\Pi_{x^0}\cap\partial B^{++}} A_i\left(\nabla_x v\big|_{x=x^0} + \nabla_\xi N_1\big|_{x=x^0},\, x^0,\, t\right) \langle\varphi\rangle_{\Pi_{x^0}} n_i ds + \Delta_{x^0}, \qquad (27)$$

where

$$\left|\Delta_{x^0}\right| \leqslant c_{10}\left(\int\limits_{\partial\Pi_{x^0}\cap\partial B^{++}} \left|\varphi - \langle\varphi\rangle_{\Pi_{x^0}}\right| ds + \varepsilon\mu \int\limits_{\partial\Pi_{x^0}\cap\partial B^{++}} |\varphi|\, ds\right),$$

with the constant c_{10} being independent of ε, μ, φ, and x^0. Property 2) (see 25b) implies that the first integral in (27) vanishes.

From Lemma 1,

$$\int\limits_{\partial\Pi_{x^0}\cap\partial B^{++}} \left|\varphi - \langle\varphi\rangle_{\Pi_{x^0}}\right| ds \leqslant$$

$$\leqslant \sqrt{\operatorname{mes}\left(\partial\Pi_{x^0}\cap\partial B^{++}\right)}\, \left\|\varphi - \langle\varphi\rangle_{\Pi_{x^0}}\right\|_{L_2\left(\partial\Pi_{x^0}\cap\partial B^{++}\right)} \leqslant$$

$$\leqslant c_8\sqrt{\varepsilon\mu}\sqrt{\operatorname{mes}\left(\partial\Pi_{x^0}\cap\partial B^{++}\right)}\left\|\sqrt{\frac{\partial\varphi}{\partial x_i}\frac{\partial\varphi}{\partial x_i}}\right\|_{L_2(\Pi_{x^0})} \leqslant$$

$$\leqslant c_{11}\sqrt{\varepsilon^s\mu^s}\left\|\sqrt{\frac{\partial\varphi}{\partial x_i}\frac{\partial\varphi}{\partial x_i}}\right\|_{L_2(\Pi_{x^0})},$$

with c_{11} being independent of ε, μ, x^0, and φ.

From Lemmas 1 and 2,

$$\int\limits_{\partial\Pi_{x^0}\cap\partial B^{++}} |\varphi|\, ds \leqslant \int\limits_{\partial\Pi_{x^0}\cap\partial B^{++}} \left|\langle\varphi\rangle_{\Pi_{x^0}}\right| ds + \int\limits_{\partial\Pi_{x^0}\cap\partial B^{++}} \left|\varphi - \langle\varphi\rangle_{\Pi_{x^0}}\right| ds \leqslant$$

$$\leqslant \frac{c_9}{\sqrt{\varepsilon\mu}}\sqrt{\operatorname{mes}\left(\partial\Pi_{x^0}\cap\partial B^{++}\right)}\, \|\varphi\|_{L_2(\Pi_{x^0})} + c_{11}\sqrt{\varepsilon^s\mu^s}\left\|\sqrt{\frac{\partial\varphi}{\partial x_i}\frac{\partial\varphi}{\partial x_i}}\right\|_{L_2(\Pi_{x^0})} \leqslant$$

$$\leqslant \frac{c_{12}}{\varepsilon\mu}\sqrt{\varepsilon^s\mu^s}\, \|\varphi\|_{W_2^1(\Pi_{x^0})}.$$

Hence $|\Delta_{x^0}| \leqslant c_{13}\sqrt{\varepsilon^s\mu^s}\|\varphi\|_{W_2^1(\Pi_{x^0})}$.

According to the Cauchy–Buniakowski–Schwarz inequality for sums,

$$\sum_{x^0:\,\Pi_{x^0}\cap G=\varnothing} |\Delta_{x^0}| \leqslant c_{14}\sqrt{\frac{\varepsilon^s\mu^s}{\varepsilon^s}\sum_{x^0:\,\Pi_{x^0}\cap G\neq\varnothing}\|\varphi\|^2_{W_2^1(\Pi_{x^0})}} \leqslant$$
$$\leqslant c_{15}\sqrt{\mu}\sqrt{\operatorname{mes}(B\cap G)}\,\|\varphi\|_{W_2^1(B\cap G)}.$$

Thus,

$$I_4 = O(\sqrt{\mu})\sqrt{\operatorname{mes}(B\cap G)}\,\|\varphi\|_{W_2^1(B\cap G)},$$

$$I_2 = \int_{B^{++}\cap G}\left(R\frac{\partial^2 v}{\partial t^2}\varphi + S\frac{\partial v}{\partial t}\varphi - \frac{\bar{\partial}}{\partial x_i}A_i\left(\nabla_x v + \nabla_\xi N^1,\, x,\, t\right)\varphi\right)dx +$$
$$+ O(\sqrt{\varepsilon}+\sqrt{\mu})\,\|\varphi\|_{W_2^1(B\cap G)}\sqrt{\operatorname{mes}(B\cap G)}.$$

We denote by $Q_{k_1\ldots k_s}$ the cube $\{x \in \mathbf{R}^s | \varepsilon k_i \leqslant x_i \leqslant \varepsilon(k_i+1)\}$. The number of cubes, such that $Q_{k_1\ldots k_s}\cap B\cap\partial G \neq \emptyset$, does not exceed $2^s k_0/\varepsilon^{s-1}$, and consequently, the total measure of these sets is $O(\varepsilon)\operatorname{mes}(B\cap G)$; therefore

$$I_5 = -\int_{B^{++}\cap G}\frac{\bar{\partial}}{\partial x_i}A_i\left(\nabla_x v + \nabla_\xi N^1,\, x,\, t\right)\varphi\, dx =$$
$$= -\sum_{Q_{k_1\ldots k_s}\cap B\subset G}\ \int_{Q_{k_1\ldots k_s}\cap B^{++}}\frac{\bar{\partial}}{\partial x_i}A_i\left(\nabla_x v + \nabla_\xi N^1,\, x,\, t\right)\varphi\, dx +$$
$$+ O(\sqrt{\varepsilon})\,\|\varphi\|_{W_2^1(B\cap G)}\sqrt{\operatorname{mes}(B\cap G)}.$$

Next, by using on each set $Q_{k_1\ldots k_s}\cap B$ the representation

$$\varphi = \langle\varphi\rangle_{k_1\ldots k_s} + \left(\varphi - \langle\varphi\rangle_{k_1\ldots k_s}\right),$$
$$\langle\varphi\rangle_{k_1\ldots k_s} = \frac{1}{\operatorname{mes}\left(Q_{k_1\ldots k_s}\cap B\right)}\int_{Q_{k_1\ldots k_s}\cap B}\varphi(x,t)\,dx, \tag{28}$$

the Poincaré–Friedrichs inequality

$$\|\varphi - \langle\varphi\rangle_{k_1\ldots k_s}\|_{L_2(Q_{k_1\ldots k_s}\cap B)} \leqslant c_{16}\varepsilon\,\|\varphi\|_{W_2^1(Q_{k_1\ldots k_s}\cap B)}$$

(c_{16} does not depend on ε and μ), and the Cauchy–Buniakowski–Schwarz inequality for sums: $\sum_{i=1}^J a_i \leqslant \sqrt{J\sum_{i=1}^J a_1^2}$, we get

$$I_5 = - \sum_{(k_1,\dots,k_s):\,Q_{k_1\dots k_s}\cap B\subset G} \int_{Q_{k_1\dots k_s}\cap B^{++}} \left(\frac{1}{M}\widehat{L}_x v - R\frac{\partial^2 v}{\partial t^2} - \right.$$

$$\left. - S\frac{\partial v}{\partial t}\right)\langle\varphi\rangle_{k_1\dots k_s} dx + O(\sqrt{\varepsilon}+\sqrt{\mu})\,\|\varphi\|_{W_2^1(B\cap G)}\sqrt{\operatorname{mes}(B\cap G)}.$$

From representation (28) we have again

$$-\int_{B^{++}\cap G} \frac{\bar{\partial}}{\partial x_i} A_i\left(\nabla_x v + \nabla_\xi N^1, x, t\right)\varphi\, dx =$$

$$= - \int_{B\cap G}\left(\frac{1}{M}\widehat{L}_x v - R\frac{\partial^2 v}{\partial t^2} - S\frac{\partial v}{\partial t}\right)\varphi\, dx +$$

$$+ O(\sqrt{\varepsilon}+\sqrt{\mu})\,\|\varphi\|_{W_2^1(B\cap G)}\sqrt{\operatorname{mes}(B\cap G)}.$$

This implies

$$I_1 = \int_{B\cap G} f(x,t)\varphi\, dx + O(\sqrt{\varepsilon}+\sqrt{\mu})\,\|\varphi\|_{W_2^1(B\cap G)}\sqrt{\operatorname{mes}(B\cap G)} =$$

$$= \int_{B\cap G}\left(R\frac{\partial^2 u}{\partial t^2}\varphi + S\frac{\partial u}{\partial t}\varphi + A_i(\nabla_x u, x, t)\frac{\partial \varphi}{\partial x_i}\right) dx +$$

$$+ O(\sqrt{\varepsilon}+\sqrt{\mu})\,\|\varphi\|_{W_2^1(B\cap G)}\sqrt{\operatorname{mes}(B\cap G)}.$$

Theorem 4 is proved. Note that Corollary 1 of Theorem 1 still holds.

REMARK 3: We emphasize that Theorems 1 and 4 do not refer to all bar structures. The solvability condition for systems (19)–(21) and conditions 1)–4) impose certain restrictions on the structure of a framework. Besides, there sometimes is an ambiguity in defining the vectors X_i^ε from 1)–4), which does not affect the form of $\hat{L}_x$.

REMARK 4: If some bars of the framework are in compression, an additional question arises concerning their stability. A detailed consideration of the question results in boundedness of μ from below.

9. Higher approximations in ε of the splitting principle for the averaged operator. In what follows, the f.a.s. is constructed by analogy with §5.1, with each coefficient of the expansion being approximated with accuracy $O(\sqrt{\mu})$ in order to obtain explicit expressions.

We seek an asymptotic solution of problems (9), (10), (12) for $G = \mathbf{R}^s$ as the series

$$u^{(\infty)} \sim v + \varepsilon N_1(\xi, \nabla_x v, x, t) + \varepsilon^2 N_2(\xi, \nabla_x v, x, t) + \dots, \quad \xi = x/\varepsilon. \quad (29)$$

We substitute (29) into A_i and use Taylor's formula. After simplifying, we get

$$A_i(\nabla_x v + \nabla_\xi N_1 + \varepsilon \overline{N}, x, t) \sim A_i(\nabla_x v + \nabla_\xi N_1, x, t) +$$
$$+ \sum_{l=1}^{\infty} \frac{1}{l!} \left(\sum_{k=1}^{n} \sum_{j=1}^{s} \varepsilon \overline{N}^{kj} \frac{\partial}{\partial y_{kj}} \right)^l A_i \Bigg|_{(\nabla_x v + \nabla_\xi N_1, x, t)} \sim A_i(\nabla_x v + \nabla_\xi N_1, x, t) +$$
$$+ \sum_{l=2}^{\infty} \varepsilon^{l-1} \left(\frac{\partial A_i}{\partial y^q} (\nabla_x v + \nabla_\xi N_1, x, t) \frac{\partial N_l}{\partial \xi_q} + \right.$$
$$\left. + W_{l,i}\left(N_1, \ldots, N_{l-1}, \nabla_x v, \ldots, \nabla_x^{(K_l)} v, \xi, x, t\right)\right);$$

here $y = (y^1, \ldots, y^s) = \|y_{kj}\|$, $A_i = A_i(y, x, t)$, $\nabla_x^{(k)} v$ is the totality of all possible derivatives v of order k, and $W_{l,i}$ is a polynomial of a power r_l in the derivatives of the elements $N_2, \ldots, N_{l-1}$; $W_{l,i}$ depends on $\nabla v, \ldots, \nabla^{(K_l)} v$ for some K_l, $\overline{N} = \|\overline{N}^{kj}\| = (\nabla_x^{(\infty)} u - \nabla_x v - \nabla_\xi N_1)/\varepsilon$. In order to be able to expand by Taylor's formula, it is necessary to require additionally that A_i, R be infinitely differentiable as functions of y, x, t. The algorithm for constructing N_1 and its principal part in m was described in the preceding subsection.

We fix $l > 1$. To each vector $e_h^\alpha \subset B$, we assign a polynomial of the $(r_l + 2)$th degree in ξ:

$$N_l^{e_h^\alpha} = \sum_{r=0}^{r_l+2} X_r^{e_h^\alpha, l}(y^{(l)}, x, t)\, \tilde{\xi}_1^r + \sum_{r=0}^{r_l+1} \sum_{i=2}^{s} Y_{r,i}^{e_h^\alpha, l}(y^{(l)}, x, t)\, \tilde{\xi}_1^r \tilde{\xi}_i +$$
$$+ \sum_{r=0}^{r_l} \sum_{i=2}^{s} Z_{r,i,j}^{e_h^\alpha, l}(y^{(l)}, x, t)\, \tilde{\xi}_1^r \tilde{\xi}_i \tilde{\xi}_j,$$

where $y^{(l)}$ is a matrix-parameter corresponding to $(\nabla_x v, \ldots, \nabla_x^{(K_l)} v)$, the $N_l^{e_h^\alpha}$ are sought under the following conditions (in the variables ξ):

$$1) \quad \left| \nu_i^j \left(\frac{\partial A_i}{\partial y^q} \left(y + \nabla_\xi N_1^{e_h^\alpha}, x, t\right) \frac{\partial N_l^{e_h^\alpha}}{\partial \xi_q} + W_{l,i} \right) \right| \leqslant c_{17} \mu^2,$$

where the vectors $\nu^1 = (\nu_1^1, \ldots, \nu_s^1), \ldots, \nu^{s-1} = (\nu_1^{s-1}, \ldots, \nu_s^{s-1})$ form the basis in $(s-1)$-dimensional subspace orthogonal to the vector e_h^α, y is an $n \times s$ matrix-parameter corresponding to $\nabla_x v$, and $c_{17} > 0$.

$$2) \quad \left| \frac{\partial}{\partial \xi_i} \left(\frac{\partial A_i}{\partial y^q} \left(\nabla_x v + \nabla_\xi N_1^{e_h^\alpha}, x, t\right) \frac{\partial N_l^{e_h^\alpha}}{\partial \xi_q} \right) + \right.$$
$$+ \frac{\partial}{\partial \xi_i} W_{l,i} + \frac{\bar{\partial}}{\partial x_i} W_{l-1,i} + \frac{\bar{\partial}}{\partial x_i} \left(\frac{\partial A_i}{\partial y^q} (\nabla_x v + \nabla_\xi N_1, x, t) \frac{\partial N_{l-1}^{e_h^\alpha}}{\partial \xi_q} \right) -$$
$$\left. - R \frac{\partial^2 N_{l-2}^{e_h^\alpha}}{\partial t^2} - S \frac{\partial N_{l-2}^{e_h^\alpha}}{\partial t} + h_l \right| \leqslant c_{18} \mu,$$

where $\overline{\partial}/\partial x_i$ is a full derivative with respect to x_i, provided the variables ξ are considered independent to x; for $l=2, W_{l-1,i}=0$; $\frac{\partial A_i}{\partial y^q}\frac{\partial N_{l-1}^{e_h^\alpha}}{\partial \xi_q}$ should be replaced by A_l;

$$h_l = -\frac{1}{\sum\limits_{e_h^\alpha \cap Q\neq\varnothing} \text{length}\,(e_h^\alpha \cap Q)\,\text{mes}\,\beta^{e_h^\alpha}} \sum_{e_h^\alpha\cap Q\neq\varnothing} \text{length}\,(e_h^\alpha\cap Q)\times$$

$$\times\,\text{mes}\,\beta^{e_h^\alpha}\left(\frac{\overline{\partial}}{\partial x_i}\left(\frac{\partial A_i}{\partial y^q}(\nabla_x v+\nabla_\xi N_1^{e_h^\alpha}, x, t)\frac{\partial N_{l-1}^{e_h^\alpha}}{\partial \xi_q}+\right.\right.$$

$$\left.\left.+W_{l-1,i}\right)-R\frac{\partial^2 N_{l-2}^{e_h^\alpha}}{\partial t^2}-S\frac{\partial N_{l-2}^{e_h^\alpha}}{\partial t}\right),$$

$\nabla_x v, \dots, \nabla_x^{(K_l)} v$ are parameters.

3) Let ξ_0 be an arbitrary nodal point $\mathcal{B}$ which is the end point for the vectors $e_1, \dots, e_q$; then

$$\left|\sum_{j=1}^{q}\sum_{i,p=1}^{s}\frac{\partial A_i}{\partial y^p}(y+\nabla_\xi N_1^{e_j}, x, t)\frac{\partial N_l^{e_i}}{\partial \xi_p}\right|_{\xi=\xi^0}\eta_i^j\,\text{mes}\,\beta^{e_j}\right|\leqslant c_{19}\mu,$$

where $(\eta_1^j, \dots, \eta_s^j)$ are the direction cosines e_j, $\beta^e(\beta^e \in \{\beta_1, \dots, \beta_I\})$ is the cylinder cross-section with the axis e in the variables $z = x/\varepsilon\mu$, y is a parameter.

4) On each of the segments e, we define the functions

$$\widetilde{N}_l(\xi) = N_l^e \quad \text{for} \quad \xi \in e,$$

$$\widetilde{N}_l^{(j)}(\xi) = \frac{\partial N_l^e}{\partial \xi_j} \quad \text{for} \quad \xi \in e.$$

Let $\widetilde{N}_l(\xi)$, $\widetilde{N}_l^{(j)}(\xi)$ be functions 1-periodic on $\mathcal{B}$, with $N_l(\xi)$ being continuous in $\mathcal{B}$.

We set

$$u^{(K)} = v^{(K)} + \sum_{l=1}^{K+1}\varepsilon^l\left(N_l^1\chi\left(\frac{\xi}{\mu}\right)+N_l^2\left(1-\chi\left(\frac{\xi}{\mu}\right)\right)\right),$$

where

$$N_l^1(\xi) = N_l^e(\xi, y^{(l)}, x, t) \quad \text{for} \quad \varepsilon\xi \in B_{hj}^{\alpha+}, \qquad \text{if} \quad B_{hj}^{\alpha+} \supset e_\varepsilon,$$

$$N_l^2(\xi) = N_l^e(\xi^0, y^{(l)}, x, t) \quad \text{for} \quad \varepsilon\xi \in \Pi_{x^0},$$

x^0 is the end node for $e_\varepsilon (x^0 = \varepsilon\xi^0)$.

Assume that there exists a collection of functions $X_r^{e_h^\alpha,l}$, $Y_{r,i}^{e_h^\alpha,l}$, $Z_{r,i,j}^{e_h^\alpha,l}$ with properties 1)–4), and each of the above-listed functions depends smoothly on its arguments $y^{(l)}$, x, t.

It is advisable that the coefficients $X_r^{e,l}$, $Y_{r,i}^{e,l}$, and $Z_{r,i}^{e,l}$ be defined in the following order (with l fixed): For each e, first from properties 1) and 2), we find $X_r^{e,l}$ for $r > 1$, $Y_{r,i}^{e,l}$ for $r > 0$, and $Z_{r,i,j}^{e,l}$; then by analogy with the functions $X_r^{e_m}$ from the preceding subsection, we find $X_r^{e_m,l}$ $(r = 0, 1)$ and $Y_{0,i}^{e_m,l}$.

We shall assume that R, S, A_j, and f are periodic in x_l with period $T \gg \varepsilon$, which is a factor of ε, and that the solution u is sought in the class of vector-functions 1-periodic in x_l.

THEOREM 5. *Let u be a solution of problems (9), (10), (12) and let $v^{(K)}(x,t) = \sum_{j=0}^{K} \varepsilon^j v_j(x,t)$ satisfy the conditions*

$$v_j(x, t) \in C^\infty (\mathbf{R}^s \times [0, \tau]),\ \tau > 0,$$

$$\hat{L}_x^{(K)} v^{(K)} \equiv \sum_{l=2}^{K+1} \varepsilon^{l-2} h_l \left(\nabla_x^{(Kl)} v^{(K)}, x, t\right) = f(x, t) + O\left(\varepsilon^K + \sqrt{\mu}\right)$$

(in the C-norm),

$$\left.\frac{\partial^l v_j}{\partial t^l}\right|_{t=0} = 0, \quad l = 0, 1, \ldots, R > 0$$

$$\text{or} \qquad S > 0,$$

where $v^{(K)}$ is periodic in all x_i with period T. Then for any infinitely differentiable function $\varphi(x,t)$ periodic in $x_1, \ldots, x_s$ with period T, the inequality

$$|\Delta| \leqslant c_{20} \left(\frac{\varepsilon^K}{\mu} + \sqrt{\mu}\right) \|\varphi\|_{W_2^1(B \cap G_T)};$$

holds (for any fixed t). Here Δ is defined by (24),

$$G = G_T = [0, T]^s, \quad \partial G_T \cap (B \backslash B^{++}) = \varnothing.$$

$u^{(K)}$ satisfies the initial conditions (in the nonstationary case) exactly.

The proof follows the pattern similar to that for the assertion of the previous subsection.

REMARK 5: This technique can be extended to the case when the coefficients and the right side are of the form $R(\xi, x, t)$, $S(\xi, x, t)$, $A_i(\xi, y, x, t)$, $f(\xi, x, t)$, are 1-periodic in ξ, and constant in ξ on each section.

§4 The Splitting Principle for the Averaged Operator for Trusses and Thin-Walled Structures

1. *Trusses*

In the definition of the framework B given in Subsection 9, §3, we shall require only the periodicity along the x_1-axis with period ε, and we set $\Phi = G_\varepsilon \cap B$, where G_ε is the parallelepiped $\{x \in \mathbf{R}^s | x_i \in (0, \varepsilon),\ i = 2, \dots, s, \text{ and } x_1 \in (d_1, d_2)\}$. When $s = 2$, the truss Φ is a strip of width ε and length $d_2 - d_1$ "cut-out" of B. The set Φ can be interpreted as a geometric model of the trusses of bridges, power transmission line supports, etc. (see Figs. 24 and 25).

Consider the problem

$$L_x u = f(x_1, t), \quad x \in \Phi, \tag{1}$$

$$\frac{\partial u}{\partial \nu} = 0, \quad x \in \partial\Phi \setminus (\{x_1 = d_1\} \cup \{x_1 = d_2\}), \tag{2}$$

$$u = 0, \quad x \in \partial\Phi \cap (\{x_1 = d_1\} \cup \{x_1 = d_2\}), \tag{3}$$

$$\begin{aligned} u\,\big|_{t=0} &= \frac{\partial u}{\partial t}\Big|_{t=0} = 0, \quad R > 0, \\ u\,\big|_{t=0} &= 0, \quad R \equiv 0, \quad S > 0, \end{aligned} \tag{4}$$

where R, S, and A_i do not depend on $x_2, \dots, x_s$. The dependence of A_i on x_1 makes it possible to use the model (3.9)–(3.12) to describe trusses with the structure slowly changing in x_1 (e.g., power transmission line supports; see Fig. 24a). We shall assume that ∂G_ε has the empty intersection with the nodal domain $B \setminus B^{++}$.

The procedure of constructing the averaged operator is similar to that described in Subsection 3.8. To each segment $e \in \mathcal{B} \cap G_1$, where G_1 is the beam $\{x \in \mathbf{R}^s | x_i \in (0, 1),\ i = 1, \dots, s,\ x_1 \in R\}$, we assign a collection of n-dimensional vectors $X_0^e(y, x_1, t), \dots, X_2^e(y, x_1, t)$ satisfying the conditions 1)–4) in Subsection 3.8. And we consider $\mathcal{B} \cap G$ instead of $\mathcal{B}$ throughout; the points where $\mathcal{B}$ intersects the boundary ∂G_1 of the beam G_1 are considered nodal and for them condition 2) must be satisfied; under conditions 3) and 4), no periodicity in $\xi_1, \dots, \xi_s$ is required; and we call the operator $\hat{L}_x^j$,

$$\hat{L}_x^j v = M_j \Bigg(R(x_1, t) \frac{\partial^2 v}{\partial t^2} + S(x_1, t) \frac{\partial v}{\partial t} - \\ - \frac{\partial}{\partial x_i} \left(A_i \left(X_1^{ej} (\nabla_x v, x_1, t), \dots, X_s^{ej} (\nabla_x v, x_1, t), x_1, t \right) \right) \Bigg),$$

the *averaged operator along* e_j. Here $M_j = \operatorname{mes} \widetilde{\beta}_j \varepsilon^{-s} \times \text{length}\,(e_j \cap Q)$.

The principal part of the averaged operator $\hat{L}_x$ of problem (1)–(4), can be represented as

$$\hat{L}_x = \sum_{j=1}^{I} \hat{L}_x^j.$$

To problems (1)–(4), we assign the averaged problem

$$\hat{L}_x v = Mf(x_1, t), \quad x_1 \in (d_1, d_2), \qquad t > 0, \tag{5}$$

$$v = 0, \quad x_1 = d_1, \quad x_1 = d_2, \tag{6}$$

$$\begin{aligned} v\big|_{t=0} = \frac{\partial v}{\partial t}\bigg|_{t=0} = 0, \quad R > 0, \\ v\big|_{t=0} = 0, \quad R = 0, \quad S > 0, \end{aligned} \tag{7}$$

where $M = \sum_{j=1}^{I} M_j$. The vector-function v does not depend on $x_2, \dots, x_s$.

Let $\Psi(x_1/\varepsilon)$ be a function equal to zero in the ε-neighbourhood of the points $x_1 = d_1$ and $x_1 = d_2$, and equal to 1 outside the 2ε-neighbourhood of $x_1 = d_1$, $x_1 = d_2, |\partial\Psi(\xi_1)/\partial\xi_q| < c_1$, $0 \leqslant \Psi \leqslant 1$, $\Psi \in C^\infty$, and let N^1, N^2, χ be the functions from Subsection 3.8 (remember that the points of $\mathcal{B} \cap \partial G_1$ are regarded as nodal and $\chi = 0$ in their neighbourhood), with $u^{(1)}$ being defined by (3.26).

By analogy with Theorem 3.4, the following theorem can be proved.

THEOREM 1. *Let u be a solution of problems (1)–(4), $v(x,t)$ a solution of problems (5)–(7), and v have all partial derivatives through third order, which are bounded by a constant c. Then for any vector-function $\varphi(x,t)|_{t=t_0} \in W_2^1(\Phi)$ ($\varphi = 0$ for $x_1 = d_1, d_2$),*

$$\Delta \equiv \frac{1}{\sqrt{\operatorname{mes}\Phi}} \left| \int_\Phi \left(R\frac{\partial^2 (u^{(1)} - u)}{\partial t^2} \varphi + S\frac{\partial (u^{(1)} - u)}{\partial t}\varphi + \right.\right.$$

$$\left.\left. + \left(A_i(\nabla_x u^{(1)}, x_1, t) - A_i(\nabla_x u, x, t)\right)\frac{\partial\varphi}{\partial x_i}\right) dx \right| \leqslant$$

$$\leqslant c_2(\sqrt{\varepsilon} + \sqrt{\mu})\|\varphi\|_{W_2^1(\Phi)}, \quad \varepsilon, \mu \to 0.$$

REMARK 1: Suppose that the latter relation implies the following approximate equality,

$$u^{(1)} \approx u, \quad A_{ij}\frac{\partial u^{(1)}}{\partial x_j} \approx A_{ij}\frac{\partial u}{\partial x_j}. \tag{8}$$

We make specific calculations of the coefficients of (5) for the truss described in the introduction to this chapter. Then (5)–(7) take the form (0.1)–(0.3), and (8) implies (0.4).

By using the method similar to the justification of Corollary 1 of Theorem 3.1, we can justify the validity of (8) in the scalar case.

2. *Thin-walled structures*

Let S_0 be the layer $\{x \in \mathbf{R}^s | x_1 \in (0, \mu\varepsilon)\}$, and let S_h^α be the layer obtained from S_0 by orthogonal transformation with the matrix α^T and by translation by h.

As a geometric model of a thin-walled structure, we take the periodic domain $\Sigma = \bigcup_{\alpha \in \Delta} \bigcup_{h \in H_\alpha} S_h^\alpha$ with the periodicity cube Q_0, while

$$\Sigma \cap Q_0 = \bigcup_{p=1}^{I} S_{h_p}^{\alpha_p} \cap Q_0. \tag{9}$$

Here Δ is a set of $s \times s$ orthogonal matrices, $H_\alpha \subset \mathbf{R}^s$; Δ, H_α are independent of ε, and $\mu\Sigma$, $\Sigma \cap Q_b$ are connected.

Fix an arbitrary pair α, h from (9). We make the transformation $\tilde{x} = \alpha x$ to convert S_h^α into S_h^E. We introduce an unknown vector-function X_1 and $s-1$ vector-parameters $Y_x, \dots, Y_s$ and solve the system of equations

$$\tilde{A}_1(X_1, Y_2, \dots, Y_s, \tilde{x}, t) = 0 \tag{10}$$

(x, t are also parameters). We set

$$\hat{L}_{\tilde{x}}^{\alpha,h} v = M^{\alpha,h} \Bigg(\tilde{R}(\tilde{x}, t) \frac{\partial^2 v}{\partial t^2} + \tilde{S}(\tilde{x}, t) \frac{\partial v}{\partial t} -$$
$$- \sum_{i=2}^{s} \frac{\partial}{\partial \tilde{x}_i} \tilde{A}_i \left(X_1 \left(\frac{\partial v}{\partial \tilde{x}_2}, \dots, \frac{\partial v}{\partial \tilde{x}_s}, \tilde{x}, t \right), \frac{\partial v}{\partial \tilde{x}_2}, \dots, \frac{\partial v}{\partial \tilde{x}_s}, \tilde{x}, t \right) \Bigg), \tag{11}$$

where $X_1(Y_2, \dots, Y_s, \tilde{x}, t)$ is a solution of (10), and $M^{\alpha,h}$ is the principal part of $\mathrm{mes}(S_h^\alpha \cap Q_0)/(\varepsilon^s \mu)$.

The principal part of the averaged operator $\hat{L}_x$ is represented as the sum

$$\hat{L}_x = \sum_{p=1}^{I} \hat{L}_x^{\alpha_p, h_p},$$

where $\hat{L}_x^{\alpha,h}$ is the operator $\hat{L}_{\tilde{x}}^{\alpha,h}$ in the variables x (under the inverse transformation $x = \alpha^T \tilde{x}$).

For thin-walled structures, an analogue of Theorem 3.1 can be formulated, with system (3.9) being averaged into (3.19), where $\hat{L}_x^{\alpha,h}$ is defined from (10), (11).

To illustrate this, we consider the domain $\Sigma = \bigcup_{h=-\infty}^{+\infty} \bigcup_{j=1}^{3} S_k^j$ where the S_k^j are thin flat films parallel to the coordinate planes:

$$S_k^j = \{x \in \mathbf{R}^3 \,|\, (k - \mu/2)\,\varepsilon \leqslant x_j \leqslant (k + \mu/2)\,\varepsilon\}.$$

Equation (1.1) is averaged into

$$\widehat{A}_{ij}\frac{\partial^2 v}{\partial x_i \partial x_j} = \overline{\text{mes}}\,(\Sigma_\xi \cap Q)\,f,$$

where

$$\widehat{A}_{ij} = \frac{1}{3}\,\overline{\text{mes}}\,(\Sigma_\xi \cap Q)\left(2A_{ij} - \sum_{l \neq j} A_{il}A_{ll}^{-1}A_{lj}\right),$$
$$\overline{\text{mes}}\,(\Sigma_\xi \cap Q) = 3\mu.$$

In the particular case $A_{ij} = \delta_{ij}K$, we have $\hat{A}_{ij} = 2\mu\delta_{ij}K$; the effective thermal conductivity coefficient is equal to $2K\mu + O(\mu^2)$, that is $(2/3)K(1-\eta)$, where η is porosity; the formula can be used to determine the effective conductivity coefficient of highly porous polymer materials with their structure close to $\Sigma = \bigcup_{k=-\infty}^{+\infty} \bigcup_{j=1}^{3} S_k^j$.

By analogy with Subsection 3.8, a more general definition of a thin-walled structure can be given, and analogues of Theorems 3.1 and 3.4 can be formulated.

§5 On Refining the Splitting Principle for the Averaged Operator

As was mentioned above, the splitting principle is based on an approximate solution of the "problem on a cell,"

$$\begin{aligned} &\frac{\partial}{\partial \xi_i} A_i(\nabla_\xi N_1 + \nabla_x v,\, x,\, t) = 0, \quad \xi \in B_\xi \cap Q, \\ &A_i(\nabla_\xi N_1 + \nabla_x v,\, x,\, t)\, n_i(\xi) = 0, \quad \xi \in \partial B_\xi, \end{aligned} \tag{1}$$

in the class of the vector-valued functions $N_1(\xi, \nabla_x v, t, \mu)$, 1-periodic in ξ, where $\nabla_x v$, x, t, and μ are parameters of the problem, B_ξ, Q are the sets B and Q_0 in the variables $\xi = x/\varepsilon$ and $n_i,(\xi)$ is the cosine of the angle between the exterior normal to ∂B_ξ and the axis $0\xi_i$. The zeroth approximation in ε of the averaged equation is of the form

$$M_\xi R\frac{\partial^2 v}{\partial t^2} + M_\xi S\frac{\partial v}{\partial t} - \frac{\partial}{\partial x_i}\langle A_i(\nabla_\xi N_1 + \nabla_x v,\, x,\, t)\rangle = M_\xi f, \tag{2}$$

where

$$\langle F(\xi)\rangle = \int_{B_\xi \cap Q} F(\xi)\,d\xi, \quad M_\xi = \text{mes}\,(B_\xi \cap Q).$$

As an approximate solution of problem (1), we have chosen the vector-function N_χ constructed in §3, which is close to N_1 in an integral norm (as $\mu \to 0$). However, N_χ differs from N_1 in the proximity of the intersections of various cylinders forming B_ξ by a quantity of order 1 in absolute terms, and hence the error $|\nabla_x(u - u^{(1)})|$ has order 1 near the "crosses."

In some cases, ignoring the terms of order $O(\mu M_\xi)$ in (2) results in the type of equation (3.9) not being maintained, or an incorrect problem arises in transition from the original problem to (3.19)–(3.21). We have already encountered this effect in §7.5. Thus, in the case of the linear system of equations of elasticity theory defined on a rectangular framework made of an isotropic material, for $s = n = 3$, $R = \rho E$, system (3.19) takes the form

$$3\pi\mu^2\rho\,\frac{\partial^2 v_i}{\partial t^2} - \pi\mu^2\mathscr{E}\,\frac{\partial^2 v_i}{\partial x_i^2} = 3\pi\mu^2 f_i(x,t), \quad i = 1, 2, 3$$

(no summation over i) where $\mathcal{E}$ is Young's modulus.

To remove the undesirable effects described above, it may be useful to introduce the boundary-layer function $\mathcal{M}(x/\varepsilon\mu, \nabla_x v, x, t)$ that tends exponentially quickly to zero in $x/\varepsilon\mu$ as it recedes from the crosses. The sum $\mathcal{M} + \chi N$ approximates N_1 more exactly. These functions are constructed in the next section as a model example (one equation).

Another approach to refining the splitting principle (see Subsection 3.9) is to construct an asymptotic expansion of a solution in powers of ε to be followed by the zeroth approximations of asymptotic solutions of cell problems as $\mu \to 0$. This approach is applicable in the case $\mu \ll \varepsilon$, and it can be used to investigate transverse vibrations of a truss (see §4). The asymptotic solutions of cell problems as $\to 0$ is sought in the class of piecewise polynomial functions satisfying conditions 1)–4) from Subsection 3.8. The problem of constructing the asymptotics of the solution of a problem on a cell reduces to an algebraic system of equations.

§6 Asymptotic Expansion of a Solution of a Linear Equation in Partial Derivatives for a Rectangular Framework

This section presents a full a.e. in powers of ε and μ of a solution of a linear equation for the simplest s-dimensional framework. The construction of higher approximations involves numerical solution of some boundary-layer problems and therefore it is rather cumbersome and laborious. Nevertheless, the a.e. of the solution enables one to investigate in detail the qualitative behaviour of the latter, as ε, $\mu \downarrow 0$.

By an s-dimensional rectangular framework with square cross-sections of bars, we shall mean the domain

$$B = \bigcup_{k_1 \ldots k_s, k_l \in Z}\ \bigcup_{j=1}^{s} \overline{B}^j_{k_1 \ldots k_s},$$

consisting of thin beams $\overline{B}^j_{k_1 \ldots k_s}$ running through the nodes of the mesh $\varepsilon(k_1, \ldots, k_s)$ ($k_l \in \mathbf{Z}$) and oriented in the jth coordinate direction:

$$\overline{B}^j_{k_1 \ldots k_s} = \{x \in \mathbf{R}^s \mid (k_l^j - \mu)\,\varepsilon < x_l < k_l\varepsilon,\ l = 1, \ldots, s,\ l \neq j\},$$

$\mathbf{Z}$ is the set of integers, $x = (x_1, \ldots, x_s)$, and ε, μ are small parameters.

Let G be a parallelepiped in $\mathbf{R}^s$, $G = \{x \in \mathbf{R}^s | -d_l \leqslant x_l \leqslant d_l\}$, where $d_l = (k_l + \theta)_\varepsilon$ ($k_l \in Z, 0 < \theta < 1$), with θ being independent of ε and μ.

Suppose that in $B \cap G$, the boundary value problem is given,

$$L_x u \equiv R\frac{\partial^2 u}{\partial t^2} + S\frac{\partial u}{\partial t} - \frac{\partial}{\partial x_i}\left(A_{ij}\frac{\partial u}{\partial x_j}\right) = f(x, t), \qquad (1)$$

$$x \in B \cap G,$$

with boundary conditions

$$u\,|_{\partial G \cap B} = 0, \quad \frac{\partial u}{\partial \nu}\bigg|_{\partial B \cap G} = 0, \qquad (2)$$

where R, S, and A_{ij} are constants satisfying the relations $R \geqslant 0$, $S \geqslant 0$, $A_{ij} = A_{ji}$,

$$A_{ij}\eta_i\eta_j \geqslant \varkappa_1 \sum_{i=1}^{s} \eta_i^2 \quad \forall\,(\eta_1, \ldots, \eta_s) \in \mathbf{R}^s, \quad \varkappa_1 > 0.$$

By the symbol $\delta/\partial\nu$, we mean the derivative with respect to the conormal: $\partial/\partial\nu = n_i A_{ij}\partial/\partial x_j$, where n_i is the cosine of the angle between the exterior normal and the axis $0x_i$, $f(x,t) \in C^\infty(\overline{G} \times [0, T_0])$.

Three cases are considered: a) hyperbolic ($R > 0$, $S > 0$); b) parabolic ($R = 0$, $S > 0$); and c) elliptic ($R = S = 0$). In case a), we impose the initial values

$$u\,|_{t=0} = \frac{\partial u}{\partial t}\bigg|_{t=0} = 0; \qquad (3a)$$

in case b),

$$u|_{t=0} = 0 \qquad (3b)$$

and assume that $f(x,t)$ satisfies the condition $\partial^l f/\partial t^l|_{t=0} = 0$ for any $l = 0, 1, \ldots$; in case c), f is supposed to be independent of t. For $s = 2$, we get the example in §1.

For $s = 2$, we set $\hat{A}_{ij} = (A_{i\bar{j}}A_{\bar{j}\bar{j}}^{-1}A_{\bar{j}j})/2$, where $\bar{j} = 3 - j$, and for $s = 3$, we assume $\hat{A}_{ij} = (A_{ij} + \sum_{l \neq j} A_{il}C_l^j/3$, where C_l^j ($l \neq j$) is a solution of the algebraic system

$$-\sum_{l \neq j} A_{il}C_l^j = A_{ij}, \quad i \neq j \qquad (4)$$

(j is a parameter). Note that $\hat{A}_{ij} = 0$ for $i \neq j$, $s = 2, 3$. To problems (1)–(3), we assign the averaged problem

$$\widehat{L}_0 v \equiv R\frac{\partial^2 v}{\partial t^2} + S\frac{\partial v}{\partial t} - \frac{\partial}{\partial x_i}\left(\widehat{A}_{ij}\frac{\partial v}{\partial x_j}\right) = f(x, t), \quad x \in G, \tag{5}$$

$$v|_{\partial G} = 0, \tag{6}$$

$$\begin{aligned} v|_{t=0} = \frac{\partial v}{\partial t}\Big|_{t=0} = 0 \qquad &\text{(in case a)),} \\ v|_{t=0} = 0 \quad &\text{(in case b)).} \end{aligned} \tag{7}$$

THEOREM 1. *Let $\mu = O(\varepsilon^\alpha)$ $(\alpha > 0)$, let u be a solution of problem (1)–(3), and let v be a solution of problem (5)–(7), which belongs to $C^4(\overline{G} \times [0, T_0])$, $M = \mathrm{mes}(B \cap G)$. Then*

$$\frac{1}{\sqrt{M}}\|u - v\|_{L_2(B\cap G\times(0,T_0))} = O(\varepsilon + \mu).$$

There exists an asymptotic expansion of u into a series in powers of ε, μ such that the partial sums $u^{(K,J)}$ satisfy the estimate

$$\|u^{(K,J)} - u\|_W = O(\varepsilon^K + \mu^J).$$

where in case a), $W = W_2^1(B\cap G\times(0,t_0))$; in case b), $W = W_2^{1,0}(B\cap G\times(0,T_0))$; and in case c), $W = W_2^1(B\cap G)$.

The existence of the a.e. can be proved under some additional condition A formulated below after formula (38). The necessary and sufficient conditions for v to belong to the class $C^4(\overline{G})$ in the case of the Poisson equation in a rectangular can be found in [41, 67].

The asymptotic solution of problem (1)–(3) of the Kth and Jth order is constructed in two stages as in §1. At the first stage, the dependence of the solution on the small parameter μ is ignored, and $u^{(K)}$ is sought in the form

$$u^{(K)} = \sum_{\substack{p,q,l\geqslant 0\\ p+q+l\leqslant K+1}} \varepsilon^{p+q+l}\sum_{|i|=l} N^p_{qi}\left(\frac{x}{\varepsilon}\right)D^{q,i}v; \tag{8}$$

here v is a solution of the averaged equation, which is independent of the fast variables $\xi_i = x_i/\varepsilon$, the $N^p_{qi_1\ldots i_l}(\xi)$ are functions depending only on $\xi = (\xi_1, \ldots, \xi_s)$ and the parameter μ, which can be represented as

$$N^p_{qi}(\xi) = N^{p,1}_{qi}(\xi) + N^{p,2}_{qi}(\xi),$$

where $N^{p,1}_{qi}(\xi)$ is a 1-periodic function, and $N^{p,2}_{qi}(\xi)$ is a boundary-layer addition (see Chapter 9), which is a quantity of order $O(\varepsilon^\beta\mu^\beta)$ for any β outside the $\theta\varepsilon$-neighbourhood of the boundary of G, and $N^{p,2}_q(\xi)$ is periodic along the coordinate directions parallel to the adjoining face ∂G and exponentially decays along the normal to that face.

We substitute (8) into (1) and (2), considering formally N^p_{qi} and v independent of μ. We simplify those of the derivatives v, that differ only in the order of differentiation with respect to x_l being considered distinct. We get

$$L_x u^{(K)} = \sum_{\substack{p,q,l\geqslant 0\\ p+q+l\leqslant K+1}} \varepsilon^{p+q+l-2} \sum_{|i|=l} H^p_{qi}(\xi)\, D^{q,i}v\,|_{\xi=x/\varepsilon} + O(\varepsilon^K), \tag{9}$$

$$\left.\frac{\partial u^{(K)}}{\partial \nu}\right|_{\partial B} = \sum_{\substack{p,q,l\geqslant 0\\ p+q+l\leqslant K+1}} \varepsilon^{p+q+l-1} \sum_{|i|=l} G^p_{qi}\,|_{\partial B}\, D^{q,i}v\,|_{\partial B} + O(\varepsilon^{K+1}),$$

$$u^{(K)}\,|_{\partial G\cap B} = \sum_{\substack{p,q,l\geqslant 0\\ p+q+l\leqslant K+1}} \varepsilon^{p+q+l} \sum_{|i|=l} N^p_{qi}\,|_{\partial G\cap B}\, D^{q,i}v\,|_{\partial G\cap B}; \tag{10}$$

here $p+q+l \leqslant K+1$,

$$H^p_{qi} = -LN^p_{qi} + T^{p,1}_{qi} + T^{p,2}_{qi}, \quad L = \frac{\partial}{\partial \xi_r}\left(A_{rm}\frac{\partial}{\partial \xi_m}\right),$$

$$T^{p,\rho}_{qi_1\ldots i_l} = -A_{i_1 j}\frac{\partial N^{p,\rho}_{qi_2\ldots i_l}}{\partial \xi_j} - \frac{\partial}{\partial \xi_j}\left(A_{ji_1}N^{p,\rho}_{qi_2\ldots i_l}\right) -$$
$$- A_{i_1 i_2}N^{p,\rho}_{qi_3\ldots i_l} + RN^{p,\rho}_{q-2,i_1\ldots i_l} + SN^{p-1,\rho}_{q-1,i_1\ldots i_l}, \quad \rho = 1, 2, \tag{11}$$

$$G^p_{qi_1\ldots i_l} = \frac{\partial}{\partial \nu_\xi}N^p_{qi_1\ldots i_l} + A_{ji_1}N^p_{qi_2\ldots i_l}n_j, \quad \frac{\partial}{\partial \nu_\xi} = n_r A_{rm}\frac{\partial}{\partial \xi_m}.$$

As usual, $N^0_0 = 1$ and the N with negative indices are zeroes; then $H^0_0 = 0$. We require that the following conditions be satisfied: Each of the expressions H^p_{qi} is equal to a constant h^p_{qi}, the coefficients of terms of order ε^{-1} in (9) are zeroes, and $G^p_{qi}|_{\partial B} = 0$ and N^p_{qi} in (10) are constants on each of the faces ∂G, that is,

$$H^p_{qi} = h^p_{qi}, \quad G^p_{qi}\,|_{\partial B} = 0, \tag{12}$$

$$h^0_1 = h^0_{01} = \ldots = h^0_{0s} = h^1_0 = 0, \tag{13}$$

$$N^p_{qi}\,|_{\partial B\cap\{x_\gamma=\pm d_\gamma\}} = n^{p,\pm\gamma}_{qi} = \text{const}, \quad \gamma = 1, \ldots, s. \tag{14}$$

If we can select N^p_{qi}, h^p_{qi}, and $n^{p,\pm\nu}_{qi}$ such that (12)–(14) hold, then the equation

$$\sum_{\substack{p,q,l\geqslant 0\\ 2\leqslant p+q+l\leqslant K+1}} \varepsilon^{p+q+l-2} \sum_{|i|=l} h^p_{qi} D^{q,i}v = f(x, t), \tag{15}$$

the boundary condition

$$\sum_{\substack{p,q,l\geqslant 0\\ p+q+l\leqslant K+1}} \varepsilon^{p+q+l} \sum_{|i|=l} n^{p,\pm\gamma}_{qi} D^{q,i}v\,|_{\partial G} = 0 \tag{16}$$

and the initial conditions

$$v\,|_{t=0} = \frac{\partial v}{\partial t}\bigg|_{t=0} = 0 \quad \text{(in case a))},$$
$$v\,|_{t=0} = 0 \quad \text{(in case b))} \tag{17}$$

can be considered as the following averaged problem: If v satisfies (15)–(17) up to terms of order ε^K, then $u^{(K)}$ from (8) satisfies problems (1)–(3) up to terms of order ε^K.

The construction of N_{qi}^p, h_{qi}^p, and $n_{qi}^{p,\pm\mu}$ is carried out by the following algorithm. First construct all the $N_{qi}^{p,1}$ and h_{qi}^p recurrently in $p+q+|i|$: $N_0^{0,1} = 1$; suppose all the $N_{qi}^{p,1}$ and h_{qi}^p are found for $p+q+|i| \leqslant K_0 - 1$. Define $T_{qi}^{p,1}(\xi)$ by (11) for $\rho = 1$ and calculate

$$h_{qi_1\ldots i_l}^p = \frac{1}{\operatorname{mes}\mathcal{M}}\left(\int\limits_{\mathcal{M}} T_{qi_1\ldots i_l}^{p,1}(\xi)\,d\xi + \int\limits_{\partial\mathcal{M}} A_{ji_1} N_{qi_2\ldots i_l}^{p,1} n_j ds\right), \tag{18}$$

where $\mathcal{M}$ is a domain on the s-dimensional torus O^s: $\mathcal{M} = \bigcup_{j=1}^s \{\xi \in O^s \,|\, \xi_l \in (-\mu, 0),\ l \neq j,\ \xi_j \in O\}$, and O is the circle of length 1. To find $N_{qi}^{p,1}$, with $p+q+|i| = K_0$, solve the boundary value problem

$$-LN_{qi}^{p,1} + T_{qi}^{p,1} = h_{qi}^{p,1}, \quad \xi \in \mathcal{M}, \tag{19}$$

with boundary conditions

$$\frac{\partial}{\partial \nu_\xi} N_{qi_1\ldots i_l}^{p,1} + A_{ji_1} N_{qi_2\ldots i_l}^{p,1} n_j = 0, \quad \xi \in \partial\mathcal{M}. \tag{20}$$

Problem (19), (20) under condition (18) has a solution defined up to a constant term.

After finding all the $N_{qi}^{p,1}$ and h_{qi}^p, define $N_{qi}^{p,2}$ and $n_{qi}^{p,\pm\nu}$ for each of the faces ∂G. For definiteness, we fix the face $\partial G \cap \{x_1 = -d_1\}$ and consider any of the parallelepipeds

$$\overline{B}_{k_1\ldots k_s}^1 \cap \{x_1 \in (-d_1, -d_1 + \varepsilon\theta/2)\}$$

in the variables $\xi_j = xj/\varepsilon$. This is the parallelepiped

$$\pi_{k_1\ldots k_s}^1 = \left\{\xi \in \mathbb{R}^s \,|\, \xi_1 \in \left(-\frac{d_1}{\varepsilon}, -\frac{d_1}{\varepsilon} + \frac{\theta}{2}\right),\ \xi_j \in (k_j - \mu,\ k_j) \quad j > 1\right\}.$$

We set $N_0^{0,2} = 0$, $n_0^{0,-1} = 1$. By increasing $p+q+|i|$ successively by 1, we construct $N_{qi}^{p,2}$ and $n_{qi_1\ldots i_l}^{p,-1}$ in order to satisfy the equation

$$-LN_{qi}^{p,2}+T_{qi}^{p,2}=0,\quad \xi\in\pi^1_{k_1\ldots k_s},\tag{21}$$

the boundary conditions

$$\frac{\partial}{\partial\nu_\xi}N_{qi}^{p,2}+A_{ji_1}N_{qi_2\ldots i_l}^{p,2}n_j=0,\quad \xi\in\partial\pi^1_{k_1\ldots k_s},\quad \xi_1\neq-\frac{d_1}{\varepsilon},\ -\frac{d_1}{\varepsilon}+\frac{\theta}{2},\tag{22}$$

$$N_{qi}^{p,2}=-N_{qi}^{p,1}+n_{qi}^{p,-1},\quad \xi\in\partial\pi^1_{k_1\ldots k_s},\quad \xi_1=-\frac{d_1}{\varepsilon},\tag{23}$$

and the condition of the exponential decay in ξ_1:

$$|N_{qi}^{p,2}|<c_1e^{-c_2(\xi_1+d_1/\varepsilon)/\mu},\qquad c_1>0,\quad c_2>0.\tag{24}$$

By passing to the new variables $z_j=\xi_j/\mu$ and by using the methods of [81, 82, 108], we can see that problem (21)–(24) has the unique solution $N_{qi}^{p,2}$, $n_{qi}^{p,-1}$. Similar constructions are performed at the other faces ∂G. Relations (19)–(24) imply (12)–(14). Problem (19), (20) is an "equation on a cell" (see Chapter 4), and (21)–(24) is a problem for the boundary-layer function $N_{qi}^{p,2}$ (see Chapter 9). This completes the first step of constructing the asymptotic expansion.

The second step involves seeking an asymptotic expansion in μ of the periodic solutions $N_{qi}^{p,1}$ of (19), (20) as well as of the solutions $N_{qi}^{p,2}$ of the boundary-layer equations. For convenience, we make the change of coordinates $y_j=\xi_j(1+\mu)$ and pass to the new small parameter $\mu_1=\mu(1+\mu)$. We introduce the notation $\mathcal{M}_y$ for $\mathcal{M}$ in the new variables.

Problem (19), (20) for the function N with an arbitrary collection of indices is rewritten as

$$-L_yN\equiv-\frac{\partial}{\partial y_r}\left(A_{rm}\frac{\partial}{\partial y_m}N\right)=f_1(y),\quad y\in\mathcal{M}_y,\tag{25}$$

$$\frac{\partial}{\partial\nu_y}N=f_2(y),\quad y\in\partial\mathcal{M}_y,\tag{26}$$

where f_1, f_2 are the known right sides found from (11), (19), (20) according to a specific collection of indices. We seek the asymptotic solution in the form

$$N^{(J)}=\sum_{j=0}^{J+2}\mu_1^j\left(N_j^{(0)}\left(\frac{y}{\mu_1}\right)+\sum_{r=1}^{s}N_j^{(r)}\left(\frac{\tilde y_r}{\mu_1}\right)\rho\left(\frac{y_r}{\mu_1}\right)\right),\tag{27}$$

where

$$\frac{\tilde y_r}{\mu_1}=\left(\frac{y_1}{\mu_1},\ldots,\frac{y_{r-1}}{\mu_1},\ y_r,\ \frac{y_{r+1}}{\mu_1},\ldots,\frac{y_s}{\mu_1}\right),$$

$$\left|N_j^{(0)}\left(\frac{y}{\mu_1}\right)\right|\leqslant c_1e^{-c_2|y/\mu_1|},\quad c_1>0,\quad c_2>0,\quad \rho(t)=\tilde\rho(t+0.5),$$

$\tilde{\rho}(t)$ is a function satisfying the conditions

$$\tilde{\rho}(t)=\begin{cases}0 & \text{for } |t|\leqslant 1,\\ 1 & \text{for } |t|\geqslant 2,\end{cases}\qquad \tilde{\rho}\in C^\infty(R),\quad \tilde{\rho}(t)=\tilde{\rho}(-t).$$

The right sides $f_\gamma(y)$ of (25) and (26) have the structure similar to (27):

$$f_\gamma(y)=\sum_{j=0}^{J}\mu_1^j\left(f_{\gamma,j}^{(0)}\left(\frac{y}{\mu_1}\right)+\right.$$
$$\left.+\sum_{r=1}^{s}f_{\gamma,j}^{(r)}\left(\frac{\tilde{y}_r}{\mu_1}\right)\rho\left(\frac{y_r}{\mu_1}\right)\right)+O\left(\mu_1^{J+1}\right),\quad \left|f_{\gamma,j}^{(0)}\left(\frac{y}{\mu_1}\right)\right|<c_1e^{-c_2|y/\mu_1|}.$$

By substituting (27) into (25), (26), we get

$$\sum_{j=0}^{J+2}\mu_1^{j-2}H_j+O\left(\mu_1^{J+1}\right)=0,\quad y\in\mathcal{M}_y,$$
$$\sum_{j=0}^{J+1}\mu_1^{j-1}G_j+O\left(\mu_1^{J+1}\right)=0,\quad y\in\partial\mathcal{M}_y,$$

where

$$H_j=H_j^{(0)}+\sum_{\gamma=1}^{s}H_j^{(\gamma)}\rho\left(\frac{y_i}{\mu_1}\right),\qquad G_j=G_j^{(0)}+\sum_{\gamma=1}^{s}G_j^{(\gamma)}\rho\left(\frac{y_j}{\mu_1}\right).$$

Here for each $\gamma=1,2,\dots s$, $H_j^{(\gamma)}$, $G_j^{(\gamma)}$ are functions defined on the set $\mathcal{M}(\gamma)=\{y\in\mathcal{M}_y|y_j\in(-1,0),\ j\neq\gamma\}$ and dependent on the fast "transverse" variables y_r/μ $(r\neq\gamma)$ and the slow "longitudinal" variable y_γ. The functions are of the form

$$H_j^{(\gamma)}=L_{z_{\bar{\gamma}}}N_j^{(\gamma)}+2\sum_{k=1}^{s}A_{\gamma k}\frac{\partial^2N_{j-1}^{(\gamma)}}{\partial y_\gamma\partial z_k}+L_{y_\gamma}N_{j-2}^{(\gamma)}-f_{1,j-2}^{(\gamma)},$$
$$G_j^{(\gamma)}=\sum_{k,l\neq\gamma}A_{kl}\frac{\partial}{\partial z_l}N_j^{(\gamma)}n_k+\sum_{k\neq\gamma}A_{k\gamma}\frac{\partial}{\partial y_\gamma}N_{j-1}^{(\gamma)}n_k-f_{2,j-1}^{(\gamma)},$$
$$\gamma>0,\quad z_j=\frac{y_j}{\mu_1},\quad z=\frac{y}{\mu_1},$$

where

$$L_{z_{\bar{\gamma}}}=\sum_{k,l\neq\gamma}A_{kl}\frac{\partial^2}{\partial z_k\partial z_l},\quad L_{y_\gamma}=A_{\gamma\gamma}\frac{\partial^2}{\partial y_\gamma^2},$$

n_k is the cosine of the angle between the x_k-axis and the exterior normal to ∂M_y; $H_j^{(0)}$ and $G_j^{(0)}$ are the boundary-layer functions defined on $\mathcal{M}_y$, depending on the fast variables $z_r = y_r/\mu$ and tending to zero exponentially fast as $z_r \to \pm\infty$,

$$H_j^{(0)} = L_z N_j^{(0)} + W_{1,j}(z) - f_{1,j-2}^{(0)}(z),$$
$$G_j^{(0)} = \frac{\partial}{\partial \nu_z} N_j^{(0)} + W_{2,j}(z)\big|_{\partial \mathscr{M}_y} - f_{2,j-1}^{(0)}(z),$$

where

$$W_{1,j}(z) = \sum_{\gamma=1}^{s} \left(L_z \left(\sum_{q=0}^{j} \frac{1}{q!} \frac{\partial^q N_{j-q}^{(\gamma)}}{\partial y_\gamma^q} \bigg|_{y_\gamma=\sigma} \bar{z}_\gamma^q \, (\rho(z_\gamma) - 1) \right) + \right.$$
$$\left. + (1 - \rho(z_\gamma)) \, L_z \left(\sum_{q=0}^{j} \frac{1}{q!} \frac{\partial^q N_{j-q}^{(\gamma)}}{\partial y_\gamma^q} \bigg|_{y_\gamma=\sigma} \bar{z}_\gamma^q \right) \right),$$

$$W_{2,j}(z) = \sum_{\gamma=1}^{s} \left(\frac{\partial}{\partial \nu_z} \left(\sum_{q=0}^{j} \frac{1}{q!} \frac{\partial^q N_{j-q}^{(\gamma)}}{\partial y_\gamma^q} \bigg|_{y_\gamma=\sigma} \bar{z}_\gamma^q \, (\rho(z_\gamma) - 1) \right) + \right.$$
$$\left. + (1 - \rho(z_\gamma)) \frac{\partial}{\partial \nu_z} \left(\sum_{q=0}^{j} \frac{1}{q!} \frac{\partial^q N_{j-q}^{(\gamma)}}{\partial y_\gamma^q} \bigg|_{y_\gamma=\sigma} \bar{z}_\gamma^q \right) \right);$$

here

$$L_z = \frac{\partial}{\partial z_r} \left(A_{rm} \frac{\partial}{\partial z_m} \right), \quad \frac{\partial}{\partial \nu_z} = n_r A_{rm} \frac{\partial}{\partial z_m},$$
$$\sigma = \sigma(z_\gamma) = (1 - \operatorname{sign} z_\gamma)/2, \quad \bar{z}_\gamma = z_\gamma - \sigma.$$

Let $\tilde{z}_\gamma = (z_1, \ldots, z_{\gamma-1}, y_\gamma, z_{\gamma+1}, \ldots, z_s)$. We represent $N^{(\gamma)}(\tilde{z}_\gamma)$ $(\gamma > 0)$ as

$$N_j^{(\gamma)}(\tilde{z}_\gamma) = \overline{N}_j^{(\gamma)}(y_\gamma) + \widetilde{N}_j^{(\gamma)}(\tilde{z}_\gamma),$$

where

$$\int_{-1}^{0} \cdots \int_{-1}^{0} \widetilde{N}_j^{(\gamma)}(\tilde{z}_\gamma) \, dz_\gamma' = 0; \tag{28}$$

here $z_\gamma' = (z_1, \ldots, z_{\gamma-1}, z_{\gamma+1}, \ldots, z_s)$. Then the conditions $H_j^{(\gamma)} = 0$, $G_j^{(\gamma)} = 0$ assume the form

$$L_{z_{\bar{\gamma}}} \widetilde{N}_j^{(\gamma)} + 2 \sum_{k=1}^{s} A_{\gamma k} \frac{\partial^2 \widetilde{N}_{j-1}^{(\gamma)}}{\partial y_\gamma \partial z_k} + L_{y_\gamma} \left(\widetilde{N}_{j-2}^{(\gamma)} + \overline{N}_{j-2}^{(\gamma)} \right) = f_{1,j-2}^{(\gamma)}, \tag{29}$$
$$z_\gamma' \in Q' = \{ z_\gamma' \mid z_l \in (-1, 0), \quad l \neq \gamma \},$$
$$\frac{\partial}{\partial \nu_z} \widetilde{N}_j^{(\gamma)} = - \sum_{k \neq \gamma} A_{k\gamma} \frac{\partial}{\partial y_\gamma} \left(\widetilde{N}_{j-1}^{(\gamma)} + \overline{N}_{j-1}^{(\gamma)} \right) n_k + f_{2,j-1}^{(\gamma)}, \quad z_\gamma' \in \partial Q'. \tag{30}$$

The condition for the existence of the solution $\widetilde{N}_j^{(\gamma)}$ to problem (29), (30) (here y is a parameter) is written as (Green's formula)

$$\int_{Q'} f_{1,j-2}^{(\gamma)} dz'_\gamma - L_{y_\gamma} \int_{Q'} \widetilde{N}_{j-2}^{(\gamma)} dz'_\gamma - L_{y_\gamma} \overline{N}_{j-2}^{(\gamma)} - \\ - \sum_{k\neq\gamma} \int_{\partial Q'} A_{k\gamma} \frac{\partial \widetilde{N}_{j-1}^{(\gamma)}}{\partial y_\gamma} n_k dS = \int_{\partial Q'} f_{2,j-1}^{(\gamma)} dS. \tag{31}$$

Let us remove the function $\widetilde{N}_{j-1}^{(\gamma)}$ in condition (31). For this purpose, we note that the solution $\widetilde{N}_j^{(\gamma)}$ of (29), (30) can be represented as

$$\widetilde{N}_j^{(\gamma)} = \widetilde{\widetilde{N}}_j^{(\gamma)} + \sum_{q\neq\gamma} C_q^\gamma (z_q + 0{,}5) \frac{\partial}{\partial y_\gamma} \overline{N}_{j-1}^{(\gamma)}, \tag{32}$$

where C_q^γ is a solution of the algebraic system (4), and $n\widetilde{\widetilde{N}}_j^{(\gamma)}$ is a solution of the problem (y is a parameter)

$$L_{z_{\bar\gamma}} \widetilde{\widetilde{N}}_j^{(\gamma)} + 2 \sum_{k=1}^{s} A_{\gamma k} \frac{\partial^2 \widetilde{N}_{j-1}^{(\gamma)}}{\partial y_\gamma \partial y_k} + L_{y_\gamma} \left(\widetilde{N}_{j-2}^{(\gamma)} + \overline{N}_{j-2}^{(\gamma)} \right) = f_{1,j-2}^{(\gamma)}, \quad z'_\gamma \in Q', \tag{33}$$

$$\frac{\partial}{\partial \nu_z} \widetilde{\widetilde{N}}_j^{(\gamma)} = - \sum_{k\neq\gamma} A_{k\gamma} \frac{\partial}{\partial y_\gamma} \widetilde{N}_{j-1}^{(\gamma)} n_k + f_{2,j-1}^{(\gamma)}, \quad z'_\gamma \in \partial Q', \tag{34}$$

$$\int_{Q'} \widetilde{\widetilde{N}}_j^{(\gamma)} dz'_\gamma = 0. \tag{35}$$

Considering (28) and (32)–(35), condition (31) can be transformed ($r = j-2$) into

$$\left(\sum_{k\neq\gamma} A_{k\gamma} C_k^\gamma + A_{\gamma\gamma} \right) \frac{\partial^2 \overline{N}_r^{(\gamma)}}{\partial y_\gamma^2} (y_\gamma) = \int_{Q'} f_{1,r}^{(\gamma)} dz'_\gamma - \int_{\partial Q'} f_{2,r+1}^{(\gamma)} dS - \\ - \sum_{k\neq\gamma} \int_{\partial Q'} A_{k\gamma} \frac{\partial \widetilde{\widetilde{N}}_{r+1}^{(\gamma)}}{\partial y_\gamma} dS, \quad y_\gamma \in (0, 1). \tag{36}$$

Relation (36) will be used to define $\overline{N}_j^{(\gamma)}$. We rewrite the problem $H_j^{(0)} = 0$, $G_j^{(0)} = 0$ in the variables z:

$$\begin{aligned} &L_z N_j^{(0)} + W_{1,j}(z) - f_{1,j-2}^{(0)} = 0, \qquad z \in X, \\ &\frac{\partial}{\partial \nu_z} N_j^{(0)} + W_{2,j}(z) - f_{2,j-1}^{(0)} = 0, \quad z \in \partial X, \end{aligned} \tag{37}$$

where

$$X = \bigcup_{l=1}^{s} \{z \in \mathbf{R}^s \,|\, z_q \in (-1, 0), \quad q \neq l, \quad z_l \in \mathbf{R}\}.$$

It can be verified that the solvability condition for problem (37);

$$\int_X \left(W_{1,j}(z) - f^{(0)}_{1,j-2}(z)\right) dz = \int_{\partial X} \left(W_{2,j}(z) - f^{(0)}_{2,j-1}(z)\right) dS$$

follows from (18) which can be represented as $\int_{\mathcal{M}_y} f_1(y)dy = \int_{\partial\mathcal{M}_y} f_2(y)ds$. By using the method given in [81, 82, 108], we can prove the exponentially quick stabilization of the solution of problem (37) as $z_l \to \pm\infty$ to the constants $c_l^{\pm}$ and make the $c_l^{\pm}$ equal to zero by selecting indefinite constants in solving (36).

Thus, the solution of the problem on a cell in constructed by the following algorithm: Let $N_j^{(0)}$, $\tilde{N}_j^{(\gamma)}$, and $\tilde{\tilde{N}}_j^{(\gamma)}$ for $j \leqslant 0$ as well as $N_j^{(\gamma)}$ and $\overline{N}_j^{(\gamma)}$ for $\nu > 0$, $j \leqslant -1$ be zero, and suppose that $N_j^{(\gamma)}$ for $j \leqslant k-1$, $\gamma \geqslant 0$ and $\overline{N}_{k-1}^{(\gamma)}$, $\tilde{N}_k^{(\gamma)}$, $\tilde{\tilde{N}}_k^{(\gamma)}$, $N_{k+1}^{(\gamma)}$ for $\gamma > 0$ are defined.

1) Define $\overline{N}_{k,0}^{(\gamma)}(y_\gamma)$ $(\gamma = 1, \ldots, s)$ from (36) with the boundary values $\overline{N}_{k,0}^{(\gamma)}$ for $y_\gamma = 0, 1$.

2) Define $W_{1,k}(z)$, $W_{2,k}(z)$ for $\overline{N}_k^{(\gamma)} = \overline{N}_{k,0}^{(\gamma)}$ and solve problem (37); let $N_{k,0}^{(0)}$ be the solution of (37), $\lim_{z_\gamma \to \pm\infty} N_{k,0}^{(0)} = c_\gamma^{\pm}$; and set

$$\overline{N}_k^{(\gamma)}(y_\gamma) = N_{k,0}^{(\gamma)}(y_\gamma) - \left(c_\gamma^- - c_\gamma^+\right) y_\gamma - c_\gamma^+, \quad N_k^{(0)} =$$
$$= N_{k,0}^{(0)} - \sum_{\gamma=1}^{s} \left(c_\gamma^+ \rho(z_\gamma) \frac{\operatorname{sign} z_\gamma + 1}{2} - c_\gamma^- \rho(z_\gamma) \frac{\operatorname{sign} z_\gamma - 1}{2}\right).$$

3) Define $\tilde{N}_{k+1}^{(\gamma)}$ $(\gamma = 1, \ldots, s)$ by (32) for $j = k+1$.

4) Define $\tilde{\tilde{N}}_{k+2}^{(\gamma)}$ $(\gamma = 1, \ldots, s)$ from (33), and (34), (35) for $j = k+2$.

The induction step is completed: $\overline{N}_k^{(\gamma)}$, $N_k^{(\gamma)} = \overline{N}_k^{(\gamma)} + \tilde{N}_k^{(\gamma)}$ $(\gamma > 0)$, $N_k^{(0)}$, $\tilde{N}_{k+1}^{(\gamma)}$, and $\tilde{\tilde{N}}_{k+2}^{(\gamma)}$ $(\gamma > 0)$ are defined.

We substitute the expansion for $N^{(J)}$ thus obtained into (23) for $N_{qi}^{p,1}$, and after the change of variables $z_l = \xi_l/\mu$, we find $N_{qi}^{p,2}(z)$ as a series in powers of μ. After substituting the expansions h_{qi}^p and $n_{qi}^{p,\pm\tau}$ in powers of μ into (15) and (16), we get the averaged problem

$$\sum_{\substack{p,q,l\geqslant 0\\ 2\leqslant p+q+l\leqslant K+1}} \sum_{j=0}^{J} \varepsilon^{p+q+l-2}\mu^j \sum_{|i|=l} h_{qi}^{p,j} D^{q,i} v = f(x,\ t), \tag{15'}$$

$$\sum_{\substack{p,q,l\geqslant 0\\ p+q+l\leqslant K+1}} \sum_{j=0}^{J} \varepsilon^{p+q+l}\mu^j \sum_{|i|=l} n_{qi}^{p,\pm\tau} D^{q,i} v \Big|_{\partial G} = 0 \tag{16'}$$

with homogeneous initial values (17). An asymptotic solution of problem (15′), (16′), (17), as in Chapter 7, is sought in the form

$$v = v^{(K,J)} = \sum_{k=0}^{K} \sum_{j=0}^{J} \varepsilon^k \mu^j v_{k,j}(x,\ t), \tag{38}$$

where the $v_{k,j}(x,t)$ do not depend on ε and μ. We assume that the following condition is satisfied:

Condition A. $v_{kj} \in C^{k+2}(\overline{G}x[0,t_0])$. Then we can use the a priori estimates for problem (1)–(3) (see Chapter 1) to obtain the estimate (for $\mu = O(\varepsilon^\alpha)$, $\alpha > O$),

$$\|u - u^{(K,\ J)}\|_W = O(\varepsilon^K + \mu^J). \tag{39}$$

To single out the principal part of the averaged equation (15′), find the asymptotics of the functions N_0^0 as $\mu \to 0$:

$$-L\left(N_{0\gamma}^0 + \xi_\gamma\right) = 0, \quad \xi \in \mathcal{M}, \qquad \frac{\partial}{\partial \nu_\xi}\left(N_{0\gamma}^0 + \xi_\gamma\right) = 0, \quad \xi \in \partial\mathcal{M}.$$

We set

$$M_\gamma(\xi) = \begin{cases} -\xi_\gamma & \text{for } \xi \in \mathcal{M},\ \xi_j \in (0,1),\ j \neq \gamma, \\ \sum\limits_{l\neq\gamma} C_l^\gamma \xi_l & \text{for } \xi \in \mathcal{M},\ \xi_\gamma \in (0,1), \end{cases}$$

$$M_{0\gamma}^0 = M_\gamma(\xi) \sum_{j=1}^{s} \rho(\xi_j).$$

By using the standard technique of integral identities (see §1), we can obtain the estimate $\|\nabla_\xi(N_{0\gamma}^0\ M_{0\gamma}^0\|_{L_2(\mathcal{M})}^2 = \operatorname{mes}\mathcal{M}\cdot O(\mu)$. This implies that the left side of equation (5) is the principal part of (15). Relations (21)–(23) for $p = q = l = 0$ suggest $n_0^{0,\pm\tau} = 1$. The assertion of Theorem 1 is the consequence of estimate (39) for $K = J = 0$.

In conclusion, it may be pointed out that the skeletal structures have two advantages over other problems of averaging fields in periodic media; namely the introduction of the small parameter μ, firstly, enables the coefficients of the averaged equation to be calculated by explicit formulas (see §§1–4) and, secondly, simplifies the problem of constructing boundary layers where G is not a layer. The splitting principle (see §3) can be applied to a great variety of skeletal structures. These advantages make it possible to use the latter as a model periodic media.

§7 Skeletal Structures with Random Properties

Problem (3.9)–(3.12) with random coefficients R, S, and $A = (A_1, \ldots, A_s)$ and random right side f is considered in [121]. In this case, the cylinders B^{α}_{hj} forming the framework B may be missing with probability $p_0^{\alpha,j}$. This framework B can be used as a geometric model of a system of cracks or capillaries in investigating processes of percolation of a viscous fluid and diffusion of the substances dissolved in it. Correlation of this model and the real structure of the soil is based on statistical data processing. The splitting principle for the averaged operator formulated in [118] generalizes the results of the tensor method well known in percolation theory (see [136, 167]).

Consider the simplest random skeletal structure. Suppose there is a connected system of indefinite capillaries (cracks) stretched in three (two) coordinate directions and spaced by ε along each of the directions. The radius of a capillary (width of a crack) is a random quantity of order $\mu\varepsilon \ll \varepsilon$. The porosity of the media is of order $\mu^2 \ll 1$ $(\mu \ll 1)$. We shall give a rigorous definition of a rectangular skeletal structure.

Let the mesh S_ε be given in $\mathbf{R}^s (s = 2, 3)$ with the coordinates of the nodes $(k_1\varepsilon, \ldots, k_s\varepsilon) = k\varepsilon$, where $k = (k_1, \ldots, k_s)$ is an integer vector, $\varepsilon \ll 1$, and through each such node, the straight lines $\Pi^i_{k_{\bar{i}}} = \Pi^i_{(k_1,\ldots,k_{i-1},k_{i+1},\ldots,k_s)}$ are drawn parallel to the coordinate axes Ox_i $(i = 1, \ldots, s)$. By $C^i_{k_{\bar{i}}}(r)$ we denote the cylinder (strip) consisting of points that are spaced less than $r\mu\varepsilon(\mu \ll 1)$ away from $\Pi_{k_{\bar{i}}}$ and we shall agree that $C^i_{k_{\bar{i}}}(0) = \emptyset$. Further, let the $r^i_{k_{\bar{i}}}$ be independent random variables which take values $\{0, r_1, \ldots, r_m\}$, while $P\{r^i_{k_{\bar{i}}} = r_j\} = p_{ij}$.

By a *random rectangular structure* we shall mean the domain

$$B_\omega = \bigcup_{i=1}^{s} \bigcup_{k_{\bar{i}} \in Z^{s-1}} C^i_{k_{\bar{i}}}\left(r^i_{k_{\bar{i}}}\right),$$

When $s = 2$, B_ω is a system of thin strips stretched in each of the coordinate directions. This model is suitable for describing a fissured structure with small porosity. When $s = 3$, B_ω is a system of thin round capillaries stretched in each of the coordinate directions.

Consider the Navier–Stokes system (unlike §4.7, here p is pressure)

$$-\nu\ \Delta u + (u,\ \nabla_x)u + \nabla_x p/\rho = f, \tag{1}$$

$$\operatorname{div} u = 0, \tag{2}$$

defined on $B_\omega \cap G$, where G is the parallelepiped $\{(x \in \mathbf{R}^s | a_i \leqslant x_i \leqslant b_i\}$ with the boundary value

$$u\,|_{\partial(B_\omega \cap G)} = 0. \tag{3}$$

It was shown in §4.7 that the averaging of the Navier–Stokes system of equations yielded Darcy's law, but to calculate the effective percolation tensor, it is necessary to solve numerically the appropriate problem on a cell. If the Navier–Stokes system of equations is defined in the random skeletal domain B_ω, the splitting principle for the averaged operator can be used to solve approximately the problem on a cell and to obtain (in heuristically rigorous terms) an explicit formula for the percolation coefficient. Namely, the averaged percolation problem has the form

$$\bar{w} - \frac{\overline{K}}{\rho g}\left(\nabla_x \bar{p} - \rho f\right) = 0, \quad x \in G, \tag{4}$$

$$\operatorname{div}\left(\rho \bar{w}\right) = 0, \quad x \in G, \tag{5}$$

$$\left(\overline{K}\left(\nabla_x \bar{p} - \rho f\right), n\right)|_{\partial G} = 0. \tag{6}$$

Here $\overline{K}$ is the effective percolation tensor, $\overline{K} = \varepsilon^2 \mu^2 M g K$,

$$M = 2^{3-s} \mu^{s-1} \pi^{s-2} \sum_{i=1}^{s} \sum_{j=0}^{m} p_{ij} r_j^{s-1}$$

is the average porosity of the medium, $K = \|K_{lh}\| = \delta_{lk} K_k$, $K_k = \frac{\kappa_s}{\nu} \mu^{s-1} \sum_{j=0}^{m} P_{kj} (r_j)^{s+1} / M$, $\kappa_2 = 2/3$, $\kappa_3 = \pi/8$, n is the exterior normal to ∂G, $\bar{w} = M \varepsilon^2 \mu^2 \bar{v}$ is the percolation speed, $\varepsilon^2 \mu^2 \bar{v}$ is the average flow velocity, ρ is density, and $\bar{p}$ is the average pressure. Relation (4) is Darcy's law.

In conclusion, we note that in the case when all the $r^i_{k_i}$ are equal to $\bar{r}$ with probability 1 (a periodic system of capillaries) and with $s = 3$, the result for the averaged coefficient of percolation is the well-known formula (see [170]): $\overline{K} = g\theta r^2/(8\nu)$, where $r = \varepsilon \mu \bar{r}$ is the radius of capillary, and θ is the volume fraction of the capillaries in any of the coordinate directions.

When $s = 2$, $\overline{K} = g\theta \bar{r}^2/(3\nu)$.

REMARK 1: With $s = 3$, the percolation coefficient depends to a great extent on the shape of the capillary's cross-section and can be calculated explicitly only for round capillaries.

REMARK 2: The above model can be applied to an investigation of the wood impregnation process (see [44]). Wood is regarded as a periodic or random capillary structure.

CHAPTER 9

MATHEMATICS OF BOUNDARY-LAYER THEORY IN COMPOSITE MATERIALS

All foregoing sections, except §§7.3 and 8.6, ignored the boundary effects that may occur near the boundary of a composite (the investigations covered only the zeroth approximation of the boundary value problem without oscillation under boundary condition). Yet the neighbourhoods of the boundaries and contact surfaces (when two pieces of different composites are in contact) have boundary layers where the gradient of a solution may greatly differ (by a value of order 1) from those of the asymptotics obtained above.

The estimates of order $O(\sqrt{\varepsilon})$ in the integral norms of the W_2^1-type "do not feel" this difference between an exact solution and an asymptotic one, since the contribution of the boundary layer having the width $O(\varepsilon|ln\varepsilon|)$ is just of order $O(\sqrt{\varepsilon})$. Therefore, there is a need for asymptotic expansions of the solutions to boundary value problems. This chapter proposes a method for constructing the a.e. of those in the domains of the simplest form (with the boundary parallel to the plane $\{x_1 = 0\}$ (§§2 and 4)) and of the problems on the contact of several microstructures with the contact surface parallel to $\{x_1 = 0\}$ (§§1, 3, 5), taking the boundary layers into account. For partial sums of such expansions, estimates of the closeness to an exact solution of any order in powers of ε are proved. Taking the boundary layers into account may be useful in investigating failures of composite materials, which often start from the boundary (see [174]).

The construction of the f.a.s. yields averaged equations of higher orders and corresponding averaged boundary conditions. Thus, if in the course of the procedure in §§2.3 and 2.4, we select an indefinite constant for the functions $N_i(\xi)$ such that $\langle N_i \rangle = 0$ $(i > 0)$, then the averaged equation takes the simplest form (2.3.38). Now, however, the N_i may not vanish for $\xi = 0$, and consequently, the substitution of the series (2.3.3) into the boundary conditions produces the asymptotic equalities

$$u^{(\infty)}\big|_{x=0,l} \sim \sum_{i=0} N_i(0) \frac{d^i v}{dx^i}\bigg|_{x=0,l} \sim 0. \tag{1}$$

These can be interpreted as averaged boundary conditions of infinite order. In the multi-dimensional case, similar relations result from the existence conditions for the boundary-layer functions which rectify the f.a.s. in §4.2.

Sections 1 and 3 consider problems on the contact of two different composite materials with the interface $\{x_1 = 0\}$. In this case, the averaged equation is constructed independently in each of the half-spaces $\mathbf{R}^s_+ = \{x \in \mathbf{R}^s | x_1 > 0\}$ and $\mathbf{R}^s_- = \{x \in \mathbf{R}^s | x_1 < 0\}$, and with $x_1 = 0$ averaged relations are derived on the interface. The leading term of the averaged relations on the interface of the composites turns out to be of the form $[v_0] = \left[\hat{A}_{ij}\frac{\partial v_0}{\partial x_j}n_i\right] = 0$. In the latter equality, the effective characteristics $\hat{A}_{ij}$ are taken separately from each side of the contact surface. These relations on the discontinuity of the coefficients $\hat{A}_{ij}$ correspond to the conditions for continuity of the averaged solution v_0 and of the effective flow $\hat{A}_{ij}\frac{\partial v_0}{\partial x_j}n_i$. They are derived also by means of the smoothing procedure described in §1.2. In this case the problem on the contact of two periodic structures is considered as a limit one relative to the parametric family of problems with smooth coefficients depending on both the fast and slow variables. Passage to the limit makes it possible to obtain relations on the interface of the structures. The third method for obtaining these relations is to apply the technique of G-convergence (see [53, 54, 157]). Note that the last two methods enable one to obtain the contact relations for an interface of an arbitrary shape, but they cannot help construct the a.e. of a boundary-layer solution in the contact area of the media. Neither do these approaches allow us to obtain averaged relations of higher orders on the interface. Therefore, in this chapter, we shall dwell on the first approach based on the construction of averaged boundary and contact relations of infinite order.

Section 3 considers a problem of the contact of two media divided by a thin inhomogeneous interlayer having a periodic structure. Note that problems on the contact of two homogeneous media divided by a thin homogeneous interlayer with large (small) physical characteristics were investigated in [69, 145, 166, 214]. The need to consider the inhomogeneous interlayer arises, for example, in investigating thermal processes in a blast furnace whose brick wall has a periodic inhomogeneous structure.

In §5 the methodology set forth in §2 is extended to the nonstationary system of equations of elasticity theory. For the stationary system of elasticity theory, an asymptotic expansion of a solution, which is exact up to the boundary, was constructed in [60, 109, 210]. Section 6 discusses the use of the asymptotics of the stress field to determine the ultimate strength of a composite. Section 7 deals with equations defined in a half-space and having second-type boundary conditions, and with some other boundary value problem. Section 8 describes approaches to investigating materials with the two-level hierarchy of the microstructure. Section 9 discusses special features of constructing boundary layers in time for the nonstationary parabolic Cauchy problem.

To simplify the presentation, we will construct the asymptotic expansions under the assumption that the coefficients of the equations are smooth functions, but as in §2.4, 4.1, and 4.2, the methodology can be generalized to the case of piecewise smooth coefficients (in the sense of §4.2). For equations with the piecewise smooth coefficients, the problems for the periodic and boundary-layer components N should be understood in the generalized (Sobolev) sense. We can see from direct substitution that the series $u^{(\infty)}$ constructed with respect to such functions N satisfies the relations on the coefficient discontinuity surfaces in the asymptotic sense. The formulations of the theorems are given for the piecewise smooth coefficients.

We shall first apply the method to a one-dimensional space problem (§1).

§1 Problem on the Contact of Two Layered Media

Consider the Cauchy problem for the system of equations

$$L_\varepsilon u \equiv R\left(\frac{x}{\varepsilon}\right)\frac{\partial^2 u}{\partial t^2} - \frac{\partial}{\partial x}\left(A\left(\frac{x}{\varepsilon}\right)\frac{\partial u}{\partial x}\right) = f(x, t) \tag{1}$$

with homogeneous initial conditions

$$u\,|_{t=0} = \left.\frac{\partial u}{\partial t}\right|_{t=0} = 0. \tag{2}$$

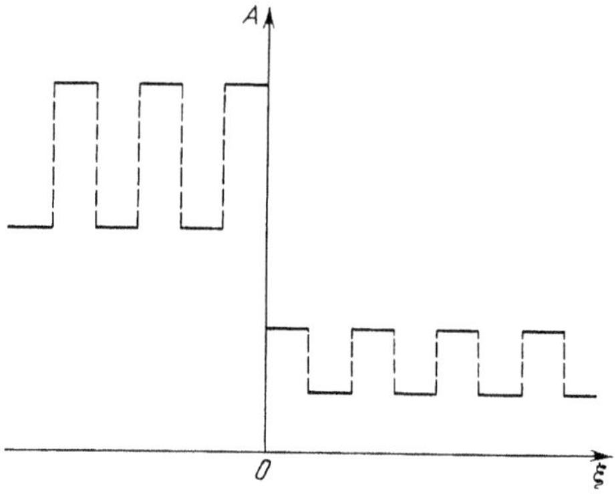

Figure 30

Here u, f are n-dimensional vectors and $R(\xi)$, $A(\xi)$ are the following piecewise smooth $n \times n$ matrix-valued functions (see Fig. 30):

$$R(\xi) = \begin{cases} R^1(\xi) & \text{for } \xi > 0, \\ R^2(\xi) & \text{for } \xi < 0, \end{cases}$$

$$A(\xi) = \begin{cases} A^1(\xi) & \text{for } \xi > 0, \\ A^2(\xi) & \text{for } \xi < 0; \end{cases}$$

here $R^i(\xi)$, $A^i(\xi)$ $(i = 1, 2)$ are 1-periodic matrix-valued functions satisfying the conditions

$$R^i(\xi) > \varkappa_1 E, \quad A^i(\xi) > \varkappa_1 E,$$

where κ_1 is a positive number.

It is assumed that $f(x,t) \in C'^\infty(\mathbf{R} \times [0, +\infty))$ (i.e., the vector-function $f(x,t)$ is infinitely differentiable in the half-plane $\mathbf{R} \times [0, +\infty)$ and is nonzero only in some bounded domain) and that the conditions

$$\left.\frac{\partial^l f}{\partial t^l}\right|_{t=0} = 0 \quad \forall l = 0, 1, \ldots \tag{3}$$

hold.

The following conjugation conditions are given at the discontinuity points of the coefficient

$$[u] = \left[A\frac{\partial u}{\partial x}\right] = 0.$$

The f.a.s. of problem (1), (2) is sought in the form

$$u^{(\infty)}(x, t) \sim \sum_{p,q,l=0}^{\infty} \varepsilon^{p+q+l} N_{ql}^{p}\left(\frac{x}{\varepsilon}\right)\frac{\partial^{q+l} v}{\partial t^q \partial x^l}, \tag{4}$$

where the $N_{qt}^{p}(\xi)$ are $n \times n$ matrix-valued functions:

$$N_{ql}^{p}(\xi) = \begin{cases} N_{ql}^{p1}(\xi) & \text{for } \xi > 0, \\ N_{ql}^{p2}(\xi) & \text{for } \xi < 0, \end{cases}$$

and the $N_{ql}^{pi}(\xi)$ are 1-periodic matrix-valued functions.

We shall construct the N_{qi}^{pi} by using the algorithm from §4.5 (the construction was made there directly for the multi-dimensional case) independently for $i = 1$, $\xi > 0$, and $i = 2$, $\xi < 0$.

Then, after substituting (4) into (1), we obtain for v two different equations with $x > 0$ and $x < 0$:

$$\sum_{\substack{p,q,l=0 \\ 2 \leqslant p+q+l}} \varepsilon^{p+q+l-2} h_{ql}^{pi} \frac{\partial^{q+l} v}{\partial t^q \partial x^l} \sim f(x, t), \tag{5}$$

where $i = 1$ for $x > 0$ and $i = 2$ for $x < 0$, while the constant matrices h_{qi}^{pi} are defined by

$$h_{ql}^{pi} = -\left\langle A\frac{dN_{ql-1}^{pi}}{d\xi} + AN_{ql-2}^{pi} - RN_{q-2l}^{pi}\right\rangle.$$

The solution of the system of equations (1) at any point $x = a$ satisfies the conjugation conditions

$$[u]_a = \left[A\frac{\partial u}{\partial x}\right]_a = 0. \tag{6}$$

By substituting (4) into (6) with $a = 0$, we obtain the contact conditions relating the solutions of the problems (5) at the point $x = 0$:

$$[v]_0 + \sum_{p,q,l=0}^{\infty} \varepsilon^{p+q+l} \left(N_{ql}^{p1}(+0) \frac{\partial^{q+l} v}{\partial t^q \partial x^l}\bigg|_{x=+0} - N_{ql}^{p2}(-0) \frac{\partial^{q+l} v}{\partial t^q \partial x^l}\bigg|_{x=-0} \right) \sim 0, \quad (7)$$

$$\left(A^1(+0) \left(\frac{dN_{01}^{0,1}}{d\xi}(+0) + E \right) \right) \frac{\partial v}{\partial x}\bigg|_{x=+0} - \left(A^2(-0) \left(\frac{dN_{01}^{02}}{d\xi}(-0) + E \right) \right) \frac{\partial v}{\partial x}\bigg|_{x=-0} + \sum_{\substack{p,q,l=0 \\ 2 \leqslant p+q+l}}^{\infty} \varepsilon^{p+q+l-1} \left(A^1(+0) \left(\frac{dN_{ql}^{p1}}{d\xi}(+0) + N_{ql-1}^{p1}(+0) \right) \frac{\partial^{q+l} v}{\partial t^q \partial x^l}\bigg|_{x=+0} - A^2(-0) \left(\frac{dN_{ql}^{p2}}{d\xi}(-0) + N_{ql-1}^{p2}(-0) \right) \frac{\partial^{q+l} v}{\partial t^q \partial x^l}\bigg|_{x=-0} \right) \sim 0. \quad (8)$$

Equation (5) with the contact conditions (7), (8) and homogeneous initial conditions

$$v\big|_{t=0} = \frac{\partial v}{\partial t}\bigg|_{t=0} = 0 \quad (9)$$

comprise the averaged problem for v. The asymptotic solution of problem (5), (7)–(9) is sought in the form (2.3.4); auxiliary problems arise with respect to v_k of the form

$$\hat{R}^1 \frac{\partial^2 v_k}{\partial t^2} - \hat{A}^1 \frac{\partial^2 v_k}{\partial x^2} = f_k^1(x, t) \quad \text{for} \quad x > 0,$$

$$\hat{R}^2 \frac{\partial^2 v_k}{\partial t^2} - \hat{A}^2 \frac{\partial^2 v_k}{\partial x^2} = f_k^2(x, t) \quad \text{for} \quad x < 0,$$

$$[v_k]_0 = g_k^1(t),$$

$$\left(\hat{A}^1 \frac{\partial v_k}{\partial x}\bigg|_{x=+0} - \hat{A}^2 \frac{\partial v_k}{\partial x}\bigg|_{x=-0} \right) = g_k^2(t);$$

here $\hat{R}^i = \langle R^i(\xi) \rangle$, $\hat{A}^i = \langle (A^i(\xi))^{-1} \rangle^{-1}$, f_k^i, g_k^i are functions defined by $v_0, \ldots, v_{k-1}$, $f_0^i = f$ and $g_0^i = 0$. By taking advantage of the arbitrary choice of constants in defining $N_{ql}^p(\xi)$ and by analogy with §2.3, we can make (7) to take the form

$$[v]_0 = 0.$$

It follows from (3) and (9) that the partial sum $u^{(K)}$ of (4) satisfies the initial conditions (2) exactly.

From the energy identity

$$\int_{-\infty}^{+\infty} \left(0.5\, R\left(\frac{x}{\varepsilon}\right)\left(\frac{\partial u}{\partial t}\right)^2 + A\left(\frac{x}{\varepsilon}\right)\left(\frac{\partial u}{\partial x}\right)^2\right) dx \Big|_0^t = \int_0^t \int_{-\infty}^{+\infty} f \frac{\partial u}{\partial t}\, dx\, dt$$

we can get the estimate

$$\max_{t\in[0,T]} \int_{-\infty}^{+\infty} \left(\left(\frac{\partial\,(u - u^{(K)})}{\partial t}\right)^2 + \left(\frac{\partial\,(u - u^{(K)})}{\partial x}\right)^2\right) dx = O(\varepsilon^K),$$

$$u^{(K)} = \sum_{\substack{p,q,l\geqslant 0\\ p+q+l\leqslant K+1}} \varepsilon^{p+q+l} N_{ql}^p\left(\frac{x}{\varepsilon}\right) \frac{\partial^{q+l} v^{(K)}}{\partial t^q \partial x^l},$$

$$v^{(K)} = \sum_{k=0}^{K} \varepsilon^k v_k(x, t).$$

Note that the $N_{ql}^p(\xi)$ are periodic up to the interface point $\xi = 0$, i.e., no rapidly oscillating boundary layers are formed in problems with one space variable; this fact is typical only for one-dimensional problems (see §§2 and 3).

§2 The Boundary Layer for an Elliptic Equation Defined on a Half-Plane

Suppose we have the equation

$$\sum_{i=1}^{2} \frac{\partial}{\partial x_i}\left(K\left(\frac{x_1}{\varepsilon}, \frac{x_2}{\varepsilon}\right)\frac{\partial u}{\partial x_i}\right) = f(x_1, x_2), \tag{1}$$

defined for $x_1 > 0$, where $K(\xi_1, \xi_2)$ is a function 1-periodic in ξ_1 and ξ_2 and piecewise smooth in the sense of §4.1. We also assume that $\sum$, the totality of the coefficient's discontinuity lines, has no common points with the axis ξ_2. Let $0 < \kappa_1 < K(\xi_1, \xi_2) < \kappa_2$. The solution of (1) is sought in the class of bounded functions, T-periodic in x_2 (T is a number of order 1 and a multiple of ε), which satisfy the Dirichlet condition

$$u\,|_{x_1=0} = g(x_2); \tag{2}$$

see the definition of the generalized solution of (1) in [82]. It is assumed that f, g are T-periodic in x_2: $f, g \in C^\infty$, $|f| < c_1 e^{-c_2 x_1}$, $c_1 > 0$, $c_2 > 0$.

We seek the f.a.s. of problem (1), (2) as

$$u^{(\infty)} \sim \left(\sum_{l=0}^{\infty} \varepsilon^l \sum_{|i|=l} N_i(\xi_1, \xi_2) D^i v \right) \Bigg|_{\xi_j = x_j/\varepsilon}, \quad j = 1, 2. \tag{3}$$

Here $v(x_1, x_2)$ is a solution of an averaged problem with constant coefficients, which will be constructed below; v is independent of ξ_1 and ξ_2; it will be shown below that $v(x_1, x_2)$ can be represented as (see §4.2)

$$v(x_1, x_2) \sim \sum_{j=0}^{\infty} \varepsilon^j v_j(x_1, x_2), \tag{4}$$

where the $v_j(x_1, x_2)$ are independent of ε.

The functions of fast variables, $N_i(\xi_1, \xi_2)$, have the following structure:

$$N_i(\xi_1, \xi_2) = N_i^1(\xi_1, \xi_2) + N_i^2(\xi_1, \xi_2), \tag{5}$$

where N_i^1 is a function 1-periodic in ξ_1 and ξ_2, and N_i^2 is a function 1-periodic in ξ_2 which tends to zero exponentially fast (as $\xi_1 \to +\infty$), that is, $|N_i^2| \leqslant c_1 e^{-c_2 \xi_1}$ for $\xi_1 > 0$. The functions N_i^2 with such properties are called the *boundary-layer-type functions*. It is also assumed that $N_\emptyset = 1$, that is, $N_\emptyset^1 = 1$, $N_\emptyset^2 = 0$. So the f.a.s. of problem (1), (2) is sought as a series in powers of $\varepsilon^{|i|}$ and the derivatives D^i of a slowly changing function v. The coefficients of this series, N_i, are rapidly oscillating functions of the form (5). Note that if we set $N_1^2 = 0$, then (3) coincides with the expansion of the solution of (1) without boundary conditions (see §4.2). To satisfy condition (2) with any order of accuracy in ε requires to introduce the additional term, N_i^2.

Substitute (3) into (1) and (2), formally considering $K(\xi_1, \xi_2) \in C^\infty(\mathbf{R}^2)$. We combine the terms with the same powers of ε and the same derivatives $D^i v$ by considering, as above, that the operators of the differentiation are not interchangeable with respect to x_q. Equation (1) and condition (2) will be written as

$$\sum_{l=0}^{\infty} \varepsilon^{l-2} \sum_{|i|=l} H_i(\xi_1, \xi_2) D^i v \sim f(x_1, x_2), \tag{6}$$

$$\sum_{l=0}^{\infty} \varepsilon^l \sum_{|i|=l} N_i(0, \xi_2) D^i v \Big|_{x_1=0} \sim g(x_2). \tag{7}$$

Here $H_i(\xi_1, \xi_2) = H_i^1(\xi_1, \xi_2) + H_i^2(\xi_1, \xi_2)$, H_i^1 is a function 1-periodic in ξ_1 and ξ_2, and H_i^2 is a function 1-periodic in ξ_2, which tends exponentially to zero as $\xi_1 \to +\infty$, or more precisely, H_i^2 is of the form

$$H_i^2 = F_0 + \sum_{j=1}^{2} \frac{\partial}{\partial \xi_j} F_j, \quad |F_j| < c_1 e^{-c_2 \xi_1}, \quad j = 0, 1, 2, \tag{8}$$

the expressions H_i^r $(r = 1, 2)$ are defined in terms of the coefficient K in (1) and the function N_i^r by

$$H_i^r(\xi_1, \xi_2) = \sum_{j=1}^{2} \frac{\partial}{\partial \xi_j}\left(K \frac{\partial N_i^r}{\partial \xi_j}\right) + \\ + \frac{\partial}{\partial \xi_{i_1}}\left(KN_{i_2 \ldots i_l}^r\right) + K \frac{\partial}{\partial \xi_{i_1}} N_{i_2 \ldots i_l}^r + \delta_{i_1 i_2} K N_{i_3 \ldots i_l}^r, \qquad (9)$$

where $i = (i_1, \ldots, i_l)$, and N_i^r with the negative length i is assumed to be zero.

By introducing the notation

$$L_{\xi\xi} = \sum_{j=1}^{2} \frac{\partial}{\partial \xi_j}\left(K \frac{\partial}{\partial \xi_j}\right),$$
$$T_i^r = \frac{\partial}{\partial \xi_{i_1}}\left(KN_{i_2 \ldots i_l}^r\right) + K \frac{\partial}{\partial \xi_{i_1}} N_{i_2 \ldots i_l}^r + \delta_{i_1 i_2} K N_{i_3 \ldots i_l}^r,$$

we can rewrite (9) as $H_i^r(\xi_1, \xi_2) = L_{\xi\xi} N_i^r + T_i^r(\xi_1, \xi_2)$. We order the N_i^r, namely we set $N_i^r \succ N_j^r$ if $|i| > |j|$. Note that T_i^r is defined by the functions N_j^r with the priority $|j| < |i|$. In the same way as in §4.1, this feature can be used for a recurrent definition of N_i^r in terms of N_j^r of lower priority if some requirements are imposed upon $H_i^r(\xi_1, \xi_2)$, $N_i(0, \xi_2)$.

We impose the following requirement:

1. $H_i^1(\xi_1, \xi_2) \equiv \text{const}$.
2. $H_i^2(\xi_1, \xi_2) \equiv 0$.
3. $N_i(0, \xi_2) \equiv \text{const}$.

We denote the constant in requirement 1 by h_i^1 and the constant in requirement 3 by h_i^2. Note that requirement 1 was introduced for the problem in the entire space in §4.2. If we can find the N_i^r and h_i^r to satisfy the requirements 1–3, then problem (6), (7) reduces to the problem with constant coefficients for v (13), (14). This problem can be solved by the standard method of expanding v in powers of ε.

Thus to construct the f.a.s. of the original problem, it suffices to construct the functions N_i^r and constants h_i^r satisfying the requirements 1–3.

By proceeding from requirement 1, we define all the N_i^1 successively by induction. Let all the N_i^1 of priority $|i| \leqslant k$ be found. We find $T_i^1(\xi_1, \xi_2)$ for all i: $|i| = k + 1$ (see the definition of T_i^r) and select $h_i^1 = \langle T_i^1 \rangle$. Then there exists a generalized solution from the class W_2^1, unique up to a constant term and 1-periodic in ξ_1 and ξ_2, of the equation

$$L_{\xi\xi} N_i^1 + T_i^1 = h_i^1 \qquad (10)$$

(see the Supplement). By solving the equation, we get N_i^1 for each multi-index i of length $|i| = k + 1$. The induction step is completed. The basis of the induction is $k = 0$, $N_\emptyset^1 = 1$.

After all the N_i^1 have been defined, we successively seek all the N_i^2 on the basis of requirements 2, 3. Namely, let all the N_i^2 of priority $|i| \leqslant k$ be found. We define $T_i^2(\xi_1, \xi_2)$ for all i: $|i| = k+1$, and select the constant h_i^2 ($|i| = k+1$) such that there exists a unique solution from the class W_2^1, 1-periodic in ξ_2 and exponentially tending to zero as $\xi_1 \to +\infty$, of the problem

$$\begin{gathered} L_{\xi\xi} N_i^2 + T_i^2(\xi_1, \xi_2) = 0, \quad \xi_1 > 0, \\ N_i^2(0, \xi_2) = -N_i^1(0, \xi_2) + h_i^2, \\ |N_i^2| \leqslant c_2 e^{-c_4 \xi_1}, \quad \xi_1 > 0, \quad c_3, \ c_4 > 0. \end{gathered} \tag{11}$$

This choice of the constant is known to be possible and unique (see [81, 82, 110]).

In order to define h_i^2, we first solve the problem

$$L_{\xi\xi} \tilde{N}_i^2 + T_i^2(\xi_1, \xi_2) = 0, \quad \xi_1 > 0, \quad \tilde{N}_i^2(0, \xi_2) = -N_i^1(0, \xi_2) \tag{12}$$

in the class of bounded functions 1-periodic in ξ_2. A solution of the problem is known to exist and be unique, and there exists the constant C_i^2 such that $|\tilde{N}_i^2 - C_i^2| < c_5 e^{-c_6 \xi_1}$. The latter proposition follows from the theorem of the Phragmén–Lindelöf type obtained in [82]. We set $h_i^2 = -\lim_{\xi_1 \to +\infty} \tilde{N}_i^2$. Then problem (11) has the required solution, and $N_1^2 = \tilde{N}_i^2 - \lim_{\xi_1 \to +\infty} \tilde{N}_i^2 |i| = k+1$, is the solution of (11).

Due to the exponential way of decay of N_i^2, problem (12) can be solved numerically in the strip $0 \leqslant \xi_1 \leqslant a$, under the condition $\partial \tilde{N}_i^2 / \partial \nu (a, \xi_2) = 0$ when a is not too large. This completes the induction step. The basis is $k = 0$, $N_\emptyset^2 = 0$.

REMARK 1: To be able to apply the results of [81, 82, 110], we must justify that T_i^2 tends to zero exponentially fast in the sense of (8). The justification is carried out using the a priori estimate of [139, 171] and by taking into account the piecewise smoothness of K.

The constants h_i^r were defined along with the N_i^r. Going back to relations (6), (7), we note that they become the problem with constant coefficients

$$\sum_{l=0}^{\infty} \varepsilon^{l-2} \sum_{|i|=l} h_i^1 D^i v \sim f(x_1, x_2), \quad x_1 > 0, \tag{13}$$

$$\sum_{l=0}^{\infty} \varepsilon^{l} \sum_{|i|=l} h_i^2 D^i v \Big|_{x_1=0} \sim g(x_2). \tag{14}$$

By using the definition of T_i^1, we can demonstrate that $h_i^1 = 0$ for $|i| = 0, 1$ and for $|i| = 2$,

$$h^1_{i_1 i_2} = \widehat{K}^{i_1 i_2} = \left\langle K \frac{\partial}{\partial \xi_{i_1}} (N^1_{i_2} + \xi_{i_2}) \right\rangle;$$

here $N^1_{i_2}(\xi_1, \xi_2)$ is a solution, 1-periodic in ξ_1 and ξ_2, of the equation

$$L_{\xi\xi}(N^1_{i_2} + \xi_{i_2}) = 0, i_2 = 1, 2$$

It can be also shown that $h^2_\emptyset = 1$, that is (13), (14) take the form

$$\sum_{q,k=1}^{2} \widehat{K}^{qk} \frac{\partial^2 v}{\partial x_q \partial x_k} + O(\varepsilon) = f(x_1, x_2),$$
$$v|_{x_1=0} + O(\varepsilon) = g(x_2),$$

which implies that v can be sought as the series $\sum_{j=0}^{\infty} \varepsilon^j v_j(x_1, x_2)$, where $v_j(x_1, x_2)$ are solutions of the recurrent chain of problems

$$\sum_{q,k=1}^{2} \widehat{K}^{qk} \frac{\partial^2 v_j}{\partial x_q \partial x_k} = f_j(x_1, x_2), \quad x_1 > 0,$$
$$v_j|_{x_1=0} = g_j(x_2). \tag{15}$$

Here f_j and g_j are determined by defining $v_0, v_1, \ldots, v_{j-1}$, $f_0 = f$, and $g_0 = g$. From Lemma 4.1 for any numbers $\eta_1, \eta_2 \in \mathbf{R}$

$$0 \leqslant \varkappa_1(\eta_1^2 + \eta_2^2) \leqslant \sum_{q,k=1}^{2} \widehat{K}^{qk} \eta_q \eta_k, \quad \varkappa_1 > 0.$$

By induction in j, it can be shown that f_j, g_j are periodic in x_2 with period T, f_j and $g_j \in C^\infty$, and $|f_j| < c_7 e^{-c_8 x_1}$, where c_7 and c_8 are independent of ε. Results of [81, 82, 110] imply that for any j a solution of problem (15) exists and is unique in the class of functions periodic in x_2 with period T, which stabilizes exponentially quickly to a constant as $x_1 \to +\infty$.

In the case of problem (15), the constant can be calculated, and it turns out that

$$\lim_{x_1 \to +\infty} v_j(x_1, x_2) = \frac{1}{T} \left(\int_0^T g_j(x_2)\, dx_2 - \frac{1}{\widehat{K}^{11}} \int_0^{+\infty} \int_\theta^{+\infty} \int_0^T f_j(x_1, x_2)\, dx_2 dx_1 d\theta \right).$$

By defining v_j successively from (15), we find $v(x_1, x_2)$ and substitute it into (3). The f.a.s. of (3) satisfies problem (1), (2) up to terms of order $O(\varepsilon^J)$ for any integer J. By $u^{(J)}$ we shall mean the partial sum of series (3) containing all the terms for $l \leqslant J + 1$:

$$u^{(J)} = \left(\sum_{l=0}^{J+1} \varepsilon^l \sum_{|i|=l} N_i(\xi_1, \xi_2)\, D^i v^{(J)} \right) \Bigg|_{\xi = x/\varepsilon}, \tag{16}$$

where

$$v^{(J)} = \sum_{j=0}^{J} \varepsilon^j v_j(x_1, x_2). \tag{17}$$

The approximation $u^{(J)}$ satisfies (1), (2) with accuracy $O(\varepsilon^J)$. The a priori estimates of [110] for problem (1), (2) imply

$$\| \nabla (u^{(J)} - u) \|_{L_2(\mathbf{R}\times(0,T))} = O(\varepsilon^J).$$

Then (16) and (17) suggest an even stronger estimate:

$$\| \nabla (u^{(J)} - u) \|_{L_2(\mathbf{R}\times(0,T))} = O(\varepsilon^{J+1}), \tag{18}$$

and [139, 147, 171] imply

$$\| u^{(J)} - u \|_{C(\mathbf{R}\times[0,T])} = O(\varepsilon^{J+2}), \tag{19}$$

$$\left| \frac{\partial u^{(J)}}{\partial x_i} - \frac{\partial u}{\partial x_i} \right| \leqslant c\varepsilon^{J+1}. \tag{20}$$

In the case of the piecewise smooth coefficient, (20) holds everywhere except on the discontinuity surfaces of the coefficient K up to these surfaces.

We shall formulate the results in the form of a theorem.

THEOREM 1. *There exists a f.a.s. of problem (1), (2) of the form (3), where N_i has structure (5), the N_i^r ($r = 1, 2$) are solutions of problem (10) for $r = 1$ and of (11) for $r = 2$, v can be represented as (4), and v_j can be defined recurrently from (15). The partial sum $u^{(J)}$ of series (3) of the form (16), (17) satisfies (1), (2) with accuracy $O(\varepsilon^J)$. The estimates (18)–(20) of the closeness of $u^{(J)}$ and u, where u is the exact solution of (1), (2) are valid.*

REMARK 2: It follows from (20) that the zeroth approximation of the gradient of the solution is of the form

$$\left(\frac{\partial u}{\partial x_i} \right)^{(0)} = \frac{\partial v_0}{\partial x_i} + \sum_{j=1}^{2} \frac{\partial}{\partial \xi_i} (N_j^1 + \overline{N}_j^2) \frac{\partial v_0}{\partial x_j},$$

where N_j^1 is a solution of the equation $L_{\xi\xi}(N_j^1 + \xi_j) = 0$ in the class of functions 1-periodic in ξ_1 and ξ_2, $\overline{N}_j^2$ is a bounded solution, 1-periodic in ξ_2, of the problem

$$L_{\xi\xi} \overline{N}_j^2 = 0, \quad \xi_1 > 0, \quad \overline{N}_j^2(0, \xi_2) = -N_j^1(0, \xi_2),$$

v_0 is a bounded solution, T-periodic in x_2, of the problem

$$\sum_{q,r=1}^{2} \widehat{K}^{qr} \frac{\partial^2 v_0}{\partial x_q \partial x_r} = f(x_1, x_2), \quad x_1 > 0,$$

$$v_0(0, x_2) = g(x_2), \quad \widehat{K}^{qr} = \left\langle K \frac{\partial}{\partial \xi_q} (N_r^1 + \xi_r) \right\rangle.$$

In this case, the estimate

$$\left| \left(\frac{\partial u}{\partial x_i} \right)^{(0)} - \frac{\partial u}{\partial x_i} \right| \leqslant c\varepsilon$$

holds.

Thus, in order to determine the gradient of the solution with accuracy $O(\xi)$, it is necessary to solve one-scale problems of the three types indicated above. And the summand $\sum_{j=1}^{2} \frac{\partial}{\partial \xi_i} \overline{N}_j^2 \left(\frac{x_1}{\varepsilon}, \frac{x_2}{\varepsilon}\right) \frac{\partial \nu_0}{\partial x_j}$ is a boundary-layer addition of order 1 near the boundary, exponentially small at the distance ε^γ away from the boundary $(0 < \gamma < 1)$.

REMARK 3: The entire reasoning can be extended unchanged to a problem in the strip $0 < x_1 < \beta$ with two boundary conditions: for $x_1 = 0$ and $x_1 = \beta$.

REMARK 4: The foregoing technique can be generalized to nonstationary and multi-dimensional problems.

REMARK 5: The Neumann conditions or the third-type boundary conditions can be considered instead of the Dirichlet condition (2) (see [112]).

REMARK 6: The constructions made above using a model example can be extended to objects corresponding to actual structures. Consider heat propagation in a toroidal layer (the body produced by rotating a ring K about the axis Oy as shown in Fig. 31) made of a composite material. Figure 32 shows the cross-section of the layer by the plane Oxy.

We pass to the curvilinear system of coordinates shown in Fig. 33. Then the process of heat propagation in new variables is described by the equation

$$S\left(\frac{x}{\varepsilon}, x\right) \frac{\partial u}{\partial t} - \frac{\partial}{\partial x_i} \left(A_{ij} \left(\frac{x}{\varepsilon}, x \right) \frac{\partial u}{\partial x_j} \right) = f(x),$$

defined in the product of the layer $\{x \in \mathbf{R}^3 | x_1 \in (r_1, r_2), (x_2, x_3) \in \mathbf{R}^2\}$ by the half-line $\{t > 0\}$. The coefficients $S(x/\varepsilon, x)$, $A_{ij}(x/\varepsilon, x)$, and the right side $f(x)$ are periodic functions in x_2 (with period $2\pi r_2$) and in x_2 (with period $2\pi r$). The boundary conditions for $x_1 = r_1$ and $x_2 = r_2$ are also periodic in x_2 and x_3.

Here the coefficients S, A_{ij} depend on both fast and slow variables. This problem can be investigated (see [112]) by analogy with the above example; the slow variable x appears in the "problem on a cell" (10), and the "problem on a strip" (11) as a parameter.

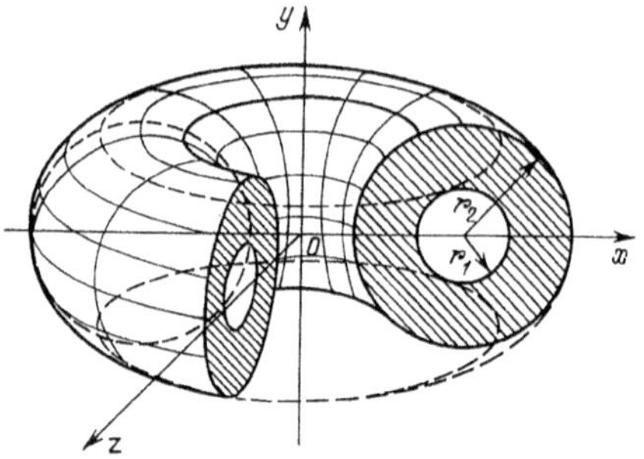

Figure 31

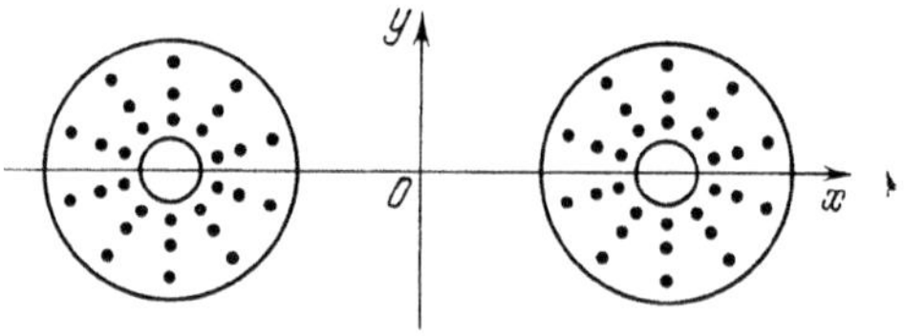

Figure 32

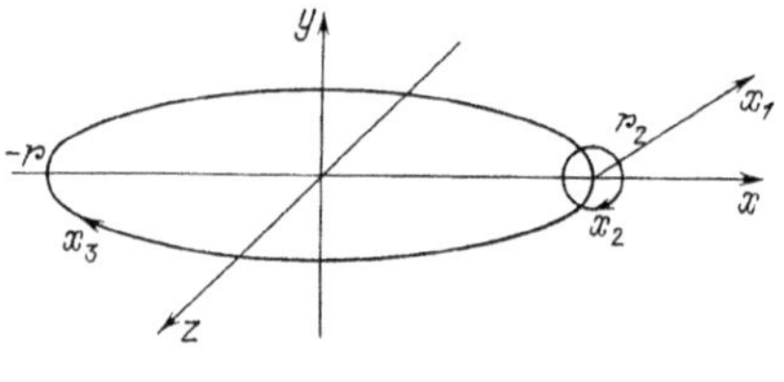

Figure 33

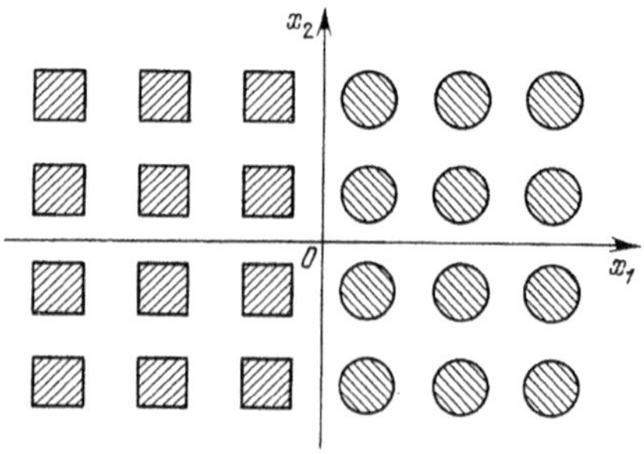

Figure 34

§3 The Boundary Layer Near the Interface of Two Periodic Structures

Consider equation (2.1) for $(x_1, x_2) \in \mathbf{R}^2$, under the assumption that the function $K(\xi_1, \xi_2)$ has the following form:

$$K(\xi_1, \xi_2) = \begin{cases} K^1(\xi_1, \xi_2) & \text{for } \xi_1 > 0, \\ K^2(\xi_1, \xi_2) & \text{for } \xi_1 < 0, \end{cases}$$

where the $K^i(\xi_1, \xi_2)$ are 1-periodic functions piecewise smooth in the above-indicated sense (see §2), the $K^i(\xi_1, \xi_2)$ satisfying the inequalities $0 < \kappa_1 < K^i(\xi_1, \xi_2) < \kappa_2$. We emphasize that, in general, $K^1 \neq K^2$ (Fig. 34), i.e., the coefficient K has different periodic structures when $x_1 > 0$ and $x_1 < 0$.

Let $f(x_1, x_2)$ be periodic in x_2, $f \in C^\infty$, $|f| < c_1 e^{-c_2|x_1|}$, where $c_1 > 0$, $c_2 > 0$,

$$\int_{-\infty}^{+\infty} \left(\int_0^T f(x_1, x_2)\, dx_2 \right) dx_1 = 0.$$

It is assumed as usual that the conjugation conditions

$$[u] = [K\, \partial u / \partial n] = 0.$$

are given on the discontinuity line of the coefficient.

The solution of equation (2.1) is sought in the class of bounded functions T-periodic in x_2 and satisfying the condition $\lim\limits_{x_1 \to -\infty} u(x_1, x_2) = 0$. The existence and uniqueness of this solution are proved in [81, 82, 110].

We seek the f.a.s. of (2.1) in the form

$$u^{(\infty)} \sim \sum_{l=0}^{\infty} \varepsilon^l \sum_{|i|=l} \Big(N_i^1(\xi_1, \xi_2) D^i v + N_i^2(\xi_1, \xi_2) (D^i v)\Big|_{x_1=+0} + \\ + N_i^3(\xi_1, \xi_2) (D^i v)\big|_{x_1=-0} \Big), \quad \xi_r = x_r/\varepsilon, \quad r = 1, 2, \quad (1)$$

where $v(x_1, x_2)$ is a solution of the averaged problem, which can be represented as (2.4). The function $N_i^1(\xi_1, \xi_2)$ is a 1-periodic in the variables ξ_1 and ξ_2 on each of the half-planes $R_+^2 = \{\xi_1 > 0\}$ and $R_-^2 = \{\xi_1 < 0\}$, $N_i^2(\xi_1, \xi_2)$, $N_i^3(\xi_1, \xi_2)$ are functions 1-periodic in ξ_2 which stabilize exponentially quickly to zero as $|\xi_1| \to \infty$, that is, $|N_i^r| \leqslant c_3 e^{-c_4|\xi_1|}$, $c_3 > 0$ and $c_4 > 0$. In addition, it is assumed that $N_\emptyset = 1$, that is, $N_\emptyset^1 = 1$, $N_\emptyset^2 = N_\emptyset^3 = 0$. If we set $N_i^r = 0$ for $r = 2, 3$, then series (1) coincides with the expansion in §4.2 of the solution of (2.1) considered on each of the half-planes R_+^2 and R_-^2 independently and without boundary conditions for $x_1 = 0$. The boundary-layer functions $N_i^2(\xi_1, \xi_2)\ (D^i v)|_{x_1=+0}$, $N_i^3(\xi_1, \xi_2)(D^i v)|_{x_1=-0}$ are introduced to take into account the contact conditions on the boundary line between two periodic media:

$$[u] = [K\,\partial u/\partial x_1] = 0 \quad \text{for} \quad x_1 = 0. \tag{2}$$

We substitute (1) into (2.1) independently on each of the half-planes R^2_+ and R^2_-. After simplifying, we get

$$\sum_{l=0}^{\infty} \varepsilon^{l-2} \sum_{|i|=l} \left(H^1_{i+}(\xi_1, \xi_2)\, D^i v + H^2_{i+}(\xi_1, \xi_2)\left(D^i v\right)\big|_{x_1=+0} + \right.$$
$$\left. + H^3_{i+}(\xi_1, \xi_2)\left(D^i v\right)\big|_{x_1=-0}\right) \sim f(x_1, x_2) \quad \text{for } x_1 > 0, \tag{3}$$

$$\sum_{l=0}^{\infty} \varepsilon^{l-2} \sum_{|i|=l} \left(H^1_{i-}(\xi_1, \xi_2)\, D^i v + H^2_{i-}(\xi_1, \xi_2)\left(D^i v\right)\Big|_{x_1=+0} + \right.$$
$$\left. + H^3_{i-}(\xi_1, \xi_2)\left(D^i v\right)\big|_{x_1=-0}\right) \sim f(x_1, x_2) \quad \text{for } x_1 < 0; \tag{4}$$

here

$$H^1_{i\pm}(\xi_1, \xi_2) = \sum_{j=1}^{2} \frac{\partial}{\partial \xi_j}\left(K \frac{\partial N^1_i}{\partial \xi_j}\right) + \frac{\partial}{\partial \xi_{i_1}}\left(K N^1_{i_2 \ldots i_l}\right) +$$
$$+ K \frac{\partial}{\partial \xi_{i_1}} N^1_{i_2 \ldots i_l} + \delta_{i_1 i_2} K N^1_{i_3 \ldots i_l},$$
$$H^r_{i\pm}(\xi_1, \xi_2) = \sum_{j=1}^{2} \frac{\partial}{\partial \xi_j}\left(K \frac{\partial N^r_i}{\partial \xi_j}\right) + \delta_{i_1 2} \frac{\partial}{\partial \xi_2}\left(K N^r_{i_2 \ldots i_l}\right) +$$
$$+ \delta_{i_1 2} K \frac{\partial N^r_{i_2 \ldots i_l}}{\partial \xi_2} + \delta_{i_1 i_2} \delta_{i_1 2} K N^r_{i_3 \ldots i_l}, \qquad r = 2, 3. \tag{5}$$

Similarly, after substitution into (2), we have

$$\sum_{l=0}^{\infty} \varepsilon^{l} \sum_{|i|=l} \left(G^2_i(\xi_2)\left(D^i v\right)\Big|_{x_1=+0} + G^3_i(\xi_2)\left(D^i v\right)\Big|_{x_1=-0}\right) \sim 0, \tag{6}$$

$$\sum_{l=0}^{\infty} \varepsilon^{l-1} \sum_{|i|=l} \left(J^2_i(\xi_2)\left(D^i v\right)\Big|_{x_1=+0} + J^3_i(\xi_2)\left(D^i v\right)\Big|_{x_1=-0}\right) \sim 0; \tag{7}$$

where

$$G^2_i(\xi_2) = N^1_i\big|_{\xi_1=+0} + [N^2_i], \quad G^3_i(\xi_2) = -N^1_i\big|_{\xi_1=-0} + [N^3_i], \tag{8}$$

$$J^2_i(\xi_2) = K^1\left(\frac{\partial N^1_i}{\partial \xi_1} + \delta_{i_1 1} N^1_{i_2 \ldots i_l}\right)\Bigg|_{\xi_1=+0} + \left[K \frac{\partial N^2_i}{\partial \xi_1}\right],$$
$$J^3_i(\xi_2) = -K^2\left(\frac{\partial N^1_i}{\partial \xi_1} + \delta_{i_1 1} N^1_{i_2 \ldots i_l}\right)\Bigg|_{\xi_1=-0} + \left[K \frac{\partial N^3_i}{\partial \xi_1}\right]. \tag{9}$$

Since the N^j_i are yet indefinite functions, they can be submitted to certain conditions. Namely, by analogy with §2, we require the following equalities to be satisfied:

1) $H^1_{i+}(\xi_1, \xi_2) = \text{const}, \quad H^1_{i-}(\xi_1, \xi_2) = \text{const};$

2) $H^r_{i\pm}(\xi_1, \xi_2) = 0, \quad r = 2, 3;$

3) $G^r_i(\xi_2) = \text{const}, \quad r = 2, 3;$

4) $J^r_i(\xi_2) = \text{const}, \quad r = 2, 3.$

We denote the constants on the right sides chosen from requirement 1) by $h_{i\pm}$, from 3) by g^r_i, and from 4) by j^r_i. If we can manage to find N^r_i, $h_{i\pm}$, g^r_i, j^r_i satisfying the requirements 1)–4), problem (3), (4), (6), (7) reduces to an equation for the function v with the coefficients constant on each of the half-planes R^2_+ and R^2_-:

$$\sum_{l=0}^{\infty} \varepsilon^{l-2} \sum_{|i|=l} h_{i+} D^i v \sim f(x_1, x_2) \quad \text{for} \quad x_1 > 0, \tag{10}$$

$$\sum_{l=0}^{\infty} \varepsilon^{l-2} \sum_{|i|=l} h_{i-} D^i v \sim f(x_1, x_2) \quad \text{for} \quad x_1 < 0 \tag{11}$$

and with the contact conditions for $x_1 = 0$:

$$\sum_{l=0}^{\infty} \varepsilon^{l} \sum_{|i|=l} \left(g^2_i (D^i v) \Big|_{x_1=+0} + g^3_i (D^i v) \Big|_{x_1=-0} \right) \sim 0, \tag{12}$$

$$\sum_{l=0}^{\infty} \varepsilon^{l} \sum_{|i|=l} \left(j^2_i (D^i v) \Big|_{x_1=+0} + j^3_i (D^i v) \Big|_{x_1=-0} \right) \sim 0. \tag{13}$$

The functions N^1_i as well as h_{i+} and h_{i-} are constructed independently on each of the half-planes by the algorithm from the preceding section. After the N^1_i, h_{i+}, h_{i-} are found, the functions N^2_i and N^3_i are defined by the following recurrent procedure. Suppose all the N^2_i and N^3_i are found for $|i| \leqslant k$. We define $T^r_i(\xi_1, \xi_2)$, $V^r_i(\xi_2)$, and $W^r_i(\xi_2)$ for all i such that $|i| = k + 1$ from (see (5), (8), (9))

$$\begin{gathered} T^r_i(\xi_1, \xi_2) = \delta_{i_1 2} \left(\frac{\partial}{\partial \xi_2} (K N^r_{i_2 \dots i_l}) + \right. \\ \left. + K \frac{\partial}{\partial \xi_2} N^r_{i_2 \dots i_l} + K N^r_{i_3 \dots i_l} \delta_{i_1 i_2} \right), \quad r = 2, 3, \\ V^2_i(\xi_2) = N^1_i \big|_{\xi_1=+0}, \quad V^3_i(\xi_2) = -N^1_i \big|_{\xi_1=-0}, \\ W^2_i(\xi_2) = K^1 \left(\frac{\partial N^1_i}{\partial \xi_1} + \delta_{i_1 1} N^1_{i_2 \dots i_l} \right) \Bigg|_{\xi_1=+0}, \\ W^3_i(\xi_2) = -K^2 \left(\frac{\partial N^1_i}{\partial \xi_1} + \delta_{i_1 1} N^1_{i_2 \dots i_l} \right) \Bigg|_{\xi_1=-0}. \end{gathered} \tag{14}$$

Let us select the constants g^r_i and j^r_i so that for any i of length $|i| = k+1$ and $r = 2, 3$, there exists a unique solution of class $W^1_2(R^2_+) \cap W^1_2(R^2_-)$, 1-periodic in ξ_2 and exponentially tending to zero as $|\xi_1| \to \infty$, of the problem

$$\begin{aligned} L_+N_i^r + T_i^r(\xi_1, \xi_2) &= 0 \quad \text{for} \quad \xi_1 > 0, \\ L_-N_i^r + T_i^r(\xi_1, \xi_2) &= 0 \quad \text{for} \quad \xi_1 < 0, \\ [N_i^r] + V_i^r(\xi_2) &= g_i^r \quad \text{for} \quad \xi_1 = 0, \\ \left[K\frac{\partial N_i^r}{\partial \xi_1}\right] + W_i^r(\xi_2) &= j_i^r \quad \text{for} \quad \xi_1 = 0, \\ |N_i^r| \leqslant c_5 e^{-c_6|\xi_1|}, \quad c_5 > 0, \quad c_6 > 0; \end{aligned} \tag{15}$$

here

$$L_+ = \sum_{q=1}^{2} \frac{\partial}{\partial \xi_q}\left(K^1(\xi_1, \xi_2)\frac{\partial}{\partial \xi_q}\right),$$

$$L_- = \sum_{q=1}^{2} \frac{\partial}{\partial \xi_q}\left(K^2(\xi_1, \xi_2)\frac{\partial}{\partial \xi_q}\right).$$

It is shown in [82] that this choice of q_i^r and j_i^r is possible and unique, and moreover,

$$j_i^r = \int_0^1 \left(\int_{-\infty}^{0} T_i^r d\xi_1\right) d\xi_2 + \int_0^1 \left(\int_0^{+\infty} T_i^r d\xi_1\right) d\xi_2 + \int_0^1 W_i^r d\xi_2; \tag{16}$$

whereas to determine g_i^r, we must first solve the problem

$$\begin{aligned} L_+\widetilde{N}_i^r + T_i^r(\xi_1, \xi_2) &= 0 \quad \text{for} \quad \xi_1 > 0, \\ L_-\widetilde{N}_i^r + T_i^r(\xi_1, \xi_2) &= 0 \quad \text{for} \quad \xi_1 < 0, \\ [\widetilde{N}_i^r] + V_i^r(\xi_2) &= 0 \quad \text{for} \quad \xi_1 = 0, \\ \left[K\frac{\partial \widetilde{N}_i^r}{\partial \xi_1}\right] + W_i^r(\xi_2) &= j_i^r \quad \text{for} \quad \xi_1 = 0 \end{aligned}$$

in the class of bounded functions 1-periodic in ξ_2. The solution of the problem exists and is unique, provided condition (16) is satisfied [110], and moreover, there is a constant C_i^r such that $|\widetilde{N}_i^r - C_i^r| < c_7 e^{-c_8\xi_1}$, $c_7, c_8 > 0$. The latter assertion follows from a theorem of the Phragmén–Lindelöf type (see [82]). The constant g_i^r is defined by the equality $g_i^r - \lim_{\xi_1 \to +\infty} \widetilde{N}_i^r$. The recurrency step is completed. The basis is $N^r = 0$ $(r = 2, 3)$.

From (14) we can get

$$g^r = (-1)^r, \quad j^r = 0, \quad h_{i\pm} = 0 \quad \text{for} \quad |i| = 0,\ 1,$$
$$j^2_{i_1} = \widehat{K}^{1i_1}_{+}, \quad j^3_{i_1} = -\widehat{K}^{1i_1}_{-}, \quad h_{i_1 i_2 \pm} = \widehat{K}^{i_1 i_2}_{\pm},$$
$$\widehat{K}^{i_1 i_2}_{+} = \left\langle K^1 \frac{\partial}{\partial \xi_{i_1}} \left(N^1_{i_{2+}} + \xi_{i_2}\right)\right\rangle,$$
$$\widehat{K}^{i_1 i_2}_{-} = \left\langle K^2 \frac{\partial}{\partial \xi_{i_1}} \left(N^1_{i_2 -} + \xi_{i_2}\right)\right\rangle,$$

the $N^1_{i_2\pm}(\xi_1, \xi_2)$ are solutions, 1-periodic in ξ_1 and ξ_2, of the equations

$$L_\pm \left(N^1_{i_2 \pm} + \xi_{i_2}\right) = 0, \qquad i_2 = 1, 2.$$

Thus, by separating the principal part of problem (10)–(13), we get

$$\sum_{q,p=1}^{2} \widehat{K}^{qp}_{+} \frac{\partial^2 v}{\partial x_q \partial x_p} + O(\varepsilon) = f(x_1, x_2), \quad x_1 > 0,$$
$$\sum_{q,p=1}^{2} \widehat{K}^{qp}_{-} \frac{\partial^2 v}{\partial x_q \partial x_p} + O(\varepsilon) = f(x_1, x_2), \quad x_1 < 0,$$
$$[v] + O(\varepsilon) = 0, \quad x_1 = 0,$$
$$\sum_{p=1}^{2} \left(\widehat{K}^{1p}_{+} \frac{\partial v}{\partial x_p}\bigg|_{x_1=+0} - \widehat{K}^{1p}_{-} \frac{\partial v}{\partial x_p}\bigg|_{x_1=-0}\right) + O(\varepsilon) = 0, \quad x_1 = 0.$$

This implies that by analogy with §2, we can seek v in the form of series (2.4) in powers of ε, and v_l can be defined from the recurrent chain of problems of the form

$$\sum_{q,p=1}^{2} \widehat{K}^{qp}_{\pm} \frac{\partial^2 v_l}{\partial x_q \partial x_p} = f_l(x_1, x_2), \quad x \in R^2_\pm,$$
$$[v_l] = g_l(x_2), \quad x_1 = 0,$$
$$\sum_{q=1}^{2} \left(\widehat{K}^{1q}_{+} \frac{\partial v_l}{\partial x_q}\bigg|_{x_1=+0} - \widehat{K}^{1q}_{-} \frac{\partial v_l}{\partial x_q}\bigg|_{x_1=-0}\right) = j_l(x_2), \quad x_1 = 0.$$

In the same way as in §2, we can calculate the constant to which v_l tends as x_1 tends to $+\infty$ and obtain the estimates of the closeness to the exact solution for the partial sum $u^{(J)}$ of series (1), as well as formulate an analogue of Theorem 2.1.

The zeroth approximation of the gradient of the solution has the form

$$\left(\frac{\partial u}{\partial x_i}\right)^{(0)} = \frac{\partial v_0}{\partial x_i} + \sum_{p=1}^{2} \frac{\partial}{\partial \xi_i} N_p^1 \frac{\partial v_0}{\partial x_p} + \\ + \sum_{p=1}^{2} \frac{\partial}{\partial \xi_i} \overline{N}_p^2 \frac{\partial v_0}{\partial x_p}\bigg|_{x_1=+0} + \sum_{p=1}^{2} \frac{\partial}{\partial \xi_i} \overline{N}_p^3 \frac{\partial v_0}{\partial x_p}\bigg|_{x_1=-0};$$

here

$$N_p^1(\xi_1, \xi_2) = \begin{cases} N_{p+}^1(\xi_1, \xi_2) & \text{for} \quad \xi_1 > 0, \\ N_{p-}^1(\xi_1, \xi_2) & \text{for} \quad \xi_1 < 0, \end{cases}$$

the $N_{p\pm}^1(\xi_1, \xi_2)$ are solutions, 1-periodic in ξ_1 and ξ_2, of the equations $L_\pm(N_{p\pm}^1 + \xi_p) = 0$, $p = 1, 2$, $\overline{N}_p^r(\xi_1, \xi_2)$ are solutions, 1-periodic in ξ_2, of the equations $L_+ \overline{N}_p^r = 0$, $\xi_1 > 0$, $L_- \overline{N}_p^r = 0$, $\xi_1 < 0$, with the contact conditions for $\xi_1 = 0$:

$$[\overline{N}_p^r] = -(-1)^r N_p^1 |_{\xi_1=(-1)^r \cdot 0},$$

$$\left[K \frac{\partial \overline{N}_p^r}{\partial \xi_1}\right] = -(-1)^r K \left(\frac{\partial N_p^1}{\partial \xi_1} + \delta_{p1}\right)\bigg|_{\xi_1=(-1)^r \cdot 0},$$

where v_0 is a bounded solution, T-periodic in x_2, of the problem

$$\sum_{q,p=1}^{2} \widehat{K}_\pm^{qp} \frac{\partial^2 v_0}{\partial x_q \partial x_p} = f(x_1, x_2), \quad x \in R_\pm^2, \tag{17}$$

$$[v_0] = \sum_{p=1}^{2} \left(\widehat{K}_+^{1p} \frac{\partial v_0}{\partial x_p}\bigg|_{x_1=+0} - \widehat{K}_-^{1p} \frac{\partial v_0}{\partial x_p}\bigg|_{x_1=-0} \right) = 0, \quad x_1 = 0. \tag{18}$$

Here the symbol $(-1)^r \cdot 0$ means $+0$ for even r and -0 for odd r. Note that equation (17) for v_0 with relations (18) for $x_1 = 0$ can be written as an elliptic equation with discontinuous coefficients in the divergent form

$$\sum_{q,p=1}^{2} \frac{\partial}{\partial x_q} \left(\widehat{K}^{qp}(x_1) \frac{\partial v_0}{\partial x_p} \right) = f(x_1, x_2),$$

where

$$\widehat{K}^{qp}(x_1) = \begin{cases} \widehat{K}_+^{qp} & \text{for} \quad x_1 > 0, \\ \widehat{K}_-^{qp} & \text{for} \quad x_1 < 0. \end{cases}$$

§4 Problem on the Contact of Two Media Divided by a Thin Interlayer

The above-described method is applicable to problems on the contact of media divided by a thin inhomogeneous interlayer with large or small physical characteristics. Due to the awkwardness of the calculations, we shall formulate only the main results for the Poisson equation. An investigation of more general problems (including nonstationary and nonlinear ones) is made in [111–113].

Suppose there is an equation

$$\sum_{i=1}^{s} \frac{\partial}{\partial x_i}\left(\overline{K}_\varepsilon\left(\frac{x}{\varepsilon}\right)\frac{\partial u}{\partial x_i}\right) = f(x), \tag{1}$$

$$x = (x_1, \dots, x_s), \quad x \in \Pi^s_{(d_1, d_2)} = \{x \in \mathbf{R}^s \mid x_1 \in (d_1, d_2)\},$$

where the d_i are multiples of $\varepsilon (d_1 < 0 < d_2)$,

$$\overline{K}_\varepsilon(\xi) = \begin{cases} K_1(\xi) & \text{for } \xi_1 < 0, \\ K_2(\xi) & \text{for } \xi_1 > 1, \\ \varepsilon^\gamma K_3(\xi) & \text{for } \xi_1 \in (0, 1), \end{cases}$$

where $\xi = (\xi_1, \dots, \xi_s)$, γ is an integer, and the $K_i(\xi)$ are 1-periodic functions piecewise smooth in the sense of §4.1. We also assume that $\sum$, the totality of the coefficient discontinuity surfaces, has no points in common with the hyperplanes $\{\xi_1 = 0\}$ and $\{\xi_1 = 1\}$. Let $0 < \kappa_1 < K_i(\xi)$, let $f(x)$ be an infinitely differentiable function of x for $x_1 \neq 0$, and suppose that when $x_1 = 0$, the function $f(x)$ and its derivatives of all orders may have the first-order discontinuity.

Let us formulate the problem of finding the generalized solution of equation (1), which would satisfy the boundary conditions

$$u\,|_{x_1 = d_i} = g_i(x_2, \dots, x_s), \qquad i = 1, 2, \tag{2}$$

T-periodic in $x_2, \dots, x_s$. It is also assumed that the functions f and g have the same periodicity, while $g_i \in C^\infty$. For $\gamma > 1$, we will assume f to be equal to zero for $x_1 \in (0, \varepsilon)$.

In [112], for any integer $J > 0$, the asymptotics $u^{(J)}$ is constructed by analogy with §§2 and 3 to satisfy (1), (2) with accuracy $0(\varepsilon^J)$. From the a priori estimates for equation (1), the closeness of $u^{(J)}$ and the exact solution u is derived in the norms in W_2^1 and C:

$$\begin{aligned} \| u^{(J)} - u \|_{W_2^1((d_1, d_2)\times(0,T)^{s-1})} &= O(\varepsilon^{J+1}), \\ \| u^{(J)} - u \|_{C([d_1,d_2]\times[0,T]^{s-1})} &= O(\varepsilon^{J+2}). \end{aligned} \tag{3}$$

The averaged problem with piecewise smooth coefficients is constructed. Its principal part (as $\varepsilon \downarrow 0$) describes the contact of two homogeneous media without the dividing interlayer but with special contact conditions, and it approximates the original problem in the sense that

$$|u - v_0| \leqslant C\varepsilon, \quad x_1 \notin (0,\ \varepsilon) \text{ for } \gamma \geqslant 1. \tag{4}$$

Estimate (4) follows from (3).

We shall now give the form of the zeroth-order averaged problem. The averaged equation is

$$\begin{aligned} &\sum_{i,j=1}^{s} \widehat{K}_1^{ij} \frac{\partial^2 v_0}{\partial x_i \partial x_j} = f \quad \text{for} \quad x \in \Pi^s_{(d_1,0)}, \\ &\sum_{i,j=1}^{s} \widehat{K}_2^{ij} \frac{\partial^2 v_0}{\partial x_i \partial x_j} = f \quad \text{for} \quad x \in \Pi^s_{(0,d_2)}, \end{aligned} \tag{5}$$

where the $\widehat{K}_q^{ij}$ are the same as in §3.

The averaged boundary condition is of the form

$$v_0 \big|_{x_1 = d_i} = g_i(x_2, \ldots, x_s), \qquad i = 1,\ 2, \tag{6}$$

where v_0 is sought in the class of functions T-periodic in $x_2, \ldots, x_s$.

We shall consider in detail the averaged contact conditions given for $x_1 = 0$. These conditions depend substantially on the parameter γ, which characterizes the conductivity of the interlayer. Namely, the conductivity of the interlayer is low when $\gamma > 0$ and high when $\gamma < 0$. The contact relations for various values of γ are given below. For brevity, we set

$$\frac{\partial v_0}{\partial \nu^q} = \sum_{j=1}^{s} \widehat{K}_q^{1j} \frac{\partial v_0}{\partial x_j}\bigg|_{x_1 = (-1)^q \cdot 0}, \qquad q = 1,\ 2.$$

We have:

1) $\gamma \geqslant 2$ ("very poorly conducting" interlayer):

$$\frac{\partial v_0}{\partial \nu^1} = \frac{\partial v_0}{\partial \nu^2} = 0, \tag{7}$$

the averaged problem splits into two independent problems in $\Pi^s_{(d_1,0)}$ and $\Pi^s_{(0,d_2)}$.

2) $\gamma = 1$ ("poorly conducting" interlayer):

$$\frac{\partial v_0}{\partial \nu^1} = \frac{\partial v_0}{\partial \nu^2} = \lambda\, [v_0], \tag{8}$$

where $\lambda = \langle K_3(\xi) \partial N^3 / \partial \xi_1 \rangle$, N^3 is a solution of the equation

$$\sum_{i=1}^{s} \frac{\partial}{\partial \xi_i} \left(K_3(\xi) \frac{\partial}{\partial \xi_i} N^3 \right) = 0,$$

$\xi \in \mathbf{R}^s$, $\xi_1 \in (0,1)$ in the class of functions from $W_2^1(0,1)^s)$, 1-periodic in $\xi_2, \dots, \xi_s$, which satisfy the conditions

$$N^3 \big|_{\xi_1=0} = 0, \quad N^3 \big|_{\xi_1=1} = 1.$$

3) $\gamma = 0$ ("conducting" interlayer): in this case the contact conditions are of the form

$$\frac{\partial v_0}{\partial \nu^1} = \frac{\partial v_0}{\partial \nu^2}, \quad [v_0] = 0. \tag{9}$$

4) $\gamma = -1$ ("well-conducting" interlayer):

$$\sum_{i,j=2}^{s} \eta^{ij} \frac{\partial^2 v_0}{\partial x_i \partial x_j} + \frac{\partial v_0}{\partial \nu^2} - \frac{\partial v_0}{\partial \nu^1} = 0, \quad [v_0] = 0, \tag{10}$$

where $\eta^{ij} = \left\langle K_3(\xi)\left(\frac{\partial}{\partial \xi_i}(N_j^4 + \xi_j)\right)\right\rangle$, N_j^4 $(j = 2, \dots, s)$ is a solution, 1-periodic in $\xi_2, \dots, \xi_s$ of the problem

$$\sum_{i=1}^{s} \frac{\partial}{\partial \xi_i}\left(K_3(\xi)\frac{\partial}{\partial \xi_i}(N_j^4 + \xi_j)\right) = 0, \qquad j = 1, \dots, s,$$

$$\xi_1 \in (0, 1), \quad \xi_i \in \mathbf{R}, \qquad i = 2, \dots, s,$$

$$\left(K_3(\xi)\frac{\partial}{\partial \xi_1} N_j^4\right)\Big|_{\xi_1=0} = \left(K_3(\xi)\frac{\partial}{\partial \xi_1} N_j^4\right)\Big|_{\xi_1=1} = 0.$$

5) $\gamma \leqslant -2$ ("highly conducting" interlayer):

$$\begin{gathered} v_0 \big|_{x_1=-0} = v_0 \big|_{x_1=+0} = \text{const}, \\ \int_0^T \cdots \int_0^T \left(\frac{\partial v_0}{\partial \nu^2} - \frac{\partial v_0}{\partial \nu^1}\right) dx_2 \dots dx_s = 0. \end{gathered} \tag{11}$$

We shall now discuss the question of correctness of the averaged problem. The correctness of equation (5) with conditions (6) and contact relations (7) or (9) can be investigated by using the standard technique. Averaged problem (5), (6), (8) can be investigated by the methods given in [76]. Problem (5), (6), (10) was investigated in [145]. Finally, problem (5), (6), (11) was considered in [112].

These results can be generalized to the case of the anisotropic Poisson equation

$$\sum_{i,j=1}^{s} \frac{\partial}{\partial x_i}\left(\overline{A}_{ij}\frac{\partial u}{\partial x_j}\right) = f$$

and to the nonstationary case (parabolic and hyperbolic). In the quasi-linear case, it has been proved that the procedure of constructing an asymptotic solution could be infinitely continued [112].

§5 The Boundary Layer for the Nonstationary System of Equations of Elasticity Theory

This section deals with the system of equations

$$L_\varepsilon u \equiv R\left(\frac{x}{\varepsilon}\right)\frac{\partial^2 u}{\partial t^2} - \frac{\partial}{\partial x_m}\left(A_{mk}\left(\frac{x}{\varepsilon}\right)\frac{\partial u}{\partial x_k}\right) = f(x, t), \tag{1}$$

given for $t > 0$, $x \in \Pi^s_{(d_1,d_2)}$. Here $\Pi^s_{(d_1,d_2)}$ is the s-dimensional layer $\{x \in \mathbf{R}^s | x_1 \in (d_1, d_2)\}$, and $R(\xi)$, $A_{mk}(\xi)$ are $s \times s$ matrices whose elements ρ^{pj}, a^{pj}_{mk} are infinitely differentiable 1-periodic functions satisfying the conditions (4.4.1) and 1) from §4.5. The f.a.s. is constructed for (1) with the Dirichlet boundary conditions for $x_1 = d_1$, d_2 and homogeneous initial conditions. The procedure is similar to the formalism described in §2 and fits into the framework of a more general procedure (see [112]).

Suppose that

$$f(x, t) = f\left(x_1, x_1', t\right) \in C^\infty\left(\Pi^s_{(d_1,d_2)} \times [0, +\infty)\right),$$

$$g_i\left(x_1', t\right) \in C^\infty\left(\mathrm{R}^{s-1} \times [0, +\infty)\right), \qquad i = 1, 2, \quad x_1' = (x_2, \ldots, x_s),$$

where f, g_1, g_2 are T-periodic in x_1 $(1 = 2, \ldots, s)$,

$$\frac{\partial^l f}{\partial t^l}(x, 0) \equiv 0, \quad \frac{\partial^l g_i}{\partial t^l}\left(x_1', 0\right) \equiv 0, \qquad i = 1, 2, \quad l = 0, 1, \ldots \tag{2}$$

It is assumed in what follows that $T = k_0\varepsilon$, $d_1 = k_1\varepsilon$, $d_2 = k$, where k_0, k_1, k_2 are positive integers.

THEOREM 1. *The problem*

$$\begin{gathered} L_\varepsilon u = f(x, t), \quad x \in \Pi^s_{(d_1,d_2)}, \quad t > 0, \\ u\left(d_i, x_1', t\right) = g_i\left(x_1', t\right), \qquad i = 1, 2, \quad t > 0, \\ u(x, 0) = \frac{\partial u}{\partial t}(x, 0) = 0, \quad x \in \Pi^s_{(d_1,d_2)}, \end{gathered} \tag{3}$$

has a unique solution $u(x, t)$ T-periodic in $x_2, \ldots, x_s$. For the solution u of (3) with the homogeneous boundary conditions $g_i = 0$, the estimate

$$\| u \|_{W_2^1(Q_{t_0})} \leqslant C(t_0) \| f \|_{L_2(Q_{t_0})}$$

holds with the constant $C(t_0)$ independent of ε. Here

$$Q_{t_0} = \{(x,t) \mid x_i \in (0,T),\quad i=2,\ldots,s,\quad x_1 \in (d_1,d_2),\quad t\in(0,t_0)\}.$$

The proof of the theorem is similar to that of Theorems 3.1, 3.2, 4.1 of Chapter 4 in [76]. In this case, the first Corn's inequality is used instead of the positive definiteness of the form $A_{ij}\xi_i\xi_j$ (see [155]).

Consider now constructing the asymptotic expansion of the solution of problem (3). We shall seek the asymptotics as the formal series

$$u^{(\infty)} \sim \sum_{r=0}^{\infty} \varepsilon^r \sum_{q=0}^{r} \sum_{|i|=r-q} N_{qi}(\xi)\, D^{qi} v(x,t), \tag{4}$$

where $\xi = x/\varepsilon$, the $N_{qi}(\xi)$ are $s\times s$ matrices 1-periodic in $\xi_2,\ldots\xi_s$, and $v(x,t) = (v^1(x,t),\ldots,v^s(x,t))^T$ is a vector-function T-periodic in $x_2,\ldots,x_s$.

By substituting (4) into (1), we get

$$f \sim \sum_{r=0}^{\infty} \varepsilon^{r-2} \sum_{q=0}^{r} \sum_{|i|=r-q} H_{qi}(\xi)\, D^{qi} v,$$

where $\xi = x/\varepsilon$, $H_{qi}(\xi) = -L_{\xi\xi}N_{qi} + T_{qi}$, $T_{qi}(\xi)$ are defined from (4.5.15) if we formally set $S \equiv 0$.

The substitution of (4) into the boundary and initial conditions yields the asymptotic equalities

$$\begin{gathered}\sum_{r=0}^{\infty} \varepsilon^r \sum_{q=0}^{r} \sum_{|i|=r-q} N_{qi}\left(\frac{d_l}{\varepsilon}, \xi_1'\right) D^{qi} v\Big|_{x_1=d_l} \sim g_l(x_1', t),\\ \sum_{r=0}^{\infty} \varepsilon^r \sum_{q=0}^{r} \sum_{|i|=r-q} N_{qi}(\xi)\, D^{qi} v\,\Big|_{t=0} \sim 0,\end{gathered} \tag{5}$$

$$\sum_{r=0}^{\infty} \varepsilon^r \sum_{q=0}^{r} \sum_{|i|=r-q} N_{qi}(\xi)\, D^{q+1i} v\,\Big|_{t=0} \sim 0, \quad \xi_1' = (\xi_2,\ldots,\xi_s). \tag{6}$$

We seek $N_{qi}(\xi)$ as the sum

$$N_{qi}(\xi) = N_{qi}^0(\xi) + N_{qi}^1(\xi_1 - k_1, \xi_1') + N_{qi}^2(\xi_1 - k_2, \xi_1'),$$

where $k_l = d_l/\varepsilon$, $N_{qi}^0(\xi)$ is 1-periodic in $\xi_1,\ldots,\xi_s$, and $N_{qi}^1(\xi)$, $N_{qi}^2(\xi)$ correspond to the boundary layers; they are 1-periodic in $\xi_2,\ldots,\xi_s$ and exponentially decay in ξ_1 as $|\xi_1| \to \infty$. We set $N_0^0 = E$, $N_0^l = 0$ $(l=1,2)$.

Introduce the notation

$$T^l_{qi}(\xi) \equiv T^l_{qi_1\ldots i_{r-q}}(\xi) \equiv R(\xi)\, N^l_{q-2i_1\ldots i_{r-q}} - A_{i_1 i_2}(\xi)\, N^l_{qi_3\ldots i_{r-q}} -$$
$$- A_{i_1 k}(\xi) \frac{\partial N^l_{qi_2\ldots i_{r-q}}}{\partial \xi_k} - \frac{\partial}{\partial \xi_m}\left(A_{mi_1}(\xi)\, N^l_{qi_2\ldots i_{r-q}}\right),$$
$$l = 0,\ 1,\ 2, \quad h^0_{qi} = \langle T^0_{qi} \rangle.$$

We define the matrices $N^0_{qi}(\xi)$, as in §4.5, from the recurrent sequence of equations with $q + |i| = 1, 2, \ldots$,

$$- L_{\xi\xi} N^0_{qi} = - T^0_{qi} + h^0_{qi}, \quad \langle N^0_{qi} \rangle = 0.$$

The matrix-valued functions N^1_{qi}, N^2_{qi} are defined successively from the recurrent chain of problems

$$\begin{aligned} - L_{\xi\xi} N^1_{qi} &= - T^1_{qi}, \quad \xi \in R^s_+ = (0, +\infty) \times \mathbf{R}^{s-1}, \\ N^1_{qi}(0, \xi_1') &= - N^0_{qi}(0, \xi_1') + h^1_{qi}, \end{aligned} \tag{7}$$

$$\begin{aligned} - L_{\xi\xi} N^2_{qi} &= - T^2_{qi}, \quad \xi \in R^s_- = (-\infty, 0) \times \mathbf{R}^{s-1}, \\ N^2_{qi}(0, \xi_1') &= - N^0_{qi}(0, \xi_1') + h^2_{qi}. \end{aligned} \tag{8}$$

The constant matrices h^l_{qi} ($l = 1, 2$) are chosen so that problems (7) and (8) would have solutions $N^l_{qi} = \|n^{l,pj}_{qi}(\xi)\|$, 1-periodic in $\xi_2, \ldots, \xi_s$, which satisfy the inequalities

$$\left| n^{l,pj}_{qi}(\xi) \right| \leqslant c^{l1}_{qi} e^{-c^{l2}_{qi} |\xi_1|}, \quad \left| \frac{\partial}{\partial \xi_k} n^{l,pj}_{qi}(\xi) \right| \leqslant c^{l1}_{qi} e^{-c^{l2}_{qi} |\xi_1|},$$

where the c^{lm}_{qi} are positive constants. The possiblity of this choice of the constants h^l_{qi} and the existence and uniqueness of the solutions of problems (7), (8) follow from [109].

Now we can continue the transformation of the asymptotic series obtained by substituting $u^{(\infty)}$ in (1), namely,

$$f(x, t) \sim \sum_{r=0}^{\infty} \varepsilon^{r-2} \sum_{q=0}^{r} \sum_{|i|=r-q} h^0_{qi} D^{qi} v, \tag{9}$$

$$g_l(x_1', t) \sim \sum_{r=0}^{\infty} \varepsilon^{r} \sum_{q=0}^{r} \sum_{|i|=r-q} h^l_{qi} D^{qi} v \big|_{x_1 = d_l}, \quad l = 1, 2, \tag{10}$$

and it turns out that

$$h^0_0 = h^0_{0i_1} = h^0_{10} = h^0_{1i_1} = 0, \quad h^l_0 = E, \quad l = 1, 2, \quad h^0_2 = \widehat{R}, \quad h^0_{0i_1i_2} = - \widehat{A}_{i_1 i_2},$$

with the matrices $\widehat{R}$, $\hat{A}_{i_1 i_2}$ having been defined in §4.5.

We seek the f.a.s. of the averaged problem (9), (10) in the form

$$v(x, t) \sim \sum_{j=0}^{\infty} \varepsilon^j v_j(x, t). \tag{11}$$

By substituting expansion (11) into (9), (10) and by equating terms of the same order in ε, we get the recurrent chain of problems for v_j:

$$\hat{L} v_j = f_j(x, t), \quad x \in \Pi^s_{(d_1, d_2)}, \quad t > 0, \tag{12}$$

$$v_j|_{x_1 = d_l} = g_{l,j}(x_1', t); \tag{13}$$

here the v_j are vector-functions T-periodic in $x_2, \dots, x_s$, and f_j, $g_{l,j}$ depend on v_k for $k \leqslant j-1$,

$$\hat{L} = \hat{R}\frac{\partial^2}{\partial t^2} - \frac{\partial}{\partial x_m}\left(\hat{A}_{mk}\frac{\partial}{\partial x_k}\right)$$

in the principal part of the averaged operator from (9). Theorem 6.1.1 implies the following theorem.

THEOREM 2. *$\hat{L}$ is the operator of nonstationary theory of elasticity, i.e., the matrices $\hat{A}_{mk}$ satisfy the conditions of type (4.4.1) and $\hat{R} > 0$.*

We supplement (12), (13) with the homogeneous initial conditions

$$v_j(x, 0) = \frac{\partial v_j}{\partial t}(x, 0) = 0. \tag{14}$$

THEOREM 3. *There exists a unique, infinitely differentiable vector-function v_j ($j = 0, 1, \dots$), T-periodic in $x_2, \dots, x_s$ and satisfying relations (12)–(14) as well as the condition $\partial^l D^{q_i} v_j / \partial t^l(x, 0) = 0$ for any q, i and $l = 0, 1, \dots$.*

The proof of Theorem 3 makes use of the circumstance that due to (2), for any j we can get by induction

$$\frac{\partial^l f_j}{\partial t^l}(x, 0) = \frac{\partial^l g_{r,j}}{\partial t^l}(x_1', 0) = 0, \qquad r = 1, 2, \ l = 0, 1, \dots,$$

and, consequently, $\partial^l D^{q_i} v_j / \partial t^l(x, 0) = 0$ ($l = 0, 1, \dots$), and v satisfies the asymptotic equalities (5), (6). Thus, the vector-function $u^{(\infty)}$ defined by (4), (11) is a f.a.s. of (3).

Suppose that

$$v^{(K)}(x, t) = \sum_{j=0}^{K} \varepsilon^j v_j(x, t),$$

$$u^{(K)}(x, t) = \sum_{r=0}^{K+1} \varepsilon^r \sum_{q=0}^{r} \sum_{|i|=r-q} N_{qi}(\xi) D^{q_i} v^{(K)}\Big|_{\xi = x/\varepsilon}.$$

THEOREM 4. *The vector-function $u^{(K)}$ is a solution of the problem*

$$L_\varepsilon u^{(K)} = f(x, t) + \varepsilon^K \Theta(x, t, \varepsilon),$$
$$u^{(K)}\left(d_l, x_1', t\right) = g_l\left(x_1', t\right) + \varepsilon^{K+1}\theta_l\left(x_1', t, \varepsilon\right),$$
$$u^{(K)}(x, 0) = \frac{\partial u^{(K)}}{\partial t}(x, 0) = 0,$$

$u^{(K)}$ is a vector-function T-periodic in $x_2, \ldots, x_s$, and the estimate

$$\| u^{(K)} - u \|_{W_2^1(Q_{t_0})} \leqslant c_1(t_0)\, \varepsilon^K$$

holds. Here

$$|\Theta(x, t, \varepsilon)| \leqslant c_2(t_0), \quad \left|\theta_l\left(x_1', t, \varepsilon\right)\right| \leqslant c_3(t_0),$$

$c_1(t_0)$, $c_2(t_0)$, $c_3(t_0)$ are independent of ε.

The proof is analogous to that given in [109].

The averaged zeroth-order problem of elasticity theory is of the form

$$\begin{aligned} &\hat{R}\frac{\partial^2 v_0}{\partial t^2} - \sum_{m,k=1}^{s} \hat{A}_{mk}\frac{\partial^2 v_0}{\partial x_m \partial x_k} = f(x, t), \\ &v_0\big|_{t=0} = \frac{\partial v_0}{\partial t}\bigg|_{t=0} = 0, \quad v_0\big|_{x_1=d_l} = g_l\left(x_1', t\right). \end{aligned} \tag{15}$$

The following zeroth approximation is written for stresses

$$\sigma_q = A_{qj}\frac{\partial u}{\partial x_j}, \qquad q = 1, \ldots, s,$$

$$\sigma_q \sim \sum_{p=1}^{s}\Bigg(\sum_{j=1}^{s} A_{qj}\frac{\partial}{\partial \xi_j}\left(N_{0p}^0(\xi) + \overline{N}_{0p}^1(\xi_1 - k_1, \xi_1') + \right.$$
$$\left. + \overline{N}_{0p}^2(\xi_1 - k_2, \xi_1') + \xi_p E\right)\Bigg)\frac{\partial v_0}{\partial x_p}, \tag{16}$$

where v_0 is a solution of problem (15), N_{0p}^0 is a solution of equation (4.4.4), and the $\overline{N}_{0p}^l$ $(l = 1, 2)$ are bounded solutions, 1-periodic in $\xi_2, \ldots, \xi_s$, of the problems

$$\frac{\partial}{\partial \xi_m}\left(A_{mk}(\xi)\frac{\partial}{\partial \xi_k}\overline{N}_{0p}^l\right) = 0, \quad \xi \in R_\pm^s,$$
$$\overline{N}_{0p}^l\big|_{\xi_1=0} = -N_{0p}^0\big|_{\xi_1=0}$$

(R_+^s corresponds to $l = 1$, R_-^s corresponds to $l = 2$).

Formula (16) makes it possible to calculate the field of microstresses near the boundary of a composite.

The foregoing method of constructing the a.e. of problem (3) can be extended to the case of piecewise smooth coefficients as well as to the case of the system of equations of elasticity theory defined in a medium with periodically arranged voids (see the stationary case in [109]).

§6 On the Ultimate Strength of a Composite

It is assumed in investigations of failure of a homogenous material that the failure starts at a given point if the function P of the stress tensor $\sigma = (\sigma_1, \dots, \sigma_s)$, depending on the material, is $P(\sigma) \geqslant 1$. In the case of a composite material, the function P is specific for each of the components; it is natural, therefore, to introduce the function $P(\xi, \sigma)$ 1-periodic in all ξ_q and depending on the point ξ. The condition for failure takes the form

$$P(x/\varepsilon, \sigma) \geqslant 1. \tag{1}$$

It is often necessary in the applications to determine the values of the effective stresses which correspond to the start of failure under a particular type of loading the composite. Then the problem usually contains a parameter b, and one must find its lowest value $b_{\min}$ for which condition (1) is satisfied at least at one point x. It is this value of the parameter that corresponds to the start of failure of the composite under the given type of loading.

To illustrate, consider the system

$$L_\varepsilon u \equiv -\frac{\partial}{\partial x_m}\left(A_{mk}\left(\frac{x}{\varepsilon}\right)\frac{\partial u}{\partial x_k}\right) = 0, \tag{2}$$

given for $x \in \Pi^s_{(d_1, d_2)}$. A parameter b appears in the boundary conditions

$$u\left(d_1, x_1'\right) = 0, \quad u\left(d_2, x_1'\right) = (b, 0, 0)^T. \tag{3}$$

The solution of problem (2), (3) is sought in the class of functions T-periodic in $x_2, \dots, x_s$. It is assumed that the coefficients $A_{mk}(\xi)$ are piecewise smooth (in the sense of §4.1) $s \times s$ matrix-valued functions satisfying the conditions (4.4.1), and T, d_i are multiples of ε.

We determine $b_{\min}$ correspondig to problem (2), (3). From the linearity of (2), (3) for any b, we have $u(x) = b\bar{u}(x)$, where $\bar{u}(x)$ is a solution of the problem

$$\begin{gathered} L_\varepsilon \bar{u} = 0, \quad x \in \Pi^s_{(d_1, d_2)}, \\ \bar{u}\left(d_1, x_1'\right) = 0, \quad \bar{u}\left(d_2, x_1'\right) = (1, 0, 0)^T, \end{gathered} \tag{4}$$

and consequently,

$$\sigma_m(x) = A_{mk}\left(\frac{x}{\varepsilon}\right)\frac{\partial u}{\partial x_k} = bA_{mk}\left(\frac{x}{\varepsilon}\right)\frac{\partial \bar{u}}{\partial x_k} \equiv b\bar{\sigma}_m(x).$$

Thus, condition (1) is equivalent to the condition $P(\xi, b\overline{\sigma}) \geqslant 1$. Next we suppose that $P(\xi,\sigma)$ is a nonnegative homogeneous function in σ of order $p > 0$, that is, $P(\xi), b\sigma) = b^p P(\xi,\sigma)$ for any positive b and any ξ. Then (1) assumes the form $b^p P(x/\varepsilon, \overline{\sigma}(x)) \geqslant 1$. Denote $P_0 = \sup_{x\in\overline{\Pi}^s_{(d_1,d_2)}} P(x/\varepsilon, \overline{\sigma}(x))$. Let $0 < P_0 < \infty$ and suppose there exists a point $x_0 \in \overline{\Pi}^s_{(d_1,d_2)}$ such that $P(x_0/\varepsilon, \overline{\sigma}(x_0)) = P_0$. Then, in order to determine $b_{\min}$, it suffices to solve problem (4), find $\overline{\sigma}(x)$, calculate P_0, and set $b_{\min} = 1/\sqrt[p]{P_0}$.

Formula (5.16) can be used to calculate $\overline{\sigma}(x_0)$ approximately. In the given case $v_0 = ((x_1 - d_1)/(d_2 - d_1), 0, \dots, 0)^T$; therefore only the first columns of the matrices N^0_{0p}, $\overline{N}^1_{0p}$, $\overline{N}^2_{0p}$ are requried to determine $\overline{\sigma}(x_0)$.

The effective stresses corresponding to $b = b_{\min}$ are calculated by the formula $\hat{\sigma}^l_k = b_{\min}\hat{a}^{l1}_{k1}(d_2 - d_1)^{-1}$. The stresses can be measured as a result of an experiment in extending the layer $\Pi^s_{(d_1,d_2)}$ along the x_1-axis.

§7 Boundary Conditions of Other Types

As has been mentioned in Chapter 1, the presentation of the averaging method is illustrated as a rule by the first boundary-value problem. However, the constructions are applicable to other types of boundary conditions. Consider the equation

$$L_\varepsilon u \equiv R\left(\frac{x}{\varepsilon}\right)\frac{\partial^2 u}{\partial t^2} - \frac{\partial}{\partial x_m}\left(A_{mk}\left(\frac{x}{\varepsilon}\right)\frac{\partial u}{\partial x_k}\right) = f(x,\ t), \quad x \in \Pi^s_{(d_1,d_2)}, \quad t > 0, \tag{1}$$

where $R(\xi)$ and $A_{mk}(\xi)$ are piecewise smooth (in the sense of §4.1), 1-periodic scalar functions satisfying the conditions (4.1.3) and > 0. In just the same way as in §2, we assume that the union of the coefficient's discontinuity surfaces Σ has no common points with the hyperplane $\{\xi_1 = 0\}$.

Suppose for $x_1 = d_1$ the first-type boundary condition

$$u\left(d_1, x_1', t\right) = g_1\left(x_1', t\right), \tag{2}$$

is given, and for $x_1 = d_2$, that of the second-type:

$$\frac{\partial u}{\partial \nu} \equiv A_{1k}\frac{\partial u}{\partial x_k} = g_2\left(x_1', t\right), \tag{3}$$

where $g_i(x_1', t)$, $f(x,t)$ are infinitely differentiable functions, T-periodic in $x_2, \dots,$ x_s, and satisfying condition (5.2). The initial conditions are homogeneous:

$$u(x, 0) = \frac{\partial u}{\partial t}(x, 0) = 0, \quad x \in \Pi^s_{(d_1,d_2)}. \tag{4}$$

The asymptotics of the solution of problem (1)–(4) is sought in the form (5.4). The substitution of $u^{(\infty)}$ into equation (1) and conditions (2), (4) results in the same relations as in §5. The substitution of $u^{(\infty)}$ into the boundary condition (3) yields

$$\sum_{r=0}^{\infty} \varepsilon^{r-1} \sum_{q=0}^{r} \sum_{|i|=r-q} G_{qi}(\xi_1') D^{qi} v \Big|_{x_1=d_2} \sim g_2(x_1', t), \tag{5}$$

where

$$G_{qi}(\xi_1') = \left(\frac{\partial}{\partial \nu_\xi} N_{qi} + A_{1i_1} N_{qi_2 \ldots i_{r-q}} \right) \Big|_{\xi_1=0}, \quad \frac{\partial}{\partial \nu_\xi} \equiv A_{1h}(\xi) \frac{\partial}{\partial \xi_h}.$$

As in §5, we seek $N_{qi}(\xi)$ as a sum of a periodic component and two components exponentially tending to zero. The procedure of constructing the functions N^0_{qi} and N^1_{qi} remains the same, whereas the N^2_{qi} are defined successively from the recurrent chain of problems

$$- L_{\xi\xi} N^2_{qi} = - T^2_{qi} \quad \text{for} \quad \xi \in R^s_-, \tag{6}$$

$$\frac{\partial}{\partial \nu_\xi} N^2_{qi} = - \frac{\partial}{\partial \nu_\xi} N^0_{qi} - A_{1i_1} \left(N^0_{qi_2 \ldots i_{r-q}} + N^2_{qi_2 \ldots i_{r-q}} \right) + h^2_{qi} \quad \text{for} \quad \xi_1 = 0.$$

The constants h^2_{qi} are chosen on the condition that there exists a solution 1-periodic in $\xi_2, \ldots, \xi_s$ of problem (6), which satisfies the inequalities

$$\left| N^2_{qi}(\xi) \right| \leqslant c^{2,1}_{qi} e^{-c^{2,2}_{qi} |\xi_1|}, \quad \left| \frac{\partial}{\partial \xi_h} N^2_{qi}(\xi) \right| \leqslant c^{2,1}_{qi} e^{-c^{2,2}_{qi} |\xi_1|},$$

where $c^{2,j}_{qi} > 0$. It is proved in [112] that

$$h^2_{qi_1 \ldots i_{r-q}} = \int\limits_{R^s_-} \left(R(\xi) N^2_{q-2 i_1 \ldots i_{r-q}} - A_{i_1 j}(\xi) \frac{\partial N^2_{qi_2 \ldots i_{r-q}}}{\partial \xi_j} - \right.$$

$$\left. - A_{i_1 i_2}(\xi) N^2_{qi_3 \ldots i_{r-q}} \right) d\xi + \int_0^1 \cdots \int_0^1 \left(\frac{\partial}{\partial \nu_\xi} N^0_{qi_1 \ldots i_{r-q}} + \right.$$

$$\left. + A_{1i_1}(\xi) \left(N^0_{qi_2 \ldots i_{r-q}} + N^2_{qi_2 \ldots i_{r-q}} \right) \right) d\xi_1.$$

We set $N^2_0 \equiv 0$ and note that then $T^2_{0i_1} \equiv 0$, $h^2_{10} = 0$,

$$h^2_{0i_1} = \int_0^1 \cdots \int_0^1 \left(\frac{\partial}{\partial \nu_\xi} N^0_{0i_1} + A_{1i_1}(0, \xi_1') N^0_0 \right) d\xi_2 \ldots d\xi_s =$$

$$= \int_0^1 \cdots \int_0^1 A_{1h}(0, \xi_1') \frac{\partial}{\partial \xi_h} \left(N^0_{0i_1} + \xi_{i_1} \right) d\xi_2 \ldots d\xi_s = \langle \tilde{A}_{1h}(\xi) \rangle = \tilde{A}_{1h}$$

(see §6.2). Now from (5), we obtain the averaged boundary condition for $x_1 = d_2$:

$$\widehat{A}_{1h}\frac{\partial v}{\partial x_h}+\sum_{r=2}^{\infty}\varepsilon^{r-1}\sum_{q=0}^{r}\sum_{|i|=r-q}h_{qi}^{2}D^{qi}v\Big|_{x_1=d_2}\sim g_2(x_1',t). \tag{7}$$

The f.a.s. of problem (5.9), (5.10), (7) is sought in the form (5.11). By analogy with Theorem 5.3, we can prove the solvability of the problems for v_j as well as the estimates of the closeness of the values of the sum $u^{(K)}$ to the exact solution of problem (1)–(4).

The formalism of constructing the asymptotics $u^{(\infty)}$, exact up to the boundary, can be extended also to the third-type boundary-value problems, to the linear and nonlinear systems of equation (see [112]), and to the systems of equations given in domains with periodic voide (see [109]).

Remark 1: Some systems given in a bounded domain reduce to the problems in a layer. Consider the system of equations of elasticity theory,

$$-\frac{\partial}{\partial x_m}\left(A_{mh}\left(\frac{x}{\varepsilon}\right)\frac{\partial u}{\partial x_h}\right)=f(x),\quad x\in Q^s_{(d_1,d_2)},$$
$$Q^s_{(d_1,d_2)}=(d_1,d_2)\times(0,T)^{s-1},$$

with the boundary conditions

$$u\,|_{x_1=d_1}=g_1(x_1'),\quad \frac{\partial u}{\partial \nu}\Big|_{x_1=d_2}=g_2(x_1');$$

the condition of rigid contact

$$u^l=0,\quad a_{lh}^{ij}\frac{\partial u^j}{\partial x_h}=0\quad \text{for}\quad i\neq l$$

being given on the other faces $x_l=0,\ T(l=2,\dots,s)$.

Let the coefficients $a_{mk}^{ij}(\xi)$ satisfy conditions B from §6.3 for the hyperplanes $\xi_h=0$ $(h=2,\dots,s)$ and suppose the functions $f(x)$, $g_j(x_1')$ have infinitely differentiable continuations $\tilde f(x)$, $\tilde g_j(x_1')$, $2T$-periodic in $x_2,\dots,x_s$, and satisfying the following *condition C*:

The lth components of the vectors $\tilde f$, $\tilde g_j$ for $l=2,\dots,s$ are odd with respect to the planes $x_1=0,\ T$, and even with respect to the remaining planes $x_q=0$, T, $(q\neq 1,l)$, with the first components of these vectors being even for all the planes $x_q=0,\ T$, $(q=2,\dots,s)$.

We continue u periodically with period $2T$ to the entire layer $\Pi^s_{(d_1,d_2)}$ so that the continuation $\tilde u$ satisfies condition C. The vector-function $\tilde u$ is a solution, $2T$-periodic in $x_2,\dots x_s$, of the system

$$-\frac{\partial}{\partial x_m}\left(A_{mh}\left(\frac{x}{\varepsilon}\right)\frac{\partial u}{\partial x_h}\right)=\tilde f(x),\quad x\in \Pi^s_{(d_1,d_2)},$$

with the boundary conditions

$$u\,|_{x_1=d_1} = \widetilde{g}_1\left(x_1'\right), \quad \left.\frac{\partial u}{\partial \nu}\right|_{x_1=d_2} = \widetilde{g}_2\left(x_1'\right).$$

The f.a.s. of this problem can be constructed and justified by the method described in §5. All the terms of the a.e., including v_0, are $2T$-periodic in $x_2, \dots, x_s$ and satisfy condition C. It follows that on the faces $x_l = 0, T$ $(l = 2, \dots, s)$, the vector v_0 satisfies the condition of rigid contact

$$v_0^l = 0, \quad \widehat{a}_{lk}^{ij} \frac{\partial v_0^j}{\partial x_k} = 0, \qquad i \neq l. \tag{8}$$

Due to (5.13) for $l = 1$ and (7), the boundary conditions for $x_1 = d_1, d_2$ are of the form

$$v_0\,|_{x_1=d_1} = g_1\left(x_1'\right), \quad \widehat{A}_{1j} \left.\frac{\partial v_0}{\partial x_j}\right|_{x_1=d_2} = g_2\left(x_1'\right). \tag{9}$$

§8 On the Averaging of Fields in Layered Media with Layers of Composite Materials

Theoretical studies have been widely conducted on materials with the two-level hierarchy of the microstructure: The material consists of a periodic system of layers, whereas each layer in turn is a periodic microstructure. If the latter (interior) microstructure is also layered, the material is called a material of the C_2 class. The interest in such materials was aroused by the following circumstance. Suppose there is a composite material with an effective heat-conductivity tensor $\hat{A}_{ij}$. It turns out that, with the same concentration of the components, a material of the C_2 class can be engineered with practically the same tensor $\hat{A}_{ij}$. Consider, for example, the heat equation in a periodic medium

$$\frac{\partial}{\partial x_i}\left(K\left(\xi_1, \xi_2\right) \frac{\partial u}{\partial x_i}\right) = 0, \quad \xi_q = \frac{x_q}{\varepsilon}, \tag{1}$$

with K being a 1-periodic function.

We assume that the region of periodicity satisfies the symmetry condition (6.3.2), and consequently the equation takes the form

$$\frac{\partial}{\partial x_i}\left(\widehat{K}_i \frac{\partial u}{\partial x_i}\right) = 0. \tag{2}$$

Let the coefficient K take the value λ_1 on the subset G_1 of the periodicity cube Q and λ_2 on $G_2 = Q \backslash G_1$, $\operatorname{mes} G_1 = \theta_1$, $\operatorname{mes} G_2 = \theta_2$. For fixed λ_1, λ_2, θ_1, θ_2, we consider the set of all possible pairs $(\widehat{K}_1, \widehat{K}_2)$ that can be coefficients of the averaged equation (2); this set was effectively constructed in [89, 90].

Let $(\widehat{K}_1^0, \widehat{K}_2^0)$ be an arbitrary point of the set; in [90] the following statement was proved: For any $\eta > 0$, a composite material of the C_2 class can be indicated with its effective coefficients $(\widehat{K}_1, \widehat{K}_2)$ such that $|\widehat{K}_1 - \widehat{K}_1^0|$, $|\widehat{K}_2 - \widehat{K}_2^0| \leqslant \eta$.

Consider the stationary temperature field in a material with the two-level hierarchy

$$L_\varepsilon u \equiv -\frac{\partial}{\partial x_m}\left(A_{mk}\left(\frac{x_1}{\varepsilon}, \frac{x}{\varepsilon\mu}\right)\frac{\partial u}{\partial x_k}\right) = f(x), \quad x \in \mathbf{R}^s; \tag{3}$$

here ε, μ are small parameters, $\varepsilon = 1/n_1$, $\mu = 1/n_2$, n_1, n_2 are positive integers, and the $A_{mk}(y_1, z)$ are 1-periodic functions of z_q and y_1 having the following structure:

$$A_{mk}(y_1, z) = \begin{cases} A^1_{mk}(z), & y_1 \in [0, d_1] \cup [d_2, 1], \\ A^2_{mk}(z), & y_1 \in (d_1, d_2), \end{cases}$$

$0 < d_l < l$, d_1 being a multiple of μ.

The $A^l_{mk}(z)$ are assumed to be functions piecewise smooth in the sense of §4.1, and 1-periodic in z_q, which satisfy the conditions for ellipticity and symmetry (4.1.3).

The f.a.s. of equation (3) is constructed in two stages. At the first step, u is sought in the form (4.2.4), where $v(x)$ and $N_i(\xi)$ depend on μ as on a parameter. The second step involves constructing an asymptotic expansion in $\mu \to 0$ of the solution N_i of the problem on a cell (4.2.13), which is a problem on the contact of two periodic structures. The problem on a cell is solved by the boundary-layer technique described in §3. A detailed solution of (3) is presented in [176].

§9 The Time Boundary Layer for the Cauchy Parabolic Problem

(The results of the section follow from [(132]).

So far we only considered the Cauchy problem with zero initial values.

This section demonstrates the possibility of extending the above method to the case of nonzero initial data. For simplicity we consider the case, rather rare in practice, when the right side and the initial function are independent of the fast variable.

Consider the Cauchy problem for a parabolic equation of the following form

$$\frac{\partial u}{\partial t} - \frac{\partial}{\partial x_m}\left(A_{mk}\left(\frac{x}{\varepsilon}\right)\frac{\partial u}{\partial x_k}\right) = f(x, t), \tag{1}$$

$$u\,|_{t=0} = g(x); \tag{2}$$

the $A_{mk}(\xi)$ are 1-periodic coefficients satisfying the conditions for uniform ellipticity,

$$\varkappa_1 \eta_j \eta_j \leqslant A_{mk}(\xi)\,\eta_m \eta_k \leqslant \varkappa_2 \eta_j \eta_j \quad \forall (\eta_1, \ldots, \eta_s) \in \mathbf{R}^s, \quad \xi \in \mathbf{R}^s,$$

where $0 < \kappa_1 < \kappa_2$; and the symmetry conditions

$$A_{mk}(\xi) = A_{km}(\xi), \quad \xi_i = x_i/\varepsilon.$$

By analogy with §4.5 we construct the f.a.s. of (1) of the form

$$\overline{u}^{(\infty)} \sim \sum_{p,q,l=0}^{\infty} \varepsilon^{p+q+l} \sum_{|i|=l} N^p_{qi}(\xi)\, D^{q_i} v. \tag{3}$$

The formal asymptotic solution $\overline{u}^{(\infty)}$ does not satisfy the initial condition (2); therefore, in order to obtain the f.a.s. of problem (1), (2), it is necessary to add the boundary-layer correction $\tilde{u}$ to $\overline{u}^{(\infty)}$ to correct the initial condition

$$\tilde{u}(x, \xi, t, \tau) \sim \sum_{r=1}^{\infty} z_r(x, \xi, t, \tau), \quad \tau = t/\varepsilon^2,$$
$$u \sim \overline{u}^{(\infty)} + \tilde{u}. \tag{4}$$

The function $\tilde{u}(x, x/\varepsilon, t, t/\varepsilon^2)$ satisfies the equation

$$\frac{\partial \tilde{u}}{\partial t} - \frac{\partial}{\partial x_m}\left(A_{mk}\left(\frac{x}{\varepsilon}\right)\frac{\partial \tilde{u}}{\partial x_k}\right) = 0 \tag{5}$$

with the initial condition

$$\tilde{u}\,|_{t=0} = -\overline{u}^{(\infty)}\,|_{t=0} + g(x). \tag{6}$$

By substituting (4) into (5), (6), we get the equations:

$$\left(\frac{\partial}{\partial \tau} - \frac{\partial}{\partial \xi_m} A_{mj}(\xi)\frac{\partial}{\partial \xi_j}\right) z_r - \left(\frac{\partial}{\partial \xi_m} A_{mj}(\xi)\frac{\partial}{\partial \xi_j} + A_{mj}(\xi)\frac{\partial^2}{\partial x_m \partial \xi_j}\right) z_{r-1} +$$
$$+ \left(\frac{\partial}{\partial t} - A_{mj}(\xi)\frac{\partial^2}{\partial x_m \partial x_j}\right) z_{r-2} = 0, \tag{7}$$

$$z_r(x, \xi, t, \tau)\,|_{t,\tau=0} = -\sum_{|i|+q+p=r} N^p_{qi}(\xi)\, D^{q_i} v\,|_{t=0} +$$
$$+ \delta_{r0}\, g(x), \quad z_r = 0 \ \text{for} \ r < 0. \tag{8}$$

Theorem 1. *The system (7), (8) is successively solvable.*

The proof of Theorem 1, as well as a generalization for the case of A_{mk} depending on the fast variable in time $A_{mk} = A_{mk}(x/\varepsilon, t/\varepsilon^2)$, can be found in [132].

SUPPLEMENT

EXISTENCE AND UNIQUENESS THEOREMS FOR THE PROBLEM ON A CELL

This supplement contains the well-known results of the investigation of the solution to the problem on a cell, which were used many times throughout the book.

1. Let the $A_{ij}(\xi)$ $(i, j = 1, \dots, s)$ be the $n \times n$ matrix-valued functions, let the $f_k(\xi)$ $(k = 0, 1, \dots, s)$ be n-dimensional vector-functions, and suppose the elements of $A_{ij}(\xi)$, $f_k(\xi)$ are bounded, measurable functions 1-periodic in ξ, and $A_{ij}(\xi) = A_{ji}^T(\xi)$.

Consider the system of equations in $\mathbf{R}^s$

$$\frac{\partial}{\partial \xi_i}\left(A_{ij}(\xi)\frac{\partial u}{\partial \xi_j}\right) = f_0(\xi) + \frac{\partial}{\partial \xi_k} f_k(\xi). \tag{1}$$

If the coefficients $A_{ij}(\xi)$ and the right sides $f_0(\xi)$, $f_k(\xi)$ are smooth functions, then by the 1-periodic solution of (1) we mean a 1-periodic vector-function satisfying (1) at each point $\xi \in \mathbf{R}^s$. After identifying the hyperplanes $\{\xi_i = 0\}$ and $\{\xi_i = 1\}$ for each $i = 1, \dots, s$, we get the s-dimensional torus T_s which is the product of s circles of unit length, while the functions $A_{ij}(\xi)$, $f_0(\xi)$, $f_k(\xi)$, and the equation (1) itself will be defined on this torus. Obviously, the inverse transition is possible, too: The function defined on the torus can be "spread" periodically over the entire space $\mathbf{R}^s$.

By multiplying (1) by the smooth function $\varphi(\xi)$ defined on T_s and integrating over the torus, we obtain the integral identity

$$-\int_{T_s}\left(A_{ij}(\xi)\frac{\partial u}{\partial \xi_j}, \frac{\partial \varphi}{\partial \xi_i}\right)d\xi = \int_{T_s}(f_0(\xi), \varphi)\,d\xi - \int_{T_s}\left(f_k(\xi), \frac{\partial \varphi}{\partial \xi_k}\right)d\xi. \tag{2}$$

It is natural that this integral identity should be used to define the generalized solutions in the case of the coefficients that are not smooth.

The vector-function $u(\xi) \in W_2^1(T_s)$ satisfying (2) for any trial vector-function $\varphi(\xi) \in W_2^1(T_s)$ is called a *generalized 1-periodic solution* of system (1).

THEOREM 1. *Let the inequality*

$$\left\langle\left(A_{ij}(\xi)\frac{\partial u}{\partial \xi_j}, \frac{\partial u}{\partial \xi_i}\right)\right\rangle \geqslant \varkappa\left\langle\left(\frac{\partial u}{\partial \xi_i}, \frac{\partial u}{\partial \xi_i}\right)\right\rangle, \quad \varkappa > 0. \tag{3}$$

hold for any vector-function $u(\xi) \in W_2^1(T_s)$. In this case, the system of equations (1) has a generalized solution if and only if

$$\langle f_0(\xi)\rangle = 0. \tag{4}$$

The solution can be defined up to an arbitrary constant vector c,

$$u(\xi) = u_0(\xi) + c;$$

here $u_0(\xi)$ is a solution of (1) satisfying the condition

$$\langle u_0(\xi)\rangle = 0.$$

Proof: Necessity. We set in (2),

$$\varphi(\xi) = \varphi_r, \qquad r = 1, 2, \ldots, n,$$

where $\varphi_r = (\delta_{1r}, \delta_{2r}, \ldots, \delta_{nr})^T$, δ_{qr} is the Kronecker symbol; then (2) implies that the rth component of the vector $\langle f_0\rangle$ is zero. Since r is arbitrary, relation (4) is satisfied.

Proof: Sufficiency. By $H(T_s)$ we denote the subspace of vector-valued functions from $L_2(T_s)$ with the zero mean, that is,

$$H(T_s) = \{u \in L_2(T_s) \,|\, \langle u(\xi)\rangle = 0\}.$$

Prove first that for any measurable bounded f_k, there exists a unique element $u \in H(T_s) \cap W_2^1(T_s)$ satisfying identity (2) for any $\varphi(\xi) \in H(T_s) \cap W_2^1(T_s)$. The Poincaré inequality (see [76]) for a scalar function $v(\xi) \in W_2^1(T_s)$

$$\langle v^2\rangle \leqslant \langle v\rangle^2 + \frac{s}{2}\sum_{k=1}^{s}\left\langle\left(\frac{\partial v}{\partial \xi_k}\right)^2\right\rangle$$

implies the inequality for a vector-function $u(\xi) \in W_2^1(T_s)$:

$$\langle u^2\rangle \leqslant \langle u\rangle^2 + \frac{s}{2}\left\langle\left(\frac{\partial u}{\partial \xi_k}, \frac{\partial u}{\partial \xi_k}\right)\right\rangle. \tag{5}$$

It follows from (5) that for any vector-function $u \in H(T_s) \cap W_2^1(T_s)$, the estimate

$$\| u \|_{L_2(T_s)}^2 \leqslant \frac{s}{2}\sum_{k=1}^{s}\left\|\frac{\partial u}{\partial \xi_k}\right\|_{L_2(T_s)}^2 \tag{6}$$

holds. Due to inequalities (3), (6) and the boundedness of $A_{ij}(\xi)$, the expression

$$[u, \varphi]_1 = \left\langle\left(A_{ij}(\xi)\frac{\partial u}{\partial \xi_j}, \frac{\partial \varphi}{\partial \xi_i}\right)\right\rangle$$

determines the scalar product on the space $H(T_s) \cap W_2^1(T_s)$, the norm $\|\varphi\|_1 = \sqrt{[\varphi, \varphi]_1}$ being equivalent to the norm of W_2^1, that is, to the norm

$$\|\varphi\|_{W_2^1(T_s)} = \left(\|\varphi\|^2_{L_2(T_s)} + \sum_{k=1}^{s}\left\|\frac{\partial\varphi}{\partial\xi_k}\right\|^2_{L_2(T_s)}\right)^{1/2}.$$

Write identity (2) as

$$[u, \varphi]_1 = -\langle(f_0, \varphi)\rangle + \sum_{k=1}^{3}\left\langle\left(f_k, \frac{\partial\varphi}{\partial\xi_k}\right)\right\rangle. \tag{7}$$

Since $f_k \in L_2(T_s)$, the right side of (7) defines a bounded linear functional on $H(T_s) \cap W_2^1(T_s)$ in φ, and by the Riesz Theorem, there exists the unique element

$$u \in H(T_s) \cap W_2^1(T_s),$$

satisfying (7), and consequently (2), for any vector-function $\varphi(\xi) \in H(T_s) \cap W_2^1(T_s)$.

Let now $\langle f_0\rangle = 0$. Let us show that identity (2) is valid for any $\varphi \in W_2^1(T_s)$. Represent φ as $\varphi = \Psi + \langle\varphi\rangle$; then $\langle\Psi\rangle = 0$, $\Psi \in H(T_s) \cap W_2^1(T_s)$, and consequently, considering (4), we have

$$\begin{aligned}
&\langle(f_0, \varphi)\rangle - \left\langle\left(f_i, \frac{\partial\varphi}{\partial\xi_i}\right)\right\rangle = \\
&\quad = \langle(f_0, \psi)\rangle + (\langle f_0\rangle, \langle\varphi\rangle) + \left\langle\left(f_i, \frac{\partial\psi}{\partial\xi_i}\right)\right\rangle = \langle(f_0, \psi)\rangle - \\
&\quad - \left\langle\left(f_i, \frac{\partial\psi}{\partial\xi_i}\right)\right\rangle = -\left\langle\left(A_{ij}\frac{\partial u}{\partial\xi_j}, \frac{\partial\psi}{\partial\xi_i}\right)\right\rangle = -\left\langle\left(A_{ij}\frac{\partial u}{\partial\xi_j}, \frac{\partial\varphi}{\partial\xi_i}\right)\right\rangle.
\end{aligned}$$

Thus, u satisfies (2) for any $\varphi(\xi) \in W_2^1(T_s)$.

Suppose there exist $u_1, u_2 \in W_2^1(T_s)$ satisfying (2), and the difference $u_1 - u_2$ is not a constant vector. Then $v = u_1 - u_2 - \langle u_1 - u_2\rangle \not\equiv 0$ satisfies (7) for $f_0 = f_k = 0$ and any $\varphi \in H(T_s) \cap W_2^1(T_s)$. Hence $[v, v]_1 = 0$, $v \equiv 0$. This contradiction proves Theorem 1. □

2. Let Q be the cube $\{\xi\mathbf{R}^s | 0 < \xi_i < 1, i = 1, \ldots, s\}$. Put a set $\overline{A}$ consisting of the finite number of domains into Q and continue the former 1-periodically onto the entire space $\mathbf{R}^s$. By $\tilde{A}$ we denote the union of all the sets thus obtained. We assume that the boundary $\partial\tilde{A}$ is an infinitely smooth manifold and the domain $\mathbf{R}^s\backslash\overline{\tilde{A}}$ is connected. Let the torus T_s be obtained from Q by identifying the faces $\{\xi_i = 0\}$ and $\{\xi_i = 1\}$ $(i = 1, \ldots, s)$. By A, we shall mean the set on T_s, which corresponds to the set $\overline{A} \subset Q$, and $n_i(\xi)$ are the direction cosines of the vector of the normal to $\partial\tilde{A}$ at the point ξ.

By a generalized 1-periodic solution of the system (1) on $\mathbb{R}^s\backslash\tilde{A}$ with the boundary condition

$$\frac{\partial u}{\partial \nu} \equiv A_{ij} \frac{\partial u}{\partial \xi_j} n_i(\xi) = f_i(\xi) n_i(\xi), \quad \xi \in \partial \widetilde{A}, \tag{8}$$

we call a vector-function $u(\xi) \in W_2^1(T_s \setminus A)$ satisfying the integral identity

$$-\left\langle \left(A_{ij}(\xi) \frac{\partial u}{\partial \xi_j}, \frac{\partial \varphi}{\partial \xi_i} \right) \right\rangle^A = \langle (f_0(\xi), \varphi) \rangle^A - \left\langle \left(f_k(\xi), \frac{\partial \varphi}{\partial \xi_k} \right) \right\rangle^A$$

for any $\varphi(\xi) \in W_2^1(T_s)$. Here $\langle f(\xi) \rangle^A = \int\limits_{T_s \setminus A} f(\xi) d\xi$.

In complete analogy with Theorem 1, we can prove the following theorem.

THEOREM 2. *Let for any vector-function $u(\xi) \in W_2^1(T_s \setminus A)$ the inequality*

$$\left\langle \left(A_{ij}(\xi) \frac{\partial u}{\partial \xi_j}, \frac{\partial u}{\partial \xi_i} \right) \right\rangle^A \geqslant \varkappa \left\langle \left(\frac{\partial u}{\partial \xi_i}, \frac{\partial u}{\partial \xi_i} \right) \right\rangle^A, \varkappa > 0. \tag{9}$$

hold. In this case, the system (1) with the boundary conditions (8) has a generalized solution if and only if

$$\langle f_0(\xi) \rangle^A = 0.$$

This solution is defined up to an arbitrary constant vector

$$u(\xi) = u_0(\xi) + c,$$

where $u_0(\xi)$ is a solution of problem (1), (8) with the zero mean.

In particular, the conditions (3), (9) are satisfied for the coefficients of an elliptic equation, since in this case the quadratic form $A_{ij}(\xi)\eta_i\eta_j$ is positive-definite. For the elasticity tensor ($n = s$), the properties (3), (9) follow from the Cauchy inequalities for the torus T_s and the domain $T_s \setminus A$ (see [155]):

$$\left\langle \frac{\partial v_i}{\partial \xi_k} \frac{\partial v_i}{\partial \xi_k} \right\rangle \leqslant c_1 \left\langle \left(\frac{\partial v_i}{\partial \xi_k} + \frac{\partial v_k}{\partial \xi_i} \right) \left(\frac{\partial v_i}{\partial \xi_k} + \frac{\partial v_k}{\partial \xi_i} \right) \right\rangle$$

for any vector-function $v = (v_1, \ldots, v_s)$ with the components v_i from $W_2^1(T_s)$ and

$$\left\langle \frac{\partial v_i}{\partial \xi_k} \frac{\partial v_i}{\partial \xi_k} \right\rangle^A \leqslant c_2 \left\langle \left(\frac{\partial v_i}{\partial \xi_k} + \frac{\partial v_k}{\partial \xi_i} \right) \left(\frac{\partial v_i}{\partial \xi_k} + \frac{\partial v_k}{\partial \xi_i} \right) \right\rangle^A$$

for any vector-function $v = (v_1, \ldots, v_s)$ with the components $v_i \in W_2^1(T_s \setminus A)$. So the assertions of Theorems 1 and 2 are valid for the elliptic equations and for the systems of equations of elasticity theory (1).

In the case of piecewise smooth coefficients, it can be demonstrated that the generalized solutions of problems (1) and (2), (8) are also classic, i.e., when continued from T_s ($T_s \backslash A$) onto $\mathbf{R}^s$ ($\mathbf{R}^s \backslash A$), they satisfy (1) everywhere outside the discontinuity surfaces of the coefficients and the relations

$$[u] = 0, \quad \left[\frac{\partial u}{\partial \nu}\right] = 0$$

on the discontinuity surfaces. Besides, the generalized solution of problems (1), (8) satisfies condition (8).

BIBLIOGRAPHY

The following abbreviations are used:
Diff. uravneniya = Differential Equations.
DAN SSSR = USSR Doklady.
MZhG = Mechanics of fluids and gases.
PMM = Applied Mathematics and Mechanics.
Izv. AN SSSR = USSR Izvestiya.
ZHVM and MF = Journal of computer mathematics and mathematical physics.
UMN = Russian Mathematical Surveys.
Matem Sb. (ornik) = Math USSR Sbornik
ZhTF = Journal of theoretical physics
Zapiski nauchn. seminarov LOMI = see translations in JOSMAR.
Trudy seminara I.G. Petrovskogo = Proc. of I.G. Petrovsky seminar.
ZhETF = Journal of experimental and theoretical physics.
FTI = Physical Technical Institute.
Vestnik MGU = Reports of Moscow University.
Vestnik KhGU = Reports of Kharkov University.
(In Russian except Refs. 32, 35, 50, 55, 56, 61, 71, 103, 148, 158, 159, 177-221)

1. Amenadze, Yu. A., *Elasticity Theory*. Vysshaia Shkola, Moscow (1976).
2. Anderson, N. S., Grin, V. Kh., and Nur, A. K., "Continual models of lattice structures of girder and plate type." *Raketnaia tekhnika i kosmonavtika* **16** (1975), 6–17.
3. Andrianov, I. V., and Manevich, L. I., "On classification of approximate equations of theory of reinforced shells taking into account the discrete arrangement of the ribs." *Gidroaeromekhanika i teoriia uprugosti* **15** (1972), 115–122.
4. Andrianov, I. V., and Manevich, L. I., "On the calculations of the stress-strain state of an orthotropic strip reinforced with stiffening ribs." *Izv. AN SSSR. Ser. Mekhanika tverdogo tela* **4** (1975), 135–140.
5. Andrianov, I. V., and Manevich, L. I., "On the design of folded shells." *Soprotivlenie materialov i teoriia sooruzhenij* **26** (1975), 11–14.
6. Andrianov, I. V., Manevich, L. I., and Nalivaiko, L. A., "On the design of shells of revolution reinforced with meridional ribs." *Stroitelnaia mekhanika i raschet sooruzhenii* (1975), 16–20.
7. Andrianov, I. V., and Diskovskii, A. A., "On the investigation of the stress-strain state of corrugated plates." *Izvestiya vuzov. Ser. Stroitelstvo i arkhitektura* **1** (1981), 36–41.
8. Bakhvalov, N. S., "Averaged characteristics of bodies with periodic structures." *DAN SSSR* **221** (1975), 516–519.
9. Bakhvalov, N. S., "Averaging partial differential equations with rapidly oscillating coefficients." *DAN SSSR* **221** (1975), 516–519.
10. Bakhvalov, N. S., "Averaging nonlinear partial differential equations with rapidly oscillating coefficients." *DAN SSSR* **225** (1975), 249–252.

11. Bakhvalov, N. S., "Averaging partial differential equations with rapidly oscillating coefficients." In *Problems of Mathematical Physics and Computing Mathematics*. Nauka, Moscow (1977), 34–51.
12. Bakhvalov, N. S., "Averaging the process of short-wave propagation in periodic media." In *Theory of Cubature Formulas and Computing Mathematics* (edited by S.L. Sobolev). Nauka, Novosibirsk (1980), 3–11.
13. Bakhvalov, N. S., "Averaging the process of heat transition in periodic media in the presence of radiation." *Diff. uravneniya* **17** (1981), 1765–1773.
14. Bakhvalov, N. S., "On the speed of sound in mixtures." *DAN SSSR* **6** (1979), 1345–1348.
15. Bakhvalov, N. S., "Averaging the transition equation for a periodic medium." *Vestnik MGU. Ser. Vychisl. matem. i kibern* **1** (1981), 14–20.
16. Bakhvalov, N. S., "Averaging partial differential equations with rapidly oscillating coefficients." In *Modern Problems of Mathematical Physics and Computing Mathematics* (edited by A. N. Tikhonov). Nauka, Moscow (1982), 38–47.
17. Bakhvalov, N. S., and Zlotnik, A. A., "The coefficient stability of differential equations and averaging equations with random coefficients." *DAN SSSR* **242** (1978), 745–748 (= Sov. Math. Dokl. 19, 1978, 1171–1175).
18. Bakhvalov, N. S., and Eglit, M. E., "Processes in periodic media which cannot be described in terms of average characteristics." *DAN SSSR* **268** (1983), 836–840.
19. Berdichevskii, A. L., "On effective thermal conductivity of media with periodically arranged inclusions."*DAN SSSR* **247** (1979), 1363–1367.
20. Berdichevskii, A. L., "On the averaged description of a liquid containing gas bubbles." *MZhG* **6** (1980), 72–79.
21. Berdichevskii, V. L., *Variational Principles of Continuum Mechanics*, Nauka, Moscow, 1983.
22. Berdichevskii, V. L., "Spatial averaging of periodic structures." *DAN SSSR* **222** (1975), 565–567.
23. Berdichevskii, V. L., "On averaging periodic structures." *PMM* **41** (1977), 993–1006.
24. Berdichevskii, V. L., and Berdichevskii, A. L., "Flow of perfect liquid about a periodic system of bodies." *Izv. AN SSSR, Ser. Mekhanika zhidkosti i gaza* **6** (1978), 3–18.
25. Berdichevskii, V. L., "The variational asymptotic method." In *Some Questions of Continuum Mechanics*. Moscow University, Moscow (1978), 271–289.
26. Berliand, L. V., "Averaging equations of linear elasticity theory in domains with fine-grained boundaries." In *Theory of Functions, Functional Analysis and Their Applications*. Kharkov University, Kharkov (1983), 16–25.
27. Bogoliubov, N. N., and Mitropolskii, Yu. A., *Asymptotic Methods in the Theory of Nonlinear Vibrations*. Nauka, Moscow, 1974.
28. Bolotin, V. V., "Basic equations of the theory of reinforced media." *Mekhanika polimerov* **2** (1965), 27–37.
29. Bolotin, V. V., "On the theory of reinforced bodies." *Izv. AN SSSR, Ser. Mekhanika* **1** (1965), 74–80.

30. Bolotin, V. V., and Moskalenko, V. N., "Macroscopic characteristics of microinhomogeneous solids." *DAN SSSR* **178** (1968), 563–565.
31. Bolotin, V. V., and Novichkov, Yu. N., *Mechanics of Laminated Structures*. Machinostroenie, Moscow, 1980.
32. Born, M., and Kun Huang, *Dynamic Theory of Crystal Lattices*. Clarendon Press, Oxford, 1956.
33. Boronkov, A. A., *The Course of Probability Theory*. Nauka, Moscow, 1972.
34. Brekhovskikh, L. M., *Waves in Laminated Media*. Nauka, Moscow, 1973.
35. Brillouin, L., and Parodi, M., *Propagation des Ondes dans les Milieux Périodiques*. Masson; Dunod, Paris, 1956.
36. Butuzov, V. F., Vasil'eva, A. B., and Fedoriuk, M. V., "Asymptotic methods in the theory of ordinary differential equations." *Itogi nauki, Matem. analiz* (1967), 5–73.
37. Van Fo Fy, G. A., *Theory of Reinforced Materials*. Nauka, Moscow, 1971.
38. Vasil'ev, F. P., *Methods of Solving Extreme-Value Problems*. Nauka, Moscow, 1981.
39. Volkov, D. B., "On averaging some boundary value problems in domains with periodic structures." *ZhVM and MF* **22** (1982), 112-122.
40. Volkov, D. B., "Asymptotic expansion for solutions of the systems of Stokes equations and of elasticity theory in domains of special form." *ZhVM and MF* **23** (1983), 1464–1476.
41. Volkov, E. A., "On the differential properties of solutions of boundary value problems for Laplace and Poisson's equations on a rectangle." *Transactions of the Steklov Mathematical Institute* **77** (1965), 89–112.
42. Vishik, M.I., and Lyusternik, L.A., "The asymptotic behaviour of solutions of linear differential equations with large or rapidly oscillating coefficients and boundary conditions." *UMN* **15**, issue 4(94) (1960), 27–95.
43. Vishik, M. I., "Solution of systems of quasi-linear equations under periodic boundary conditions." *DAN SSSR* **137** (1961).
44. Gorshin, S. N., *Preservation of Wood*. Lesnaia Promyshlennost', Moscow, 1977.
45. Grigoliuk, E. I., and Filshtinskii, L. A., *Perforated Plates and Shells*. Nauka, Moscow, 1970.
46. Dolgikh, V. N., and Filshtinskii, L. A., "Theory of linearly reinforced composite material with anisotropic components of the structure." *Izv. AN SSSR, Ser. Mekhanika tverdogo tela* **6** (1978).
47. Dolgikh, V. N., and Filshtinskii, L. A., "A model of an anisotropic medium reinforced with thin bands." *Prikladnaya mekhanika* **15** (1979), 24–30.
48. Dul'nev, G. N., *The Coefficients of Transfer in Inhomogeneous Media. Thermophysical Properties of Substances*. The Leningrad Institute of Fine Mechanics and Optics, Leningrad, 1979.
49. Dul'nev, G. N., and Zarichniak, Yu. P., *Thermal Conductivity of Mixtures and Composite Materials*. Energiia, Leningrad, 1974.
50. Duvaut, G. "Functional analysis and solid-state mechanics. Application to the study of composite materials with periodic structure." In: *Theoretical Applied Mechanics*. M. : Mir, (1979), p. 323–346.

51. Egorov, Yu. V., "On hyperbolic equations with discontinuous coefficients." *DAN SSSR* **143** (1960), 514–517.
52. Zhigun, I. G., and Poliakov, V. A., *Properties of 3D-Reinforced Plastics*. Zinatne, Riga, 1978.
53. Zhikov, V. V., Kozlov, S. M., Oleinik, O. A., and Ha Tien Ngoan. "Averaging and G-convergence of differential operators." *UMN* **34**, issue 5 (209) (1979), 65–133.
54. Zhikov, V. V., Kozlov, S. M., and Oleinik, O. A., "G-convergence of parabolic operators." *UMN* **36** issue 1 (217) (1981), 11–58.
55. Zienkievich, O. C., *Method of Finite Elements in Technology*. M. : Mir, (1975).
56. Siegel, R., and Howell, J. R., *Thermal Radiation Heat Transfer*. New York, 1973.
57. Il'iushin, A. A., *Plasticity*. Gostekhizdat, Moscow, 1948, part 1.
58. Il'iushin, A. A., *Continuum Mechanics*. Nauka, Moscow, 1979.
59. Il'iushina, E. A., "A version of moment plasticity theory for a one-dimensional continuous medium of inhomogeneous periodic structure." *PPM* **36** (1972), 1086–1093.
60. Iosif'ian, G. A., Oleinik, O. A., and Panasenko, G. P., "Asymptotic expansion of a solution of the system of elasticity theory with periodic, rapidly oscillating coefficients." *DAN SSSR* **266** (1982), 18–22.
61. Kelly, A., *Materials of High Durability*. M. : Mir, (1976).
62. Kozlov, S. M., "Averaging differential operators with almost periodic, rapidly oscillating coefficients." *Matem. sbornik* **107 (149)** (1978), 199–217.
63. Kozlov, S. M., "Averaging random structures." *DAN SSSR* **241** (1978), 1016–1019.
64. Kozlov, S. M., "Conductivity of two-dimensional random media." *UMN* **34**, issue 4 (1979), 193–194.
65. Kolpakov, A. G., "Effective rigidities of composite plates." *PMM* **46** issue 4 (1982), 666–673.
66. *Composite Superhard Materials*. Institute of Superhard Materials of the Ukranian SSR Academy of Sciences, Kiev, 1979.
67. Kondrat'ev, V. A., "On the smoothness of the solution of the Dirichlet problem for the second-order elliptic equations in the piecewise smooth domain." *Differentsial'nye uravneniua* **6** (1070), 1829–1843.
68. Kosarev, A. Yu., "The asymptotics of the averaged characteristics of periodic elastic media with strongly variable properties." *DAN SSSR* **267** (1982), 38–42.
69. Krasnov, I. P., "On solving magnetostatic problems for thin closed shells." *ZhTF* **42**, issue 8 (1972), 1545–1549.
70. Kregers, A. F., and Melbardis, Yu. G., "Determination of the deformability of 3D-reinforced composites using the method of averaging rigidities." *Mekhanika polimerov* **1** (1978), 3–8.
71. Kristensen, R. M., *Mechanics of Composite Materials*. Trans. from Engl. (Yu. M. Tarnopolskogo, ed.). M. : Mir, 1982.
72. Krylov, N. M., and Bogoliubov, N. N., *Introduction to Nonlinear Mechanics*. The Ukranian SSR Academy of Sciences, Kiev, 1937.

73. Kuzmak, G.E., "Asymptotic solutions of nonlinear second-order differential equations with variable coefficients." *PMM* **23**, issue 3 (1959), 515–526.
74. Ladyzhenskaya, O. A., *A Mixed Problem for a Hyperbolic Equation*. Gostekhizdat, Moscow, 1953.
75. Ladyzhenskaya, O. A., *Mathematics of the Dynamics of Viscous Noncompressible Liquid*. Nauka, Moscow, 1970.
76. Ladyzhenskaya, O. A., *Boundary Value Problems of Mathematical Physics*. Nauka, Moscow, 1973.
77. Ladyzhenskaya, O. A., and Solonnikov, V. A., "On some problems of vector analysis and generalized formulations of boundary value problems for Navier–Stokes equations." *Zap. nauchn. sem. LOMI AN SSSR* **59** (1976), 82–116.
78. Ladyzhenskaya, O. A., Solonikov, V. A., and Ural'tseva, N. N., *Linear and Quasi-Linear Parabolic Equations*. Nauka, Moscow, 1967.
79. Ladyzhenskaya, O. A., and Ural'tseva, N. N., *Linear and Quasi-Linear Elliptic Equations*. Nauka, Moscow, 1973.
80. Landis, E. M., *Second-Order Elliptic and parabolic equations*. Nauka, Moscow, 1971.
81. Landis, E. M., and Panasenko, G. P., "A theorem on the asymptotics of solutions of elliptic equations with coefficients periodic in all the variables except one." *DAN SSSR* **235** (1977), 125–1255.
82. Landis, E. M., and Panasenko, G. P., "On a variant of a theorem of the Phrägman–Lindelöf type for elliptic equations with coefficients periodic in all the variables except one." *Trudy seminara im.* I. G. Petrovskogo. Moscow University, Moscow (1979), 105–136.
83. Lesnichaya, V. A., and Manevich, L. I., "Asymptotic investigation of vibrations of plates reinforced with stiffening ribs." *Prikladnaia Mekhanika* **16** (1980), 67–72.
84. Lions, J. L., "Remarks on some computerizable aspects of the homogenization method in composite materials." In *Computing Methods in Applied Science and Engineering*, 1974.
85. Lifshits, I. M., and Rozentsveig, L. N.,"On the theory of elastic properties of polycrystals." *ZhETF* **16** (1946), II.
86. Lifshits, I. M., and Rozentsveig, L. N., "On the construction of Green's tensor for the basic equation of elasticity theory in the case of an unbounded, elastic anisotropic medium."*ZhETF* **17**, issue 9 (1947).
87. Lomakin, V. A., *Elasticity Theory of Inhomogeneous Bodies*. Moscow University, Moscow, 1976.
88. Lomov, S. A., *Introduction to General Theory of Singular Perturbations*. Nauka, Moscow, 1981.
89. Lur'e, K. A., and Cherkaev, A. V., "*G*-convergence of the set of anisotropically conducting media in the case of two dimensions." *DAN SSSR* **259** (1981), 328–331.
90. Lur'e, K. A., and Cherkaev, A. V., "Exact estimates of the conductivity of composites formed by two isotropic conducting media taken in a given proportion." *FTI, Leningrad*, **783** (1982), preprint.

91. Malmeister, A. K., Tamuzh, V. P., and Teters, G. A., *Strength of Polymer and Composite Materials*. Zinatne, Riga, 1980.
92. Manevich, L. I., and Pavlenko, A. V., "Asymptotic analysis of equations of the theory of eccentrically reinforced cylindrical shells." In *Theory of Plates and Shells*. Nauka, Moscow (1971), 185–190.
93. Marchenko, V. A., and Khruslov, E. Ya., "Boundary value problems with a fine-grained boundary." *Matem. sb.* **65** (107) 458–472.
94. Marchenko, V. A., and Khruslov, E. Ya., *Boundary Value Problems in Domains with Fine-Grained Boundaries*. Naukova Dumka, Kiev, 1974.
95. Marchenko, V. A., and Khruslov, E. Ya., "New results in the theory of boundary value problems with a fine-grained boundary." *UMN* **33**, issue 3 (1978), 127.
96. Marchuk, G. I., and Agoshkov, V. I., "Introduction to the projection network methods." (1981).
97. Maslov, V. P., *Operator Methods*. Nauka, Moscow, 1973.
98. Mindlin, R. D., "Microstructure of linear elasticity." In: *Mekhanika*; a collection of translations. Nauka, Moscow (1964).
99. Mindlin, R. D., and Tirsten, Kh. F., "The effect of moment stresses in linear elasticity theory." In: *Mekhanika*; a collection of translations. Nauka, Moscow (1964).
100. Mitropolskii, Yu. A., *The Averaging Method in Nonlinear Mechanics*. Naukova Dumka, Kiev, 1971.
101. Moiseev, N. N., *Asymptotic Methods of Nonlinear Mechanics*. Nauka, Moscow, 1969.
102. Mukhina, G. V., and Sobolev, S. L., "On the solution of a boundary-value problem." *PMM* **23** (1959), 534–539.
103. Nayfeh, A. H., *Perturbation Methods*. Wiley, NY, 1973.
104. Oleinik, O. A., "On the Dirichlet problem for elliptic equations." *Matem. sb.* **24** (66) (1949), 3–14.
105. Oleinik, O. A., "Boundary value problems for linear elliptic and parabolic equations with discontinuous coefficients." *Izv. AN SSSR, Ser. matem.* **25** (1961), 3–20.
106. Oleinik, O. A., "On the convergence of solutions of elliptic and parabolic equations with weak convergence of the coefficients." *UMN* **30**, issue 4 (184) (1975), 257–258.
107. Oleinik, O. A., "On heat propagation in many-dimensional dispersed media." In *Problems of Mathematical Physics and Mechanics*. Nauka, Moscow (1976), 224–236.
108. Oleinik, O. A., and Iosif'ian, G. A., "Energy estimates of the generalized solutions of boundary value problems for second-order elliptic equations and their applications." *DAN SSSR* **232** (1977) 1257–1260.
109. Oleinik, O. A., Iosif'ian, G. A., and Panasenko, G. P., "Asymptotic expansion of solutions of the system of elasticity theory in perforated domains." *Matem. sb.* **120** (162) (1983), 22–41.
110. Panasenko, G. P., "On the existence and uniqueness of the solution of some elliptic equations defined in $\mathbf{R}^s$." *Tr. Konf. molodykluchenykh f-ta vychisl. matem. i kibern.* (The proceedings of the conference of young scientists

of the department of computing mathematics and cybernetics), issue 3, Moscow University, Moscow (1976), 53–61.

111. Panasenko, G. P., "Asymptotics of higher orders of solutions of equations with rapidly oscillating coefficients." *DAN SSSR* **240** (1978), 1293–1296.
112. Panasenko, G. P., "Asymptotics of higher orders of solutions of problems on the contact of periodic structures." *Matem. sb.* **110 (152)** (1979), 505–538.
113. Panasenko, G. P., "Averaging the problem on the contact of two periodic structures divided by a thin interlayer." In *Applied Mathematics and Computer Software* (edited by A.S. Il'inskii and L. N. Korolev). Moscow University, Moscow (1979), 20–31.
114. Panasenko, G. P., "Averaging periodic structures with well-conducting inhomogeneities." *Vestnik MGU, Ser. Vychisl. matem. i kibern.* **3** (1980), 4–11.
115. Panasenko, G. P., "Asymptotics of solutions and eigenvalues of elliptic equations with strongly variable coefficients." *DAN SSSR* **252** (1908), 1320–1325.
116. Panasenko, G. P., "Averaging fields in framework and thin-walled structures." *UMN* **36**, issue 4 (1981), 224.
117. Panasenko, G. P., "Averaging fields in skeletal structures." In *Applied Mathematics and Computer Software* (edited by L.N. Korolev, A. D. Kuz'min and O. Yu. Tikhomirov). Moscow University, Moscow (1981), 13–14.
118. Panasenko, G. P., "The principle of splitting an averaged operator for a nonlinear system of equations in periodic and random skeletal structures." *DAN SSSR* **263** (1982), 35–40.
119. Panasenko, G. P., "Mathematics of the theory of 3D-reinforced media." *UMN* **37**, issue 4 (1982), 136–137.
120. Panasenko, G. P., "Averaging fields in composite materials with high-modulus reinforcement." *Vestnik MGU, Ser. Vychisl. Matem. i Kibern.* **2** (1093), 20–27.
121. Panasenko, G. P., "Averaging processes in skeletal structures with random properties." *ZhVM and MF* **23** (1983).
122. Pen'kovskii, V. I., "On the diffusion coefficient of the equations of motion of moisture in soil." *Izv. ASN SSSR, Ser. Mekh. i mashinostr.* (1964), 149–152.
123. Petrovskii, I. G., *Lectures on the Theory of Ordinary Differential Equations.* Nauka, Moscow, 1964 (= *Ordinary Differential Equations*, Prentice-Hall, Englewood Cliffs, 1966).
124. Pisanko, A. N., "Vibrations of eccentrically reinforced shells." *Problemy prochnosti* **3** (1980), 93–96.
125. Pobedria, B. E., *Numerical Methods in Theory of Elasticity and Plasticity.* Moscow University, Moscow, 1981.
126. Pobedria, B. E., and Gorbachev, V. I., "On static problems of elastic composites." *Vestnik MGU, Ser. Matem. Mekhan.* **5** (1977), 101–110.
127. Pobedria, B. E., and Sheshenin, S. V., "On the matrix of influence." *Vestnik MGU, Ser. Matem. Mekhan.* **6** (1979), 76–81.
128. Polubarinova-Kochina, P. Ya., *Theory of Motion of Ground Water.* Nauka, Moscow, 1977.

129. Preobrazhenskii, I. N., "Dynamic problems of the theory of thin-walled elements of supporting structures weakened by cutouts" (Review). *Problemy prochnosti*, part 1, **5** (1980), 82–90; part 2, **6** (1980), 99-109.
130. Preobrazhenskii, I. N., *Stability and Vibrations of Plates and Shells with Openings*. Mashinostroenie, Moscow, 1981.
131. Pshenichnov, G. I., *Theory of Thin Elastic Laced Shells and Plates*. Nauka, Moscow, 1982.
132. Piatnitskii, A. L., "A parabolic equation with rapidly oscillating coefficients." *Vestnik MGU, Ser. Matem. Mekhan.* **3** (1980), 33–39.
133. Piatnitskii, A. L., "On the limit behaviour of the dependence domain of a hyperbolic equation with rapidly oscillating coefficients." *Matem. sb.* **115** (157) (1981), 130–145.
134. Piatnitskii, A. L., "On the behaviour on the infinity of a solution of a second-order elliptic equation defined on a cylinder." *UMN* **37**, issue 2 (1982), 231–232.
135. Raitum, U. E., "On the G-convergence of quasi-linear elliptic operators with unlimited coefficients." *DAN SSSR* **261** (1981), 30-34.
136. Rats, M. V., Ivanova, N. B., and Malyshev, L. I., *Manual for Calculating the Percolation Coefficient of Fissured Rock Bodies by the Parameters of the Cracks*. Stroiizdat, Moscow, 1979.
137. Reztsov, M. V., "The averaged operator for a nonlinearly elastic skeletal structure." *Vestnik MGU, Ser. vych. matem. i kibern.* **2** (1982).
138. Rikards, R. B., and Chate, A. K., "The initial surface of strength of a unidirectionally reinforced composite under plane stress state." *Mekhanika polimerov* **4** (1976), 633–639.
139. Roitberg, Ya. A., and Sheftel, Z. G., "Energy inequalities for elliptic operators with discontinuous coefficients and for general boundary conditions and conjugation conditions." *DAN SSSR* **148** (1963), 531–533.
140. Romm, E. S., *Percolation Properties of Fissured Rock*. Nedra, Moscow, 1966.
141. Samarskii, A. A., "On a problem of heat propagation." *Vestnik MGU* **3** (1947), 85–101; **6** (1947), 119–129.
142. Samarskii, A. A., *Theory of Difference Schemes*. Nauka, Moscow, 1977.
143. Sedov, L. I., *Continuum Mechanics*. Nauka, Moscow, 1976.
144. Senatorov, P. K., "The coefficient stability of solutions of ordinary second-order differential equations and of parabolic equations on a plane." *Dif. uravneniia* **7** (1971), 754–758.
145. Simonenko, I. B., "Problems of electrostatics in an inhomogeneous medium. The case of a thin dielectric with a high dielectric constant." *I. Diff. uravneniia* **10** (1974), 301–309.
146. *Modern Composite Materials*. (ed. by Brautman, L., and Krok, R.) M. : Mir, (1970).
147. Sobolev, S. L., *Some Applications of Functional Analysis in Mathematical Physics*. Leningrad University, Leningrad, 1950.
148. Strang, G., and Fix, G. J., *An Analysis of the Finite Element Method*. Mir, Moscow, 1977.

149. Tamuzh, V. P., and Teters, G. A., "Problems of mechanics of composite materials." *Mekhankia kompoz. materialov* **1** (1979), 34–45.
150. Tarnopol'skii, Yu. M., Roze, A. V., Zhigun, I. G., and Gunyaev, G. M., "Structural features of materials reinforced with high-modulus fibers." *Mekhanika polimerov*, **4** (1971).
151. Tikhonov, A. N., and Samarskii, A. A., *Equations of Mathematical Physics*. Nauka, Moscow, 1977.
152. Tupin, R. A., "Theories of elasticity taking stresses into account." *Mekhanika*. A collection of translations (1965).
153. Fedorova, O. A., "Variational difference scheme for a homogenous equation of diffusion." *Matem. Zametki* **17** (1975).
154. Fenchenko, V. I., and Khruslov, E. Ya., "Asymptotic solutions of differential equations with strongly oscillating and degenerate coefficients." *DAN USSR* Ser. A **4** (1980), 26–30.
155. Fichera, G., *Existence Theorems in Elasticity. Boundary Problems of Elasticity with Unilateral Components*. Mir, Moscow, 1974.
156. Ha Tien Ngoan, "On the convergence of solutions of boundary value problems for a sequence of elliptic equations." *UMN* **32**, issue 3 (195) (1977), 183–184.
157. Ha Tien Ngoan, "On the convergence of solutions of boundary value problems for a sequence of elliptic systems." *Vestnik MGU, Ser. Matem., mekhan.* **5** (1977), 83–92.
158. Hashin, Z., Rozen, B. V., " Elastic modules of materials reinforced with fibers." *Applied Mechanics*, Series E (USA) **2** (1964) p. 223–232. Transl. from English.
159. Hill, R., "Elastic properties of composite media, some theoretical principles." *Mechanics*, **5** (1964) Collection of translations.
160. Khoroshun, L. P., "Methods of the theory of random functions in problems on the macroscopic properties of microinhomogeneous media." *Prikladnaia mekhanika* **14** (1978), 3–17.
161. Khruslov, E. Ya., "The Dirichlet problem in domains with random boundaries." *Vestnik KhGU, Ser. Matem. mekhan.* **34** (1970), 3–7.
162. Khruslov, E. Ya., "On the conditions for convergence of the sequence of solutions to the first boundary value problem." *Teoriia funktsii, funktsional'nyi analiz i ikh prilozheniia* **16** (1972), 28–38.
163. Khrusolv, E. Ya., "The first boundary value problem in domains with complex boundaries for the higher-order equations." *Matem. sb.* **103, (145)** (1977), 614–629.
164. Khruslov, E. Ya., "The asymptotic behaviour of solutions of the second boundary value problem under detailization of the boundary of the domain." *Matem. sb.* **106** (1978), 604–621.
165. Khruslov, E. Ya., "On the convergence of solutions of the second boundary value problem in weakly connected domains." In *Theory of Operators in Functional Spaces and its Applications*. Naukova Dumka, Kiev (1981), 129–174.
166. Tseitlin, L. A., "On the determination of magnetic and electric fields of thin layers and shells." *ZhTF* **28**, issue 6 (1958), 1326–1329.

167. Chernyshev, S. N., *Motion of Water Along the Networks of Cracks*. Nedra, Moscow, 1979.

168. Shamaev, A. S., "Spectral problems in theory of averaging." *UMN* **36**, issue 4 (1981), 224.

169. Shemergor, T. D., *Theory of Elasticity of Microinhomogeneous Media*. Nauka, Moscow, 1977.

170. Shestakov, V. M., *Dynamics of Underground Waters*. Moscow University, Moscow, 1973.

171. Sheftel' Z. G., "Estimates in L_p of solutions of elliptic equations with discontinuous coefficients satisfying general boundary conditions and conjugation conditions." *Sib. matem. zh.* **4** (1965), 636–668.

172. Sheshenin, S. V., "Averaged moduli of a composite." *Vestnik MGU, Ser. Matem. mekhan.* **6** (1980), 79–83.

173. Shokin, Yu. I., *Method of Differential Approximations*. Nauka, Novosibirsk, 1979.

174. Shorshorov, M. Kh., "The problem of compatibility in developing composite materials with a metal matrix and reinforced with high-modulus fibers." In *Fibrous and Dispersion-Ordered Composite Materials*. Nauka, Moscow (1976), 10–15.

175. Yurinskii, V. V., "On averaging an elliptic boundary value problem with random coefficients." *Sib. matem. zh.* **21** (1980), 209–223.

176. Yakovleva, E. A., "Averaging the stationary heat equation for periodic media of a special structure." *ZhVM and MF* **22** (1982) 623–633.

177. Artola, M., and Duvaut, G., "Homogeneisation d'une plaque renforcée par un système périodique de barres curvilignes." *Comp. Rend. Acad. Sci. Paris*, Ser. A **286** (1978), 659–662.

178. Artola, M., and Duvaut, G. "Homogeneisation d'une plaque renforcée." *Comp. Rend. Acad. Sci. Paris*, Ser. A **284** (1977), 707–710.

179. Artola, M., Duvaut, G., "Homogeneisation d'une classe de problèmes non lineaires." *Comp. Rend. Acad. Sci. Paris*, Ser. A **288** (1979), 775–778.

180. Babuska, I., "Homogeneisation and its applications. Mathematical and computational problems." *Proc. Symp. Numerical Solution of Part. Diff. Equations* **3**, Maryland (1975). New York, Acad. Press, 1976, 89–116.

181. Babuska, I., "Solution of interface problems by homogenisation." *SIAM J. Math. Anal.* **7** (1976), 603–634; **7** (1976), 635–645; **8** (1977), 923–937.

182. Babuska, I., "Homogenisation approach in engineering." *Lecture Notes in Economics and Math. Systems* **134** (1976), 137–153.

183. Bakhvalov, N. S., "Investigation of differential equations with rapidly oscillating coefficients." *Proceedings of the Fourth Symposium on Basic Problems of Numerical Mathematics, MFFUK v Praze* (1980), 5–14.

184. Bakhvalov, N. S., *Homogenization and Perturbation Problems. Computing Methods in Applied Science and Engineering*. Amsterdam, North Holland (1980), 645–658.

185. Becus, A. G., "Homogeneisation and random evolutions: applied to the mechanics of composite materials." *Quart. Appl. Math.* **37** (1979) 209–217.

186. Bensoussan, A., Lions, J. L., Papanicolaou, G., "Sur quelques phénomènes asymptotiques stationaires." *S. R. Acad. Sci. Paris* **281** (1975), 89–94.

187. Bensoussan, A., Lions, J. L., Papanicolaou, G., "Boundary layers and homogenization of transport processes." *Publ. RIMS Kyoto Univ.* **15** (1979), 53–157.

188. Bensoussan, A., Lions, J. L., Papanicolaou, G., *Asymptotic Analysis for Periodic Structures*. Amsterdam, North Holland, 1978.

189. Biroli, M.G.G., "Convergence for elliptic variational and quasi-variational inequalities." *Proc. Intern. Meeting on Recent Methods in Nonlinear Anal.* Ed. De Giorgi, Magenes, Mosco, Pitagora. Bologne (1979). 361–383.

190. Bourgat, J. F., "Numerical experiments of the homogenization method for operators with periodic coefficients." *Rapport de Recherche.* **277** (1978). I.R.I.A. Rocquencourt, France.

191. Cioranescu, D., Saint Jean Paulin, J. "Homogenisation in open sets with holes." *Journ. Math. Anal. Appl.* **71** (1979), 590–607.

192. Duvaut, G., "Comportement macroscopique d'une plaque perforée périodiquement." *Lecture Notes in Mathematics* **594** (1977), Springer-Verlag, 131–145.

193. Duvaut, G., "Homogeneisation des plaques à structure périodique en théorie non lineaire de Von Karman." *Lecture Notes in Mathematics.* **665** (1977). Springer-Verlag. 56–69.

194. Hill, R., "Theory of mechanical properties of fibre-strengthened materials." *Journ. Mech. Phys. Solids* **12** (1964), 199–218.

195. Huffington, N. J., "Theoretical determination of rigidity properties of orthogonally stiffened plates." *Journ. of Applied Mechanics* **23** (1956), 15–20.

196. Gibson, J. E., *Thin Shells (Computing and Theory)*. New York, Pergamon Press (1980), 289.

197. Damlamian, A., "How to homogenize a nonlinear diffusion equation: Stefan's problem." *SIAM J. Math. Anal.* **12** (1981), 306–313.

198. De Giorgi, E., Spagnolo, S. "Sulla convergenza degli integrali dell'energia per operatori ellittici del secondo ordine." *Boll. Un. Mat. Ital.* **8** (1973), 391–411.

199. De Giorgi, E., "Γ-convergenza e G-convergenza." *Boll. Un. Mat. Ital.* **5** (1977), 213–220.

200. Kesavan, S., "Homogenisation of elliptic eigenvalue problems." *Appl. Math. and Optim.* Part I, **5** (1979), 153–167; Part II, **5** (1979), 197–217.

201. Lions, J. L., "Remarks on non-local phenomena in composite materials and in perforated materials." *Proc. of the Jutam Symposium, Northwestern Univ.* (ed., Nemat). Amsterdam, North Holland, 1979.

202. Lions, J. L., "Homogeneisation non local." *Proc. Intern. Meeting on Recent Methods in Non-Linear Analysis.* (ed., De Giorgi, Magenes, Mosco, Pitagora). Bologne (1979), 189–203.

203. Lions, J. L., "Asymptotic expansions in perforated media with periodic structures." *Rocky Mount. J. Math.* **10** (1980), 125–140.

204. Lions, J. L., "Perturbations singulières dans les problèmes aux limites et en contrôle optimal." *Lecture Notes in Mathematics.* **323** Springer, 1973.

205. Lions, J. L., *Some Methods in the Mathematical Analysis of Systems and Their Control*. Beijing, Science Press, (1981).

206. Marcellini, P., "Periodic solutions and homogenization of non-linear variational problems." *Annali de Matematics* **117** (1978).
207. Marcellini, P., Sbordone, C., "Sur quelques de G-convergence et d-homogenisation non-lineaires." *C.R. Acad. Sci. Paris*. **284** (1977).
208. Mignot, F., Puel, J.-P., Suquet, P. M., "Homogenisation and bifurcation of perforated plates." *Int. J. of Eng. Sci.* **18** (1980), 409–414.
209. Mignot, F., Puel, J.-P., Suquet, P. M., "Homogeneisation d'une problème de bifurcation." *Proc. Int. Meeting on Recent Methods in Non-Linear Analysis*. Ed., De Giorgi, Magenes, Mosco, Pitagors. Bologne (1979), 281–307.
210. Oleinik, O. A., Panasenko, G. P., Yosifian, G. A., "On the asymptotic expansion for solutions of a homogenization problem in elasticity." *Applicable Analysis* **15** (1983), 15–32.
211. Papanicolaou, G., "Introduction to the asymptotic analysis of stochastic equations." *Lectures in Applied Mathematics* **16** (1977), 109–147.
212. Reuss, A., "Berechnung der Fliehgrenze von Mischkristallen auf Grund der Plastizitätsbedingung für Einkristalle." *Z. angew. Math. und Mech.* **9** (1929), 49–58.
213. Sanchez-Paleicia, E., "Comportements local et macroscopique d'un type de milieux physiques hétérogènes." *Intern. Jour. Engin. Sci.* **12** (1974), 331–351.
214. Sanchez-Palencia, E., "Non-homogeneous media and vibration theory." *Lecture Notes in Physics* **127**, Berlin, Springer-Verlag, 1980.
215. Spagnolo, S.,"Sulla convergenza di solutioni di equazioni paraboliche e ellipticke." *Ann. Scuola Norm. Sup. Pisa, cl. Sci.* **22** (1968), 577–597.
216. Tartar, L., *Homogeneisation*. Paris, Cours Peccot En Collège de France, 1977.
217. Tartar, L., "Estimation de coefficients homogeneises." *Analyse Numérique et Fonctionnelle*. CNRS et Université Paris-Sud, 364–373.
218. Vanninathan. M., "Homogeneisation des valeurs propres dans les milieux perforés." *Comp. Rend. Acad. Sci.* **287** (1978), 403–406.
219. Vanninathan, M., "Homogeneisation des problèmes de valeurs propres dans les milieux perforés. Problème de Dirichlet." *Comp. Rend. Acad. Sci.* **287** (1978), 823–825.
220. Voight, W., *Lehrbuch der Kristallphysik*. Berlin, Teubner, 1928.
221. Maxwell, C., *Treatise on Electricity and Magnetismus, V.1.* Oxford, Oxford Univ. Press, 1873.
222. Bakhvalov, N. S., "Determination of effective characteristics of a nonlinear elastic periodic medium in the case of small strains." In *Numerical Methods of Solving Problems of Elasticity and Plasticity*. Proceedings of the 7th All-Union Conference. Inst. for Theoretical and Appl. Math of Siberian Branch of the USSR Acad. of Sci. Novosibirsk (1982), 3–7.
223. Obraztsov, I. F., Vasil'ev, V.V.,and Bunakov, V. A., *Optimum Reinforcement of Shells of Revolution made of Composite Materials*. Mashinostroenie, Moscow, 1977.
224. Zlotnik, A. A., "The coefficient stability of systems of ordinary differential equations." *Diff. uravneniia.* **2** (1984), 220–229.

225. Kalugin, O. Yu., "Averaging boundary value problem of elasticity theory in perforated domains." *DAN SSSR* **5** (1982), 1097–1100.
226. Mol'kov, V. A., "The effective characteristics of a unidirectional fibrous composite." *Abstract of a thesis. Kand. Sc. (Phys-Math).* Moscow, 1983.
227. Gorbachev, V. I., "Problem of reduction for elastic space weakened by a system of cylindrical pores." *Izv. AN SSSR, ser. MTT* **5** (1983), 63–67.
228. Oleinik, O. A., and Markov, V. G., "On heat propagation in homogeneous dispersion media." *PMM* **39** (1975), 1073–1081.

SUBJECT INDEX

asymptotic expansion (a.e.) 12
averaged boundary conditions of infinite order 309
  equation xii, xxvi, xxx, 167
    of infinite order 126, 170, 184
  operator along the strip 270
    cylinder 272, 273
    system 280, 291
  segment 144, 149
  (zero-order) problem 23, 40, 66
averaging = homogenisation
boundary-layer addition 320
Darcy equations 154
effective coefficients xii, xxx
elliptic equation 6
equation on a periodic cell 109
fast variables xxvi
formal asymptotic solution (f. a. s.) xxxii
  averaged equation 21
framework 269
generalized solution 2, 3, 7, 9, 10
  formulation 3
hyperbolic equation 8
hypothesis of equivalent homogeneity 212
macroscopic variables xxvi
materials with two-level hierarchy of the microstructure 341
mean over period xxviii
medium with a periodic structure xxx
microscopic variables xxvi
multiscale method xxvi
multi-speed transition equation 165
nonstationary system of equations of elasticity theory 10
$\omega\varepsilon^2$-condition 226
operator of local averaging 206

parabolic equation 7
percolation tensor 154
periodic cell problem 21, 105
principle of frozen coefficients 58
  splitting the averaged operator 243, 275
problem on a strip 262, 270
random coefficient 83
  rectangular structure 306
rectangular periodic framework 258, 268, 295
section 269, 279
skeletal structure 279
slow variables xxvi
stationary Navier–Stokes equations 150
  Stokes equations 150
system of equations of motion of a linear visco-elastic medium 46
    the couple-stress theory 184
T-periodicity xxvii
truss 291
wave dispersion 188
Here "i" is a notation of the INTRODUCTION's pages

9 780792 300496

GPSR Compliance

The European Union's (EU) General Product Safety Regulation (GPSR) is a set of rules that requires consumer products to be safe and our obligations to ensure this.

If you have any concerns about our products, you can contact us on ProductSafety@springernature.com

In case Publisher is established outside the EU, the EU authorized representative is:

Springer Nature Customer Service Center GmbH
Europaplatz 3
69115 Heidelberg, Germany

Batch number: 10823957

Printed by Printforce, the Netherlands